Automation and Robotics

Automation and Robotics

Edited by Paula Sutton

CLANRYE INTERNATIONAL
www.clanryeinternational.com

Clanrye International,
750 Third Avenue, 9th Floor,
New York, NY 10017, USA

ISBN: 978-1-63240-788-7

Cataloging-in-Publication Data

Automation and robotics / edited by Paula Sutton.
p. cm.
Includes bibliographical references and index.
ISBN 978-1-63240-788-7
1. Automation. 2. Robotics. 3. Robots. 4. Automatic control. I. Sutton, Paula.
T59.5 .A98 2019
670.427--dc23

For information on all Clanrye International publications
visit our website at www.clanryeinternational.com

CLANRYE INTERNATIONAL

Contents

Preface

The ever growing need of advanced technology is the reason that has fueled the research in the field of robotics in recent times. Robotics is the branch of engineering that utilizes methods and systems from multiple industries for designing, building and operating of robots. Automation on the other hand is the design and development of a range of computer software to perform tasks that are commonly carried out by humans. This book aims to showcase how automation methods and technologies are applied in the field of robotics. Such technology is applied in multiple industries to reduce human labor and simplify complex industrial processes such as manufacturing, packaging, construction, etc. The various sub-fields of robotics along with technological progress that have future implications are also glanced at in this book. It includes some of the vital pieces of work being conducted across the globe, on various topics related to this area of study. As this field is emerging at a fast pace, this book will help the readers to better understand the concepts of automation and robotics.

Significant researches are present in this book. Intensive efforts have been employed by authors to make this book an outstanding discourse. This book contains the enlightening chapters which have been written on the basis of significant researches done by the experts.

Finally, I would also like to thank all the members involved in this book for being a team and meeting all the deadlines for the submission of their respective works. I would also like to thank my friends and family for being supportive in my efforts.

Editor

1

Acquisition of Contact Force and Slippage Using a Vision-Based Tactile Sensor with a Fluid-Type Touchpad for the Dexterous Handling of Robots

Yuji Ito[1*], Youngwoo Kim[2] and Goro Obinata[3]

[1]*Graduate School of Engineering, Nagoya University, Nagoya, Japan*

[2]*Korea Institute of Machinery & Materials (KIMM), Daegu Research Center for Medical Devices and Green Engergy, Dalseo-gu, Korea*

[3]*EcoTopia Science Institute, Nagoya University, Nagoya, Japan*

Abstract

This paper presents a new approach for estimating contact force and slippage by using a vision-based tactile sensor with a fluid-type touchpad for the dexterous handling of robots. The sensor consists of a CCD camera, LED lights, a transparent acrylic plate and a deformable touchpad. The sensor can obtain a variety of tactile information, such as the contact force, shape, contact region, position and orientation of an object in contact with the fluid-type touchpad. The previous method for measuring contact force requires extensive calculation; this paper proposes a new method based on lookup tables for measuring normal force, tangential force and moment, using a fluid-type touchpad. Additionally, we clarify the mechanism of slippage between a fluid-type touchpad and a contacted object. This mechanism is efficient for applying the proposed slippage estimation method to a fluid-type touchpad. The validation of the proposed methods is confirmed in the experimental results.

Keywords: Contact Force Measurement; Dexterous Handling; Robot Hands; Slippage Estimation; Tactile Sensors

Introduction

Tactile receptors in the skin allow humans to sense multimodal tactile information such as the contact force, slippage, shape, position and orientation of a contacted object. Humans easily control their muscles by feeding back information from tactile receptors. Therefore, tactile sensing is a key factor in enabling robots to imitate skilled human behaviors. Precise control and dexterity in robots are due to information feedback from tactile sensors.

When considering practical applications, tactile sensors should meet three specific requirements. First, flexible sensor surfaces are optimal, as sensors should fit the object geometrically. Second, a simple structure is required for a compact robot. Third, in order to achieve multifunctional and dexterous robots, we need a sensor that allows the simultaneous acquisition of multiple types of tactile information.

Various tactile sensors have been developed using resistive, capacitive, piezoelectric, ultrasonic or electromagnetic sensing elements [1,2]. A sensor with strain gauges embedded in an elastic body has been proposed [3] for estimating the slippage of a contacted object. Noda et al. have developed a sensor including standing cantilevers and piezo resistors arrayed in orthogonal directions for detecting shear stress [4]. Schmitz et al. have arrayed twelve capacitance-to-digital converter (CDC) chips in an array on the body of each robot finger; these provide twelve 16-bit measurements of capacitance [5]. Hakozaki et al. have arrayed sensing elements on conductive rubber at regular intervals to measure three components of stress based on a capacitive method [6]. However, the key practical issues remain unresolved. Since these sensors require multiple sensing elements and complicated wiring, their structures are complex and cannot satisfy the second important requirement as described in the previous paragraph. A wire-free tactile sensor based on transmitters/receivers [7], and a sensor based on micro coils for changing impedance by contact force [8], were housed in complex structures. Although compact sensors using microelectromechanical system (MEMS) can be manufactured, the surfaces of these sensors are minimally deformable [9-12] and cannot satisfy the first requirement as described above.

However, vision-based approaches are extremely suitable for tactile sensors [13-15]. Typical vision-based sensors include the following two components: a deformable contact surface made of elastic material to fit its shape to contacted objects; and a camera to capture the deformation of the contact surface. Tactile information is acquired by analyzing the deformation of the surface. Compact vision-based sensors can be easily fabricated, as they do not require multiple elements or complex wiring in the contact region. Moreover, using elastic material does not decrease sensitivity in this type of sensor. K. Kamiyama et al. estimated a three-axis contact force by detecting the sensor body's deformation, using a two-layered dotted pattern and a charge-coupled device (CCD) camera [16,17]. The sensor reported in [18,19] consists of rubber sheets with nubs, a transparent acrylic plate, a light source and a CCD camera. Light travels through the transparent plate and is diffusely reflected where the nubs come in contact with the plate. The three-axis contact forces are obtained based on the intensity of the reflected light as captured by the CCD camera. Yamada arranged reflector chips on the surface of an elastic body [20]. In this case, deformation of the contact surface is estimated using the four corner positions of the reflector chips. However, these sensors can only detect one form of tactile reception. In order to achieve the dexterous handling of robots in a dynamically changed environment, various types of tactile information should be acquired simultaneously.

We have proposed a vision-based tactile sensor that can sense

***Corresponding author:** Yuji Ito, Graduate School of Engineering, Nagoya University, Nagoya, Japan, E-mail: ito_yuji@nagoya-u.jp multiple types of tactile information simultaneously [21,22].The sensor consists of a CCD camera, LED lights, a transparent acrylic plate and a transparent hemispherical elastic touchpad for contacting the object. Here, the touchpad is deformable and the structure of the sensor is simple. Our proposed sensor satisfies the above requirements: simultaneous acquisition of various kinds of tactile information; simple structure; and a deformable surface for the sensor. This sensor measures normal force, tangential force and rotational moment based on preliminarily-constructed lookup tables, which relate the deformation of the elastic touchpad to force and moment [21]. This sensor can also estimate slippage between the touchpad and a contacted object [22].

Next, a tactile sensor using a fluid-type touchpad has been proposed for estimating the shape, contact region, position, orientation and contact force of an object in contact with the touchpad [23-25]. The surface of the fluid-type touchpad is made of a silicon rubber elastic membrane. The inside of the membrane is filled with translucent red-colored water. Implementation of the fluid-type touchpad has extended the sensing ability of our tactile sensor. However, the issue of how to estimate slippage using the fluid-type touchpad has not yet been addressed in these papers [23-25]. The difficulty is that the mechanism of slippage in the case of a fluid-type touchpad differs significantly from that of an elastic touchpad. Moreover, the method for measuring contact force by using a fluid-type touchpad requires extensive calculation [25], while the elastic touchpad can measure the contact force rapidly with the use of look-up tables [21]. In consideration of applications requiring rapid sensor response, the methods using lookup tables are much more efficient.

The purpose of this study is to estimate the contact force and slippage of an object by using a fluid-type touchpad. Firstly, we will propose a new method based on lookup tables for measuring normal force, tangential force and moment. Secondly, we will clarify the mechanism of slippage in the case of a fluid-type touchpad, comparing it to an elastic touchpad. Finally, the validation of our proposed methods will be confirmed in the experimental results.

Vision-Based Tactile Sensors

Figure 1 shows the configuration of a vision-based tactile sensor consisting of a CCD camera, LED lights, a transparent acrylic plate and a touchpad. The dimensions of the CCD camera and the LED lights are 8 × 8 × 40 mm and 60 × 60 × 60 mm, respectively. The touchpad is hemispherical, with a curvature radius and height of 20 mm and 13 mm. The two types of touchpads are presented in Figure 2: one is a transparent elastic touchpad, and the other is a semitransparent fluid-type touchpad. The elastic touchpad consists of a simple, transparent elastic body. The surface of the fluid-type touchpad is made of an elastic membrane constructed of silicon rubber; the inside of the membrane is filled with translucent, red-colored water. A dotted pattern is printed on the inside of the touchpad surface for observing the touchpad's deformation. Analysis of the deformations formed when the touchpad comes in contact with objects yields multimodal tactile information, using an image of the inside of the deformed touchpad captured by the CCD camera. Figure 3a and 3b show the captured images, sized 640 × 480 effective pixels, in the cases of the elastic touchpad and the fluid-type touchpad, respectively. This sensor can obtain multiple types of tactile information, including the shape, contact region, position and orientation of an object [23-25].

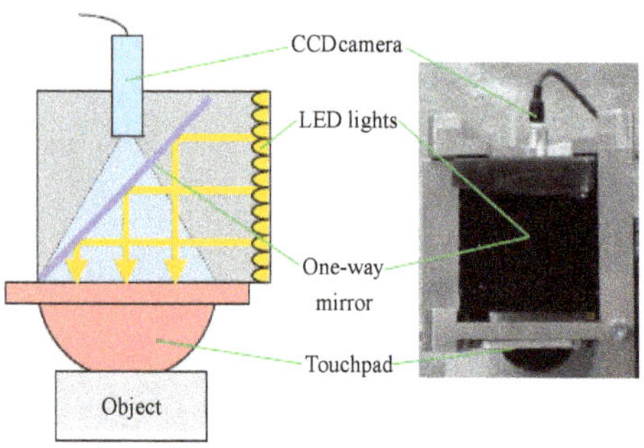

Figure 1: Configuration of the vision-based tactile sensor.

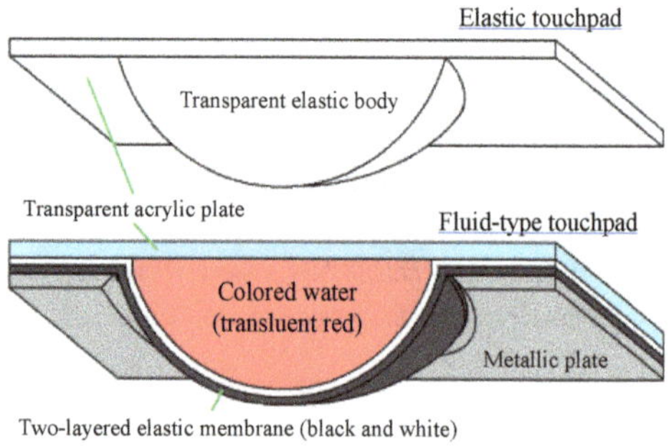

Figure 2: Configuration of the elastic touchpad and the fluid-type touchpad.

V-Measurement of Contact Force

Method for measuring normal force

The previous method in [21] regarded the elastic touchpad and a contacted flat object as an elastic spherical object and a rigid flat object, respectively. The contact region in this case is a circle. Xydas et al. and Kao et al. analyzed the stiffness and contact mechanics of soft fingers based on the relationship of the Hertzian contact model [22,23]. Their approach calculates the relationship between the radius of a contact region and normal force, which is given as follows:

$$r_c = cf_n^{\gamma} \quad (1)$$

Here, r_c, c and f_n are the radius of the contact region, the constant coefficient and normal force, respectively. The constant coefficient c depends on the mechanical properties of the touchpad, such as shape and stiffness. However, the above relation cannot be applied to the fluid-type touchpad because of the difference in structure.

Therefore, we newly consider the contact between a fluid-type touchpad and a flat object based on the following approach. When the fluid-type touchpad comes in contact with the flat object in a normal direction, the object is subject only to the inner pressure of the contact region by the touchpad. Therefore, the equilibrium equation of the flat surface in a normal direction can be expressed as follows:

$$F_z = p_n S_c . \quad (2)$$

Here, F_z signifies normal force. p_{in} and S_c represent the inner pressure of the touchpad, and the contact area between the touchpad surface and the object, respectively. Note that the contact area S_c and the inner pressure p_{in} increase along with the depth D of the contact between the touchpad and the object, as shown in Figure 4. There is a non-linear relationship between the contact area S_c, the contact depth D and the inner pressure p_{in}. Moreover, the acquisition of inner

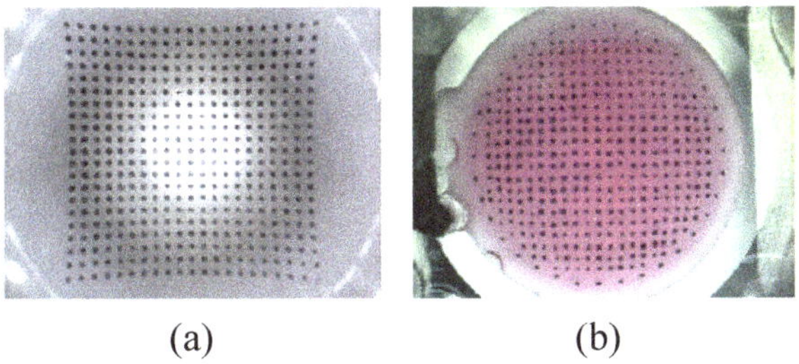

Figure 3: Captured images; (a) The elastic touchpad. (b) The fluid-type touchpad.

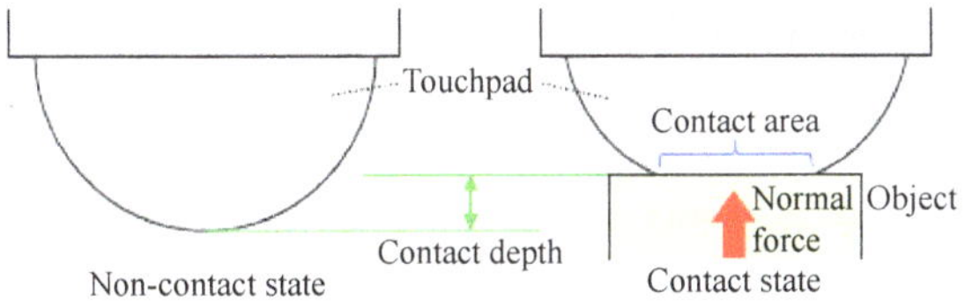

Figure 4: Relationship between the contact area, the contact depth and normal force.

pressure p_{in} is not straightforward, while the previous methods can obtain the contact area S_c and the contact depth D in [23,24]. Therefore, we express p_{in} as the function of S_c or D, and thus normal force F_z can also be given by the function of S_c or D as follows:

$$F_z = \begin{cases} F_S(S_c) \\ F_D(D) \end{cases} . \quad (3)$$

Here, F_S and F_D are the functions of S_c and D, respectively. We approximate the functions F_S and F_D as quadratic functions.

Method for measuring tangential force

Tangential force was measured in the previous method by the displacement of the central dot and the radius of the contact region, using the elastic touchpad [21]. Figure 5a shows the displacement of the central dot when an object is in contact with the touchpad.

In the case of the fluid-type touchpad, we measure tangential force by using the displacement of the central dot, the contact area S_c and the depth of the contact D with tuning parameters. Adapting the complex characteristics of the fluid-type touchpad is achieved by the tuning parameter which are newly introduced. Here, this relationship depends on normal force, transformed into the contact area S_c or the depth of the contact D. When normal force is large, a tangential force is gained with reference to the displacement. In reverse, a smaller tangential force is generated with reference to the displacement when normal force is small. Therefore, we normalize the displacement of the central dot by using the contact area S_c or the depth of the contact D with the tuning parameters as follows:

$$d_0^{nor} = \begin{cases} \left(\dfrac{S_c}{S_{ref}}\right)^{T_S} d_0 \\ \left(\dfrac{D}{D_{ref}}\right)^{T_D} d_0 \end{cases} . \quad (4)$$

Here, d_0^{nor}, d0, S_{ref} and D_{ref} are the normalized displacement of the central dot, the displacement before the normalization, the reference contact area and the reference depth, respectively. T_S and T_D are the tuning parameters for adapting the complex characteristics of the fluid-type touchpad, not used by the previous method in [21].

Method for measuring moment around normal direction

In a manner similar to the measurement of tangential force, the previous method using the elastic touchpad also measured moment in a normal direction perpendicular to the contact surface [21]. The rotation angle of the touchpad's surface yields the moment after identifying the relationship between the rotation angle of the contact surface and the moment. Figure 5b shows the rotation angle when an object is in contact with the touchpad.

When we use the fluid-type touchpad, this relationship also depend on normal force, transformed into the contact area S_c or the depth of the contact D. Moment increased or decreased with reference to the rotation angle according to S_c and D. Therefore, we normalize the rotation angle of the contact surface by using S_c and D with the tuning parameters as follows:

$$\theta^{nor} = \begin{cases} \left(\dfrac{S_c}{S_{ref}}\right)^{R_S} \theta \\ \left(\dfrac{D}{D_{ref}}\right)^{R_D} \theta \end{cases} . \quad (5)$$

Here, θ^{nor} and θ are the normalized rotation angle of the contact surface, and the rotation angle before the normalization, respectively. R_S and R_D are the tuning parameters for adapting the complex characteristics of the fluid-type touchpad, not used by the previous method in [21].

Estimation of Slippage

Definition of stick ratio

In order to achieve dexterous handling by robot hands, it is necessary to obtain not only contact force, but also information on slippage between a touchpad and a grasped object. Moreover, it is difficult for a robot to control grip force before macroscopic slippage occurs unless the sensor can predict the slippage. The grasped object may drop due to a delay in controlling the grip force. If macroscopic slippage occurs, a bigger grip force is required to stop slippage because the friction coefficient decreases in a dynamic situation.

The slippage degree between a touchpad and a grasped object has been focused on in [26,27]. Slippage degree refers to the various slip phases, such as perfect sticking, incipient slippage and macroscopic

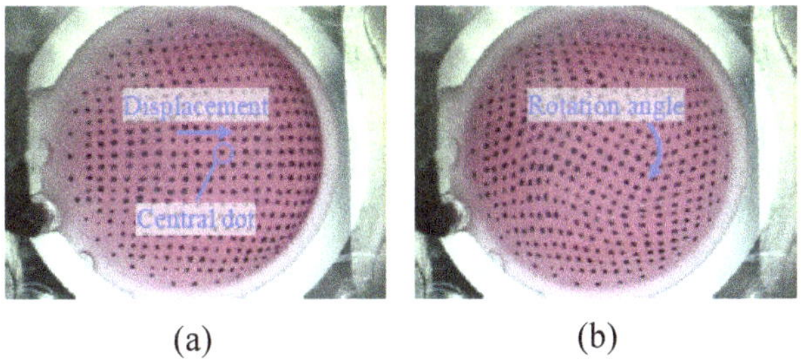

Figure 5: Deformation of the surface of the touchpad; (a) The displacement of the central dot. (b) The rotation angle of the contact surface.

slippage. The contact state between an elastic sensor and an object is in a state of continuous change from a sticking phase to a slipping phase until macroscopic slippage occurs. When tangential force is small, the elastic sensor surface is partially deformed and incipient slippage occurs in the contact region. Incipient slippage increases along with tangential force until macroscopic slippage finally occurs. If incipient slippage is detected, a robot can keep grasping an unknown object and prevent it from slipping, even if the mass and friction coefficients are unknown. Watanabe et al. proposed a method to calculate and maintain appropriate grip force on an unknown object by feeding back the slippage degree, based on proportional control [28]. When the slippage degree indicates perfect sticking, increasing the grip force should be avoided because it may crush the object. When incipient slippage becomes evident, the grip force should be increased before macroscopic slippage occurs.

In order to evaluate the method for estimating slippage degree, and to apply it to various systems, we first must represent the slippage degree quantitatively. The area ratio of the stick region to the total contact region is called the stick ratio ϕ, which is defined in [21,22] as follows:

$$\phi \equiv \frac{S_s}{S_c} \,. \tag{6}$$

Here, S_s is the area of the stick region, and S_c is the area of the contact region. When ϕ = 1, the contact region exists perfectly in the sticking state. As we increase the ratio of tangential force divided by normal force, the stick ratio then decreases. A slippage region appears in the contact region and gradually increases. This state is called incipient slippage. When incipient slippage progresses and the entire contact region becomes the slippage region, the slippage of the object is macroscopically visible and ϕ = 0. This state is called macroscopic slippage. The stick region and the slippage region are graphically represented as shown in Figure 6. In the following section, the estimation of the stick ration is presented.

Method for estimating stick ratio

We obtained the stick ratio from captured images of the surface of the touchpad. We used the relative displacement of each dot to the object between the reference image and the current image. The reference image is a previously captured image, such as the initial image taken before testing, which is appropriately updated [22]. Figure 7 shows the displacement of the dots between the reference and current images. If the relative displacement of a certain dot to the object is zero, the dot remains in its original position relative to the object. Otherwise, the dot slips on the object. Here, the displacement of the object can be approximated as the displacement of the central dot in the dot pattern [21]. Therefore, the dot *k* satisfies the following inequality as regarded in the stick region:

$$|d_0 - d_k| < \delta \,. \tag{7}$$

Here, d_0 and d_k are the displacement of the central dot and the dot *k*, respectively, between the reference image and the current image. δ is the minute threshold value. Next, we approximate the two regions Ss and *Sc* as the number of contacting and sticking dots. Therefore, the stick ratio is obtained in [22] as follows:

$$\phi \cong \frac{N_s}{N_c} \,. \tag{8}$$

Here, N_c and N_s represent the number of the dots in the contact region, and the number of the dots satisfying (7), respectively. When estimating the stick ratio, each dot in the contact region is evaluated based on whether it satisfies (7) between the reference image and the current image.

In the following section, we consider the mechanism of slippage in order to apply this estimation method to the fluid-type touchpad.

Mechanism of slippage degree

In order to estimate slippage degree, tactile sensors must satisfy the structure generating incipient slippage. This means that the stick region and slippage region occur simultaneously in the contact region. For example, if rigid objects are in contact with each other, macroscopic slippage occurs immediately, without incipient slippage, since the objects are barely deformed. When we used the elastic touchpad, we confirmed that the touchpad's structure generated incipient slippage based on Hertzian contact [29]. Here, Hertzian contact addresses the contact between an elastic spherical object and a rigid flat surface. Therefore, Hertzian contact is inefficient when considering fluid-type touchpads, as a fluid-type touchpad consists of a membrane surface and inner liquid.

In order to understand the mechanism of slippage occurring with the fluid-type touchpad, we compartmentalize the surface of the touchpad into many small segments, as shown in Figure 8. We consider the theory of small segments on the surfaces of both the elastic and fluid-type touchpads from a mechanics viewpoint. We also focus on the differences in the small segments between the elastic and fluid-type touchpads.

We formulate the balance of force of a small segment from the viewpoint of the static mechanics, dividing the sensor surface into an I × J array of segments. Figure 9 shows the balance of the segment *i, j (i=1,2,..., I, j=1,2,..., J)*, which is subjected to forces on the four-

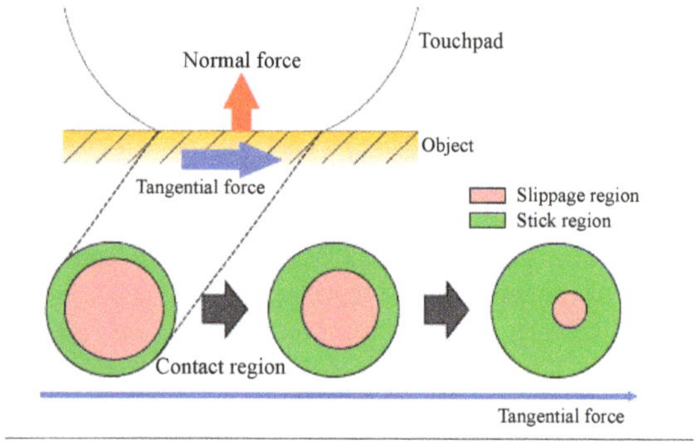

Figure 6: The stick region and the slippage region.

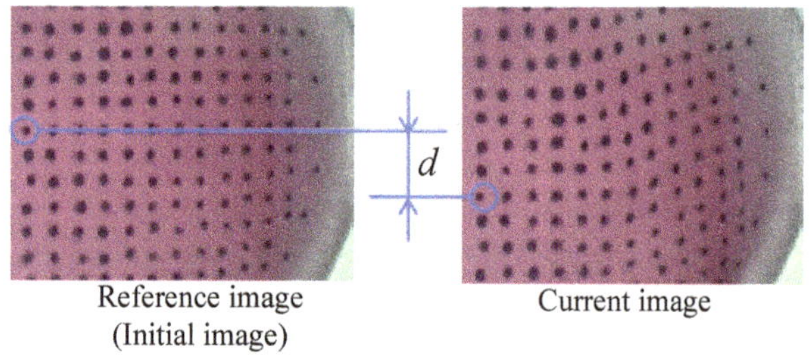

Figure 7: Displacements of dots between the reference image and current image.

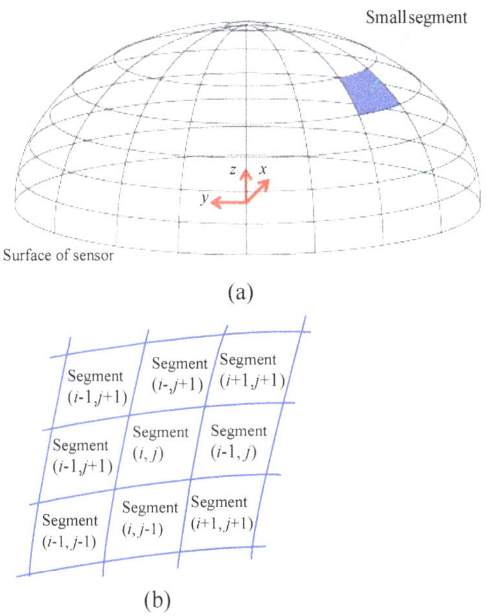

Figure 8: Small segments on the surface of the touchpad; (a) Compartmentalization of the surface. (b) Array of small segments.

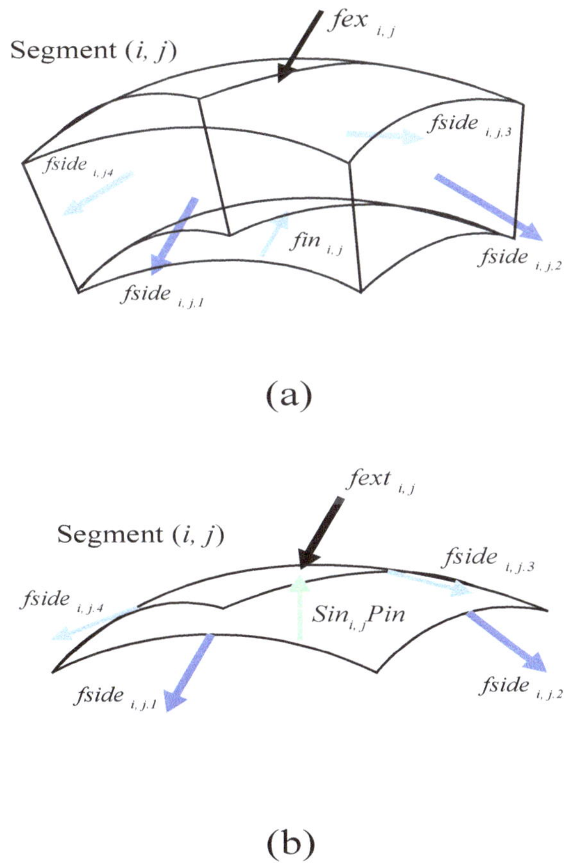

Figure 9: Balance of the small segment for formulation; (a) The small segment of the elastic touchpad. (b) The small segment of the fluid-type touchpad

sided cross-section surface $f^{side}_{i,j}k$ $(k = 1, 2, 3, 4)$, and a force on the inner cross-section surface $f^{ini,}_{j}$. External force is also applied when the touchpad is in contact with an object. The segment of the sensor surface is balanced as follows:

$$f^{ext}_{i,j} + f^{in}_{i,j}\left(\mathring{a}^{in}_{i,j}\right) + \sum_{k=1}^{4} f^{side}_{i,j,k}\left(\mathring{a}^{side}_{i,j,k}\right) = 0 \cdot \tag{9}$$

Here, $f^{ext}_{i,j}$ is the external force applied to the segment *i, j*. $f^{side}_{i,j}k$ *(k =1, 2, 3, 4)* and fini,j represent the functions of the strains of each cross-section surface $\varepsilon^{side}_{i,j}k$ *(k =1, 2, 3, 4)*, and the function of the strain of the inner cross-section surface $\varepsilon^{in}_{i,j}$, respectively. Here, the strain vector ε consists of normal strain and two-directional shear strains. The strains $\varepsilon^{side}_{i,j,k}$ *(k =1, 2, 3, 4)* and $\varepsilon^{in}_{i,j}$ are the functions of the gradient of the displacement distribution $u_{dis}(p^{seg}_{i,j})$ at the point $p^{seg}_{i,j}$ of the sensor surface, where $p^{seg}_{i,j}$ is the three-dimensional position of each small segment *i, j* based on the orthogonal coordinate system, as shown in Figure 8. Therefore, the forces on the cross-section surface are given as follows:

$$f^{side}_{i,j,k}\left(\mathring{a}^{side}_{i,j,k}\right) \equiv g^{side}_{i,j,k}\left(u_{dis}\left(p^{seg}_{i,j}\right)\right) \; (k = 1,2,3,4), \tag{10}$$

$$f^{in}_{i,j}\left(\mathring{a}^{in}_{i,j}\right) \equiv g^{in}_{i,j}\left(u_{dis}\left(p^{seg}_{i,j}\right)\right) \cdot \tag{11}$$

Substituting (10) and (11) into (9) expresses the external force for balancing the segment statically as follows:

$$f^{ext}_{i,j} = -g^{in}_{i,j}\left(u_{dis}\left(p^{seg}_{i,j}\right)\right) - \sum_{k=1}^{4} g^{side}_{i,j,k}\left(u_{dis}\left(p^{seg}_{i,j}\right)\right) \cdot \tag{12}$$

Here, we consider the relation between the external force and the position of the segment $p^{seg}_{i,j}$. We also define the orthogonal coordinate system *X(i, j, k)-Y(i, j, k)-Z(i, j, k)*, such that the *X(i, j, k)* axis and *Z(i, j, k)* axis are perpendicular to each cross-section surface and the outer surface of the segment i, j, respectively. When the segment *i, j* is in contact with the object and does not slip, the *X(i, j, k)* and *Y(i, j, k)*-directional gradients of the displacement distribution $u_{dis}(p^{seg}_{i,j})$ are not changed regardless of $u_{dis}(p^{seg}_{i,j})$, as the deformation on the *X(i, j, k)-Y(i, j, k)* plane is fixed. Therefore, the following equation is satisfied:

$$\left[\frac{\partial g^{side}_{i,j,k}\left(u_{dis}\left(p^{seg}_{i,j}\right)\right)}{\partial u_{dis}\left(p^{seg}_{i,j}\right)}\right]_X = 0 \cdot \tag{13}$$

Here, $[U^*]_{XY}$ represents the *X(i, j, k)* and *Y(i, j, k)*-directional components of *U**. Therefore, the relation between the external force and the segment position $p^{seg}_{i,j}$ satisfies the following equation:

$$\left[\frac{\partial f^{ext}_{i,j}}{\partial u_{dis}\left(p^{seg}_{i,j}\right)}\right]_X = \left[-\frac{\partial g^{in}_{i,j}\left(u_{dis}\left(p^{seg}_{i,j}\right)\right)}{\partial u_{dis}\left(p^{seg}_{i,j}\right)}\right]_X \cdot \tag{14}$$

Here, the size of the external force increases with the displacement $u_{dis}(p^{seg}_{i,j})$ in reference to the initial state, which in most cases is the non-contact state of the touchpad. This result, when taken with the elastic touchpad, is straightforward.

On the other hand, in the case of the segment of the fluid-type touchpad, $f^{in}_{i,j}$ is not the function of the strain of the segment, but also the inner pressure of the membrane p_{in} as follows:

$$\begin{cases}\left[g^{in}_{i,j}\left(u_{dis}\left(p^{seg}_{i,j}\right)\right)\right]_Z = S^{in}_{i,j}\, p_n \\ \left[g^{in}_{i,j}\left(u_{dis}\left(p^{seg}_{i,j}\right)\right)\right]_X = 0\end{cases} \cdot \tag{15}$$

Here, $[U^*]Z$ represents the *Z(i, j, k)*-directional component of *U**. $S^{in}_{i,j}$ is the area of the inner surface of the segment *i, j*, and the direction of the inner pressure p_{in} is vertical to the inner surface. Inner pressure is independent of segment position psegi,j as follows:

$$\frac{\partial \boldsymbol{g}^{in}_{i,j}\left(\boldsymbol{u}_{dis}\left(\boldsymbol{p}^{seg}_{i,j}\right)\right)}{\partial \boldsymbol{u}_{dis}\left(\boldsymbol{p}^{seg}_{i,j}\right)} = \boldsymbol{0}. \tag{16}$$

From (14) and (16), $f^{ext}_{i,j}$ is also independent of segment position $pseg^{i,j}$ as follows:

$$\left[\frac{\partial \boldsymbol{f}^{ext}_{i,j}}{\partial \boldsymbol{u}_{dis}\left(\boldsymbol{p}^{seg}_{i,j}\right)}\right]_X = -\left[\frac{\partial \boldsymbol{g}^{in}_{i,j}\left(\boldsymbol{u}_{dis}\left(\boldsymbol{p}^{seg}_{i,j}\right)\right)}{\partial \boldsymbol{u}_{dis}\left(\boldsymbol{p}^{seg}_{i,j}\right)}\right]_X = \boldsymbol{0}. \tag{17}$$

This characteristic of the fluid-type touchpad is different from that of the elastic touchpad. This indicates that the external force applied to the segments that do not slip does not increase by the displacement of the segment, even if the touchpad's surface is significantly deformed. For example, tangential force applied to the sticking segment does not change when a flat object, in contact with the touchpad, moves in a tangential direction. This characteristic is helpful for understanding the mechanism of incipient slippage when using the fluid-type touchpad.

Now, we consider the mechanism by which incipient slippage occurs when using the fluid-type touchpad. When the spherical surface of a fluid-type touchpad is in contact with an object, the pressure is uniformly distributed across the contact surface. This occurs because the applied normal force is balanced by the uniform inner pressure of the fluid-type touchpad, in contrast to the elastic touchpad, as shown in Figure 10. As described in the previous paragraph, tangential force applied to the sticking segment does not increase even when a flat object, in contact with the touchpad, moves in a tangential direction. Tangential force applied in the slippage region remains almost constant, and tangential force increases only at the border between the slippage region and stick region. Therefore, incipient slippage also occurs gradually in the border between the stick region and the slippage region. In our analysis, we demonstrated that the method for estimating slippage degree can be successfully applied to the fluid-type touchpad.

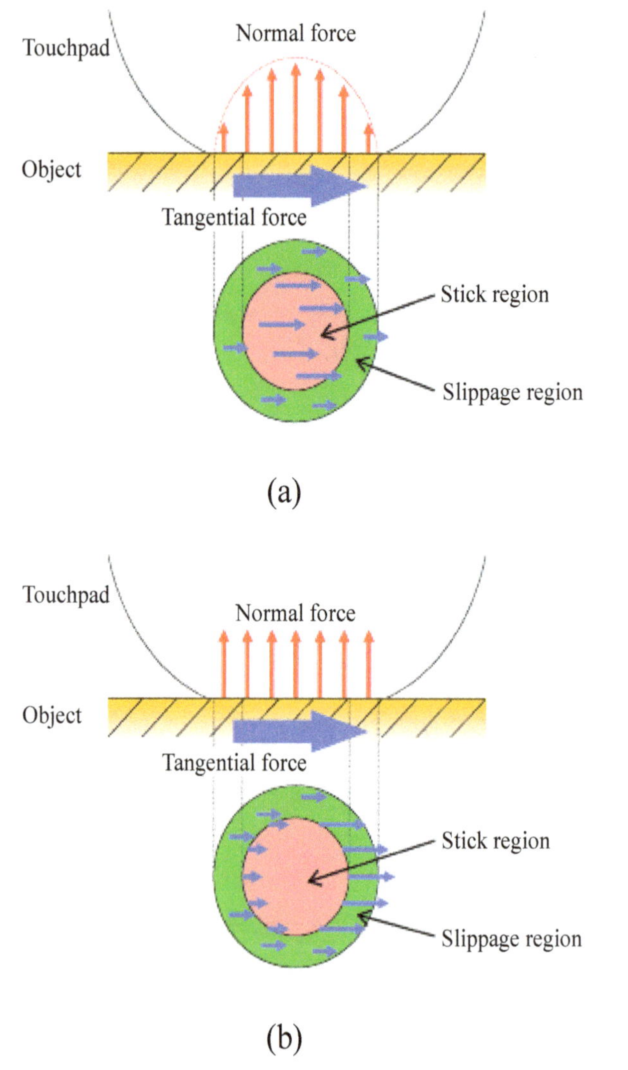

Figure 10: Mechanism of incipient slippage; (a) The elastic touchpad. (b) The fluid-type touchpad.

Experimental Results

In this chapter, the proposed methods for using the fluid-type touchpad are confirmed by the experiment's results. The proposed sensor was fixed on a movable stage, in contact with a flat object in a normal direction. When we moved the object and sensor on the movable stage, normal force, tangential force, moment and slippage were simultaneously generated. A laser displacement meter measured the displacement of the object to estimate and evaluate the slippage degree.

Measurement results of normal force

We measured normal force when the fluid-type touchpad was in contact with the object in a normal direction, altering the magnitude of the force. Figure 11 shows the measurement result of normal force. We can thus determine the relationship between normal force and the contact area S_c. The relationship between normal force and the contact depth D can also be successfully calculated. These relationships, defined as the functions F_S and F_D in (3), were identified as follows:

$$F_z = \begin{cases} F_S(S_c) = 0.00008S_c^2 + 0.0191S_c + 0.6242 \\ F_D(D) = 0.776D^2 + 0.5563D + 0.3435 \end{cases}. \tag{18}$$

Next, the above relationships changed when normal and tangential forces were simultaneously applied. In this case, we compensated the relationships by using the displacement of the central dot d_0, based on

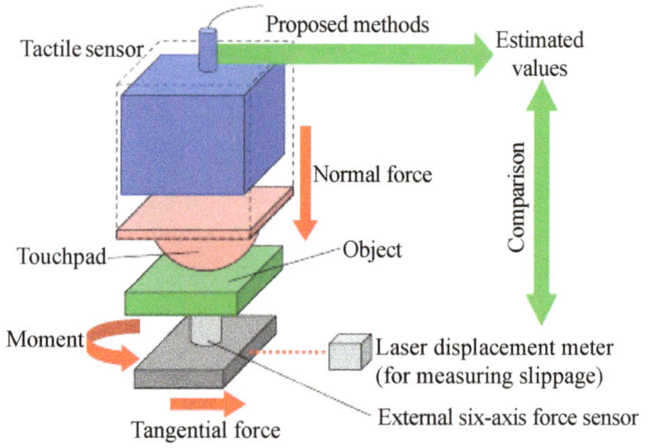

Figure 11: Configuration of the experimental setup.

the results of regression analysis as follows:

$$F_z = \begin{cases} 0.748F_S(S_c) - 0.637|d_c| + 1.7 \\ -0.356F_D(D) + 0.548|d_c| + 3\ .2 \end{cases} . \quad (19)$$

Figure 12 shows the results of this compensation. We can see its success when normal and tangential forces were applied, as the average absolute measurement errors in Figure 12a and 12b were 0.54 (N) and 0.46 (N), respectively.

Finally, we also compensated the relationships change when normal force and moment were simultaneously applied. Regression analysis yielded the following compensation:

$$F_z = \begin{cases} 1.2\ \ F_S(S_c) + 0.00171|\theta| - 1.8 \\ -0.272F_D(D) - 0.0074|\theta| + 2\ .5 \end{cases} . \quad (20)$$

Figure 13 shows the results of this compensation and its success when normal force and moment were applied. The average absolute measurement errors of the results in Figure 13a and 13b were 0.20 (Nm) and 0.090 (Nm), respectively. In all results, we confirmed that using the contact depth D obtained more accurate results. This is because the contact depth D is calculated more precisely than the contact area S_c, when using the previous methods in Ito et al. [23,24].

$$\{F_z\}_{\mathrm{N}} = \begin{cases} 0.748\{F_S(S_c)\}_{\mathrm{N}} - 0.637\{|d_c|\}_{\mathrm{m}} \ + 1.7 \\ 1.2\ \ \{F_D(D)\}_{\mathrm{N}} + 0.509\{|d_c|\}_{\mathrm{m}} \ - 3.5 \end{cases} . \quad (21)$$

$$\{F_z\}_{\mathrm{N}} = \begin{cases} 1.2\ \ \{F_S(S_c)\}_{\mathrm{N}} + 0.00171\{|\theta|\}_{\mathrm{deg}} - 1.8 \\ 0.6\ \ \{F_D(D)\}_{\mathrm{N}} - 0.00102\{|\theta|\}_{\mathrm{deg}} - 0.0518 \end{cases} . \quad (22)$$

The average absolute measurement errors in Figure 12a and 12b were 0.54 N and 0.27 N, respectively. The maximum errors in Figure 12a and 12b were 1.49 N and 0.72 N, respectively.

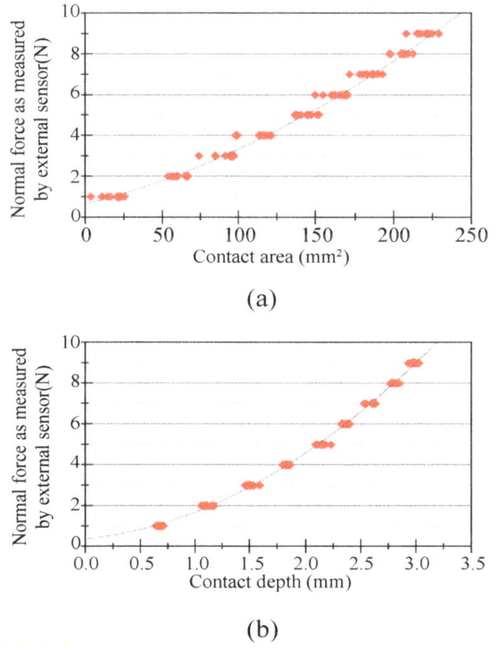

Figure 12: Measurement results of normal force; (a) Measurement using the contact area. (b) Measurement using the contact depth.

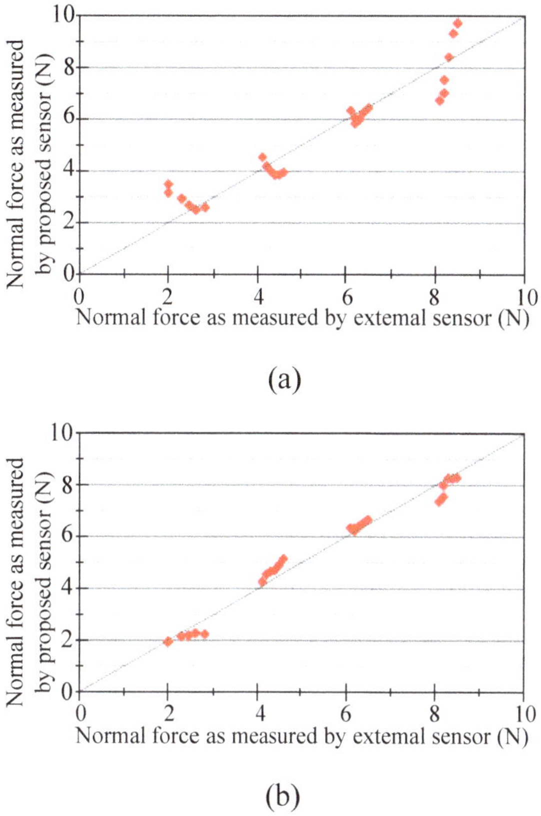

Figure 13: Measurement results of normal force when tangential force is applied; (a) Measurement using the contact area. (b) Measurement using the contact depth

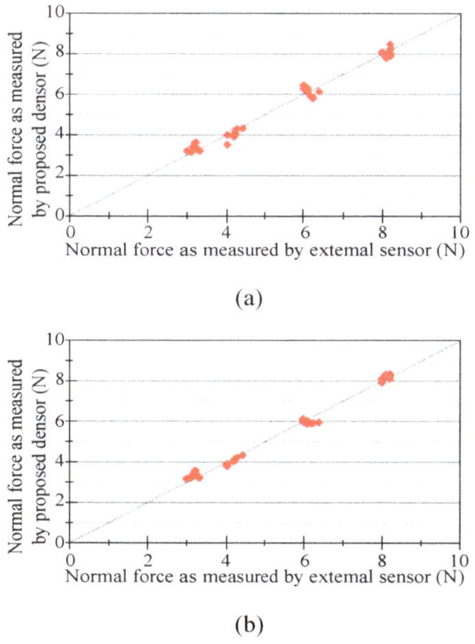

Figure 14: Measurement results of normal force when moment is applied; (a) Measurement using the contact area. (b) Measurement using the contact depth.

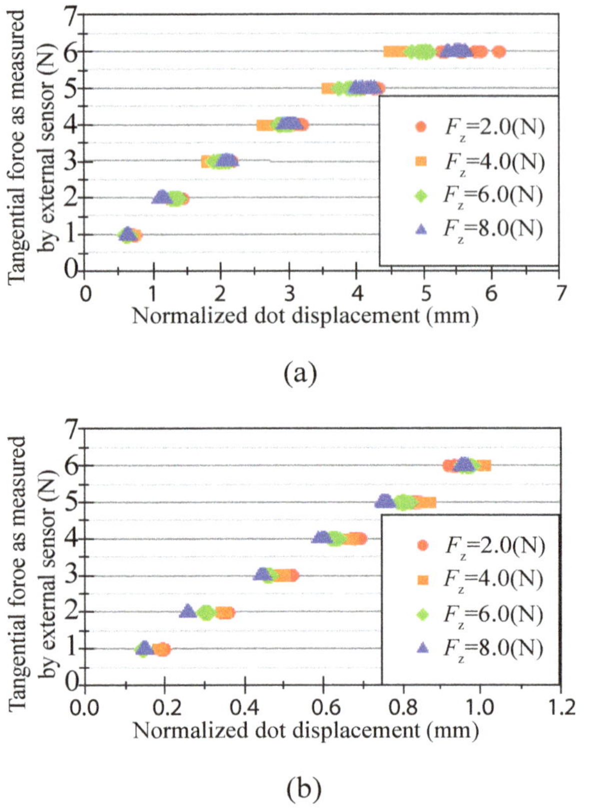

Figure 15: Measurement results of tangential force; (a) Measurement using the contact area. (b) Measurement using the contact depth.

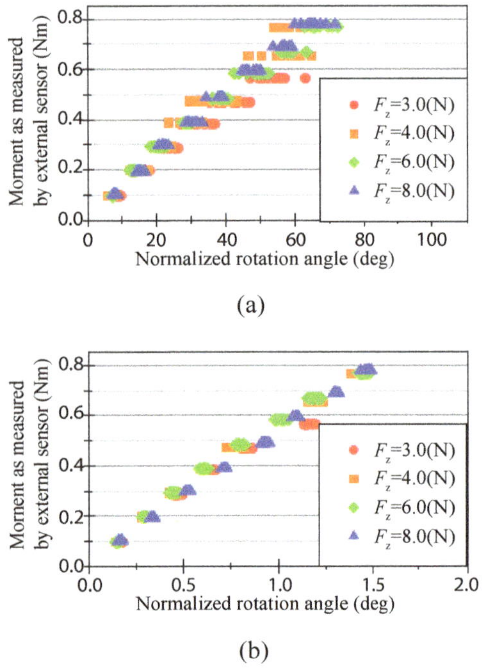

Figure 16: Measurement results of moment; (a) Measurement using the contact area. (b) Measurement using the contact depth.

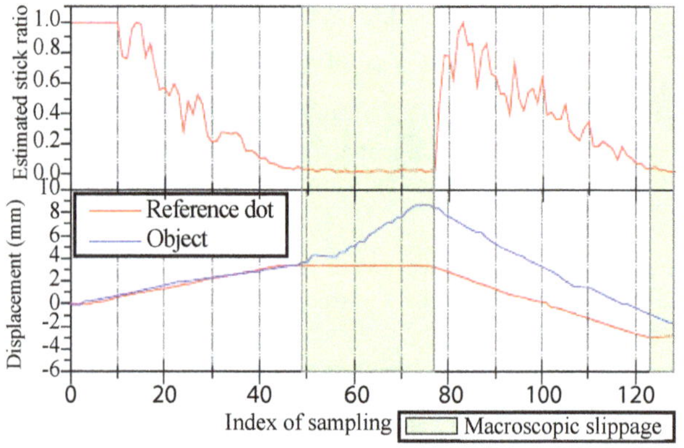

Figure 17: Results of the estimated stick ratio.

The average absolute measurement errors in Figure 13a and 13b were 0.20 N and 0.16 N, respectively. The maximum errors in Figure 13a and 13b were 0.49 N and 0.43 N, respectively.

Measurement results of tangential force

We measured tangential force when the object moves in a tangential direction while keeping in contact with the touchpad, as reported in Figure 14. The proposed method using the normalized displacement of the central dot d0nor in (4) can be successfully applied to tangential force as well, since the relationship between the normalized displacement and tangential force is almost linear. The parameters S_{ref}, D_{ref}, TS and T_D were identified as 100 mm^2, 10 mm, 0.65 and 0.75, respectively. In this experiment, using the contact depth D also obtained more accurate results.

Measurement results of rotational moment around normal direction

When we simultaneously applied normal force and moment in a normal direction, moment was measured by the proposed method using a normalized rotation angle of the contact surface θ^{nor} in (5), as reported in Figure 15. Figure 15 shows the linear relationship between the normalized rotation angle and moment. As demonstrated, the fluid-type touchpad can successfully measure moment. The parameters S_{ref}, D_{ref}, R_S and R_D were identified as 100 mm2, 10 mm, 1.7 and 2.2, respectively. Using the contact depth D also obtained more accurate results in this experiment.

Results of Estimation of Object Slippage: In this section, we demonstrate the results achieved when the slippage estimation method was applied to the fluid-type touchpad. Tangential force was applied when the touchpad was in contact with a flat object. Figure 16 shows the result of the estimated stick ratio when using the fluid-type touchpad. The upper and lower figures show the stick ratio and the displacements of the central dot and the object, respectively. We can see that the stick ratio was successfully estimated after macroscopic slippage occurred. The rapid increase of the stick ratio after initial macroscopic slippage occurred because a section of the incipient slippage region remained in a deformed state and joined the stick region. This shows that we can estimate the stick ratio in spite of a deformation in the stick region. When macroscopic slippage occurs in the opposite direction, the proposed method can still estimate the stick ratio. Since these results are equivalent to the results achieved when using the elastic touchpad [22], this demonstrates that the slippage estimation method can be successfully applied to the fluid-type touchpad. We confirmed that the fluid-type touchpad can also estimate slippage degree using the proposed method.

Conclusion

We have achieved an estimation of contact force and slippage by using a tactile sensor with a fluid-type touchpad. Consequently, we have proposed a new method based on look-up tables with newer, more efficient compensations for measuring normal force, tangential force and moment. Secondly, we have clarified the mechanism of slippage when using the fluid-type touchpad. We have shown that the fluid-type touchpad is able to generate and estimate slippage degree accurately. Finally, the validation of our proposed methods using the fluid-type touchpad has been confirmed in the experimental results.

Our developed sensor can be fabricated easily and at a low cost, as the sensor has a simple structure and does not require complex sensing elements or wiring. Although the size of the sensor developed and used in this study is relatively large, it can be easily downsized by using a smaller CCD/CMOS camera.

In the process of contributing to this paper, our vision-based sensor with a fluid-type touchpad was developed to a greater level of practicality. Combined with the previous work [23-25], our sensor has demonstrated that it can simultaneously obtain multiple types of tactile information, including the contact force, moment, slippage degree, shape, contact region, position and orientation of an object in contact with the fluid-type touchpad. Future work involves the implementation of fluid-type tactile sensors across the industrial and medical fields and in various practical applications, such as robot hands for dexterous handling.

References

1. Lee MH, Nicholls HR (1999) Tactile sensing for mechatronics-a state of the art survey. In Proc. of the Mechatronics 1-31.
2. Dahiya RS, Metta G, Valle M, Sandini G (2010) Tactile Sensing—From Humans to Humanoids. IEEE Trans. on Robotics 26: 1-20.
3. Yamada D, Maeno T, Yamada Y (2002) Artificial Finger Skin having Ridges and Distributed Tactile Sensors used for Grasp Force Control. Journal of Robotics and Mechatronics 14: 140-146.
4. Noda K, Hoshino K, Matsumoto K, Shimoyama I (2006) A shear stress sensor for tactile sensing with the piezoresistive cantilever standing in elastic material. Sensors and Actuators A: Physical 127: 295-301.
5. Schmitz A, Maggiali M, Natale L, Bonino B, Metta G (2010) A Tactile Sensor for the Fingertips of the Humanoid Robot iCub. In Proc. of IEEE/RSJ International Conference on Intelligent Robots and Systems 2212-2217.
6. Hakozaki M, Shinoda H (2002) Digital tactile sensing elements communicating through conductive skin layers. In Proc. of IEEE/RSJ International Conference on Intelligent Robotics and Automation 3813-3817.
7. Yamada K, Goto K, Nakajima Y, Koshida N, Shinoda H (2002) Wire-Free Tactile Sensing Element Based on Optical Connection. In Proc. of 19th Sensor Symposium 433-436.
8. Yang S, Chen X, Motojima S (2006) Tactile sensing properties of protein-like single-helix carbon microcoils. Carbon 44: 3352-3355.
9. Takao H, Sawada K, Ishida M (2006) Monolithic Silicon Smart Tactile Image Sensor With Integrated Strain Sensor Array on Pneumatically Swollen Single-Diaphragm Structure. IEEE Trans. on Electron Devices 53: 1250-1259.
10. Engel J, Chen J, Liu C (2003) Development of polyimide flexible tactile sensor skin. Journal of Micromechanics and Micro Engineering 13: 359-366.
11. Mei T, Li WJ, Ge Y, Chen Y, Ni L, Chan MH (2000) An integrated MEMS three-dimensional tactile sensor with large force range. Sensors and Actuators A: Physical 80: 155-162.
12. Engel J, Chen J, Liu C (2003) Development of a multimodal, flexible tactile sensing skin using polymer micromachining. In Proc. of 12th International Conference on TRANSDUCERS, Solid-State Sensors, Actuators and Microsystems 2: 1027-1030.
13. Ferrier NJ, Brockett RW (2000) Reconstructing the Shape of a Deformable Membrane from Image Data. International Journal of Robotics Research 19: 795-816.
14. Saga S, Kajimoto H, Tachi S (2007) High-resolution tactile sensor using the deformation of a reflection image. Sensor Review 27: 35-42.
15. Johnson MK, Adelson EH (2009) Retrographic sensing for the measurement of surface texture and shape. In Proceedings of the IEEE Conference on Computer Vision and Pattern Recognition 1070-1077.
16. Kamiyama K, Vlack K, Mizota T, Kajimoto H, Kawakami N, et al. (2005) Vision-Based Sensor for Real-Time Measuring of Surface Traction Fields. IEEE Computer Graphics and Applications 25: 68-75.
17. Sato K, Kamiyama K, Nii H, Kawakami N, Tachi S (2008) Measurement of Force Vector Field of Robotic Finger using Vision-based Haptic Sensor. In Proc. of IEEE/RSJ International Conference on Intelligent Robots and Systems 488-493.
18. Ohka M, Mitsuya Y, Matsunaga Y, Takeuchi S (2004) Sensing characteristics of an optical three-axis tactile sensor under combined loading. Robotica 22: 213-221.
19. Ohka M, Takata J, Kobayashi H, Suzuki H, Morisawa N, et al. (2009) Object exploration and manipulation using a robotic finger equipped with an optical three-axis tactile sensor. Robotica 27: 763-770.
20. Yamada Y, Iwanaga Y, Fukunaga M, Fujimoto N, Ohta E, et al. (1999) Soft Viscoelastic Robot Skin Capable of Accurately Sensing Contact Location of Objects. In Proc. of IEEE/RSJ/SICE International Conference on Multisensor Fusion and Integration for Intelligent Systems 105-110.
21. Obinata G, Ashis D, Watanabe N, Moriyama N (2007) Vision Based Tactile Sensor Using Transparent Elastic Fingertip for Dexterous Handling. Mobile Robots: Perception & Navigation 137-148.
22. Ito Y, Kim Y, Obinata G (2011) Robust Slippage Degree Estimation based on Reference Update of Vision-based Tactile Sensor. IEEE Sensors Journal 11: 2037-2047.
23. Ito Y, Kim Y, Nagai C, Obinata G (2011) Contact State Estimation by Vision-based Tactile Sensors for Dexterous Manipulation with Robot Hands Based on Shape-Sensing. International Journal of Advanced Robotic Systems 8: 225-234.
24. Ito Y, Kim Y, Nagai C, Obinata G (2012) Vision-based Tactile Sensing and Shape Estimation Using a Fluid-type Touchpad. IEEE Trans. on Automation Science and Engineering 9: 734-744.
25. Ito Y, Kim Y, Obinata G (2011) Multi-axis Force Measurement based on Vision-based Fluid-type Hemispherical Tactile Sensor. In Proc. of IEEE/RSJ International Conference on Intelligent Robots and Systems 4729-4734.
26. Xydas N, Kao I (1999) Modeling of contact mechanics and friction limit surface for soft fingers in robotics with experimental results. International Journal of Robotics Research 18: 941-950.
27. Kao I, Yang F (2004) Stiffness and Contact Mechanics for Soft Fingers in Grasping and Manipulation. IEEE Trans. on Robotics and Automation 20: 132-135.
28. Watanabe N, Obinata G (2008) Grip Force Control Using Vision-Based Tactile Sensor for Dexterous Handling. In Proc. of the European Robotics Symposium 44: 113-122.
29. Johnson KL (1987) Contact Mechanics. Cambridge University Press.

Developing Serpentine Robot Control to use Lateral Affordances

William R Hutchison*, Betsy J Constantine and Jerry Pratt

MeMeMe Inc, 1470 Birchmount Rd, Scarborough, ON M1P 2G1, Canada

Abstract

This paper describes the unique challenges in developing control of complex lateral movements needed by a serpentine robot to ascend steep slopes by climbing over and around affordances/obstacles on the slope. The research extends previous serpentine robot work developing control of sagittal movements for climbing up stairs and over uneven parallel bars. Effective lateral control was developed using an iterative combination of learning, a genetic algorithm, and developer programming. The robot's many simultaneous movements were controlled mostly as a function of very local sensory inputs and little centralized coordination.

Keywords: Robots; Serpentine robot; Genetic algorithm; Local sensory inputs.

Introduction

Snakes can traverse as wide a range of terrains as almost any animal, with unique advantages in terrains with narrow passageways and wide horizontal and vertical gaps that can be spanned by their long bodies. In hopes of duplicating this range of mobility, a number of serpentine robots have been developed in recent years. Early serpentine robots tended to imitate snake locomotion, so had active joints but no wheels or tracks on the segments [1-3]. The most effective serpentine robots across a wide range of terrains have active joints and active tracks or wheels on the segments, including the Souryu-I [4], Millibots [5], Moira [6] and OmniTread OT-4 [7-9]. Much research has focused on developing methods for controlling lateral movements involving only ventral (downward) contact, but many of the terrains of interest for these robots require the use of lateral affordances, sometimes referred to as obstacles. Our previous work has focused on saggital (i.e., up and down) control to use affordances such as steps and bars, while the current study focuses on lateral movements to use affordances typically found in natural environments. Transeth et al. [10] provide a sophisticated analysis of the physics of lateral movements with affordances, but they do not provide a control strategy to enable the robot to actively locate and use affordances not appropriately located along the robot's default path.

Controlling Serpentine Robots

Like other complex robots, serpentine robots are very challenging to control because of their many degrees of freedom (DOF) as well as the complexity of sensors needed in many terrains.

Approaches to controlling serpentine robots

Three approaches that have been used to control serpentine robots are briefly described.

Manual operation: The earliest form of control for these robots was manual control, where operators could see the robot and terrain. Manual operation enables demonstrations of the robot's mobility, but the complexity of movements and the need for complete overview means it cannot be used in most situations of interest to potential users. For example, the OT-4 robot in manual mode requires three operators for six joints, and usually speed will be well below the robot's physical limits even in terrains requiring control mainly in only one plane at a time.

Aim and propagate: The most common control approach potentially usable in real applications takes advantage of the fact that the bodies of active drive serpentine robots usually follow approximately the same path as the head. The human operator controls head movements while observing the robot or viewing a display from a camera mounted on the robot, and automated control propagates the head's actions back to following segments, controlling each joint when it reaches the same location along the path. Propagation algorithms have been based on either fairly straightforward odometry or analytic methods such as clothoids adapted from rail track analyses [3,11,12]. Passive and active propagation of head movements can be adequate for certain terrains, such as maneuvering on level ground through obstacles, up easy stairs, or through a sewer line. However, for the more demanding terrains, there are several major flaws with propagation of head movements as a sole strategy:

1. The head is a special case whose joint positions often cannot or should not be copied by subsequent joints;

2. Deformation of the terrain from compression or displacement may require different movements for different segments;

3. Changes in robot orientation from rolling, slipping, and pitching often alter movement requirements for subsequent segments; and

4. Operators often need automated assistance with head movements, not just body movements.

Fixed gaits

Evolutionary control development methods, including genetic algorithms (GAs) [13] and direct policy search [14], have been applied successfully to a number of robot control problems. Dowling demonstrated the use of GAs to develop control for a range of serpentine movement patterns, including snakelike gaits (e.g., lateral undulation, side winding) and non snake like gaits (e.g., rolling like a wheel).

Uniform gaits can be effective on smooth surfaces or easy terrains,

***Corresponding author:** William R. Hutchison, MeMeMe Inc, 1470 Birchmount Rd, Scarborough, ON M1P 2G1, Canada, E-mail: hutchison.w@gmail.com but uniform gaits will fail in complex natural terrains because the terrain may vary along the length of the serpentine body, requiring different kinds of movements at different points along the body. Even a simple transition from one terrain type to another may require different behaviors of segments before and after the transition point in the terrain, with each behavior chosen based on local sensory input.

Problems in controlling serpentine robots: Serpentine robots are difficult to program or manually operate in real time because many coordinated movements requiring interpretation of complex sensor data from external and internal sensors are required to traverse natural terrain features.

In addition to the general problems associated with control of high-DOF robots, serpentine robots present three major control challenges.

1. Tight coupling prevents independent movement of segments, because:

- The limits on joint angles potentially constrain the positions of all coupled segments,
- Actuating each joint affects all attached segments, and
- The leverage of distal coupled segments can often prevent movement when torque is applied.

This is in stark contrast with legged robots where most target leg movements can be executed relatively independently of each other.

2. Situational awareness is much more of a challenge for both autonomous control and teleoperation because most of body is behind the head. Some solutions and their drawbacks include:

- A rear-mounted camera, which has been implemented or planned on several robots [11], but many terrains do not allow seeing the front from behind (or vice versa);
- Touch sensors, which are valuable for knowing what is around the body, but contact is often lost and 3D spatial memory is very challenging; and
- Odometry, which is essential, but existing contact and visual odometry are often poor due to longitudinal and lateral slippage, rolling, terrain gaps, turning, and 3-dimensionality.

3. The extended body of a serpentine robot makes it very likely that different kinds of movement strategies will be necessary simultaneously at different points along the body. Each movement must be suited to its local terrain conditions, making uniform gaits and propagation of head movements ineffective in many situations.

Developing control of the Omni Tread OT-4 serpentine robot

The current study aimed to develop teleoperational control that would succeed in a range of challenging terrains. In this approach, the operator sets the overall direction and the control system provides assistance to the operator for close-in head movements and completes control of all other body movements. The goal was to have a single control system handle all target terrains automatically, including transitions between them, rather than having specialized control for each terrain. Not only would the latter approach require knowing when to switch, but even the best switching would sometimes fail in transitions or in heterogeneous terrains.

The OmniTread OT-4 serpentine robot: The OmniTread OT-4 robot, shown in Figure 1, has seven segments, each 8.2 cm wide, 8.2 cm high, and 10.3 cm long (center segment is 10.9 cm long). Its total

Figure 1: The Omni Tread OT-4 serpentine robot climbing over parallel bars under automated control.

length is 94 cm and it weighs 4.0 kg. Each segment has four sides, each side nearly covered by two tracks. The tracks are driven by worm gears on a shaft from a single electric motor in the center segment. Each two-DOF joint is activated by four pneumatic bellows capable of a wide range of lengths and pressures, enabling the body to be stiff to cross wide gaps or relaxed to comply with uneven terrain. The maximum joint angle is approximately 40 degrees. Gas for the bellows can be provided by onboard cylinders or an onboard compressor. The OT-4 is equipped with accelerometers on each segment to sense pitch and roll and potentiometers on each joint to sense joint pitch and roll.

Control Developed With the Seventh Generation Control System

Control of the OmniTread OT-4 serpentine robot was developed using the Seventh Generation (7G) Control System, a software system implementing a combination of reinforcement learning and genetic algorithms along with extensive support for sensory preprocessing, simulations, and user interface components [15,16]. Control was developed in simulation in a series of iterations using the Yobotics! Simulation Construction Set [17]. In each iteration, the developer made manual changes to a set of programmed behaviors ("scripts") and used learning and a GA in repeated simulation runs to optimize the sensory and behavior parameters [16]. The robot's movements were grouped into four logical movement groups: sagittal and lateral head movements and sagittal and lateral bod movements. For example, the head sagittal movement group includes lift, push down, relax, hold, and continue previous action. Note that if every joint's sagittal and lateral movements were specified as a movement group, the agent would have at least 12 groups. Following a strategy that can often be applied to many high-DOF robots, however, a single movement group produced default sagittal behaviors for all joints behind the first one. The default behaviors could be overridden in various situations for the different segments (e.g., special tail behaviors). The same strategy was applied to lateral body behaviors.

Development of sagittal control

Initial development focused on controlling the OT-4 in two terrains, stairs and a series of uneven parallel bars that could be traversed by controlling only sagittal movements. The sagittal control program is briefly described as follows. The head was commanded to lift when obstacles were sensed within about half a segment length ahead of the robot if the top of an obstacle was above a line sloping upward ahead of the "chin." If the head was lifted nearly to its limit and the obstacle was still above it, the second joint was lifted, and similarly for the third joint. When the bottom track of the head touched an edge (defined as contact on a narrow section of track) behind its "chin," the head was commanded to push down until the segment was slightly beyond horizontal (i.e., with "chin" down). The main control strategy for body movements attempted to keep the bottom of each segment as horizontal as possible whenever its bottom track touched an edge by rotating the joints ahead and behind appropriately in opposite directions. When an edge was sensed at the very front of a segment, the joint ahead of the edge was kept closed to keep the gap at the joint from getting hung up on the edge. If a joint received conflicting commands from the segment ahead of it and the segment behind it, the command from the segment behind the joint was executed. If joints #2 and #3 received commands from both the head (e.g., to lift for obstacles) and neighboring body segments (e.g., to lower a joint), the command from the head was executed. If a joint received no command, the default action was to reduce the current torque slightly. The initial test terrains stairs and parallel bars did not require lateral movements, so the control system simply kept the joints straight laterally.

The sagittal control program enabled the simulated OT-4 to climb 20.3 cm high stairs and over a series of parallel bars spaced at random heights and gaps near the robot's physical limits. The control developed in simulation was then tested on the real OT-4 in corresponding terrains in the laboratory. Figure 1 shows the OT-4 traversing the parallel bars. For the tests, the 7G control system used input from human observers to substitute for an array of IR sensors on the head and touch sensors along the bottom and sides of each segment that were modeled in simulation but not present on the real OT-4 [18]. The autonomous control system enabled the OT-4 to climb the stairs and parallel bars terrains in the laboratory test courses, moving at higher speeds than could be demonstrated with human operators.

Extending serpentine control to lateral movements

Although sagittal control was effective for unobstructed stairs and parallel bars, lateral movements are often necessary for mobility and stability, especially for climbing where there are few surfaces with low enough slope for traction. In the sagittal plane, gravity inescapably arranges pressure between the robot and objects beneath it, but in the lateral plane such force must usually be created by intelligent movements of the robot. The physics of force resolution that dominate sagittal plane movements for climbing can be extended in large part to lateral contact with objects, because the component of the force of gravity parallel to the surface resolves into forces normal to the lateral contact and parallel to it. The lateral friction combines with the friction on the bottom tracks to prevent slipping.

Implementing aim and propagate lateral control

As a first step in developing control of lateral movements, a version of the Aim and Propagate approach was implemented. The head joint was programmed to comply with horizontal joystick positions whenever the operator was also pressing an override button on the joystick. Unlike other approaches in which commanded joint positions for the head were propagated to all following joints, each joint was programmed to propagate the actual joint position of the previous joint when it reached the corresponding location in the terrain. This approach was much better at accommodating cumulative unplanned changes in position (e.g., from slipping, rotating, or terrain deformation) as the robot moved through rough terrain. Position data obtainable in simulations permits perfect odometry for propagation of movements, but on the real OT-4 position was estimated by applying a calibration factor to the product of track speed time's elapsed time. This simple programming enabled effective manual control of the robot through both simulated and real slalom courses, as well as remote teleoperation in simulated and real courses where the operator could rely only on camera input from the front of the robot.

Combining lateral and sagittal control

The next step was to test a combination of sagittal control of the head and body joints with the Aim and Propagate lateral control. The real stairs terrain was modified by adding large irregularly shaped rocks along a winding path up a staircase, as shown in Figure 2 [18]. The operator's joystick position determined the head orientation whenever the operator pressed the override button. This hybrid control enabled the real OT-4 to climb this challenging course, given that the joystick operator could directly observe the entire robot and terrain. The most obvious control weakness was that when the tracks slipped, the simple approximation to odometry became very inaccurate, causing lateral movements to propagate before reaching the corresponding location. This produced obviously suboptimal movements, but the robot could still move forward because the lateral movements pressed the side tracks against rocks to enable forward movement, although with unnecessary stress on joints and on the terrain. The operator learned through experience to add a certain amount of continuous lateral head oscillation in order to avoid persisting in stuck positions. It was also noted that propagating each joint position from the previous joint rather than from the head enabled recovery from odometry errors within one segment length of movement after slippage ended. While accurate odometry can potentially improve performance, subsequent testing in simulation with perfect odometry showed that propagation alone is insufficient for terrains such as this. However, propagation provided a valuable default behavior that could be overridden by more intelligent control developed in the next phase.

Developing intelligent lateral control

The previous tests showed that a relatively small set of general movement strategies was useful in a wide range of terrains. Much of their effectiveness is due to movements being controlled mostly on the basis of local sensory input, producing several "subgait" level behaviors along the length of the robot at the same time. The next phase of development added intelligent lateral control. A large and important class of terrains requires much more sophisticated lateral movements than the wiggling movements that had been successful in narrow passageways between rocks. Lateral objects have potential value for mobility and stability in many terrains, but they are essential for some terrains such as climbing slopes with few narrow passageways or horizontal affordances.

Figure 2: The Omni Tread OT-4 robot on stairs terrain with rocks, requiring control of sagittal and lateral movements.

There are two general classes of lateral movement strategies for climbing slopes. One is what was observed among the rocks on the stairs, which was to arch by bending multiple joints to press against objects on both sides of the path. The other class is necessary in terrains without objects on both sides, where the best movement is to move above and behind objects along the way to push against surfaces that produce upward force components. This phase of development focused on the latter class because a physics analysis suggests that it is a direct extension of the existing sagittal control. Moreover, it is consistent with the overall development strategy of focusing first on very local control of movements based on mostly local sensory input, as opposed to the more global arching behavior for narrow passages.

Terrain for developing lateral control

The terrain designed for this phase is a ramp, shown in Figure 3, that is too steep simply to climb in a straight line (i.e., the tangent of the slope is greater than the coefficient of static friction), but the slope contains objects analogous to rocks or posts along the path on which the robot can climb. As in the sagittal control problem, touch sensors provide simple and relatively unambiguous information about the location of objects, although they require intelligent control to maintain effective contact with those objects as well as some short-term memory to remember objects when contact is temporarily lost.

Method for developing intelligent lateral control

Following the same iterative development procedure used to develop sagittal control, effective lateral control was developed for this terrain. In this terrain, lateral body movements must use sensor input to relate effectively to the objects in order to climb successfully. The robot must approach nearby affordances, move behind or above them, turn back in the goal direction at appropriate times, and avoid getting hung up on edges between joint spaces on its sides.

The new lateral climbing behaviors were triggered only if the robot sensed that it was on a slope steep enough to risk slipping. As with sagittal collision avoidance, the head turns away to avoid hard collisions with objects. However, just after passing objects, the head turns toward the object to touch it along the side of the head. When on a slope, all segments turned toward objects touching their sides. Short-term memory based on lagged touch sensor input enabled segments to remember the location of objects in order to turn toward them when contact was temporarily lost.

As with the sagittal head movements, the segments behind the head were successively recruited for lateral movements as needed: if joint #1 is bending near its limit and segment #2 is not touching anything on the side opposite to joint #1's turning direction, joint #2 is commanded to turn in the same direction as joint #1, and so on through joint #5.

Figure 3: Terrain for developing intelligent lateral control of the OT-4 model in the Yobotics! Simulation Construction Set.

When no object was detected on either side or in front, the default head behavior was to turn toward the direction parallel to the slope of the ramp (a short-range goal that would be set by the operator in a fielded teleoperation application). The default behavior for the other joints was to propagate the angle from the previous joint at the same location.

A subtler behavior was to lift the head whenever the head was changing direction so as to avoid pivoting the body, but to push the head down with moderate force whenever the head was fixed in a sideways orientation in order to pull the body to follow the head. Because the robot can get hung up on a lateral edge just as on a vertical edge, a corresponding joint-closing behavior was implemented. Explicit sensor preprocessing inferred that a joint was hung up when the current and all lagged front sensors were touching only at the front of the adjoining segment, implicitly indicating no movement. Segments #5 and #6 just in front of the tail had additional behaviors that overrode the control scripts that otherwise were the same for all body joints, as follows. In addition to segments #5 and #6 commanding the joint behind them to turn toward whichever side they touch, those two rear joints are triggered by touch on the segment behind them to turn away from, rather than toward, the touch unless the contact is at the front (in which opening the joint risks hanging up).

The developers found that developing lateral control was significantly more difficult than developing sagittal control due to:

- Gravity not automatically maintaining contact with terrain affordances in this plane,
- Conflicts between movements for mobility and movements to maintain touch,
- The need to remember the location of objects while not contacting them,
- The need to learn sensory patterns to use as cues to approach objects, and
- The need to coordinate head lifting and downward pushing with lateral head and body movements to make turning effective (see below).

Results

The new control system developed in this phase retained the previously developed effective sagittal head and body control, retained control by propagation developed previously as the default lateral behavior, and added intelligent lateral movements as overrides in appropriate situations. Figure 4 shows a sequence of snapshots from a video made in the Yobotics simulator showing the simulated OT-4 robot climbing the ramp by pressing against the available posts and rocks in the simulated terrain[1]. In Figure 4, panels h, i, and j, the rear segments of the robot can be seen maneuvering laterally to climb using a post, while the middle segments are moving sagittally to pull up over the ledge, and the head is turning sideways to avoid collision with the wall. All these movements must be executed correctly and simultaneously to succeed in climbing this terrain. The subgait control strategy succeeded in this case because each movement is in response to local terrain characteristics and movements can be executed independently of each other.

The combination of relatively simple behaviors often produces oscillation, such as turning toward and away from the objects.

[1]The complete simulation video is available at: http://www.behaviorsystems.com/files/ramp with posts and ledge.mov

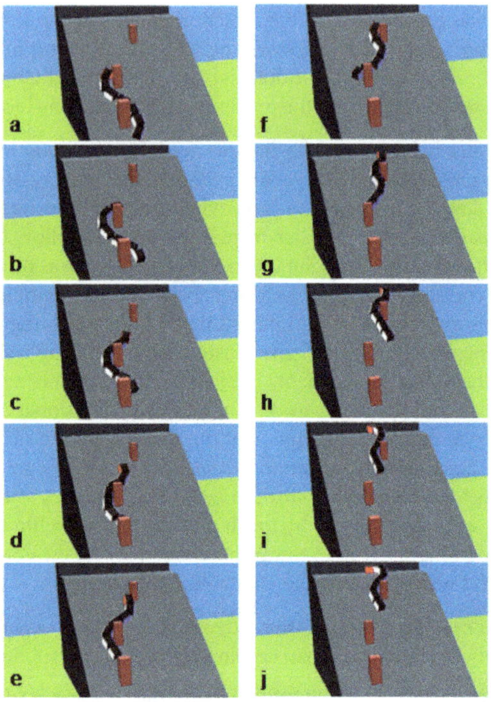

Figure 4: Snapshots from simulation of OT-4 robot climbing a steep ramp that required intelligent lateral movements to use posts to avoid slipping down the ramp, along with saggital movements to pull itself over the ledge and lateral head movements to avoid hitting the wall.

Oscillation-reducing alternatives were tested, but the simpler control based on seeking and avoiding proved more robust across the variety of situations encountered. A number of observers of the simulations have commented that the behaviors, including the natural oscillation, give the impression of being more like an animal "trying" to climb the terrain than a programmed mechanical motion.

This last study was done in simulation and has not yet been validated by controlling the real OT-4 robot climbing an equivalent ramp. Previous validation tests of control of the real OT-4, however, have shown that 7G control developed in simulation transferred well to controlling the real OT-4 robot in similar terrains in the laboratory [18].

References

1. Hirose S (1993) Biologically inspired robots: Snake-like locomotors and manipulators. Oxford: Oxford University Press.
2. Nilsson M (1997) Snake robot free climbing in Proc. International Conf. on Robotics and Automation (ICRA), Albuquerque, NM.
3. Ikeda H, Takanashi N (1987) Joint assembly moveable like a human arm. US Patent 4683406 Assignee: NEC Corp.
4. Takayama T, Hirose S (2000) Development of Souryu-I connected crawler vehicle for inspection of narrow and winding space.26th Annual Conf. of the IEEE Ind. Electronics Society (IECON), Nagoya, Aichi, Japan.
5. Brown H, Weghe J, Bererton C, Khosla P (2002) Millibot trains for enhanced mobility. IEEE/ASME Trans. Mechatronics 7: 452-461.
6. Osuka K, Kitajima H (2003) Development of mobile inspection robot for rescues activities: MOIRA. IEEE/RSJ Intl. Conf. on Intelligent Robots and Systems (IROS), Las Vegas, NV.
7. Grzegorz G, Malik GH, Borenstein J (2005) The OmniTread serpentine robot for industrial inspection and surveillance. Industrial Robot: An International Journal 32: 139-148.
8. Borenstein J, Granosik G, Hansen M (2005) The Omni Tread serpentine robot design and field performance. SPIE Defense and Security Conference: Unmanned Ground Vehicle Technology VII, Orlando, Florida.
9. Borenstein J, Hansen J, Nguyen H (2006) The Omni Tread OT-4 serpentine robot for emergencies and hazardous environments. 2006 International Joint Topical Meeting: "Sharing Solutions for Emergencies and Hazardous Environments, Salt Lake City, Utah, USA.
10. Transeth A, Leine R, Glocker C, Pettersen K, Liljeback P (2008) Snake Robot Obstacle-Aided Locomotion: Modeling, Simulations, and Experiments. IEEE Trans. Robotics 24: 88-104
11. Kamegawa T, Yamasaki T, Igarashi H, Matsuno F (2004) Development of the snake-like rescue robot KOHGA. IEEE Intl. Conf. on Robotics and Automation, New Orleans, LA, USA.
12. Klaassen B, Paap K (1999) GMD-SNAKE2: A snake-like robot driven by wheels and a method for motion control. IEEE International Conference in Robotics and Automation.
13. Dowling K (1997) Limbless locomotion: Learning to crawl with a snake robot. Ph.D. dissertation, Robotics Institute, Carnegie Mellon Univ of Pittsburgh, PA.
14. Ng AY, Coates A, Mark D, Varun G, Jamie S, et al. (2006) Autonomous helicopter flight via reinforcement learning. Experimental Robotics IX 21: 363-372.
15. Hutchison WR, The Seventh Generation (7G) System.
16. Hutchison WR, Constantine BJ, Borenstein J, Pratt J (2007) Developing control of a high-DOF robot using reinforcement learning, genetic algorithms, scripting, and simulation. IEEE/RSJ Intl. Conf. on Intelligent Robots and Systems.
17. Yobotics! Simulation Construction Set.
18. Hutchison WR,Constantine BJ, Borenstein J, Pratt J (2007) Development of control for a serpentine robot. IEEE Intl. Symposium on Computational Intelligence in Robotic Automation.

e_GRASP: Robotic Hand Modeling and Simulation Environment

Ebrahim Mattar*

College of Engineering, University of Bahrain, Kingdom of Bahrain

Abstract

Modeling and simulation of robotics hands are significant topics that have been looked into by many robotics Specialists and programming experts. This is due to a demand to build a friendly platform for analyzing proposed hand design and movements, earlier to hand physical construction. For meeting such demands, a dexterous robotic hand software simulator was synthesized. The developed code is dexterity characterized robotic hand modeling and simulation software environment. The simulator was developed for robotics hands research purposes. This manuscript is presenting a brief documentation of such a modeling and simulating environment for simulating dynamic movements of multi-finger robotics hand. The environment is named as the *e*_GRASP. To make use of other supporting environments, *e*_GRASP is a Mat lab based simulator, with a quite large number of linked functionalities and routines that helps in simulating hand movements in a defined 3D space. *e*_GRASP was built after a number of years of experience while dealing with robot hands, hence it is a comprehensive Mat lab based Toolbox that makes use of other Mat lab defined Toolboxes. *e*_GRASP can also be interfaced to real-time hand control, with an ability to be linked with even higher levels of hierarchy. This includes Mat lab AI Tools, optimization, as considered useful toolboxes for dexterous hands for grasping and manipulation.

Keywords: Dexterous robotic manipulation; Modeling; Real-time simulation; Control; Matlab

Introduction

Literature and related studies

There are number of efforts to build robotics system modeling and simulation environments using Mat lab. For example, in reference [1], a Mat lab Toolbox for the iRobot Create (MTIC) was reported. It was mentioned that, "The toolbox replaces the native low level numerical commands, with a set of high level, intuitive, Mat lab functions (aka "wrappers")" [1]. In terms of modeling complexity, furthermore, Shaoqiang et al. [2] have mentioned that, "Biped robots are often treated as inverted pendulums for its simple structure. But modeling of robot and other complex machines is a time-consuming procedure. A new method of modeling and simulation of robot based on SimMechanics is proposed", Shaoqiang et al. [2]. Over number of years, Corke [3], has been working towards a MATLAB based robotic arm simulation. Hence, Corke [3], has introduced the well-known book for MATLAB environment with associated libraries. This is known as "Robotics Toolbox for Matlab", Corke [3]. The environment was a very useful tool for modeling robotic arms. It was also an easy methodology and coding for modeling and simulation different robotics arm structures. Jambak, et al. [4], mentioned the importance and paramount of Robot Simulation Software nowadays. This is to increase the accuracy and efficiency of industrial robot. They adopted a project using "virtual reality interface design methodology and utilizes MATLAB/Simulink and V-Realm Builder as the tools". They also mentioned that, "a robot model has been developed and a Robot Simulation Software life cycle has been implemented", Jambak et al. [4].

Furthermore, in reference to Olivier [5], Cyberbotics, it was reported that, Webots TM has a number of essential features intended to make such simulation tool both easy to use and powerful:

- Models and simulates any mobile robot, including wheeled, legged and flying robots.

- Includes a complete library of sensors and actuators.

- Lets you program the robots in (*C*, *C++* and *Java*), or from third party software through TCP/IP.

- Transfers controllers to real mobile robots, including Aibo®, Lego®, Mindstorms®, Khepera®, Koala®, and Hemisson®.

- Uses the ODE (*Open Dynamics Engine*) library for accurate physics simulation.

- Creates AVI or MPEG simulation movies for web and public presentations.

- Includes many examples with controller source code and models of commercially available robots.

- Lets you simulate multi-agent systems, with global and local communication facilities", Olivier [5].

Gourdeau [6] has indicated that, "Using an object-oriented programming approach, ROBOOP, a robotic manipulator simulation package which is both platform and vendor independent, compares favorably against a package requiring similar coding effort", Gourdeau [6]. In fact, he also indicated that, the performance tests show that with ROBOOP, the routine (with a class), inverse dynamics of a 6-DOF robot was made faster, can be computed and simulated in less than (5 ms) with a Pentium (100 MHz) computer.

Ramasamy and Arshad, [7] have both developed a robotic hand simulator. They indicated that "This robotic hand simulation is divided into three main parts. The main objective is to design a three dimensional graphic of a robotic hand and its movement animation

***Corresponding author:** Ebrahim Mattar, College of Engineering, University of Bahrain 32038, Kingdom of Bahrain, E-mail: ebmattar@ieee.org that imitates the movement of a human hand". They used the graphic design as a foundation to find the kinematics and dynamic properties of the robotic hand.

Furthermore Miller and Allen [8] have both presented the GRASPIT! In reality, GRASPIT! is a dexterous robotic hand simulation code. It was reported that the work focus of the grasp analysis, has been on force-closure grasps, which are useful for pick-and-place type tasks. Miller and Allen in [8], have also reported that, "This work discusses the different types of world elements and the general robot definition, and presented the robot library".

Miller et al [9]. have furthermore developed a simulation platform, where they focused on (Design/methodology/approach), as they are unlike other simulation systems. However, the simulation system was built specifically to analyze grasps by robotics hand. Miller et al. [9] also reported that, "It can import a wide variety of robot designs by using standard descriptions of the kinematics and link geometries. Various components support the analysis of grasps, visualization of results, dynamic simulation of grasping tasks, and grasp planning".

Jagdish et al. [10], have proposed and presented in their chapter, a fast as well as automatic hand gesture detection and recognition system. It was worked towards a reorganization of a hand gesture, hence an appropriate command is to be sent to the robot hand. Jagdish et al. [10], also mentioned that, once robot receives a command, it does a pre-defined work and keeps doing until a new command arrives.

Corrales et al. [11], have developed a kinematic, dynamic and contact models of a three-fingered robotic hand. The hand name was (BarrettHand), as this to obtain a complete description of the system which is required for manipulation tasks and simulation. Corrales et al. [11] have also stated that, "The developed models have been implemented on a software simulator based on the EASY JAVA simulations platform. Several experiments have been performed in order to verify the 3 accuracy of the proposed models with regard to the real physic system".

Gourret et al. [12], have addressed the problem of simulating deformations between objects and the hand of a synthetic character during a hand grasping progression. Hence, a numerical method based on finite element theory was developed and used to allow considering into account active forces of the each finger on the object and the reactive forces of the object on the fingers. Magnus [13], stated that, "before the implementation of the controller was made on the real hand it was tested and development on a simulation created in MATLAB/ simulink with help from a graphic physics engine called GraspIt!".

In this context, Magnus [13], have also looked into movements of the hand finger, and found how this is affected by the force from a leaf spring and a tendon that bends the finger. They also exposed the hand fingers to contact forces. They used these results and all these components to create models that are used in the simulation, hence to make the finger perform accurately.

Tarmizi et al. [14], as justified by increasing demands robotics hands both modeling and simulation, and due to the increasingly gaining importance amongst researchers for industrial and medical applications. Tarmizi et al. also stated that, "a multifinger robot hand with five fingers is modeled and simulated for grasping task. This was done using a CAD tool known as SOLID WORKS, and an analytical tool known as SimMechanics of MATLAB". In fact, obtained simulation results, indicated that the improved performance of grasping functions for robotics hands.

Ramasamy and Arshad [7], used animation techniques, hence, this to facilities designs and three dimensional graphic of an robotic hand, and its movement animation that imitates movements of the human hand. Such graphic designs, are hence, used as a foundation to find the kinematics and dynamic properties of the robotic hand. The end result is a robotic hand simulation that comes with analyses of the kinematics and dynamic properties".

Boughdiri et al. [15], have mentioned that there are numerous simulation results that shown, the derived dynamic model can predict the motion of the multi-fingered hand in free motion without holding any object. Additionally they have developed dynamic model that used and led to decoupling dynamic characteristics, by which the control of different parts of the system are simulated. They have reported a model-based computed torque technique, as for tracking control of the multi-fingered hand.

Ohol and Kajale [16] have looked into a number of issues related to robotic hand simulation. This includes; required task for the robots, as this becoming more complicated. In addition, handling of objects with various properties, e.g. material, size, mass, and physical interaction between the finger and an object. Ohol and Kajale [16] have stated in their research that, "Design procedure, solid modeling, force analysis and simulation have been discussed for further dynamic analysis towards confirmation of the viability".

Chan and Yun-Hui [17], have further looked into the issue of a dynamic simulator, that can helps and facilitates developments and applications of a multi-fingered robot hand. They indicated that, the existing dynamic simulators cannot effectively simulate dexterous manipulation of a multi-fingered robot hand. This is due to the lack of capability to cope with frequent changes in contact constraints and grasping configurations as well as impulsive collision occurring during manipulation. Chan and Yun-Hui, [17] also mentioned that, "We propose a unified framework to model free motions, collisions, and different contact motions including sticking, rolling, and sliding". Hence they proposed a innovative transition model for handling transitions between these contact motions. Finally, a 3D dynamic simulator has been developed to simulate dexterous robotic manipulation tasks, while involving combination of different contacts. That was based on a unified dynamic model. Simulation results indicated and confirmed the validity of the dynamic model and the simulator efficiency.

Main article contribution

In an effort to synthesis a dexterous multi-finger robotic hand MODELING and SIMULATION environment, this research framework was thoroughly focused towards the utilization and employment of Matlab-based platform coding, for a kinematics and dynamic simulation of an (n) number of fingers hand, with (m) number of joint within each finger dexterous robot hand.

The coding software is referred to as the "*e*_Grasp". *e*_Grasp was coded using Matlab functions and routine, with some associated Matlab libraries. The main motivation for selecting Matlab as the programming environment, this due to the availability of extended libraries and the ability to link Matlab with other external libraries. The simulator was built to be even supportive for complicated model-based control algorithms for hand-object movements. Examples of which include the Computed Torque approach, and adaptive control. The simulator was also linked with MATLAB AI tools, as an attempt to make use of available AI techniques for implementing intelligent hand manipulation.

Manuscript organization

Within this manuscript, we shall be introducing *e*_Grasp simulator, where the main features of such modeling and simulation space are presented. We shall also show how this environment is linked to other internal Matlab main Toolboxes and external C++ libraries. Particularly, this manuscript is presenting a long term research framework, which is actually focused for dexterous robotics manipulation that have gained experience and achieved over a number of years, where it was totally dedicated towards building a robotic hand modeling and simulation environment. The manuscript was divided into six sections. Section 3 gives a brief introduction and literatures related to this research theme. In Section 4 we present he simulator integrated structure and blocks. Robotics hand model building, as in relation to hands kinematics and dynamics, are therefore introduced in Section 5. Section 6 gives the simulator HIERARCHCY and DATA_STRUCTURE. This focuses on coding features for dexterous hand simulation. In Section 7 we show few *e*_Grasp simulation results. Finally, in Section 8, we provide conclusions and draw few concluding remakes.

Simulator Integrated Structure and Blocks

Multi finger robotics hands do represent complicated dynamic and interrelated closed chain kinematics system. Therefore, modeling and simulation of such systems, always represent challenging tasks. This is illustrated in Figure 1. Here we show the top level building blocks of the *e*_Grasp simulator hierarchy. It composes of five building units. The first is the top level (which includes the hand motion planning and higher USER commands), within such environment task definition is developed. In this top unit an appropriate supervisory software has been developed to facilitate the communication with the various functions of the hand. Typically the user can state the needed motion parameters, like velocities and positions. The second level of hierarchy, is the controllers array box; motion and force control hardware aspects are mainly located within this unit. The third level of hierarchy is dedicated towards the mechanical hand and the drive box used for actuation. Finally, the last level of hierarchy is dedicated towards the level hand motion sensing and instrumentations. Within Section 5, the theme of hand design will be further expanded; hence, kinematic and dynamic modeling matters are therefore discussed in further details.

The e_GRASP has developed its awareness, as a result of the continuous demands to look into the most challenges effort, which is modeling and simulating dexterous robotic hands. Most computer controlled systems used for robotics control, have distributed simulators, which are dedicated to specific tasks. For a six degrees of freedom arm, six simulators they are usually dedicated for the production of digitally controlled motion, hence parallel motion of all the DOF is achieved. All of these DOF are then managed by a single upper simulator. Multi-simulators are unusual distributed simulators which may be utilized to share the computational load for the same task, for providing redundancy in computation, or for sharing the multi axis controllability load in a system.

For multi-fingered robot hands, a few dedicated simulation structures have been used in practice. For multi-fingered robot hands, and due to the large number of actuators to be controlled by a supervisory software, the design adopted here comprises of (n) linked motion simulators, simulating the digital control, for hand tendon displacement control. Furthermore, the simulator is to consider control of hand distal joints, as each distal joint in the hand is also digitally force controlled.

Due to the large number of actuators and the need to have a user friendly software interface, a TOP SUPERVISORY code has been written for the purpose of hand programming and coordinating the various units over the hand entire system. Smooth fingertip Cartesian motion is an essential capability for robot hands in general, therefore for achieving an (n) path Cartesian point of fingertip motion, the motion is required to be planned, such that, fingertips pass near to defined via points without stopping. The *e*_Grasp hand simulator should be featuring the below listed features:

(i) Model robotic hand kinematics and dynamics.

(ii) Simulate closed hand-object chain kinematics.

(iii) Simulate constrained hand-object dynamics.

(iv) Perform joint-space inverse kinematics

(v) Plotting and graphing functionalities.

(vi) Linking the simulator to others MATALB toolboxes.

(vii) Optimization, and optimal force distribution.

(viii) Linking to external tools, and others C++ libraries.

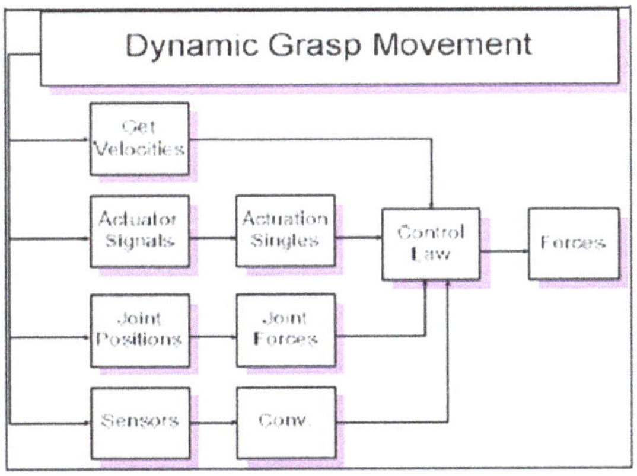

Figure 1: Hand simulator hierarchy and DATA_STRUCTURE, made it an easy procedure for interfacing different hand blocks.

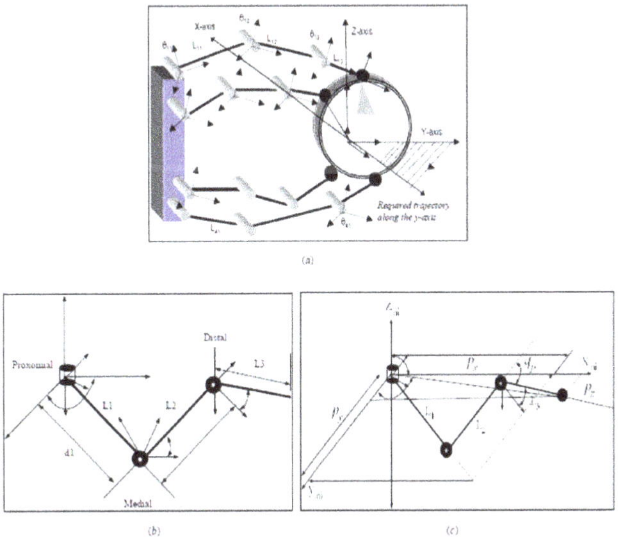

Figure 2: (a) We need to simulate a typical hand-object motion in 3D.That was made an easy task with e_Grasp, (b) A typical finger Kinematics and (c) Single finger geometry and related kinematics.

Building Simulator Models

Simulator library building: kinematics models

Hand Thumb Finger Model: In reference to Figure 2, and Table 1, $\left({}^{0}H_{3}\right)$ is formed by matrix multiplication of all the individual link matrices. While forming such a product, sub-results can be derived. This will be useful while solving hand inverse kinematics. By multiplying for simulation purposes, and for a simple situation of three joints in a finger, forward kinematics are expressed in terms of forward transformation matrix $\left({}^{0}H_{3}\right)$. Here the $\left({}^{0}H_{3}\right)$ entries have been calculated by:

$$R_1 = \begin{cases} +(C_1C_{23}) \\ -(C_1C_{23}) \\ S_1 \\ C1\left(l_3C_{23} + l_2C_2 + l_1\right) \end{cases} \qquad R_2 = \begin{cases} +(S_1C_{23}) \\ -(S_1S_{23}) \\ -C_1 \\ S_1\left(l_3C_{23} + l_2C_2 + l_1\right) \end{cases} \tag{1}$$

$$R_3 = \begin{cases} +S_{23} \\ +C_{23} \\ 0 \\ \left(l_3C_{23} + l_2C_2\right) \end{cases} \qquad R_4 = \begin{cases} 0 \\ 0 \\ 0 \\ 1 \end{cases}$$

$$\left({}^{0}H_{3}\right) = \begin{pmatrix} C_1C_{23} & -C_1C_{23} & S_1 & C_1\left(l_3C_{23} + l_2C_2 + l_1\right) \\ S_1C_{23} & -S_1S_{23} & -C_1 & S_1\left(l_3C_{23} + l_2c_2 + l_1\right) \\ S_{23} & C_{23} & 0 & \left(l_3C_{23} + l_2c_2\right) \\ 0 & 0 & 0 & 1 \end{pmatrix} \tag{2}$$

While describing fingers kinematic, it is also essential to describe their kinematics with respect to fixed Cartesian frames in the hand palm. These frames are essential elements for a grasped object motion. To define the fingertips of the four fingers to a fixed Cartesian frame, a permanent palm frame has been placed at the base of the first finger sub-system. This is illustrated in Figure 2.

Other hand digits: Each of the fingers is related to the hand palm reference coordinate (attached to the root of the first finger) by transformations which are designated by *d2*, *d3* and *d4* for second, third, and fourth fingers respectively. Due to the position of the second finger in the hand palm, the fingertip location with respect to the hand palm is defined by the kinematics equation, as obtained by multiplying the finger transformation matrix $\left({}^{0}H_{3}\right)$ by the transformation matrix which represents the displacement of that finger from the origin of the palm coordinate frame. If displacement of the second finger is defined by $\left(-d_{x2}\ \ 0\ -d_2\right)$, the location various hand fingers with respect to a fixed hand frame are:

2nd finger:

$${}^{0}H_{3} = \begin{pmatrix} C_1C_{23} & -C_1C_{23} & S_1 & \left(P_x - d_{x2}\right) \\ S_1C_{23} & -S_1S_{23} & -C_1 & \left(P_y\right) \\ S_{23} & C_{23} & 0 & \left(-d_2 + P_z\right) \\ 0 & 0 & 0 & 1 \end{pmatrix} \tag{3}$$

3rd finger:

$${}^{0}H_{3} = \begin{pmatrix} C_1C_{23} & -C_1C_{23} & S_1 & \left(P_x - d_{x2}\right) \\ S_1C_{23} & -S_1S_{23} & -C_1 & \left(P_y\right) \\ S_{23} & C_{23} & 0 & \left(P_z\right) \\ 0 & 0 & 0 & 1 \end{pmatrix} \tag{4}$$

4th finger:

$$H_f = \begin{pmatrix} \left(0.76C_2S_{23} + 0.64S_{23}\right) & \left(0.76C_1S_{23} + 0.64C_{23}\right) & 0.76S_1 & \left(0.7P_x + 0.64P_z - d_4\right) \\ S_1C_{23} & -S_1S_{23} & -C_1 & \left(P_y\right) \\ \left(-0.64C_2S_{23} + 0.76S_{23}\right) & \left(0.64C_2S_{23} + 0.76S_{23}\right) & -0.64S_1 & \left(-0.64P_x + 0.76P_z\right) \\ 0 & 0 & 0 & 1 \end{pmatrix} \tag{5}$$

Eq (2)-(5) do constitute the kinematics model of a (n) digits fingers hand. They also specify the computation of the position and orientation of frame 3 in reference to frame (0, *initial*), for each finger. These are the basic kinematics equations for simulating hand kinematics.

Simulator library building: (inverse kinematics)

In order to describe a finger position, reverse problem must be considered. Given a finger posture in terms of 3D components (P_x, P_y, P_z), what are the corresponding joint coordinates? In Section (5.1), the issue of computing position and orientation of the fingertip frame $\left({}^{0}H_{3}\right)$ was considered relative to a frame fixed at the hand palm. Here we are summing a given a set of joint angles of the finger. In this section, inverse kinematics is presented. Expressed another way, the inverse kinematics problem is given a frame or a homogeneous $\left({}^{0}H_{3}\right)$ matrix in the reachable subset, solve for the corresponding values of the joint variables that would result in similar $\left({}^{0}H_{3}\right)$ matrix numerical values. In contrast to forward kinematics, inverse kinematics problem does not have a general analytical solution. Given the desired position and orientation of the fingertip relative to the finger root, how we compute a set of joint angles which will achieve this desired result within a constrained kinematic system? The Cartesians upon which to base a decision vary, but a reasonable choice would be the closest solution. A finger will be considered kinematically solvable, if joint variables can be determined by an algorithm which allows one to determine all the sets of joint variables associated with a given position and orientation. We shall restrict the attention to closed form solution. In this context closed form means a solution method based on analytic expressions or on the solution of a polynomial. Within closed form solutions, two methods of obtaining a solution are distinguished. FIRST APPROACH: In

Finger revolute joints	Finger 1st joint proximal	Finger 2nd joint medial	Finger 3rd joint distal	---	---	Finger joint n^A (distal)
θ_i	θ_1	θ_2	θ_3	---	---	θ_n
α_i	90°	0°	0°	---	---	0°
a_i	l_1	l_2	l_3	---	---	
d_i	0	l_1	0	---	---	0°
Motion range	$\begin{pmatrix} -45^\circ \\ \leftrightarrow \\ +45^\circ \end{pmatrix}$	$\begin{pmatrix} -45^\circ \\ \leftrightarrow \\ +45^\circ \end{pmatrix}$	$\begin{pmatrix} -45^\circ \\ \leftrightarrow \\ +45^\circ \end{pmatrix}$	---	---	$\begin{pmatrix} -45^\circ \\ \leftrightarrow \\ +45^\circ \end{pmatrix}$

Table 1: Hand Kinematics and interrelated parameters.

reference to eq (2), and frame assignment given in Figure 2, we desire a solution for joint space vector $\Theta^{T}{}_{i}=(\theta_{1i}\ \theta_{2i}\ \theta_{3i})$ given a Cartesian numeric coordinate of a given fingertip posture. Equate ${}^{0}A_{3}$ with numeric matrix, gives:

$$\begin{pmatrix} \chi_{11} & \chi_{12} & \chi_{13} & \chi_{14} \\ \chi_{21} & \chi_{22} & \chi_{23} & \chi_{24} \\ \chi_{31} & \chi_{32} & \chi_{33} & \chi_{34} \\ 0 & 0 & 0 & 1 \end{pmatrix} = \begin{pmatrix} \eta_x & o_x & a_x & P_x \\ \eta_y & o_y & a_y & P_y \\ \eta_z & o_z & a_z & Pz \\ 0 & 0 & 0 & 1 \end{pmatrix} \tag{6}$$

This result in:

$$P_x=\left(C_1\left(l_3C_{23}+l_2C_2+l_1\right)\right)\ and\ Py=\left(S_1\left(l_3C_{23}+l_2C_2+l_1\right)\right)$$

$$\left(P_y/P_x\right)=\left(S_1/C_1\right) \tag{7}$$

$$\Theta_1=\tan^{-1}\left(P_y/P_x\right) \tag{8}$$

Likewise using $P_z=\left(l_3S_{23}+l_2C_2\right)$ divide by $\cos\left(\theta_2\right)$ This results in:

$$\left(P_z/C_2\right)=\left(L_3S_{23}+l_2S_2/C_2\right)\quad and\quad \tan\left(\theta_2\right)=\left(P_z-l_3S_{23}/l_2C_2\right) \tag{9}$$

Searching for a suitable value of $\cos\left(\theta_2\right)$ in Equ (9):

$$P_x=\left(l_3C_1O_z+C_1l_2C_2+C_1l_1\right)$$

$$l_2C_2=\left(\left(P_x/C_1\right)-l_3O_z-l_1\right)$$

$$\theta_2=\tan^{-1}\left(\left(P_z-l_3n_z\right)/\left(P_x/C_1\right)-l_3O_z-l_1\right) \tag{10}$$

$$\theta_3=\tan^{-1}\left(n_z/O_z\right)-\theta_2 \tag{11}$$

The initial solution is that which corresponds to a known orientation vectors. SECOND APPROACH: A geometric approach for solving the inverse kinematic problem is a technique of decomposing the spatial geometry of the finger into several plane geometry problems. Joint space vectors is solved for, using tools of plane geometry. Figure 2b also displays the kinematics and geometric configuration of one finger. Applying rules of geometry to one finger shape, this results in the following equations describing the joint space vector in terms of the given (P_x) (P_y) (P_z) fingertip vector. From Figure 2c, and for the triangle as due to finger geometry:

$$\tan\left(\theta_1\right)=\left(P_y/P_x\right)\ and\ \left(\theta_1\right)=\tan^{-1}\left(P_y/P_x\right) \tag{12}$$

Furthermore, from the figure:

$$\left(P_q=\mathrm{P}_r-l_1\right)\ and\ P_t=\sqrt{\left(P_z^{2}+P_q^{2}\right)}$$

$$\theta_z=\tan^{-1}\left(P_z/P_q\right)\qquad W_a=C^{-1}\left(l_3^{2}-P_t^{2}-l_2^{2}/-2P_tl_2\right)$$

Having found θ_z, then $\left(\theta_z-W_a\right):\theta_3=180^{\circ}-\cos^{-1}\left(\left(P_t^{2}-l_2^{2}-l_3^{2}\right)/2l_2l_3\right)$ (13)

Motion Finger. Using the geometric model of the finger for the simulator is an easy and quick approach to obtaining the joint space vector. One drawback of the geometric method is that: at a definite finger posture the geometric configuration of the finger fails to give a solution. Refer to Figure 1 to observe that certain configurations of the fingertip where the solution is not well defined, hence multiple solutions may appear. However as θ gets large enough, the fingertip location P_t goes to a position with respect to (0) and T_r angles where the cosine rule used in eq (13) is not applicable. To achieve realistic finger solutions, hand simulator source code has been designed in such a way as, to get the inverse kinematic for the finger model using the geometric approach. Subsequently, once joint space vector are evaluated for, orientation vectors are evaluated, as they constitute columns of the well-known finger Jacobian matrix of J_θ We shall use two known methods in literature to find the inverse kinematics geometric and algebraic approaches of solutions fail to give definite solution.

Simulator library building: building hand Jacobian

The Jacobian matrix plays important role in hand simulation. Differential changes in finger end locations, are caused by as results of joint differential changes. The number of rows in a Jacobian is determined by the number of degrees of freedom in Cartesian space required by the task which is, in turn, determined by the degrees of mobility of a task. In fact, the JACOBIAN is a multi-dimensional form of the derivative of a function of several variables. Since the Jacobian is a differential formula of a matrix, hence it is possible to obtain elements of that matrix using the CHAIN FORMULA as far as the functions of independent variables. For the finger Jacobian J_θ there are a number of techniques which have been developed for the Jacobian calculation by: Waldron Algorithm, Renand Algorithm, Paul's Algorithm, [18,19]. Location of a fingertip is characterized by relative positioning of frame f_{i+1} attached to the fingertips, with respect to a frame f_0. In addition, a fingertip velocity is always a vector in the tangent space of the frame space, which is related by the Jacobian matrix J_θ. Such framings are illustrated in details in Figure 3.

Simulator library building: jacobian singularities avoidance

Due to the fact the Jacobian is a position dependent matrix, in particular finger configurations, the matrix becomes not full of rank. When this occurs the Jacobian rows and column vectors are linearly dependent, thus do not span the $\in\Re^{(3\times1)}$ vector space of X. Therefore, there exists at least one direction in which the fingertip cannot be moved no matter how the joint velocities θ_1, through θ_3 are chosen. Here we shall introduce the derivation of a finger Jacobian matrix in terms of D-H matrices. In Eq (5), the Jacobian is defined with respect to the reference frame, frame (0) at the finger root. However, for our purpose it has been assumed that the points of contact are stationary on the grasped object and another algorithm can be implemented based upon the kinematic model of the finger. Such algorithm is based on the method shown by Paul [19]. From the time when all joints are revolute, the three columns of the matrix are given by:

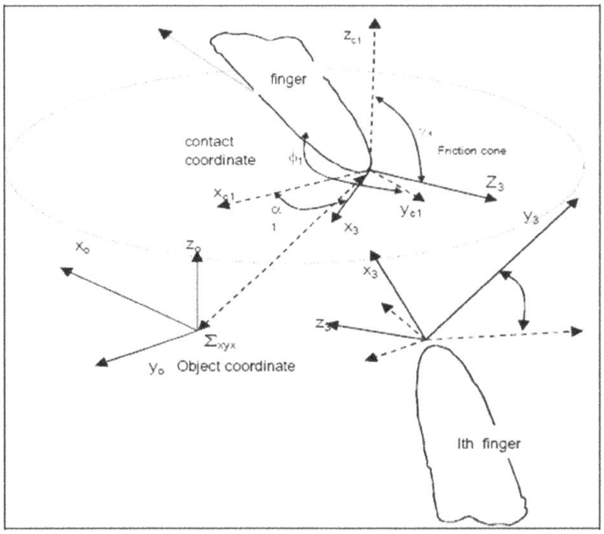

Figure 3: Simulator deals with diverse classes of fingertip contact models. The default type, is the (frictional point of contact).

$$\left.\begin{aligned} A^3 d_i &= \left(\left(-n_x P_y\right) i \left(-o_x P_y + o_y P_x\right) j \left(-a_x P_y + a_y P_x\right) k\right) \\ A^3 \delta_i &= \left(n_z i \;\; o_z j \;\; a_z k\right) \end{aligned}\right\}$$

While using this algorithm, it is necessary to evaluate the orientation vectors of a finger posture, which are assumed to be known by using the geometric finger model and the forward transformation matrix. Paul's algorithm can be used to calculate the Jacobian of an (n) degrees of freedom arm, where each degree of freedom corresponds to one joint. Thus, each column in the Jacobian is a vector which describes the differential motion of a joint. A revolute joint has a differential rotation around the axis of rotation Z axis:

$$\left.\begin{aligned} &\left(\partial P_x / \partial \theta_n\right) = J_{1n} = \left({}^{n-1}n_y \; {}^{n-1}P_x - {}^{n-1}n_x \; {}^{n-1}P_y\right) \\ &\left(\partial P_y / \partial \theta_n\right) = J_{2n} = \left({}^{n-1}o_y \; {}^{n-1}P_x - {}^{n-1}o_x \; {}^{n-1}P_y\right) \\ &\left(\partial P_z / \partial \theta_n\right) = J_{3n} = \left({}^{n-1}a_y \; {}^{n-1}P_x - {}^{n-1}a_x \; {}^{n-1}P_y\right) \\ &\left(\partial \delta_x / \partial \theta_n\right) = J_{4n} = {}^{n-1}n_z \;\; \left(\partial \delta_y / \partial \theta_n\right) = J_{5n} = {}^{n-1}o_z \;\; \left(\partial \delta_z / \partial \theta_n\right) = J_{6n} = {}^{n-1}a_z \end{aligned}\right\} \tag{15}$$

To relate a fingertip position to a set of joint angles of the fingers, one may make use of the forward solution and obtain the fingertip matrix $\left({}^0H_3\right)$. Paul [19] has presented a routine that relates differential changes in joint variables, to the differential changes in the tip position and orientation. This is summarized here as:

$${}^0H_3 = \begin{pmatrix} C_1C_{23} & -C_1C_{23} & S_1 & C_1\left(l_3C_{23} + l_2C_2 + l_1\right) \\ S_1C_{23} & -S_1S_{23} & -C_1 & S_1\left(l_3C_{23} + l_2c_2 + l_1\right) \\ S_{23} & C_{23} & 0 & \left(l_3S_{23} + l_2S_2\right) \\ 0 & 0 & 0 & 1 \end{pmatrix} \tag{16}$$

$${}^1H_3 = \begin{pmatrix} C_{23} & -S_{23} & 0 & \left(l_3C_{23} + l_2C_2\right) \\ C_{23} & -C_{23} & 0 & \left(l_3C_{23} + l_2c_2\right) \\ S_{23} & C_{23} & 1 & 0 \\ 0 & 0 & 0 & 1 \end{pmatrix} \tag{17}$$

$${}^2H_3 = \begin{pmatrix} C_3 & -S_3 & 0 & \left(l_3C_3\right) \\ S_3 & -C_3 & 0 & \left(l_3S_3\right) \\ 0 & 0 & 1 & 0 \\ 0 & 0 & 0 & 1 \end{pmatrix} \tag{18}$$

Further application of the algorithm, and using the relation defined by eq (17) gives: (1st) column of J_θ

$$\left.\begin{aligned} &\partial x_1 = \left(\psi S_1 C_1 C_{23} - S_1 C_{23}\right) \\ &\partial y_1 = \left(\psi C_1 S_1 S_{23} + S_1 C_1 S_{23}\right) \\ &\partial z_1 = \left(-C_1^2 \psi\right) - S_1^2 \psi \\ &\partial x_1 = 0 \;\; \partial y_1 = 0 \;\; \partial z_1 = 0 \end{aligned}\right\} \tag{19}$$

$$\left.\begin{aligned} &\partial x_1 = S_{23} \\ &\partial y_1 = C_{23} \\ &\partial z_1 = 0 \end{aligned}\right\} \tag{20}$$

$$(2^{nd})\text{ column of } J_\theta \left.\begin{aligned} &\partial x_2 = \left(S_{23}\left(l_3C_{23} + l_2C_2\right)\right) - C_{23}\left(l_3S_{23} + l_2S_2\right) \\ &\partial y_2 = \left(C_{23}\left(l_3C_{23} + l_2C_2\right)\right) - S_{23}\left(l_3S_{23} + l_2S_2\right) \\ &\partial z_2 = 0 \\ &\partial z_2 = 0 \;\; \partial y_2 = 0 \;\; \partial z_2 = 0 \end{aligned}\right\}$$

Leading to (2nd) column of J_θ to be $\left(\partial x_2 = l_2 S_3 \;\; \partial y_2 = l_3 + l_2 C \;\; \partial z_2 = 0\right)$ (21)

$$\text{Third Column: } \left.\begin{aligned} &\partial x_3 = \left(\left(l_3C_3S_3 - l_3S_3C_3\right)\right) \;\; \partial y_3 = l_3 \;\; \partial z_3 = 0 \\ &\delta x_3 = 0 \;\; \delta y_3 = 0 \;\; \delta z_3 = 1 \end{aligned}\right\} \tag{22}$$

Hence J_Θ is finally written as:

$$J_\Theta = \left(u_1 \; u_2 \; u_3\right), \;\; u_1^T = \left(0 \;\; 0 \;\; -\psi\right), \;\; u_2^T = \left(\left(l_2S_3\right) \;\; \left(l_3 + l_2C_3\right) \;\; 0\right), \;\; u_3^T = \left(0 \;\; l_3 \;\; 0\right) \tag{23}$$

Eq (23) has been verified using Matlab software, and using velocity cross product (VCP) technique used by Fu et al. [20]. Since each finger has a 3-DOF mechanism responsible for producing translational differentials motion at the fingertip, the Jacobian matrix can be rewritten in terms of differentials motion of the fingertip. The methodology taken here is to divide the simulation environment into (n) main blocks of simulators. Respectively, individual blocks have been dedicated for the simulation and control of the (m) joints of one finger. This is shown in Figure 4.

Simulator Hierarchicy

Top level: simulator data structure

Configuration of hand simulating software was based on an

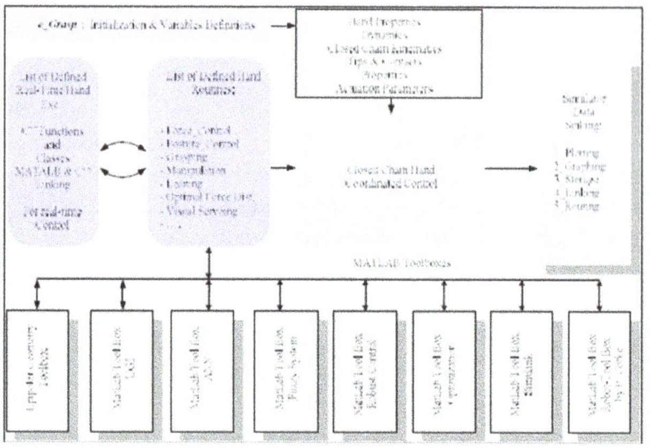

Figure 4: e_GRASP, a MULTI-FINGER robot hand comprehensive modeling and simulator MATLAB based environment.

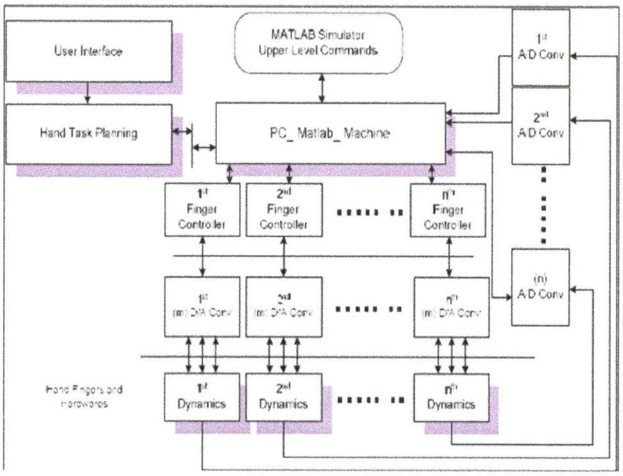

Figure 5: e_Grasp simulator cascaded levels. The simulator deals with both top level control laws, in addition to the lower level finger joints laws.

assumption, of using discrete time joint-space controllers. This is clearly illustrated in Figure 5. That was also based on using a model of MOTION CONTROLLER. There a number of available real-tome DSP controllers. Example of which is the national semiconductor product, (LM628) motion processor. For simulation purposes, there are up to ($n \times m$) motion processors that have been employed for the overall hand control. Individual single processors were dedicated control an individual DOF in the hand. Overall MatLab supervisory coding and software designing for the entire hand simulation, is to be run on high speed machine. However, it was also achieved while using an up-to-date high speed Laptops. High speed machine is needed, as this requires communicating with ($n \times m$) separated controllers blocks via ($n \times m$) processors hardware. From software viewpoint, simulating coordinated hand control, takes place at different computational levels. The hand hierarchical nature leads inherently to a bottom-up supervisory simulation design. Top level of concurrency, looks at entire hand simulation.

e_GRASP hand data structure and classes

At start, we shall define a finger data structure to contain a finger name. The "." operator, used in the case of Structure. Field tells mat lab to access the field named field in the structure.

% Creating Hand Data Structure And Classes..
Finger_Data.First='Finger_1';
Finger_1Data.Hand='Joint_1';
NameData.Last='Joint_m';
% Creating a HandData structure with a name field. HandData. Finger=FingerData;
% Initializing rest of the hand structure ...
Finger_1.Status=Finger_1_Data';
Finger_1Data.x_pos=10;
Finger_1Data.y_pos=40;
Finger_1Data.z_pos=-60;
Finger_1Data.q_1=-60°;
Finger_1Data.q_m=30°;
HandData.Fingers=linspace(60,30,45,...);
% View contents of the data_structure FingerData FingerData. Name FingerData.deatils
% Operating on Elements of the Structure.
FingersData.X_Y_Z(3)=0;
FingerData.X_Y_Z FingertData.X.First='First_Finger'; Finger_ Data.First
% Creating arrays of structures for each hand finger.
num_fingers=4; % depends on hand number of fingers
for i=1:num_Fingers
ClassData(i)=FingerData;
end
ClassData
ClassData(2)

Joints space and fingers simulation: non closed chain dynamics simulation

For simulating whole hand dynamics, either in open or closed loops, it is therefore important to have a model of the individual joint dynamics. This consists of a model of a single discrete time motion closed controller. A good model is an (n) bit controller, connected with a D/A converter, driving current amplification. The digital closed loop system also requires a simulation for measuring the individual joint displacement and position. In order to create an entire closed loop simulator model of the individual joints, and for describing the whole closed chain system, the following sub-sections will rather focus on the dynamics of a single joint controller, as this will be leading to a finger simulation, hence, leading to entire hand simulation.

Discrete time simulation of a single axis joint

Digital simulation of the lower level of individual joints, is discussed and simplified. That is because the control input for ith joint depends only on locally measured variables, not on the variables of the other joints. Moreover, computations are easy and do not involve solving the complicated derived nonlinear hand inverse dynamics. Contrary to this, for manipulation and exertion of forces points of view, an additional joint torque control has to be added, this is for an exertion of grasping forces throughout object motion. From literature, a typical PID controllers expressed by:

$$u(t) = k\left(e(t) + \frac{1}{\tau_i}\int_0^t e(t)dt + \tau_d\left(\frac{de(t)}{dt}\right)\right) \tag{24}$$

In eq (24), $u(t)$ is input to the joint controller, and $e(t)$ is the controller output. Digitizing eq (24), resulting in:

$$U(z) = \left(K_p + \frac{K_{ip}}{\left(1-z^{-1}\right)} + Kd\left(1-z^{-1}\right)\right)E(z) \tag{25}$$

A simplified dynamical model of the joint with the electric actuator may be written as:

$$\left(J_a + J_m + n_i^2 J_L\right)\ddot{\theta}_i + \left(\frac{\left(B_a + B_m + k_b k_i\right)}{R}\right)\dot{\theta}_i = \left(\frac{k_i}{Rv_i}\right) - n_i d_i \tag{26}$$

$$\tau_a(s) = \left(J_T S^2 + B_T S\right)\theta(S)$$

In eq (26), the $\left(J_a + J_m + n_i^2 J_L\right)$ terms are effective inertia and effective damping coefficients of actuators shafts. This also includes the joint inertia. The joint position trajectory is calculated by a trajectory planning routine. This is passed to the digital control as number of pulses.

A typical joint space closed loop transfer function, relating an input to the joint, to its output, (block diagram shown in Figure 6, this is expressed by:

$$\left(\frac{\theta_m(S)}{\theta_d(S)}\right) = \left(\frac{\left(nk_\theta k_{PID} k_i RJ_T\right)}{\left(S^2 + \left(\frac{RB_T + k_i k_b}{RJ_T}\right)S + \left(\frac{nk_\theta k_{PID} k_i}{RJ_T}\right)\right)}\right) \tag{28}$$

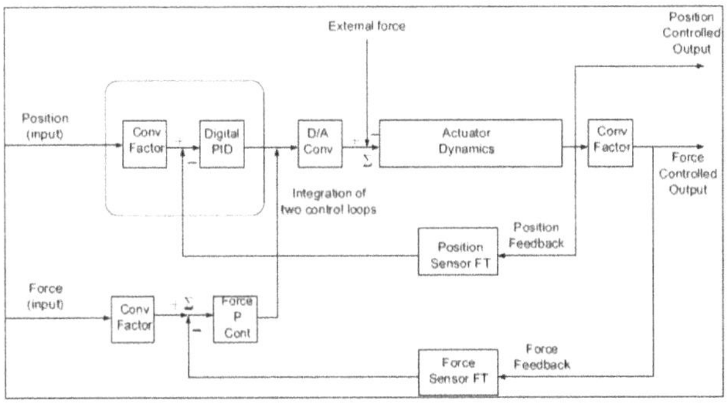

Figure 6: e_GRASP detailed discrete-time controller synthesis for a single joint closed loop control.

$$\frac{\theta_m(z)}{\theta_d(z)}=\left(\frac{z^2}{(z-\alpha_1)(z-\alpha_2)}\right) \tag{29}$$

Individual joint space controllers synthesis, was achieved using and supported by Matlab. Matlab environment, with its control toolbox synthesis and facilities, provided an aid to the controller testing and simulation. Typical design tools do include, Nyquist diagram, Modulus Frequency response, and time step response. For achieving a discreet time simulation within Matlab, the finger joints were being controlled digitally rather than in analog approach. Moving from continuous time domain to discrete time domain, Within (*Z*) domain, sampling time has an important role in determining the system stability. We start to do that by initially modeling the D/A converter in terms of sampling rate, $\left(\left(1-e^{-TS}\right)\big/_{s}\right)$, hence the corresponding (*Z*) transform of the (D/A), including actuator, is evaluated. In addition, the computed discrete-time poles of eq (29) are, as well, function of the variable factor in the system, the scalar value of the proportional gain (k_p).

Simulation of joints-space controllers

Furthermore, in Figure 6 we show the bottom low-level controller block diagram, and the associated parameters. This is to avoid a joint dynamic behavior that is not desirable, as due to the amount of joint motion disturbances. Therefore we need to overcome any undesirable effects. The use of adjusted parameters control law have the effect of making the response behavior enhanced. The location of adjusting the full PID control parameters can be seen in enhancing fingers motion. As this will be rather simulated later; the choice of low proportional gain in addition to the integral and derivative has the effect of decreasing the corresponding (rise time) as compared with the result for a purely high proportional gain.

N Configuration hand motion: (creation of arrays, pointers, and records)

The simulator has a large data structure. This is to comprehend the complicated fingers motion. System points of contact motions are generated by specifications of (*r*) direction of motion. From an object to be moved within the work-space of the hand, principle axes of motions are defined in addition to Cartesian fingertips paths. If fingertips paths are specified as (*H, K, L, U,......*), for the (*n*) fingers, then for (*n*) segments path, the Cartesian segment (localities in space), are passed and download to the entire hand controller coding. This is based on the use of such data structure:

$$\left.\begin{array}{r} \textit{configuration} \\ \textit{First finger} \\ \textit{Second finger} \\ \textit{Third finger} \\ \textit{Fourth finger} \end{array}\right\} \left.\begin{array}{c} A=(1\ 2\ 3\ 4) \\ H=(H_1\ H_2\ H_3\ H_4) \\ K=(K_1\ K_2\ K_3\ K_4) \\ L=(L_1\ L_2\ L_3\ L_4) \\ U=(U_1\ U_2\ U_3\ U_4) \end{array}\right\} \tag{30}$$

As indicated earlier, every finger segment vector consists of an ARRAY (Data_Type). This is for storage of finger joints positions and velocities. In accordance to the specified path coordinates, hand controller creates array of Matlab based records. At the ending phase of the (RECORDS) and (POINTERS) creation, the hand controller holds ($n \times n$) of records containing the Cartesian motion information of (*n*) segment paths for the individual fingers.

Simulation of joint force control

For simulating a joint torque dynamics and control, this requires a continuous measurement of the joint torque via some typical sensors. It is an essential for a joint torque sensor to be enclosed within a closed loop system. Furthermore, this is to be incorporated within the position control loop. While discussing the simulating aspects of a joint force control, the mechanism of transmission at joints do play an important role in the performance of the servo system itself. The problem of closing a joint force control feedback loop is the non-ideal transmission and drive system, i.e. presence of STICTION, COGGING, COULOMBIC friction, together with higher order resonance modes in the system. Furthermore, equivalent rotational stiffness do play an essential role in force control simulation. It relates the change in joint torque with changes in joint displacement and has to be computed for a rotational system. For a specific geometry of joint pulley and cantilever beam, a joint space rotational stiffness is mathematically expressed:

$$k_\theta=\gamma\big(c(\alpha)-s(\alpha)s(\beta)t(\alpha+\beta)\big)\chi \tag{31}$$

$$\gamma=\big(1/\big(8s(\beta)s(\alpha+\beta)\big)\big) \tag{32}$$

and χ is a parameter representing the force sensor stiffness. For an ideal actuator characteristic, the system force control transfer function is given by:

$$\tau_a=\left(J_T\left(\frac{d^2\theta}{dt^2}\right)+B_T\left(\frac{d\theta}{dt}\right)\right) \tag{33}$$

Hence, for simulation of the force proportional control system, a typical control law is considered:

$$\tau_a=k_{ps}\Delta\tau \tag{34}$$

In eq (34), (k_{ps}) is the force loop controller gain, $\Delta\tau=(\tau_r(t)-\tau_f(t))$, where (τ_r) is the desired behavior, whereas the term (τ_f) is the finger actual joint torque. Furthermore, Eq (34) represents an ideal mechanism for controlling a joint torque. While examining a real typical physical construction of a force sensor, joint transmission system, and grasped object compliance, do add complexity to force control system. There is also variation in the dynamics of the finger due to its configuration dependence, however such changes are masked by the actuator dynamics. A grasped object compliance plays an important role in the response of the force controller. There exists the following linkage between (τ_f) and $(\Delta\theta)$, the angular displacement of the driving pulley. This is expressed by:

$$\tau_f=(\delta\Delta\theta) \tag{35}$$

In eq (35), δ is the compound rotational stiffness of the finger mechanical structure, as expressed in terms of: (i) sensor rotational stiffness, (ii) the grasp rigidity characteristics, (iii) and stiffness of the tendon. This is expressed by:

$$\left(1/_{\delta}\right)=\left(\left(1/_{\upsilon}\right)+\left(1/_{v_p}\right)+\left(1/_{\eta}\right)\right) \tag{36}$$

While considering eq (34) and eq (35), a relation between τ_f and τ_r is expressed by:

$$\left(\frac{\tau_f(s)}{\tau_r(s)}\right)=\left(\frac{(\delta k_p)}{J_a s^2+B_a s+\delta k_i k_p}\right) \tag{37}$$

Eq (37) gives the torque response of the closed loop system is controllable by υ and k_p parameters. The simulation behavior of the designed force servo system, does not remain uniform even for constant gains k_p. This is due to the fact, that δ involves some varying quantities η and v_p. For instance increases from a very stiff environment while it decreases for a compliant object surface. Additionally v_p, tendon stiffness, might change if a greater stretching force is applied to its ends. The behavior could be further simplified by taking into account that the robot hand has been categorized to be grasping a rigid object, hence η is having a very high value. v_p is quite a high quantity for the routed tendon due to the short length of the tendon and the use of high stiffness tendon characteristics. An equivalent rotational stiffness υ, by itself is a function of the physical construction of the sensor and its location in the finger. Having considered such assumptions, Eq (36) is totally dominated by an equivalent rotational stiffness. Here υ is a function of the sensor physical parameters. Finally, for real time simulation of the force closed loop system (feedback control), we transform the dynamics from time domain to frequency domain, i.e. relying on Laplace transform of the entire force closed loop control transfer function. This is expressed as:

Finally, for real time simulation of the force closed loop system (feedback control), we transform the dynamics from time domain to frequency domain, i.e. relying on Laplace transform of the entire force closed loop control transfer function. This is expressed as:

$$\left(\frac{F_0(S)}{F_r(S)}\right)=\left(\frac{\left(nk_{c\Theta}P_ik_i/RJ_T\right)}{S^2+\left(RB_T+k_ik_b/RJ_T\right)S+\left(nk_\Theta k_{c\Theta}P_ik_pk_i\right)\left(RJ_T\right)}\right) \tag{38}$$

In eq (38), p_i, is a conversion ratio. It relates the a conversion from a desired fingertip force, into a volts within the loop. The expressed relation of Eq (38), does indicate the changing parameters that do affect the force closed loop dynamics. Therefore, the built simulator environment is to take care of such parameters while simulating a typical a joint closed loop system.

Closed chain hand-object simulation: advanced control laws (computed torque method)

One of the great benefits of such a simulation instrument, is an ability to simulate robotics hands, under advanced control laws. Typical example is the Cartesian computed torque control law. In addition, other advanced control laws is the sliding mode control. In reality, such control methodologies are complicated, and do require massive computational requirements. However, they were made easy while using *e*_Grasp simulator. More results will be shown over the next Section. In addition, adaptive or even ANN based nonlinear control can also be simulated through the *e*_Grasp simulator. For (*m*) joints (in a finger), with (*n*) fingers robotic hand, the individual fingers are modeled by the time varying equation of motion:

$$A_i(\theta)\ddot{\theta}+B_i(\theta,\dot{\theta})\dot{\theta}+C_i(\theta)\theta=(\tau_i+\sigma_i) \tag{39}$$

In eq (39), (θ) is a trajectory of finger joint in rad. $(\dot{\theta})$ is joint speed trajectory in (rad/sec). $(\ddot{\theta})$ is joint acceleration in (*rad/sec2*). $A_i(\theta), B_i(\theta,\dot{\theta})$ *and* $c_i(\theta)$ are the hand concatenated dynamics. Additionally, (σ_i) is equivalent of externally added forces. Hand wrenches and forces are found by:

$$f_{hand}=-\left(G^{+1}M_b+\lambda\eta\right) \tag{40}$$

M_b the grasp dynamics, and G^{+1} *1* is the hand grip transform inverse. For a balanced motion, the resulting forces and moments of the entire closed chain dynamics is equal to zero. Hence, equating hand dynamics with object dynamics, this gives:

$$A_hJ_h^{-1}\left(\dot{G}^T\dot{u}-J_h\ddot{q}_h\right)+B_h\dot{q}_h+C_h=\tau_h+\dot{J}_h^T\dot{G}^{+1}\left(A_o\ddot{u}+B_o\dot{u}+C_o\right)-J_h^T\lambda\eta \tag{41}$$

Furthermore, in eq (40), f_{hand} is a set fingertips forces. Calculating fingertip forces using pseudo inverse of grip transform *G*. In Equ (41), $\lambda\eta$ is part of the force solution, and λ is an adjusting vector. Furthermore, the vector $u^c{}_d$ is the position and orientation of the grasped object. For non-slipping contacts, there is no change at contact points, i.e. $\left(\partial\Gamma_h/\partial t\right)=0$, though there might be rotational change at each point of contact. eq (41) also represents a typical and the entire hand-object dynamics. Hand fingertips forces, f_{hand} are playing major roles in such balancing equation. Defining a Cartesian based posture error (*e*) of an object in 3D as $\in\Re^{6\times1}$, as the error between a defined posture $(u^c{}_a)$, and the real object posture $(u^c{}_a)$ as $(e\cong u^c{}_a-u^c{}_a)$. Object-hand closed chain system is described in terms of hand joint-space torques (τ_h) joint torques and Euler dynamics. In Cartesian space, and for the three terms Cartesian based controller, this is expressed by:

$$\tau_h=\left(A_hJ^{-1}{}_hX_h+T_{ex}\right) \tag{42}$$

$$T_{ex}=\left(B_k+C_k\right)+J^T{}_h\left(F_{cd}-Z_i\int_0^x\left(\phi_{cd}-\eta\lambda\right)\right),\quad X_h=\left(G^{+1}\Theta_a-J_h\Theta_k\right)\to\Re^{n\times1} \tag{43}$$

In eq (42), $X_h\in\Re^{12\times1}$ and expressed mathematically as in Equ (43). In addition, A_k, B_k *and* C_k are the entire hand augmented dynamics. F_{cd} is a commanded set of forces, $(\eta\lambda)$ is an adjustable term, J_h is hand Jacobain matrix. Each finger maps its joints torque to the object via the entire hand gasp *G* which is formulated as $G=\left(G_1\ \vdots\ G_2\ \vdots\ \to\ \vdots\ G_n\right)$, as grab sub-matrices $G_i\in\Re^{6\times3}$, for $i=\left(1\ 2\ \cdots n\right)$ are defined in terms of contact location by Eq (42):

$$G_i=\left(I_{3\times3}/\gamma_i\right) \tag{44}$$

In eq (44), (γ_i) are sub-matrices for contact configuration. They are performing a skew-matrix of position contact (γ_i) over the grasped object surface. In addition, hand fingertip force distribution depends completely on dually heavily computed matrices. The first is G^{+1} witch is an irregular matrix. The second is the hand Jacobian inverse matrix as the $(J^{-1}{}_h)\cdot(J^{-1}{}_h)$ is a large matrix, and it is a concatenated matrix, compromising all the fingers Jacobians, as $\left(J_h=diga\left(J_1\ J_2\ \cdots J_n\right)\right)$. Finally, eq (43) expresses a typical Cartesian based hand controller using PID, that will be simulated through the *e*_Grasp.

Result and Analysis

In particular, *e*_Grasp has been used successfully as a tool for motion and manipulation analysis of robotic multifinger hands. In this respect, for running this environment, this needs a Matlab environment. The code has been also tested lately on MATLAB Version 7.8.0.347 (R2009a). The simulator has been tested over a number of times. For validating the *e*_Grasp simulator potential, we shall present within this section few simulation results as related to a control for moving while grasping a grasp by a robot hand. The chosen hand is of four fingers, where each finger is having three rotational joints, i.e. $(n=4, m=3)$. The simulation environment gives the user an ability to select the hand configuration, dynamic and kinematics parameters, nature of fingertip contact, control sampling rate, law of controller, simulation time, in addition to others simulation parameters. There are large number of results to be shown, as a result of running the simulator, however, we shall show only few graphics results. In this respect, in Figure 7a, we show the simulator graphical interface with other plotting graphics and the m-editor in Matlab. Furthermore, in Figure 7b we show part of simulator interfacing with other Matlab related toolboxes. In this respect, here we show the Matlab optimization toolbox being interfaced with e_Grasp simulator to compute optimal forces and torques needed by hand fingers to make an object motion.

For a demonstration, a 3D grasp movement is shown in Figure 8. A grasp object of a known dimension and weight was manipulated by movements of fingertips, while applying a suitable set of fingertips forces. The grasp was moved in periodic sinusoidal movement. Results shown that, simulating a 3D grasp movements was made an easy task while using the *e*_Grasp simulator. Dynamics of a grasped object can be changed while alternating both Cartesian controller parameters and joints parameters. This leads to make a grasp motion stable, oscillatory, faster, or slower. Furthermore, Figure 9 displays a finger joint-space closed loop displacement performance. The simulator has taken into account fingertip rotations during a course of a grasp motion. Furthermore, in Figure 10 we show a simulation of hand torques, while grasping an object in 3-D space. The figure shows how torques are computed in response to motion. The simulator was able to transform a grasp from an initial posture to another over a minimum time. For such a simulation, the simulation environment offers a full adjustment of the three terms PID controller parameters, hence they have been selected for minimum overshoot. A grasped object movements can be made much unstable or even with less overshoot over its movement path in 3D. In Figure 11 we show a typical simulator capabilities. The figure is showing the computed error while comparing artificial neural network output with actual joint-space output. The 2nd finger joint-space motion simulations are therefore realistic in terms of displacement and motion. Joint space torques are very realistic in terms of producing an adequate amount of torques for fingertip movements. In Figure 12, we also display an important ability of the *e*_Grasp, which the ability of linking its output to other Matlab toolboxes. Here we show a link between *e*_Grasp with Matlab fuzzy toolbox. The figure shows an adaptation of fuzzy membership functions for optimal forces

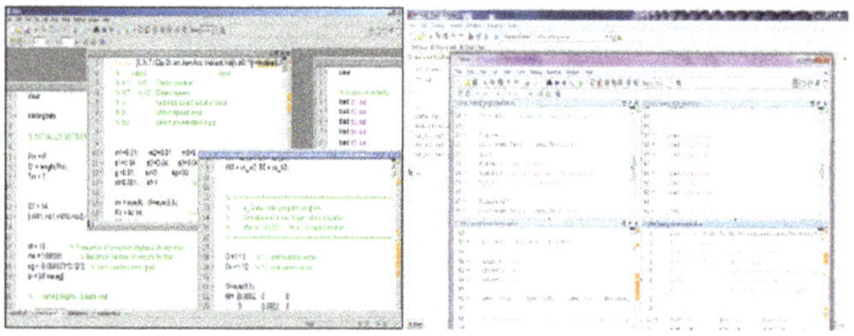

(a) (b)

Figure 7: (a) The simulator: A screen shot for e_GRASP connection, (b) Part of simulator interfacing with Mat lab related tool boxes. Here we show the Mat Lab optimization toolbox to compute optimal forces and torques needed by hand fingers to achieve an object motion.

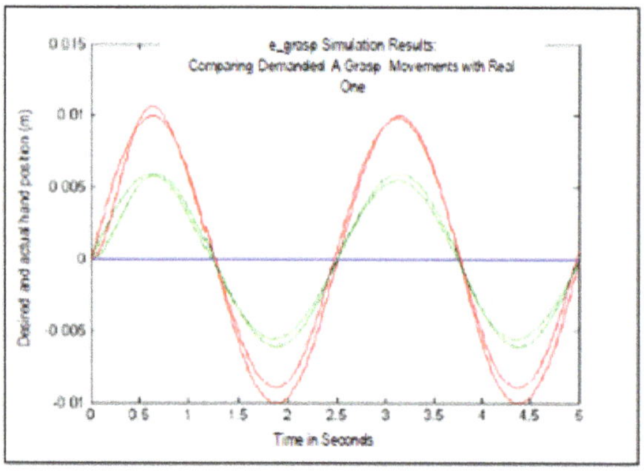

Figure 8: Creating a 3-D grasp movement is an easy task with e_GRASP.

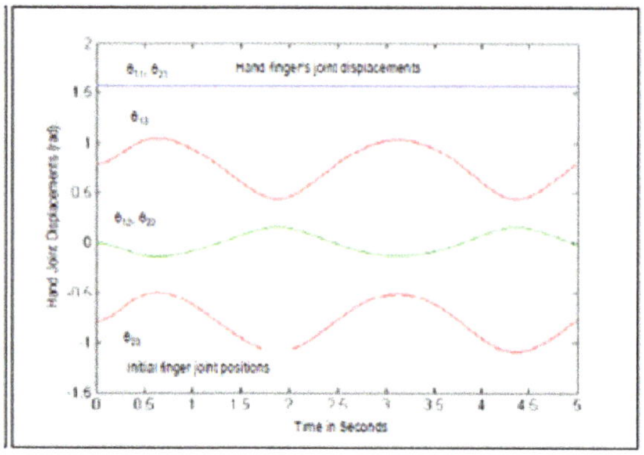

Figure 9: Due to hand motion a 1st finger closed loop simulation.

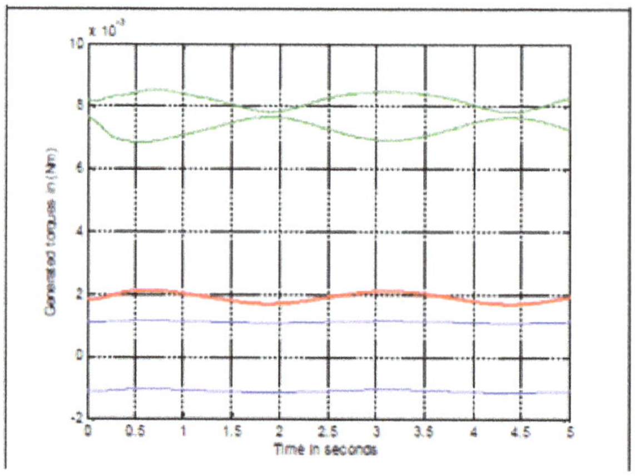

Figure 10: e-Grasp simulation of hand torques. Simulation includes plotting of individual torques while grasping in 3-D space.

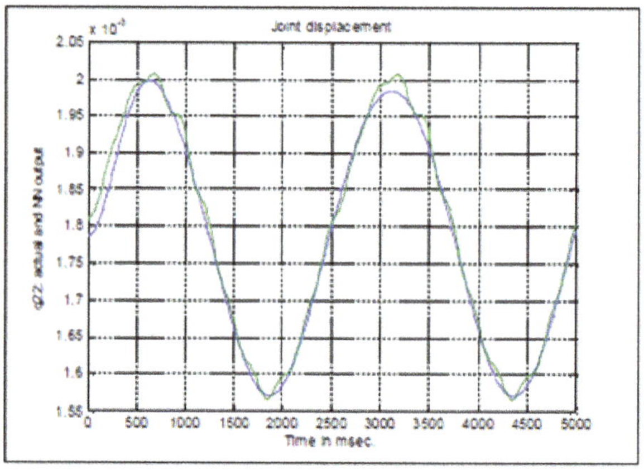

Figure 11: e-Grasp simulation capabilities: Plotting of individual joint error, while comparing ANN controlled output with actual joint output.

learning via fuzzy system. In addition, the package is also able to be linked with much advanced Matlab functionalities, which is the ANN tools. In addition to the basic functions, the simulation package can also provide more analysis of hand movement. For example, Figure 13 shows the ability to plot training patterns for hand ANN training. Such patterns were also used for learning and understanding the hand for optimization of fingertips contact forces.

Conclusions

Software simulation of an (*n*) DOF articulated robotic hand system is not an obvious task. Specifically, such simulators are truly needed for testing purposes, and for viewing hand performance before physical implementation. This manuscript has focused on a developed simulation software for supervising and controlling a fully described (kinematically and dynamically) (*n*) fingers robotic hand. The simulation environment was achieved via Matlab with a link to other available Matlab toolboxes. The developed simulator also has the ability to be linked internally to Matlab toolboxes, in addition, to be externally linked to other libraries, like (C^{++}), for a possible real-time hand control. The simulator even has the ability for linking high level commands to the low-level digital motion processor. The simulator has proved to be an effective way to look and view kinematics and dynamics models of robotic hand. These models are found to be essential elements for hand object dynamic simulation. Next stage, is to take the simulator further, with graphical interfaces and functionalities.

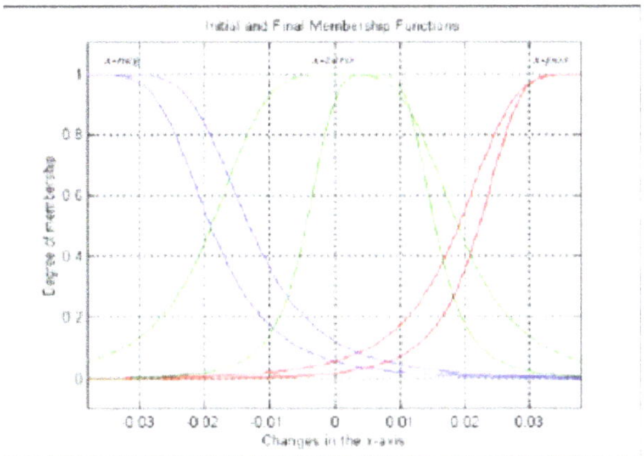

Figure 12: e_GRASP simulator capabilities: Hand learning simulator linkage to MATLAB/FUZZY Toolbox. The FUZZY Toolbox was used for learning of hand motion parameter.

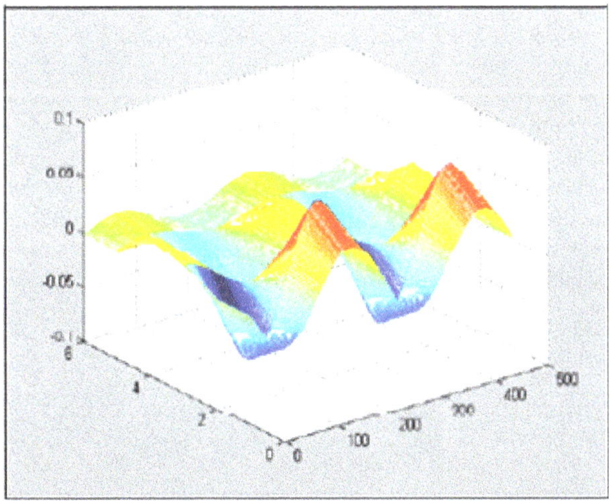

Figure 13: e_GRASP simulator capabilities: Typical integration of MATLAB/ANN Toolbox hand motion.

References

1. Matlab Toolbox for the iRobot Create (MTIC) (2011) Version 2.0.
2. Shaoqiang Y, Zhong L, Xingshan L (2008) Modeling and simulation of robot based on Matlab/SimMechanics. Control Conference, CCC 2008, 27th Chinese, Kunming.
3. Peter Corke (2008) Robotics Toolbox for Matlab.
4. Jambak MI, Haron H, Nasien D (2008) Development of Robot Simulation Software for Five Joints Mitsubishi RV 2AJ Robot Using MATLAB/Simulink and V-Realm Builder. 5th International Conference on Computer Graphics, Imaging and Visualisation, Penang.
5. Olivier M (2004) Cyberbotics Ltd WebotsTM: Professional Mobile Robot Simulation. Inernational Journal of Advanced Robotic Systems.
6. Gourdeau R (1997) Object-oriented programming for robotic manipulator simulation. Robotics and Automation Magazine, IEEE 4: 21-29.
7. Ramasamy S, Arshad R (2000) Robotic hand simulation with kinematics and dynamic analysis. TENCON 2000 3: 178-183.
8. Miller T, Allen K (2004) Graspit! A versatile simulator for robotic grasping. Robotics & Automation Magazine, IEEE 11: 110-122.
9. Miller M, Allen P, Santos V, Valero-Cuevas F (2005) From robotic hands to human hands: a visualization and simulation engine for grasping research. Industrial Robot: An International Journal 32: 55-63.
10. Jagdish R, Radhey S, Rajsekhar A, Bhanu P (2012) Real-Time Robotic Hand Control Using Hand Gestures", Robotic Systems – Applications, Control and Programming. Book. Edited by Dr. Ashish Dutta, ISBN 978-953-307-941-947.
11. Corrales A, Jara A, Torres F (2010) Modelling and simulation of a multi-fingered robotic hand for grasping tasks.11th International Conference on. Control Automation Robotics & Vision (ICARCV), Singapore.
12. Gourret J, Thalmann N, Thalmann D (1989) Simulation of object and human skin formations in a grasping task. 16th annual conference on Computer graphics and interactive techniques.
13. Magnus B (2008) Controlling a Robot Hand in Simulation and Reality. Degree Project Department of Management and Engineering LIU-IEI-TEK-A--08/00336—SE, (2008).
14. Tarmizi W, Adly A, Amirfaiz W, Elamvazuthi I, Begam M (2010) Modeling and simulation of a multi-fingered robot hand. International Conference on Intelligent and Advanced Systems (ICIAS), Kuala Lumpur, Malaysia.
15. Boughdiri R, Bezine H, Sirdi M, Naamane A, Alimi M (2011) Dynamic modeling of a multi-fingered robot hand in free motion. 8th International Multi-Conference on Systems, Signals and Devices (SSD) Sousse.
16. Ohol S, Kajale S (2008) Simulation of Multifinger Robotic Gripper for Dynamic Analysis of Dexterous Grasping. Proceedings of the World Congress on Engineering and Computer Science,San Francisco.
17. Chan C , Yun-Hui L (1999) Simulating dextrous manipulation of a multi-fingered robot hand based on a unified dynamic model . IEEE International Conference on Robotics and Automation Detroit, MI.
18. Orin E, Chao H, Olson W, Schrader W (1985) Pipeline/Parallel Algorithms for the Jacobian and Inverse Dynamics Computations. Proceedings of the IEEE International Conference on Robotics and Automation.
19. Paul RP (1981) Robot Manipulators: Mathematics, Programming, And Control. MIT Press, Cambridge, USA.
20. Fu S, Gonzalez C, Lee S (1987) Robotics Control, Sensing, Vision, and Intelligence. McGraw-Hill International Editions, Singapore.

4

Closed-Form Inverse Kinematic Solution for Anthropomorphic Motion in Redundant Robot Arms

Yuting Wang and Panagiotis Artemiadis*

Mechanical and Aerospace Engineering, School for Engineering of Matter, Transport and Energy, Arizona State University Tempe, USA

Abstract

As robots are increasingly migrating out of factories and research laboratories and into our everyday lives, they should move and act in environments designed for humans. For this reason, the need of anthropomorphic movements is of utmost importance. This paper proposes a framework for solving the inverse kinematics problem of redundant robot arms that results to anthropomorphic configurations. The swivel angle of the elbow is used as a human arm motion parameter for the robot arm to mimic. The swivel angle is defined as the rotation angle of the plane defined by the upper and lower arm around a virtual axis that connects the shoulder and wrist joints. Using kinematic data recorded from human subjects during every-day life tasks, we validate the linear relations between intrinsic and extrinsic coordinates of the human arm that estimates the swivel angle, given the desired end-effector position. Defining the desired swivel angle simplifies the kinematic redundancy of the robot arm. The proposed method is tested with an anthropomorphic redundant robot arm and the computed motion profiles are compared to the ones of the human subjects. We show that the method computes anthropomorphic configurations for the robot arm, even if the robot arm has different link lengths than the human arm and starts its motion at random configurations.

Keywords: Human arm kinematics; Robot arm model; Redundancy resolution; Anthropomorphic motion

Introduction

During the last decade, robots have successfully migrated out of factories and academic labs and into our. Every day lives, creating new families of co-robots and bionics. These robots should be able to move and act in environments designed for humans, and more importantly use tools for executing tasks designed for humans. Therefore, there is an increasing demand of robots which can interact, communicate and collaborate with humans. This requires human-like behaviour, which will allow the human subject to be able to understand robot's intentions and seamlessly collaborate with the robot. Its application fields range widely from service robotics to therapeutic devices [1]. In order for the human-robot cooperation to be intuitive, the robot configurations should be anthropomorphic [2,3]. This is not straightforward, since the human arm is redundant, i.e. it has 7 degrees of freedom (DOFs), while only 6 DOFs are required for a given position and orientation of the arm endpoint. This creates a challenge for the redundant robot inverse kinematics that need to be solved in a similar way of that of the human arm, in order to guarantee seamless integration [4-6]. The exploitation of kinematic redundancy for the generation of human-like robot arm motions has been already proposed in the literature. In Cruse [7], it is shown that it is possible to associate some cost function to each human arm joint. The arm performs movements that optimize these cost functions. A variety of cost functions have been used to explain the principles of human arm motor control, such as ones related to dynamics [8-10], neuro-physiological and psychophysical ones [11-13], as well as combinations of those [7,14]. However, the majority of these cost functions are used with global optimization methods which are computationally expensive and not suitable for real-time implementation. In [3,15], a mathematical cost functional describing the muscle fatigue is used to achieve human-like joint motions of a robot arm during writing tasks. Even though it is of a local nature, the applicability of this method to generate human-like manipulation motions is not yet clear. Human motion capture has been widely used for the generation of kinematic models describing human and humanoid robot motions [16]. There have been also efforts to generate human-like motion by imitating human arm motion as closely as possible. In Kim [17], a method to convert the captured marker data of human arm motions to robot motion using an optimization scheme is proposed. The position and orientation of the human hand, along with the orientation of the upper arm, were imitated by a humanoid robot arm.

However, this method was not able to generate human-like motions, given a desired three dimensional (3D) position for the robot end-effector. Similarly, most of the previous works on biomimetic motion generation for robots are based on minimizing posture difference between the robot and human arm, using a specific recorded data set [18,19]. Therefore, the robot configurations are exclusively based on the recorded data set. In this way, the methods can not generate new human-like motion. The latter is a major limitation for the kinematic control of anthropomorphic robot arms and humanoids, because the range of possible configurations is limited to the ones seen in the data. In [4], a method is proposed to solve the inverse kinematic problem by defining the swivel angle, i.e. the rotation angle of the plane defined by the upper and lower arm around a virtual axis that connects the shoulder and wrist joints. However, the method is only demonstrated for simple tasks, i.e. natural reaching and grasping, while the method

***Corresponding author:** Panagiotis Artemiadis, Mechanical and Aerospace Engineering, School for Engineering of Matter, Transport and Energy, Arizona State University Tempe, USA, E-mail: panagiotis.artemiadis@asu.edu requires the initial configuration of the human arm and the robot arm to be known, a feature that is not readily available in real scenarios. In Asfour [2], the upper arm joints values are first calculated for positioning the robot elbow and then the remaining joints are solved with closed-form inverse kinematics. Such an approach though cannot be easily applied to robots having a kinematic structure different from that of the human upper limb. There are also some biomimetic approaches based on the dependencies among the human joint angles [1,20]. In Edsinger A [1], the authors generalized the inverse kinematic solution by encompassing joint limitation, singularity avoidance and optimum manipulability measures. However, the proposed requires the robot arm with similar structure of the human arm, while its iterative solution method is not efficient for real-time processing. In this paper we consider the problem of generating human-like motions from the kinematic point of view taking into account data recorded during a wide variety of everyday life tasks. We rely on a hypothesis from neurophysiology and apply it for generating human-like motions of a redundant anthropomorphic robot arm. The human arm swivel angle is used as a parameter for the robot arm to mimic, and the problem of the inverse kinematics is simplified. The linear relationship between the intrinsic and extrinsic coordinates was validated and used to estimate the desired swivel angle using previous knowledge of human arm motion recordings. The proposed controller is finally applied to a 7 DOFs robot arm (LWR 4+, KUKA) for evaluation purposes.

Method

Human arm kinematics

As it shown in Figure 1, the human arm consists of a series of rigid links connected by three anatomical joints (shoulder joint, elbow joint, and wrist joint) while neglecting the scapular and clavicle motions [21]. In this study, 7 DoFs were analyzed for simplicity: shoulder exion-extension, shoulder abduction-adduction, shoulder lateral-medial rotation, elbow exion-extension, elbow pronation-supination, wrist exion-extension and wrist pronation-supination, which can be simulated by 7 corresponding joint angles, i.e. q1, q2, q3, q4, q5, q6,q7 for the human arm. The Denavit-Hartenberg (DH) parameters of the kinematic model of the arm that we used are listed in Table 1 [1,20] where L1, L2, L3 are the lengths of the upper arm, forearm and palm respectively1. In order to track the motion of the upper limb, a 3D position sensor and associated positioning markers attached to each rigid link were used to compute the joint angles of the shoulder, elbow and wrist. The position and orientation of the end effector for the human arm can be expressed as a function of all those joint angles through forward kinematics.

Robot arm model

Figure 1 also shows the reference and link coordinate systems of the 7-DoF robot arm (LWR4+, KUKA) using the DH convention [22]. The values of the DH parameters are listed in Table 2, where Lu, Lf and Lh are the link lengths of robot upper-arm, forearm and hand respectively. There are several differences between the human arm and the robot arm:

1. The human arm has a spherical wrist (q5, q6 and q7 axes intersect at a single point) while the robot arm does not.
2. The length of the KUKA arm is almost twice as much as that of the human arm.

Reduction of number of joint variables

For a given position and orientation of the end-effector with respect to the base frame, the wrist position can be easily defined. As shown in Figure 2, once the end-effector position and orientation are specified in terms of p_{ee} and $R_{ee} = [n\ s\ a]$, so that the homogeneous transformation relating the description a point in the end-effector frame to the description of the same point in base frame can be represented as

$$T^b_e = \begin{bmatrix} n\ s\ a\ p_{ee} \\ 0\ 0\ 0\ 1 \end{bmatrix} \quad (1)$$

then the wrist position in the end-effector frame can be represented as

$$P_{we} = \begin{bmatrix} d_x \\ d_y \\ d_z \end{bmatrix} = d \quad (2)$$

i	α_i	a_i	d_i	θ_i
1	90°	0	0	q_1
2	90°	0	0	q_2 + 90°
3	90°	0	L_1	q_3 + 90°
4	90°	0	0	q_4 + 180°
5	90°	0	L_2	q_5 + 180°
6	90°	0	0	q_6 + 90°
7	90°	L_3	0	q_7 + 180°

Table 1: Human arm D-H parameters.

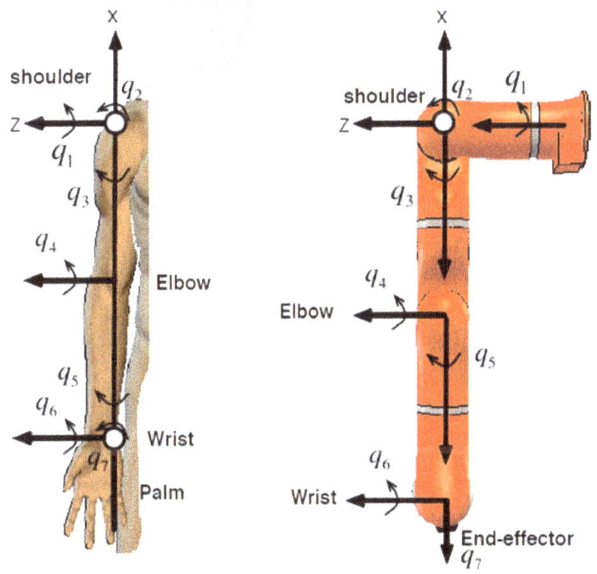

Figure 1: Kinematic models of the Human and Robot arms.

i	α_i	a_i	d_i	θ_i
1	90°	0	0	q_1
2	90°	0	0	q_2
3	90°	0	L_u	q_3
4	90°	0	0	q_4
5	90°	0	L_f	q_5
6	90°	0	0	q_6
7	0	0	L_h	q_7

Table 2: KUKA arm model D-H parameters.

Where dx, dy, dz are the coordinates of the wrist center with respect to the origin of the end-effector frame. This allows direct computation of the wrist position with respect to the base frame given the end-effector frame [22]. In our case, the base frame is located at the center of the shoulder for both arm models shown in Figure 1. Therefore, the position of the wrist Pw in the base frame can be found as:

$$\begin{bmatrix} P_w \\ 1 \end{bmatrix} = T^b{}_e \begin{bmatrix} P_{we} \\ 1 \end{bmatrix} \quad (3)$$

Therefore, given a desired position and orientation of the end-effector, we split the problem into two parts:

1. Using (3) we compute the wrist position, and since the wrist position is only a function of the first 4 joint angles (q1-q4) we define a method to solve for those joints. This step involves redundancy in joint angles and it is going to be solved in a way to guarantee anthropomorphism (see Redundancy resolution section).
2. Knowing q1-q4 and the desired position and orientation of the end-effector frame, we can then analytically solve for q5-q7. Therefore, we can neglect the structure difference of these two arms.

Redundancy resolution

There is evidence from previous works that the position of the elbow joint in space is an important parameter of anthropomorphism for arm configurations [4,2]. The redundancy of the arm is actually found on the first four joints, that need to position the wrist on a 3D position in Cartesian Space. A simple physical interpretation of the redundant degree of freedom is based on the observation that if the wrist is held fixed, the elbow is still free to swivel about a circular arc whose normal vector is parallel to the axis from the shoulder to the wrist [23]. In Veljko P (3), P_s, P_e and P_w represents the position of the shoulder, the elbow and the computed location of the wrist based on the given position and orientation of the end effector. As the swivel angle ϕ varies, the elbow traces an arc of a circle lying on a plane which is perpendicular to the wrist-to-shoulder axis. In order to measure ϕ, we define the normal vector of the plane as $\tilde{n}$, P_c as the center of the circle and unit vector $\tilde{u}$, $\tilde{v}$ as the coordinate system on the plane. Here $\tilde{u}$ is set as the projection of an arbitrary vector $\tilde{a}$ onto the plane and $\tilde{v}$ is the cross product of $\tilde{u}$ and $\tilde{n}$.

The selection of ~ will determine where $\phi = 0$[23, 4]. Therefore, the position of the elbow can be represented as:

$$P_e = R\left[\cos(\phi)\tilde{u} + \sin(\phi)\tilde{v}\right] + P_c \quad (4)$$

where R is the Euclidean distance between P_e and P_c.

The analytic expression above is an advantage when an objective function is used to select an appropriate value of ϕ since it is often necessary to express the objective function in terms of the joint angles. Once the value of ϕ is determined, the elbow position can be computed. With the position of the wrist and elbow, as well at the 3D position and orientation of the rigid body of the end effector, we are able to analytically give a unique solution to the inverse kinematic problem, and therefore compute the 7 joint angles of the upper limb.

Another benefit of the swivel angle is that it is not constrained by the limitation of the arm length. In Hyunchul [4], it has been proven that the direction of the longest axis in the manipulability ellipsoid of the wrist is only a function of the swivel angle. Therefore, the swivel angle of the robot arm should be selected close to that of the human arm with the same position and orientation of the end effector in order to maintain the manipulability which leads to an anthropomorphic configuration. This will be validated in the Results section.

The linear relationships between intrinsic and extrinsic coordinates

The idea to generate anthropomorphic robot arm motions is inspired by the results obtained in [24,25]. The arm movements are in shoulder-centered spherical coordinates and there is a linear sensorimotor transformation model that maps the extrinsic coordinates on a natural arm posture using the intrinsic coordinates. The extrinsic coordinates are the wrist position expressed in the spherical coordinates, where R denotes the radial distance, χ the azimuth, and ψ the elevation. The reason to choose the spherical coordinates rather than Cartesian or cylindrical is the former leads to a more compact representation of the linear relation [24]. The wrist position in the Cartesian coordinates can be transformed to the spherical coordinates by:

$$R2 = P_w(x)^2 + P_w(y)^2 + P_w(x)^2 \quad (5)$$

$$\tan(\chi) = \frac{P_w(z)}{-P_w(y)} \quad (6)$$

$$\tan(\varphi) = \frac{P_w(x)}{\sqrt{P_w(y)^2 + P_w(z)^2}} \quad (7)$$

Where $P_w(x), P_w(y)$ *and* $P_w(z)$ designate the Cartesian components of the wrist position with respect to the shoulder frame. The intrinsic coordinates consist of angles defined the upper arm elevation (θ) and yaw (η) and the forearm elevation (α) and yaw (β). The elevations (θ, β) define the angle between each limb segment and the vertical axis measured in a vertical plane. The yaw angles (η, α) define the angle between each of the limb segments and the anterior direction, measured in the horizontal plane, as shown in Figure 3. However, this linear relation has only been proven by one simple task, arm pointing in [24,25]. Here, we demonstrate a more accurate relationship with more daily life tasks (see Results section).

Based on this linear relationship, the orientation angles of the upper arm and the forearm $(\theta, \eta, \alpha, \beta)$ can be obtained only knowing the wrist position (R, χ, φ). Therefore, as is shown in Figure 4, the elbow position in this coordinate system can be expressed as:

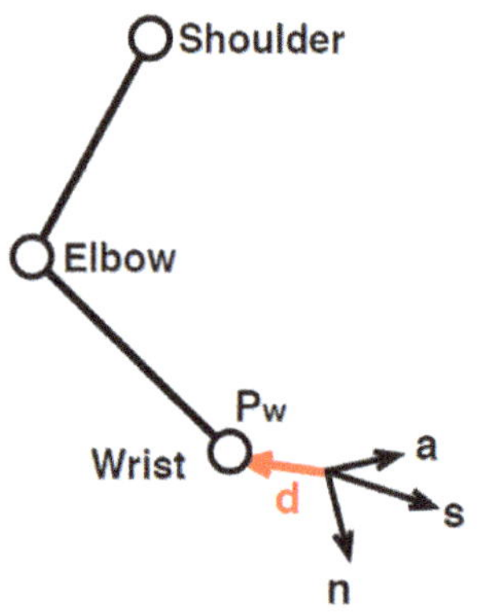

Figure 2: Wrist position with respect to the end-effector frame.

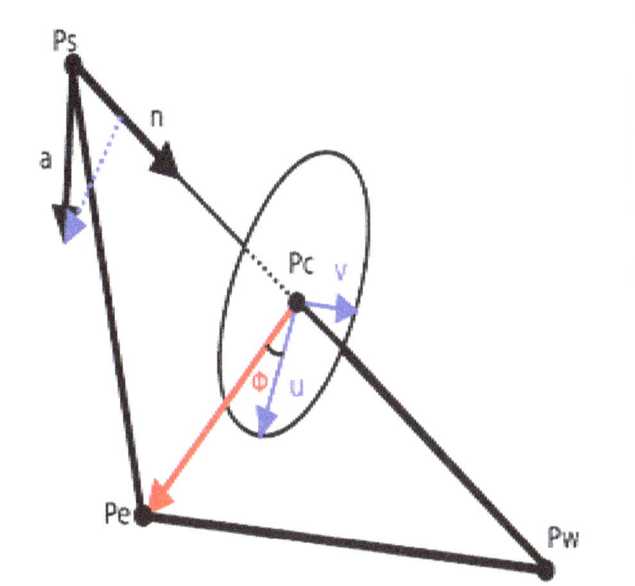

Figure 3: Coordinate frame at the center of the elbow circle and the swivel angle that allows the parameterization of the elbow position by a single variable.

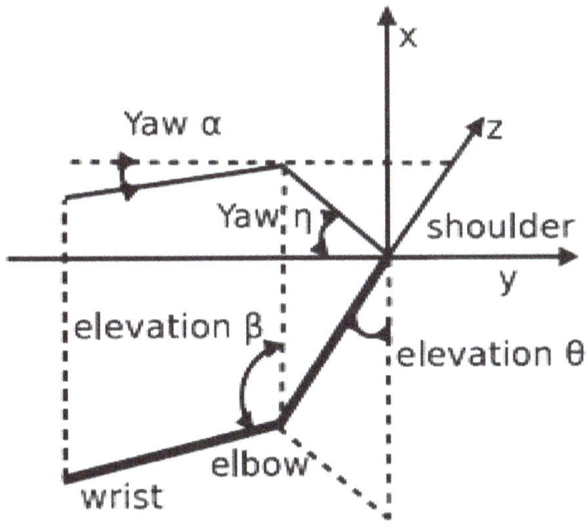

Figure 4: Orientation angles of the arm and forearm. The angles θ and η define the elevation and the yaw of the upper arm and β and α represent the elevation and yaw of the forearm.

$$P_e(x) = -U\cos(\theta) \tag{8}$$

$$P_e(y) = -U\sin(\theta)\cos(\eta) \tag{9}$$

$$P_e(z) = U\sin(\theta)\sin(\eta) \tag{10}$$

Where $P_e(x), P_e(y)$ *and* $P_e(z)$ designate the components of the elbow position with respect to the shoulder frame. U represents the length of the upper arm.

The swivel angle can be obtained knowing the elbow position computed by (8,9,10) and wrist position computed by (3). Then the swivel angle is selected using (4):

$$\phi = a\tan 2\left(\frac{(P_e - P_c).\vec{u}}{\|P_e - P_c\|}, \frac{(P_e - P_c).\vec{v}}{\|P_e - P_c\|}\right) \tag{11}$$

Solution for inverse kinematic problem

Let $P_e = [x_e\ y_e\ z_e]^T$, $P_w = [x_w\ y_w\ z_w]^T$ be the position of the elbow and wrist with respect to the base frame. The position of the elbow is computed by (4), with the estimation of the swivel angle, using (11). The position of the wrist is computed using (3). Let the desired position and orientation of the end- effector be given by

$$T = \begin{bmatrix} n_d & s_d & a_d & p_d \\ 0 & 0 & 0 & 1 \end{bmatrix} \tag{12}$$

Since the position of the wrist and elbow are computed, then the solution for the first 4 joint angles has a

closed form which is shown below:

$$q_1 = \arctan 2(y_e, x_e) \tag{13}$$

$$q_2 = \arctan 2\left(\sqrt{(x_e)^2 + (y_e)^2}, z_e\right) \tag{14}$$

$$q_3 = \arctan 2(-M_3, M_1) \tag{15}$$

$$q_4 = \arctan 2\left(\sqrt{(M_1)^2 + (M_3)^2}, M_2 - L_1\right) \tag{16}$$

Where

$$M_1 = x_w\cos(q_1)\cos(q_2) + y_w\sin(q_1)\cos(q_2)z_w\sin(q_2) \tag{17}$$

$$M_2 = x_w\cos(q_1)\sin(q_2)y_w\sin(q_1)\sin(q_2)z_w\cos(q_2) \tag{18}$$

$$M_3 = x_w\sin(q_1) + y_w\cos(q_2) \tag{19}$$

We should note that although multiple solutions could arise, they are eliminated by violating the human

joint limitations. In order to solve for q5-q7, we can articulate the inverse kinematics problem into two sub-problems. After solving the inverse kinematics for q1 to q4, we can compute the transformation matrix from the base frame to the wrist frame, $T^0{}_4(q_1, q_2, q_3, q_4)$. Then, the transformation matrix from the wrist frame to the end-effector frame can be computed as

$$T^4{}_7(q_5, q_6, q_7) = \begin{bmatrix} n_x^{(4)} & s_x^{(4)} & a_x^{(4)} & p_x^{(4)} \\ n_y^{(4)} & s_y^{(4)} & a_y^{(4)} & p_y^{(4)} \\ n_z^{(4)} & s_z^{(4)} & a_z^{(4)} & p_z^{(4)} \\ 0 & 0 & 0 & 1 \end{bmatrix} = (T^0{}_4)^T T \tag{20}$$

Where

$$n^{(4)} = \left[n_x^{(4)}\ n_y^{(4)}\ n_z^{(4)}\right]^T, s^{(4)} = \left[s_x^{(4)}\ s_y^{(4)}\ s_z^{(4)}\right]^T, a^{(4)} = \left[a_x^{(4)}\ a_y^{(4)}\ a_z^{(4)}\right]^T$$

are the orientation vectors and $p^{(4)} = \left[p_x^{(4)}\ p_y^{(4)}\ p_z^{(4)}\right]^T$ is the position vector of the end-effector reference system with respect to the one at the wrist. Since $T^4{}_7(q_5, q_6, q_7)$ is known, the joint angles q_5, q_6, q_7 can be computed using (20). The analytical solution is given by:

$$q_5 = \arctan 2(a_y^{(4)}, a_x^{(4)}) \tag{21}$$

$$q_6 = \arctan 2\left(\sqrt{(a_y^{(4)})^2 + (a_x^{(4)})^2}, a_z^{(4)}\right) \tag{22}$$

$$q_7 = \arctan 2(n_z^{(4)}, -s_z^{(4)}) \tag{23}$$

For $q_5 \varepsilon(0,\pi)$

$$q_5 = \arctan 2(-a_y^{(4)}, -a_x^{(4)}) \tag{24}$$

$$q_6 = \arctan 2\left(-\sqrt{(a_y^{(4)})^2 + (a_x^{(4)})^2}, a_z^{(4)}\right) \tag{25}$$

$$q_7 = \arctan 2(-n_z^{(4)}, s_z^{(4)}) \tag{26}$$

For $q_5 \varepsilon(-\pi,0)$

Results

Experiments

In order to collect data to demonstrate the linear relation of the orientation angles and the wrist position, as well as test the proposed method, we conducted experiments with three right-handed subjects (two male and one female). We should note that our goal is to prove this linear relationship works across all subjects and can be applied to estimate the swivel angle of the human arm for anthropomorphic control of the robot arm. A motion capture system with associated positioning markers attached to each rigid link of the human arm were used to compute the joint angles of the shoulder, elbow and wrist. Initially we demonstrated the linear relationship between the angles of the orientation of the upper arm and the forearm and the wrist position using data from a variety of daily tasks and then conducted separate validation tasks with the same human subjects. Six types of experimental tasks were selected from activities of daily living: (Type 1) arm reaching and pointing, (Type 2)placing a water bottle in discrete locations, (Type 3) placing a ping-pong ball in discrete locations, (Type 4) eating, (Type 5) face and head touching and (Type 6) writing. These tasks are chosen from the basic Activities of Daily Life (ADLs) and each task lasts 20 seconds. No initial information of the configuration of the human arm and the robot arm is needed.

Demonstration of the linear relations

Given the elbow and wrist position of the human subjects, we can solve for the actual orientation angles of the upper arm and the forearm. The spherical coordinates of the wrist position can be computed using (5, 6, 7). Therefore, the optimum linear relation is chosen by:

$$\min \int_T^0 (\sigma_{act}(t) - \sigma(t, p_w))dt \tag{27}$$

where T corresponds to 1=5 of total data recording time, $\sigma_{act}(t)$ is the actual angle of the orientation of the upper arm and the forearm, $\sigma(t, P_w)$ is the estimated orientation angle as a linear function of Pw. The combined linear relation between the wrist position and the orientation angles across all tasks and subjects is shown in Figure 5 and the statistical details are given in Table 3. However, due to the arm length difference of the human subjects, the relation will differ from subjects. In Figure 6, two types of graphical analysis are plotted. The mean errors for all the orientation angles are below 5 degrees across all subjects and tasks. It also shows that the mean errors vary from different tasks but are similar $(\pm 1^{\circ})$ for different orientation angles.

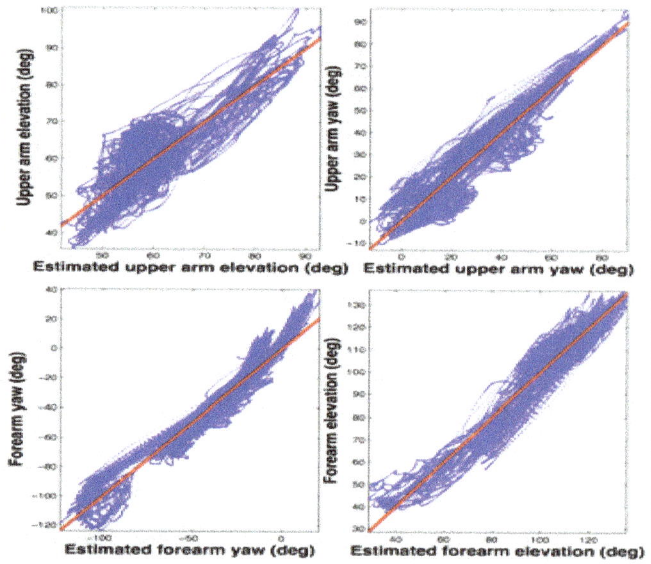

Figure 5: Dependence of intrinsic coordinate on extrinsic parameters across all tasks and subject. The vertical axis represents the actual orientation angles and the horizontal axis represents estimated values from the linear combination of target parameters which gave the best fit to the data.

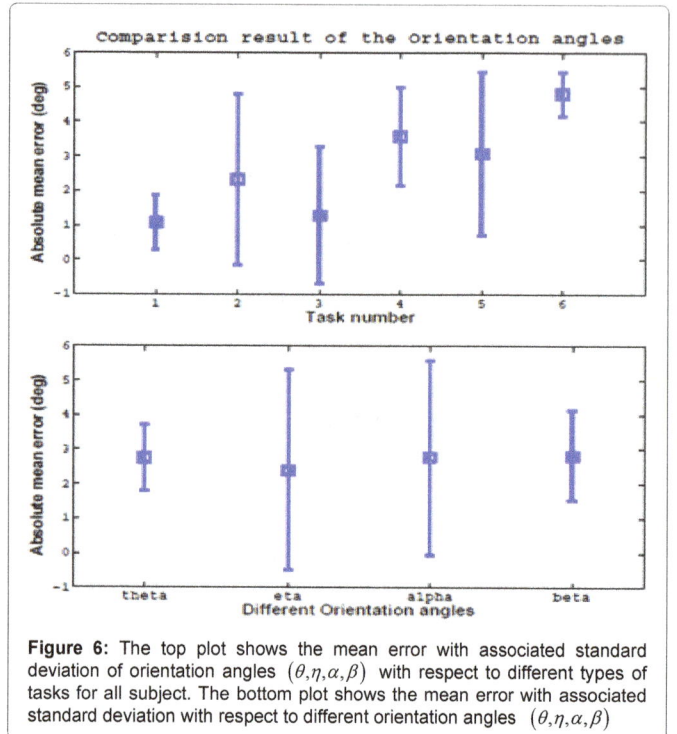

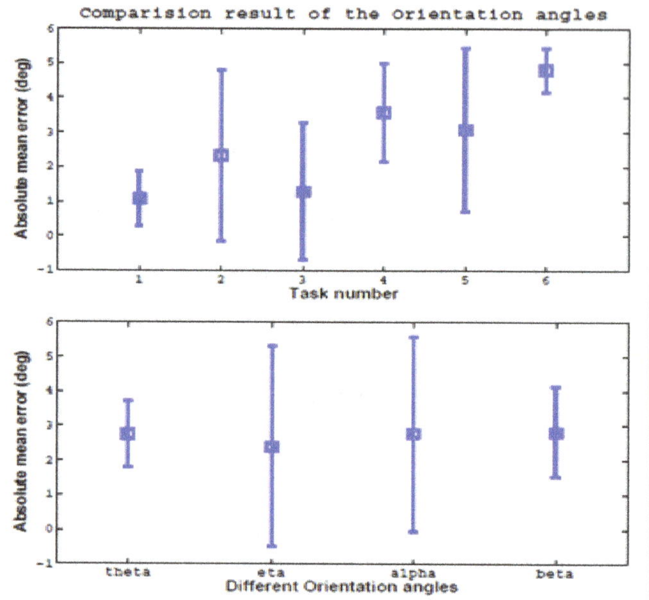

Figure 6: The top plot shows the mean error with associated standard deviation of orientation angles $(\theta,\eta,\alpha,\beta)$ with respect to different types of tasks for all subject. The bottom plot shows the mean error with associated standard deviation with respect to different orientation angles $(\theta,\eta,\alpha,\beta)$

Method evaluation

I	Orientation angle	Linear function	r²
1	θ	0.2678R+0.5735ψ+66.46	0.80
2	η	-0.1516R +1.052χ +80.51	0.91
3	α	0.3248R +0.9347χ -154.7	0.95
4	β	0.1471 R-0.981 ψ +24.49	0.95

Table 3: Linear relationship between the intrinsic and extrinsic coordinate.

The proposed method was used in order to control a robot arm (LWR4+, KUKA) to reach the desired position and orientation which was identical to that of the human subject during each trial. For the performance estimation, the mean and standard variation of the absolute difference between the the measured swivel angle collected from the subjects during the experiments and the estimated swivel angle of the robot arm based on the proposed criterion were calculated. The performance estimation results are plotted with two representative

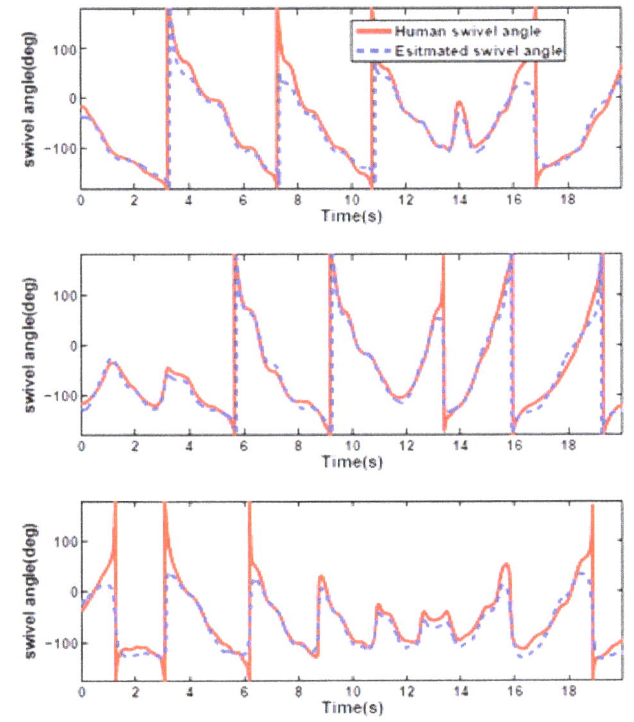

Figure 7: Comparison between the robot (estimated) swivel angle (dotted line) and the human (measured) Swivel angle (solid line) for Type 1 task across all subject. Rows correspond to subject 1,2,3 respectively.

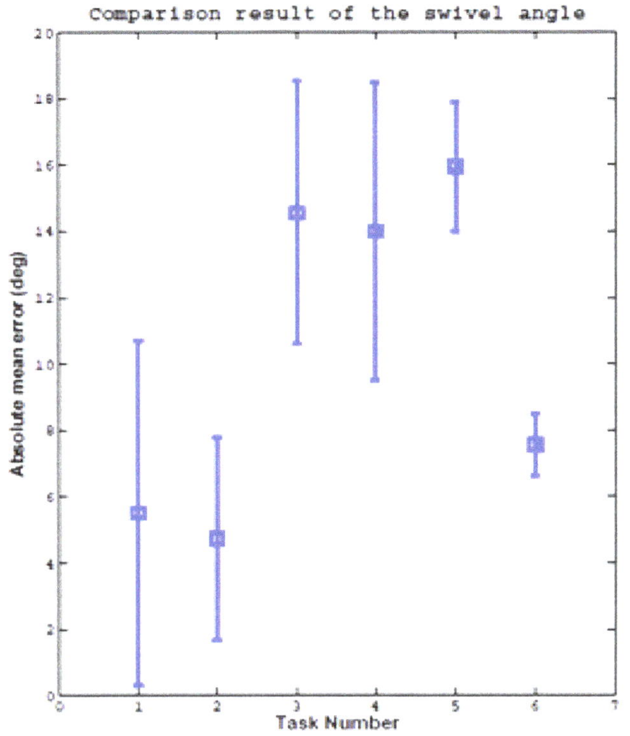

Figure 9: The mean error with associated standard deviation of the swivel angle (ϕ) with respect to types of tasks.

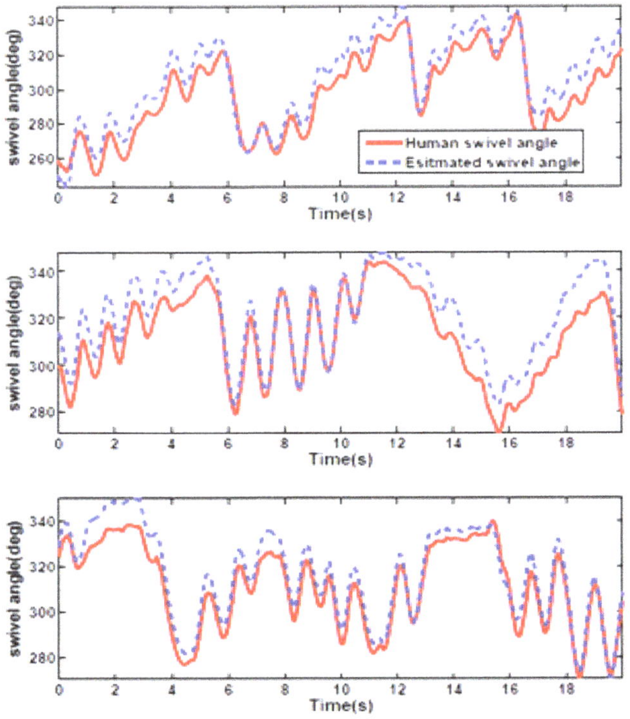

Figure 8: Comparison between the robot (estimated) swivel angle (dotted line) and the human (measured). Swivel angle (solid line) for Type 6 task across all subject. Each row is from subject 1,2,3 respectively.

types of tasks for all subjects. Figures 7 and 8 show the comparison of the human and the robot arm swivel angles for two representative types of tasks. In Figure 9, an analysis of the data indicates that the mean errors vary for different types of tasks. Here we should notice that due to the length difference from the end effector to the wrist, the wrist position of the robot arm will be slightly different from that of the human arm which can cause some error in the estimation of the swivel angle. Also, there is not any initial information for the human arm and the robot arm but the estimation value is still very close to the actual swivel angle with 11° mean error across all tasks and subjects. Comparing the results for Type 2 and Type 3 tasks, it implies that the hand orientation caused by the different types of object has affect on the estimation result (Type 2 task, passing the water bottle has more constraints on the wrist orientation than Type 3 task, passing the ping-pong ball). In order to show that the swivel angle is indeed a possible metric of anthropomorphism, we compared the human and robot arm configurations during the tested tasks. These are shown in Figure 10. It must be noted that the joint angles of the human and robot arm will be different due to the difference in the link lengths of the two. Despite this, the robot arm configurations are very similar to the ones of the human, proving that the swivel angle is representative of anthropomorphism in robot motions.

Conclusions

In this paper we consider the problem of generating human-like motions from the kinematic point of view, taking into account data recorded during a wide variety of everyday life tasks. We use findings from neurophysiology that note the importance of the elbow position and orientation in anthropomorphic arm movements. The

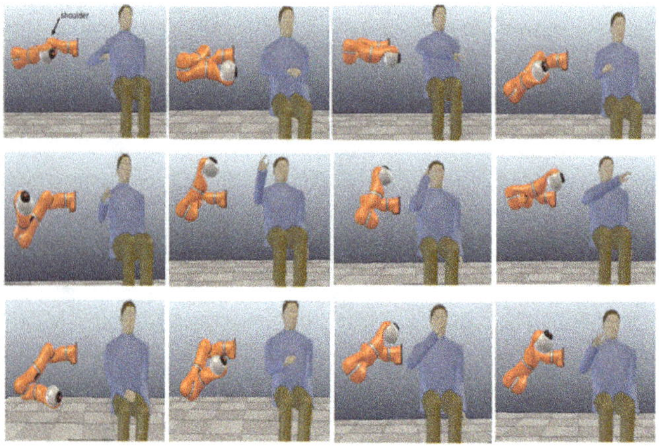

Figure 10: Snapshot of the human arm performing 3D motions and the robot arm driven by the proposed method.

swivel angle of the elbow is used as a human arm motion parameter for the robot arm to mimic. Using experimental data recorded from a human subject during every-day life tasks, we validate the linear relations between intrinsic and extrinsic coordinates that estimates the swivel angle, given the desired end-effector position. Requiring a desired swivel angle simplifies the kinematic redundancy of the robot arm. The proposed method is tested with an anthropomorphic redundant robot arm and the computed motion profiles are compared to the ones of the human subject. We show that the method computes anthropomorphic configurations for the robot arm, even if the robot arm has different link lengths than the human arm, or starts its motion from random configurations. The novelty of the proposed method can be found in two main points. First the method uses the concept that the positioning of the elbow joint is a decisive factor of anthropomorphic configurations in humans. Based on that, we define the swivel angle and design our inverse kinematic problem in order to provide similar (but not identical) robot swivel angles with those of the human during every-day life tasks. This results to an analytic closed-form solution of the inverse kinematic problem. Secondly, the method is generalizable, since it can be used in a wide variety of redundant robot arms, as long as an elbow-equivalent point is defined on the robot arm.

References

1. Edsinger A, Kemp CC (2007) Human-Robot Interaction for Cooperative Manipulation: Handing Objects to One Another. The 16th IEEE International Symposium on Robot and Human interactive Communication.
2. Asfour T, Dillmann R (2003) Human-like motion of a humanoid robot arm based on a closed-form solution of the inverse kinematics problem. Intelligent Robots and Systems 2:1407-1412.
3. Veljko P, Spyros T, Dragan K, Goran D (2001) Human-like behavior of robot arms: general considerations and the handwriting task—Part I: mathematical description of human-like motion: distributed positioning and virtual fatigue. Robotics and Computer-Integrated Manufacturing 17:305-315 .
4. Hyunchul K, Miller LM, Byl N, Abrams G, Rosen J (2012) Redundancy Resolution of the Human Arm and an Upper Limb Exoskeleton. Biomedical Engineering 59: 1770-1779.
5. Hyunchul K, Miller LM, Rosen J (2011) Redundancy and joint limits of a seven degree of freedom upper limb exoskeleton. Annual International Conference of the IEEE on Engineering in Medicine and Biology Society, Boston.
6. Hyunchul K, Miller LM, Al-Refai A, Brand M, Rosen J (2011) Redundancy resolution of a human arm for controlling a seven DOF wearable robotic system. Annual International Conference of the IEEE on Engineering in Medicine and Biology Society, Boston.
7. Cruse H, Wischmeyer E, Brüwer M, Brockfeld P, Dress A (1990) On the cost functions for the control of the human arm movement. Biological Cybernetics 62: 519-528.
8. Hollerbach JM, Ki Suh (1987) Redundancy resolution of manipulators through torque optimization. Robotics and Automation 3:308-316.
9. Khatib O, Burdick J (1986) Motion and force control of robot manipulators. Robotics and Automation 3:1381-1386.
10. Hogan H (1984) An organizing principle for a class of voluntary movements. The Journal of Neuroscience 4: 2745-2754.
11. Sief NA, Winters J(1989) Changes in musculoskeletal control strategies with loading: inertial,.isotonic, random. in ASME Biomech. Symp, pp. 355{358, 1989.
12. Temprado JJ, Swinnenb SP, Carsonc RG, Tourmenta A, Laurenta M (2003) Interaction of directional, neuromuscular and egocentric constraints on the stability of preferred bimanual coordination patterns. Human Movement Science 22:339-363.
13. Debaerea F, Wenderotha N, Sunaertb S, Van Heckeb P, Swinnen SP (2004) Cerebellar and premotor function in bimanual coordination: parametric neural responses to spatiotemporal complexity and cycling frequency. NeuroImage 21: 1416-1427.
14. Latash ML (1993) Control of human movement.
15. Potkonjak V, Kostic D, Tzafestas S, Popovic M, Lazarevic M, Djordjevic G (2001) Human-like behavior of robot arms: general considerations and the handwriting task part ii: the robot arm in handwriting. Robotics and Computer-Integrated Manufacturing 17: 317- 327.
16. Ude A, Man C, Riley M, Atkeson CG (2000) Automatic generation of kinematic models for the conversion of human motion capture data into humanoid robot motion.
17. Kim C, Kim D, Oh Y (2005) Solving an inverse kinematics problem for a humanoid robot2019s imitation of human motions using optimization," in ICINCO'05, pp. 85{92, 2005.
18. Pollard N, Hodgins JK, Riley M, Atkeson C(2002) Adapting human motion for the control of a humanoid robot .IEEE International Conference on Robotics and Automation 2: 1390-1397.
19. Caggiano V, De Santis A,Siciliano B, Chianese A (2006) A biomimetic approach to mobility distribution for a human-like redundant arm. The First IEEE/RAS-EMBS International Conference on Biomedical Robotics and Biomechatronics.
20. Artemiadis P, Katsiaris P, Kyriakopoulos K (2010) A biomimetic approach to inverse kinematics for a redundant robot arm. Autonomous Robots 29: 293-308.
21. Korein JU (1986) A geometric investigation of reach.
22. Sciavicco L, Siciliano B(1996) Modeling and control of robot manipulators.
23. Tolani D, Goswami A, Badler NI(2000) Real-time inverse kinematics techniques for anthropomorphic limbs. Graphical Models 62: 353- 388.
24. Soechting JF, Flanders M (1989) Sensorimotor representations for pointing to targets in three dimensional space. Journal of Neurophysiology 62: 582-594.
25. Soechting JF, Flanders M (1989) Errors in pointing are due to approximations in sensorimotor transformations. Journal of Neurophysiology 62: 595-608.

Development of a Versatile Robotic System with Multiple Training Modes for Upper-Limb Rehabilitation Study

Furui Wang[1]*, Duygun Erol Barkana[2] and Nilanjan Sarkar[3]

[1]*Abbott Point of Care, Abbott Laboratories, Princeton, NJ, 08540, USA*
[2]*Department of Electrical and Electronics Engineering, Yeditepe University, Istanbul, Turkey*
[3]*Department of Mechanical Engineering, Vanderbilt University, Nashville, TN, 37212, USA*

Abstract

This paper presents the development of a versatile robotic system with multiple training modes to serve as a test-bed that will facilitate the study of upper-limb rehabilitation following stroke. Seven different training modes, i.e., passive, low-impedance, assist-as-needed, resist-as-needed, visual error augmentation, viscous force field and force perturbation, have been integrated in this robotic system. A hierarchic control system is developed to coordinate these training modes. Initial experiments on unimpaired participants have verified that the robotic system is able to provide the above training modes properly.

Keywords: Rehabilitation Robotics; Multiple Training Modes; Upper-limb movement; Motor learning and adaptation

Introduction

Stroke is a highly prevalent condition [1], especially among the elderly, that results in high costs to the individual and society [2]. According to the American Heart Association (2013), in the U.S., approximately 795 000 people suffer a first or recurrent stroke each year [1]. Of these, 60-75% will live beyond one year after incidence, resulting in a current stroke population of 6.5 million [1,3,4]. Arm function is acutely impaired in a large majority of those diagnosed with stroke [5]. It is a leading cause of disability, commonly involving deficits of motor function. Recent clinical results have indicated that movement assisted therapy can have a significant beneficial impact on a large segment of the population affected by stroke or other motor deficit disorders. In the last few years, robot-assisted rehabilitation for the stroke patients has been an active research area, which provides repetitive movement exercise and standardized delivery of therapy with the potential of enhancing quantification of the therapeutic process [6-13].

Various robot-assisted rehabilitation systems are developed for the upper-limb rehabilitation such as MIT-MANUS [6,7], Assisted Rehabilitation and Measurement (ARM) Guide [8,9]. Mirror Image Movement Enabler (MIME) [10-12] and GENTLE/s [13]. Studies with these robotic systems verified that robot-assisted rehabilitation results in improved performance of functional tasks.

The promising results of robot-assisted rehabilitation systems indicate that robots could be used as effective rehabilitation tools. Different strategies for robot-assisted rehabilitation therapies have been developed, including passive [14,15] active-assistance[8,9,14-18], active-constrained [15], counterpoise control [19] resistive[18], error-amplifying [20-23] , and bimanual modes [15,24,25]. In spite of all these notable studies, questions remain on how best to retrain movement following stroke. Should movement be assisted, assisted only as needed, resisted, manipulated such that movement error is augmented, etc.? Do these training strategies need to be applied separately, or are they more effective when integrated in a specific manner? How do we understand the kinematic and dynamic adaptation of upper-limb movement? In order to investigate such questions a versatile robotic system is needed which can deliver different training therapies thus allow researchers and therapists to evaluate and compare various training strategies implemented with the same platform.

In this work, we develop a versatile robotic system integrated with multiple training modes, i.e., low impedance, passive, assist-as-needed, resist-as-needed, visual error augmentation, viscous force field and force perturbation, to provide a test-bed for the evaluation of these training strategies in upper-limb rehabilitation and the understanding of the motor adaptation in upper-limb movement. Furthermore, the developed robotic system, which is called the PRSAR (Puma Robotic System for Arm Rehabilitation), can combine some of these training modes together so that one can investigate the effect of administering multiple modes simultaneously. The PRSAR is developed based on a PUMA robot and augmented with hand attachment device. An intelligent control framework is designed with a high-level supervisory controller to coordinate different training modes and monitor the task execution and safety, a middle level task planner to generate the reference tasks, and a low-level assistive controller to provide the robotic assistance/resistance in an accurate manner.

This paper is organized as follows. It first introduces the available training modes integrated in the robotic system in Section II. Then Section III presents the development of the robotic system. Section IV shows the results of the validating experiments on unimpaired participants with different training modes. Section V discusses the potential contributions of this work and Section VI concludes the paper and proposes future research directions.

Rehabilitation Training Strategies in PRSAR

In this section, we will introduce the training task designed for

***Corresponding author:** Furui Wang, Abbott Point of Care, Abbott Laboratories, Princeton, NJ, 08540, USA, E-mail: furui.wang@gmail.com the upper-limb rehabilitation, the background and implementation of rehabilitation training strategies integrated in the PRSAR.

The training task

The training task designed in our work is a reaching task of the arm movement involving cognitive processing that could contribute to a variety of functional daily living activities. The reaching task is commonly used for rehabilitation training of upper extremity after stroke. In this task, the participant is asked to move his/her arm in the forward direction to reach a desired point in the horizontal plane and then bring it back to the starting position repetitively. The reaching task requires a combination of shoulder and elbow motion which could increase the active range of motion (AROM) in the shoulder and the elbow in preparation for later functional reaching activities in rehabilitation. The allowable motion is either restricted in one direction only or in both directions of the 2D horizontal plane, depending on the implementation of specific training modes. The participant is required to follow the tip of a certain position trajectory so that the speed of the arm motion is dictated. The idea here is to improve the ability of the participant's arm movement by helping him/her complete a daily living task at a desirable speed. Improving the speed of movement for such tasks is an important criterion to measure the success of a therapy [26]. Note that the PRSAR system is not limited to the reaching task alone - it can implement other rehabilitation tasks involving arm movements. However, in order to describe the training modes we need to choose a task, and in this paper we have chosen the reaching task.

Passive training mode: The passive training mode refers to the training where the user is passive as the robot takes over and moves his/her arm towards a target following a predefined position profile. In passive training mode of the PRSAR, the predefined position profile is a smooth trajectory in X direction, i.e., moving the arm forward, with a bell-shaped velocity profile (Figure 1).

Low impedance training mode: In the low impedance training mode, the participant actively moves his/her arm while the robot passively follows the arm motion where the impedance of the robot is minimized. In this training mode, the inertia and friction of the robot is compensated by the motor torques of the robot, the participant moves his/her arm in a low impedance environment.

At the beginning of a rehabilitation training session, the participant will participate in the low impedance training mode to learn the manipulation of the robot. The baseline and after-training performance of a participant are also measured during this training mode. Moreover, the participant may need to take extra low impedance training following some deceptive training sessions, e.g., visual error augmentation and viscous force field, to washout any sensory-motor distortion.

Assist-as-needed (ANN) training mode: The assist-as-needed training mode defined in this work is similar to the active assistance strategy in literatures [8,9,14-18]. Active assistance is the primary therapy strategy tested so far, where the patient attempts a movement (active) and in which a therapist manually helps complete the movement if the patient is unable to move (assistance) [27]. It is intuitive that in comparison to passive training mode, where the robotic system provides continuous assistance without considering the patient's actual performance, the performance-based active assistance training mode will be more effective. Assisting every movement of a patient has been shown to be not beneficial compared to no assistance or assistance as needed [27]. It has also been suggested [8,15] that performance-based therapy showed better results in improving patients' impairment scores than conventional therapies. Thus, a robot-assisted rehabilitation system could be more efficient if the assistance provided to the patient is given only as and when needed.

In the assist-as-needed training mode of the PRSAR, the participant is required to move his/her arm following the tip of a desired position trajectory as shown in Figure 1. The arm motion is restricted in X direction only. By tracking the tip of the desired position trajectory, the participant will be actually moving their arm with a predefined velocity profile. The idea here is to improve the motor ability of participant's arm by helping him/her to improve his/her speed of movement, which is an important criterion to measure the success of a therapy.To make the robot assists only as needed, an acceptable position band (Figure 2) is defined with the upper bound x_{upper} and the lower bound x_{lower} calculated by Equation (1).

$$X_{upper} = x_d + (x_d * \text{percentage})$$
$$X_{lower} = x_{d -} (x_d * \text{percentage}) \quad (1)$$

Where percentage is the value chosen to set the upper and lower bounds for the defined position trajectory. If the actual position x (t) lies within the acceptable band, then the participant is considered to be able to track the trajectory without robotic assistance. If the actual position x(t) is not between the upper bound x_{upper} and the lower bound x_{lower}, then the robot is activated to provide assistance to bring the arm position back into the acceptable range.

Note that each participant requires a certain amount of time (settling time) to generate the desired motion. The robot should not be

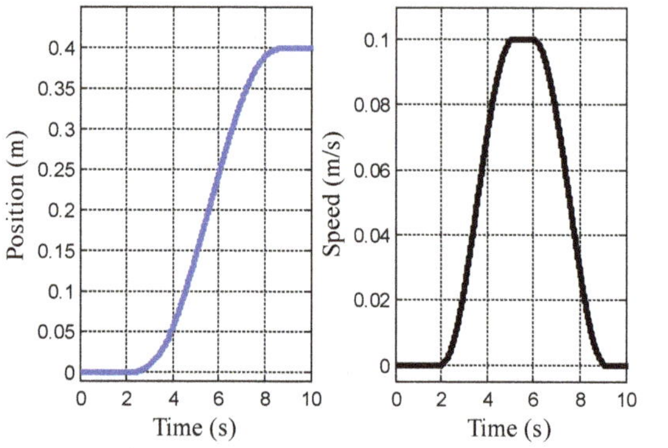

Figure 1: Position and velocity profiles. Here, the distance of the task is 0.4m and the maximum speed is 0.1 m/s.

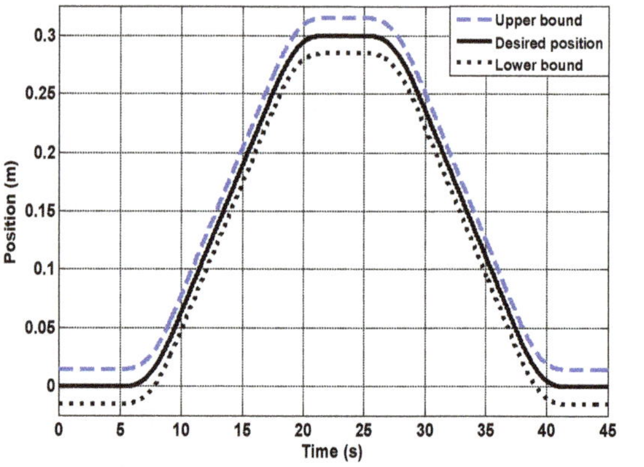

Figure 2: The Acceptable Position Band.

activated until it is determined that the participant is not able to generate the required motion by his/her own effort. Thus, in our application, the average position is used to determine the activation of the robotic assistance instead of the instant position. During a given time interval, the averages of the actual arm position, x_{ave}, the upper bound, x_{upper_ave}, and the lower bound, x_{lower_ave}, are calculated by Equation (2):

$$x_{ave} = \frac{t_s}{(t_f - t_i)} \bullet \sum_{t_i}^{t_f} x(t), \quad x_{lower_ave} = \frac{t_s}{(t_f - t_i)} \bullet \sum_{t_i}^{t_f} x_{lower}(t)$$
$$x_{upper_ave} = \frac{t_s}{(t_f - t_i)} \bullet \sum_{t_i}^{t_f} x_{upper}(t) \quad (2)$$

Where t_i and t_f are the starting time and final time of a time interval, respectively; t_s is the sampling time of the data acquisition; x(t) is the actual arm position at any given time t.

For any time interval, if condition: $x_{lower_ave} < x_{ave} < x_{upper_ave}$ is satisfied, the assistive controller is not activated and the participant continues the tracking task without robotic assistance. If condition is not satisfied, the robot is activated to provide assistance.

Resist-as-needed (RAN) training mode: Stroke patients may have excessive co-activation or increased stiffness in their affected limbs, so it becomes difficult for them to move their arm in an accurate manner. In the resist-as-needed training mode, the participant is required to move his/her arm on his/her effort from an initial position O to a target position T in a straight line following a position profile while his/her arm is able to move in the 2D horizontal plane. The robot provides resistance to the participant's arm movement if the arm motion deviates from the straight line path. The purpose of this training is to improve the motor ability of the patient's arm move in an accurate manner. The idea of resist-as-needed training comes from the virtual channel training strategies [28,29].

A virtual channel is defined as shown in Figure 3. No robotic resistance will be generated if the arm position remains inside the virtual channel. However, if the arm position goes out of the channel, a spring-like restoring force will be applied to the arm to enforce it back to the right path. The restoring force is,

$$F = \begin{cases} 0, & |y(t)| \le \frac{W}{2} \\ -K \cdot (y(t) + \frac{W}{2}), & y(t) < -\frac{W}{2} \\ -K \cdot (y(t) - \frac{W}{2}), & y(t) > \frac{W}{2} \end{cases} \quad (3)$$

Where W is the width of the virtual channel in mm and K is the stiffness of the spring-link force field in N/mm. W and K are adjustable to meet participant's actual motor condition.

Visual error augmentation (VEA) training mode: The latest research in many models and artificial learning systems such as neural networks suggest that error drives sensorimotor learning, so that one can learn adaptation more quickly if the error is larger [30]. Such error-driven learning processes are believed to be central to adaptation and the acquisition of skill in human movement [31,32]. Visual error augmentation can improve the rate and extent of motor learning in healthy participants and may facilitate neuro-rehabilitation strategies that restore function in brain injuries such as stroke [21]. Feedback distortion is shown to be able to elicit functional improvements in patients with chronic stroke and traumatic brain injury [23].

In the VEA training mode, the task is a 1D reaching task, which is the same with the AAN training mode. However, the arm position trajectory displayed to the participant is amplified by a certain gain factor K so that makes the position error more noticeable to the participant and stimulate him/her to make faster response to correct the position error Figure 4.

The position error

$$e = x_d(t) - x(t) \quad (4)$$

where, $x(t)$ is the actual arm position, $x_d(t)$ is the desired arm position.

The arm position displayed to the participant on the monitor, x*, is

$$x^* = x_d(t) + K \cdot (x(t) - x_d(t)) \quad (5)$$

Viscous force field training (VFF) mode: This training mode is designed to help the understanding of the motor adaptation and motor learning process of the upper-limb during the reaching movement. Stroke patients may have excessive co activation or increased stiffness in their affected limbs. These conditions need to be considered in understanding the stroke rehabilitation. An undesirable environment is created for healthy participants by applying a force field to their arms during the upper-limb movement to simulate these conditions. Examining the adaptation process and after-adaptation effect of the arm movement in the dynamic environment will help understand the kinematic and dynamic influence in motor learning during the goal-directed arm movement as an alternative to study directly with the more vulnerable stroke patients [28,33].

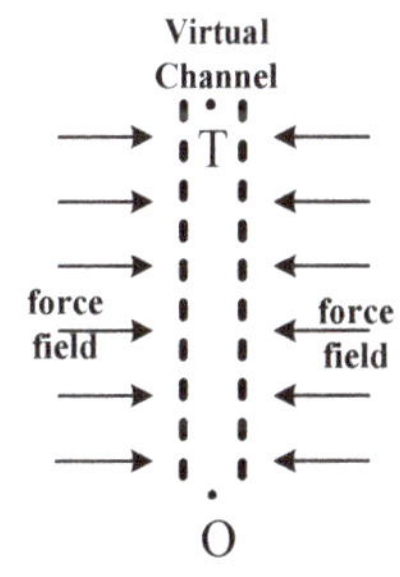

Figure 3: Illustration of the Resist-As-Needed Training.

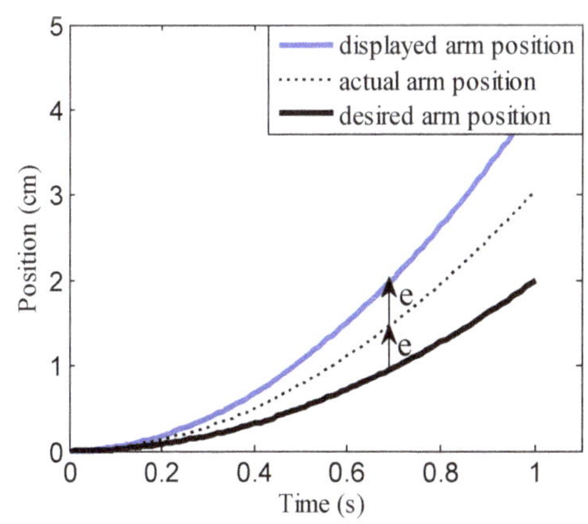

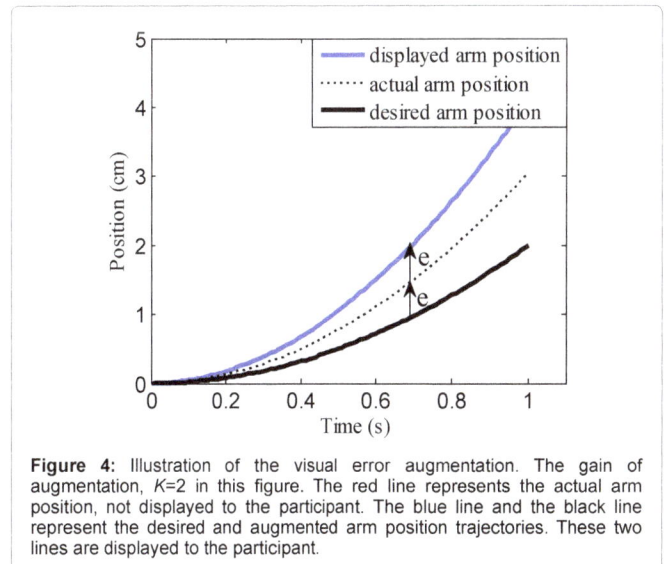

Figure 4: Illustration of the visual error augmentation. The gain of augmentation, K=2 in this figure. The red line represents the actual arm position, not displayed to the participant. The blue line and the black line represent the desired and augmented arm position trajectories. These two lines are displayed to the participant.

In the viscous force field training mode, the participant is required to do the reaching task in the 2D horizontal plane following a position profile along a straight line path in X direction. A dynamic environment is created so that a viscous force, which is proportional to the arm movement, will be applied perpendicular to the direction of the arm movement (Figure 5). The viscous force is defined as,

$$\begin{bmatrix} F_x \\ F_y \end{bmatrix} = \begin{bmatrix} 0 & 0 \\ K_v & 0 \end{bmatrix} \cdot \begin{bmatrix} v_x \\ v_y \end{bmatrix} = \begin{bmatrix} 0 \\ K_v \cdot v_x \end{bmatrix} \quad (6)$$

Force perturbation (FB) training mode: This training mode, similar to the viscous force field training mode, is designed to examine how the upper-limb movements are modified during adaptation to external force perturbations applied on the arm during the reaching task. Instead of creating a dynamic force field, force perturbation is randomly applied to the opposite direction as resistance against the arm movement to simulate the excessive stiffness in the affected arm of stroke patients. The arm motion is restricted in one direction only in the force perturbation training mode.

The Robotic System Development

Hardware

A PUMA 560 robotic manipulator is used as the main hardware platform in this work. The PUMA robot is augmented with a force-torque sensor and a hand attachment device (Figure 6).

The PUMA 560 robot is a 6 degrees-of-freedom (DOF) device consisting of six revolute axes. Each major axis (joints 1, 2 and 3) is equipped with an electromagnetic brake, which can be activated when power is removed from the motors, thereby locking the robot arm in a fixed position. The technical specifications of this robot can be found in [34]. In order to record the force and torque applied by the robot, an ATI Gamma force/torque sensor is used. The robot has been interfaced with Matlab Simulink/Realtime Workshop to allow fast and easy system development. The force values recorded from the force/torque sensor are obtained using a National Instruments PCI-6031E ADC board. The joint angles of the robot are measured by encoders and received by the computer through a Measurement Computing PCIQUAD04 encoder board. The torque output calculated by the Simulink model is sent to the robot through a Measurement Computing PCIM-DDA06/16 DAC board. The sampling rate of all boards is set as 1 kHz. A computer monitor is placed in front of the participant to provide visual feedback about his/her motion trajectory during the execution of the task.

In order to provide robotic assistance to the participant's upper arm, a hand attachment device (Figure 6 bottom) is designed to couple the arm to the robot. The hand attachment device consists of an aluminum plate with two small flat-faced electromagnets (Magnetool Inc.), and a steel forearm padded splint (Moore Medical Inc.), which is attached to the aluminum plate by the magnetic power. The participant's arm is strapped on the splint. An ATI Gamma force/torque sensor is placed between the robot and hand attachment device for force measurement. A handheld controller, which controls the power to magnetize and demagnetize the electromagnets, is given to the participant to remove the hand attachment device from the robot in a safe and quick manner if needed.

Control system

The control architecture of the PUMA robotic system consists of a high-level supervisory controller to coordinate and switch between different training modes, and monitor the task execution and the participant's safety; a middle level task planner to modify task parameters and generate task trajectories; and a low-level assistive controller to execute the task accurately (Figure 7). A graphic user interface (GUI) is developed to provide easy mode selection and task parameter setting. The control system is developed in Matlab/Simulink Real-time Windows Target to interface with the hardware and execute the control program in real-time.

The high-level supervisory controller: The high-level supervisory

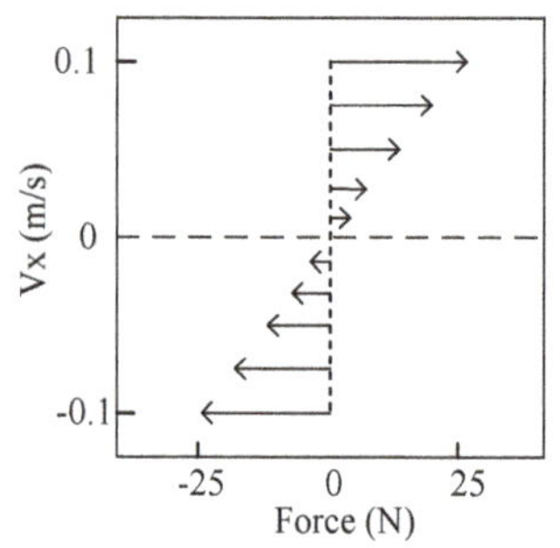

Figure 5: Illustration of Viscous Force Field The viscosity of the force field is 250 N/(m/s).

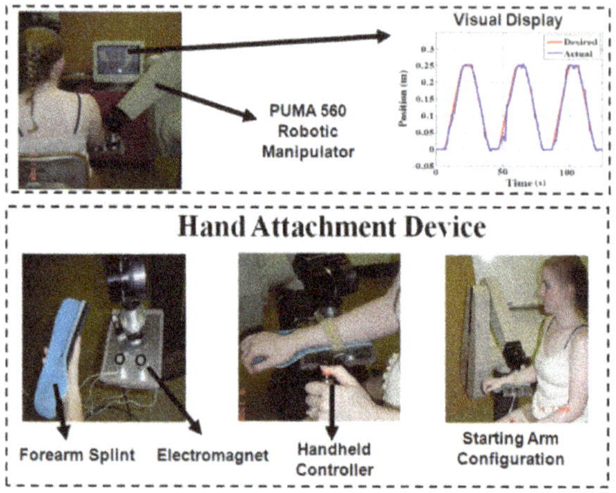

Figure 6: The PUMA Robotic System.

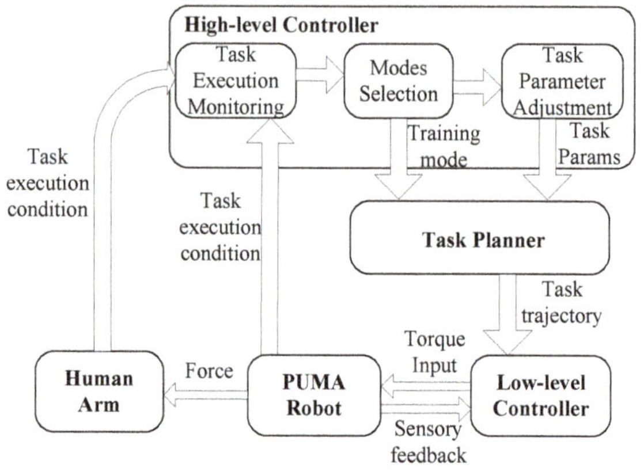

Figure 7: Control Architecture.

controller is a discrete event system that can be modeled as a deterministic finite automaton. We use a state chart to represent the high-level controller as shown in Figure 8. From the statechart, a vector *S* is defined to make the training mode selection by setting each bit of the vector to 1 or 0 to enable or disable the corresponding training mode. Each training mode is represented by a sub-state. The AAN, RAN and VEA training modes are designed as parallel sub-states, which means these modes can be enabled alone or in different combinations and work as integrated training modes. Others are exclusive sub-states that cannot be enabled together. The structure of the high-level controller allows flexible and versatile training modes selection of the robotic system.

The middle-level task planner: The task planner receives the mode selection and task parameters from high-level controller and generates the reference task trajectory for the low-level controller to execute. The task planner is a trajectory generator built with Matlab/Simulink S-function. Once a training mode is selected, the total task distance, maximum speed and acceleration, and the DOF of the motion will be defined in the high-level controller. The task planner will use these parameters to generate the reference task trajectory for the low-level controller. The reference trajectory is a smooth position profile with bell-shape velocity profile as shown in Figure 8.

Graphic user interface: To provide a user-friendly interface which allows mode selection without accessing the Simulink program, a user interface is designed with Matlab GUI to make quick training mode selections and task parameter settings for the experimenter (Figure 9). The left half of the GUI is the mode selection panel. The default button is *Reset*, which corresponds to the *Off* state in the state chart. Each training mode has an exclusive selection button, while AAN, RAN and VEA training modes are grouped in a box where multiple selection is permitted. Each button represents a bit in the *S* vector. Selecting a button will set the corresponding bit to 1 and unselecting a button will set that bit to 0. The right half of the GUI is the task parameter setting panel. The experimenter can define the total task distance, maximum speed and acceleration by type in the values in text boxes.

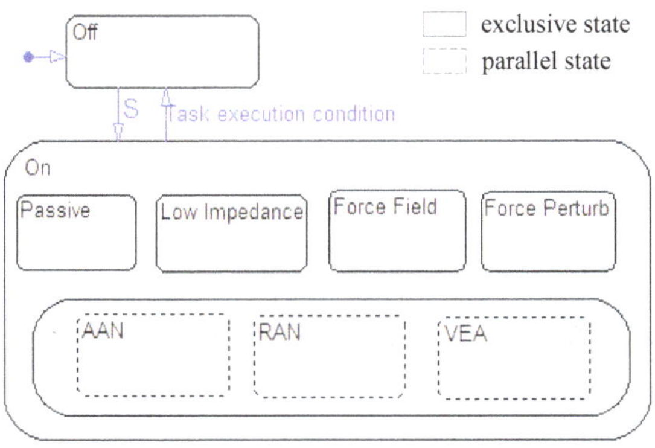

Figure 8: State chart of High-level Supervisory Controller.

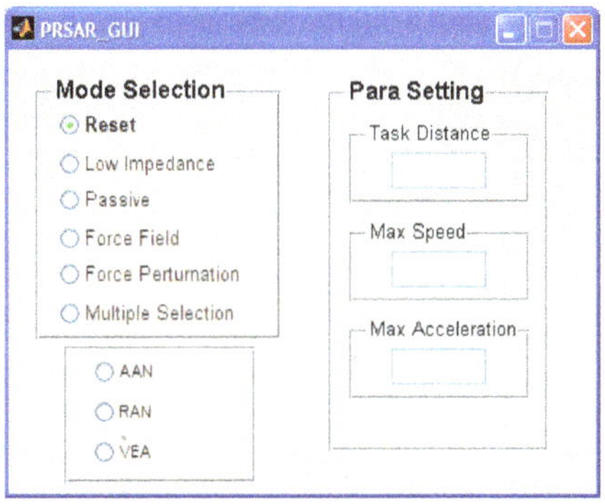

Figure 9: User interface for task selection and parameter setting in GUI.

The low-level assistive controller: The low-level controller is designed to provide robotic assistive/resistive force to the participant to complete the movement tracking task in the task-space in an accurate manner as required in different training modes. In the PRSAR, the low-level controller is an assistive controller which has an outer force feedback loop and an inner position feedback loop (Figure 10) .The tracking of the reference trajectory in X direction is guaranteed by the inner motion control. The design of the low-level controller has been reported in [35]. The contact force in X direction, which is given as a force reference to the controller, is computed by a planner. The proposed controller is similar to an impedance controller; however it allows specifying the reference time-varying force directly. A brief discussion of the controller is presented below.

The equations of the motion in X direction for the robot are given by:

$$\Gamma = u - J^T(q)\cdot F = M(q)\ddot{q} + V(q,\dot{q}) + G(q) \tag{7}$$

where M(q) represents the inertia matrix, $V(q,\dot{q})$ is the summation of the Coriolis torques and centrifugal torques, and G(q) is the vector of gravity torques. Γ is the generalized joint force which is calculated using $u-J^T(q)F$, where u is the input to the robot, J(q) is the Jacobian matrix and F is the contact force exerted by the robot.

Using inverse dynamics control, the robot dynamics are linearized and decoupled via feedback. Control output u to the PUMA robot is designed as follows:

$$u = M(q)y + V(q,\dot{q}) + G(q) + J^T F \tag{8}$$

Here, $y = \ddot{q}$ represents a new input. The new control input y is designed so as to allow tracking of the desired force F_d. For this purpose, the control law is selected as follows:

$$y = J(q)^{-1} M_d^{-1}(-K_d\dot{x} + K_p(x_f - x) - M_d\dot{J}(q,\dot{q})\dot{q}) \tag{9}$$

where X_f is a suitable reference to be related to force error; M_d (mass), K_d (damping) and K_p (stiffness) matrices specify the target impedance of the robot; X and $\dot{X}$ are the position and velocity of the end-effector in the Cartesian coordinates, respectively. The relationship between the joint space and the Cartesian space acceleration is used to determine position control equation:

$$\ddot{x} = J(q)\ddot{q} + \dot{J}(q,\dot{q})\dot{q} = J(q)y + \dot{J}(q,\dot{q})\dot{q} \tag{10}$$

By substituting Equation (9) into Equation (10),

$$\ddot{x} = J(q)\{J(q)^{-1}M_d^{-1}[-K_d\dot{x} + K_p(x_f - x) - M_d\dot{J}(q,\dot{q})\dot{q}]\} + \dot{J}(q,\dot{q})\dot{q} \tag{11}$$

$$= -M_d^{-1}K_d\dot{x} + M_d^{-1}K_p(x_f - x)$$

Thus,

$$M_d\ddot{x} + K_d\dot{x} + K_p x = K_p x_f \tag{12}$$

Equation (12) shows the position control tracking of *x* with

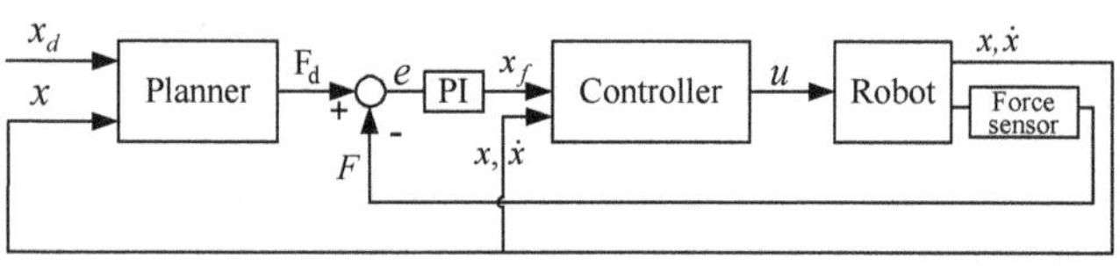

Figure 10: The Assistive controller.

dynamics specified by the choices of K_d, K_p and M_d matrices. Impedance is attributed to a mechanical system characterized by these matrices that allows specifying the dynamic behavior. Let Fd be the desired force reference, which is computed using a PID loop:

$$F_d = P_d(x_d - x) + I_d \int (x_d - x)dt + D_d \frac{d(x_d - x)}{dt} \quad (13)$$

Where X_d, x , P_d, I_d and D_d are the desired position, actual position, the proportional, integral and derivative gains of the PID position loop, respectively. The relationship between X_f and the force error is expressed in Equation (14) as:

$$x_f = P(F_d - F_i) + I\int (F_d - F_i)dt \quad (14)$$

where *P* and *I* are the proportional and integral gains, respectively. The *P* and *I* gains are tested in our previous work to guarantee a smooth and sufficient assistance to the participant [36]. Fi is the force applied by the human. Equations (12) and (14) are combined to obtain below equation:

$$M_d\ddot{x} + K_d\dot{x} + K_p x = K_p(P(F_d - F_i) + I\int (F_d - F_i)dt) \quad (15)$$

From Equation (15), the desired force response is achieved by controlling the position of the robot.

Besides the assistive/resistive force in X direction, the low-level controller also provides viscous force and resistive force in y direction in RAN and VFF training modes. In this case, an extra term F_y in the contact force tern F is calculated using Equation (3) and (6) and then plugged into Equation (8) to compute the control output to the PUMA robot.

Safety consideration: Ensuring safety of the participant is a very important consideration when designing a robot-assisted rehabilitation system. Thus, in case of emergency situations, the experimenter can press a kill switching button to stop the PUMA robot. With the quick-release hand attachment device, the participant can also quickly release his/her arm from the PUMA robot by pressing the handheld controller to deal with any physical safety related events. When the controller is pressed, the electromagnets are demagnetized instantaneously and the participant is free to remove the splint from the robot.

Moreover, rotation angle and torque limits of each joint of the robot are monitored within the control system to disable and stop the robot with its inherited brakes to prevent unexpected movement or excessive motor torque.

Validating Experiments and Results

In this section we present the validating experiments conducted on unimpaired participants to illustrate the implementation of different training modes with the proposed robotic system. Note that the experiments focus on the validation of the developed robotic system to administer different training modes. As a result, we used unimpaired participants to demonstrate the functionality of the robotic system, PRSAR. In the future, we will systematically study the impact of different training strategies on stroke rehabilitation using PRSAR.

Procedure

Three participants with no arm impairement were recruited for the experiments. One participant participated in the experiments with Low-Impedance, AAN, RAN and VEA training modes. The second participant participated in the experiments with viscous force field and force perturbation training modes. The third participant participated in the integrated training of AAN and VEA modes. The total distance and the maximum speed of the reaching movement were set as 0.25m and 0.05m/s, respectively. The participants were required to repeat the forward and backward reaching task 5 times in each training mode.

In the AAN training mode, the time interval to calculate the average position and determine the activation of the robotic assistance was chosen to be 4 second. In the RAN training mode, the width of the virtual channel was set as 0.01m, which was 5 mm deviation from the desired path on each side. The stiffness of the restoring force was 2500N/m., which was tested to be sufficient to bring the arm back into the channel quickly. In the VEA training mode, the gain of the visual error augmentation was chosen as 2, which had been shown to elicit the best training performance in literature [21]. The augmented and desired arm position trajectories were displayed to the participant on monitor screen. In the viscous force field training mode, the viscosity of the force field was 500 N/(m/s). In the force perturbation training mode, the amplitude of the perturbation force was 20N and the duration of the force perturbation was 1 second, applied at t=12, 32, 52, 72 and 92 seconds. In the integrated training, the participant received the augmented arm position trajectories displayed on the monitor while the robotic assistance was available to assist the participant when needed.

Participants were seated in a height adjustable chair as shown in Figure 6 (top left) and were required to place their forearm on the hand attachment device as shown in Figure 6 (bottom left). The height of the endpoint of the PUMA robot was adjusted for each participant to start the tracking task in the same arm configuration. The starting arm configuration was selected as shoulder at neutral 0° position and elbow at 90° flexion position. The task required moving the arm in forward flexion to approximately 60° in conjunction with elbow extension to approximately 0° and then coming back to the starting position. The release button of the hand attachment device was given to the participants in case of emergency situations during the task execution (Fig. 6- bottom middle). The participants received visual feedback of the task trajectories and their own position trajectories on a computer monitor in front of them (Figure 6-top right). The actual arm position and contact force were recorded at 1 kHz sampling rate.

Experimental results

The first participant participated in the Low Impedance, AAN, RAN and VEA training modes. In all these training modes, the participant was asked to follow a reference trajectory in X direction displayed on the monitor screen. Even though the participant had no arm impairment, it was not easy to perfectly track the desired trajectory. By comparing the actual arm position trajectory with the desired trajectory, we can evaluate the efficacy of each training mode. The position trajectories in the task direction (X direction) during all training modes are shown in Figure 11. The averages of the absolute position errors in different training modes are: e_{LI}=5.3mm, e_{AAN}=4.5mm, e_{RAN}=5.2mm and e_{VEA}=3.9mm. Note that the low-impedance mode can be considered as the baseline performance for the participant. In the Low Impedance and RAN training modes, the participant was able to move their arm in

Y direction also. By comparing the deviation from the task direction, we can examine the improvement of the participant's ability to conduct the training task in the right direction. The position trajectories of the participant in X-Y plane in Low Impedance and RAN training modes are shown in Figure 12. The averages of absolute position errors in Y direction are 3.87mm in the Low Impedance training and 2.66 mm in the RAN training.

The second participant participated in the VFF and FP training modes. The purpose of these experiments was to see how the participant would adapt to dynamic environment during arm movement thus to help understand the kinematic and dynamic influence in motor learning during a goal-directed arm movement. The participant was allowed to move his arm in the X-Y plane in the VFF training mode. The arm position trajectory and the corresponding viscous force applied in the VFF training mode are shown in Figure 13. The viscous force was applied perpendicular to the task direction properly as defined in Equation (6) during the arm movement. Although deviation from the task line was observed during the movement, the participant managed to reach the target position under the influence of the force field. Exploring the adaptation process of the arm movement under the influence of the dynamic force field will be an alternative to help understand motor learning process of the affect limbs of stroke patients. In the FP training mode, the participant was allowed to move his arm in X direction only. The position error trajectory and the force perturbation during the FP training mode are shown in Figure 14. The position error increased significantly during the periods when the force perturbation was applied. Errors in opposite direction were observed right after the force perturbation, which could be an indication of adaptation in the motor learning. This training mode can be viewed as a simulation of motor learning process of those stroke patients whose affected limbs have excessive muscle stiffness at certain configuration during arm movement.

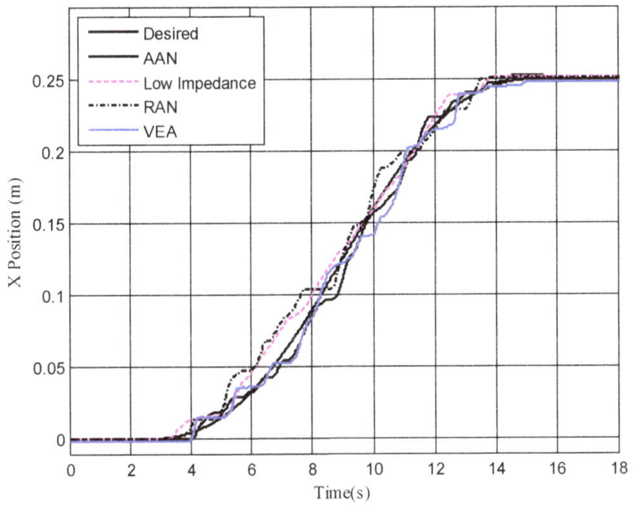

Figure 11: Position Trajectories in Different Training Modes.

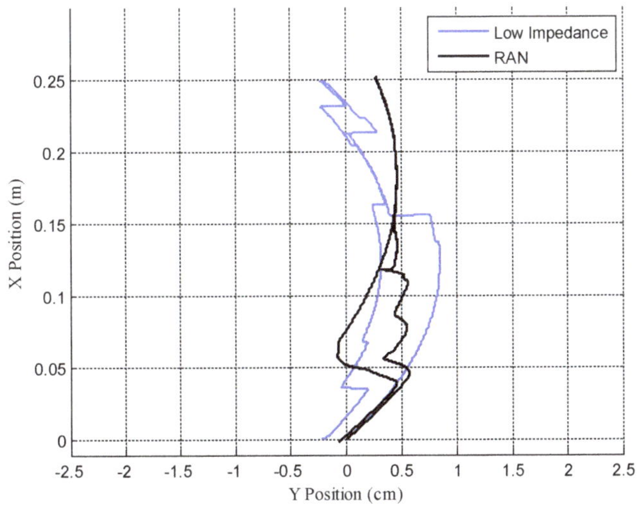

Figure 12: Position trajectories in Low Impedance and RAN trainings. The restoring force keeps the arm from moving out of the virtual channel in the RAN Training mode.

The third participant participated in the integrated training of AAN and VEA modes. In our previous work [37], we have shown that by combining some of the training modes, it is possible to improve the efficiency of the rehabilitation training. It is of great interests to see whether more training modes can be integrated to provide optimal training strategies for stroke rehabilitation. The desired, actual and displayed arm position trajectories are shown in Figure 15 up. The deviation from the displayed arm position to the desired arm position was twice as the real position error from actual arm position to the desired arm position. The corresponding position error and the activation of robotic assistant are shown in Figure 15 down. Once activated, the robotic assistance would continue for 4 seconds and then bring the arm position back into the acceptable error band. Note that, the robotic assistance was activated based on average position error not instant position error. For example, at t=20.5, where the instant position error was out of the acceptable error band, the robotic assistance was not activated. It was activated until t=24s, when the average position error in 4s interval (t=20~24s) was out of acceptable error band. The average position is computed using Equation (2) in Section 2.4. By comparing the training performance in this training mode with other training modes, we want to find optimal training strategies for stroke patients with different levels and types of arm impairments.

Discussion

With the large and ever increasing stroke population and the huge amount of stroke related medical cost, robot-assisted rehabilitation has become an active research area during recent years, aiming to deliver reliable and standardized rehabilitation therapies as well as reduce the cost of rehabilitation. Various robotic systems have been developed and different rehabilitation training strategies have been proposed to improve the rehabilitation training efficacy. The goal of our present work is to provide a general test-bed to evaluate and compare different training strategies. A PUMA robotic system for arm rehabilitation (PRSAR) is developed in this paper. The PRSAR is integrated with different training strategies for the study of upper-limb rehabilitation after stroke. A similar system, MIME, has been reported by Lum in 2002 [15], however, the implementation for assist-as-needed and resist-as-needed training strategies proposed in this work are different. Additionally, new training strategies reported in recent literatures, e.g., visual error augmentation, viscous force field and force perturbation modes, have been implemented in the PRSAR to enhance the versatility of the robotic system. Moreover, in our system, some of these training modes can be implemented together as new integrated training

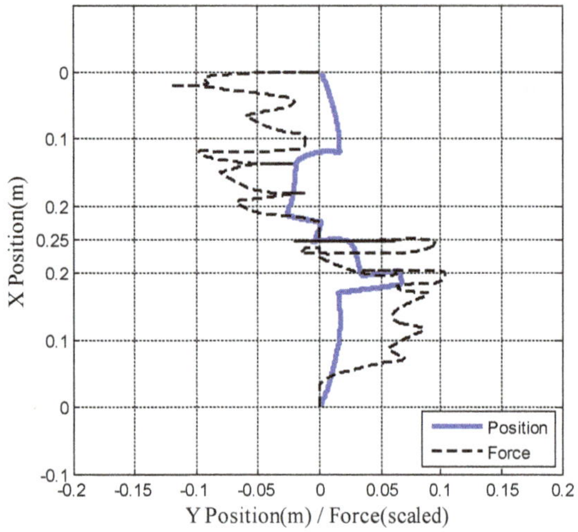

Figure 13: Position and Force Trajectory in Viscous Force Field Training. Note the backward movement is flipped around Y=0.25m line for better demonstration.

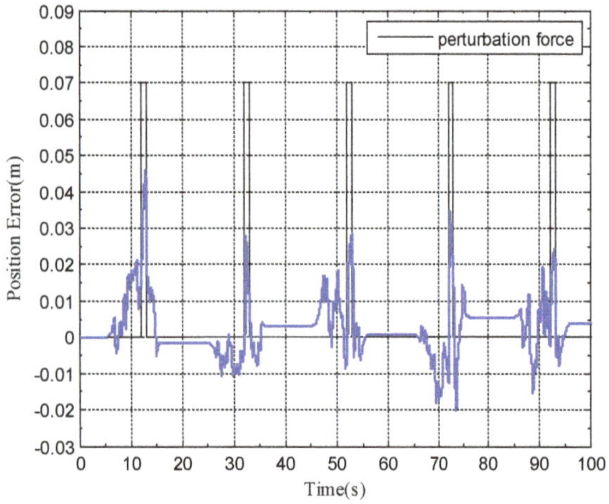

Figure 14: Position and Force Trajectory in Force Perturbation Training. The perturbation force is scaled down for readability.

modes under the coordination of the presented high-level supervisory controller (Figure 15). With all the available training modes, this PRSAR has the potential to serve as a test-bed to evaluate and compare the training efficacy of different training strategies for upper-limb rehabilitation and to understand the motor learning and adaptation process during upper-limb movement. We are aware that PUMA robot was not originally developed for rehabilitation purposes and thus have a few drawbacks in this regard. However, with improved control and planning, PUMA robot has been shown to be useful in rehabilitation purpose as well [15]. The primary purpose of our work is to develop a versatile control architecture that can provide different training modes. To implement this architecture we have used the PUMA robot because of availability. But note that our development is not restricted to PUMA robot alone.

The control system of the robotic system has a hierarchical architecture consisting of a high-level supervisory controller, a middle-level planner, and a low-level assistive controller. The high-level supervisory controller and middle-level planner, which work together to coordinate different training modes, adjust task parameters, generate task trajectories and monitor the task execution, are flexible and expandable. They can be integrated in other robotic systems to produce the same reference profiles for reaching task with minor modifications. Furthermore, new training strategies, if discovered in future works, can be integrated to the control architecture by adding these training strategies as new sub-states into the state chart of the high-level controller without changing other states in the control architecture.

The enhanced PRSAR was tested with unimpaired participants in the initial validating experiments. A reaching movement was chosen for the training task but the PRSAR is not restricted to this task only. The experimental process proved that all training modes worked properly. The robotic assistive/resistive/viscous/perturbation forces were enabled and applied as designed. The AAN and VEA modes were able to work together as an integrated training mode. The performance of the unimpaired participant showed differences among Low-Impedance, AAN, RAN and VEA training modes (Figures 11,12). The viscous force field and force perturbation showed impact to the motor adaptation during the arm movement in the VFF and FP training modes (Figures 13,14). The efficacy of these training modes in upper-limb rehabilitation and the influence in motor adaptation during arm movement will be further evaluated with formally designed experiments and adequate subject population.

Conclusion and Future Work

The presented robotic system (PRSAR) is capable of providing multiple training modes to facilitate a test-bed for the study of upper-limb rehabilitation after stroke. The control system designed for this robotic system consists of a flexible and expandable high-level supervisory controller to coordinate different training modes, modify task parameters and monitor the task execution and the participant's safety, a middle level task planner to generate task trajectories, and a low-level assistive controller to execute the task accurately. The presented robotic system has been tested in the initial experiments and all training modes work as designed.

As a future work, experiments will be conducted on both healthy

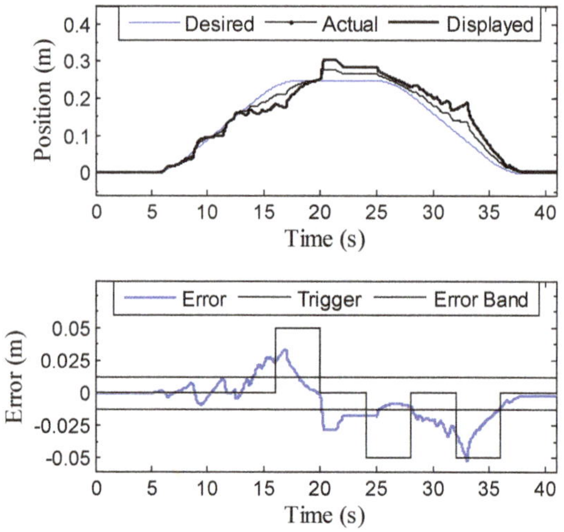

Figure 15: Up: Desired, actual and displayed (error augmented) arm position. Down: Position error (blue line) and triggering of robotic assistance in the integrated training of AAN and VEA modes.

and stroke participants with different training modes to evaluate the efficacy of different training strategies and understand the kinematic and dynamic adaptation in arm movement. Further, the robotic system will be coupled with a hand exoskeleton under development in our research group for the study of the upper-limb reach-to-grasp activity.

References

1. American Stroke Association "Statistics about Stroke".
2. Matchar DB, Duncan PW (1994) Cost of stroke . Stroke Clin Updates 5: 9-12.
3. Bonita R, Steward A, Beaglehole R (1990) International trends in stroke mortality: 1970-1985. Stroke 21: 989-992.
4. Broderick JP, Phillips SJ, Whisnant JP, O'Fallon WM, Bergstralh EJ (1989) Incidence rates of stroke in the eighties: the end of the decline in stroke? Stroke 20: 577-582.
5. Gray CS, French JM , Bates D, Cartlidge NE, James OF, Venables G (1990) Motor recovery following acute stroke. Age Ageing 19: 179-184.
6. Krebs HI, Palazzolo JJ, Dipietro L, Ferraro M, Krol J, et al. (2003) Rehabilitation Robotics: Performance-Based Progressive Robot-Assisted Therapy. Autonomous Robots 15 :7-20.
7. Krebs H, Ferraro M, et al. (2004) Rehabilitation robotics: pilot trial of a spatial extension for MIT-Manus. J Neuroeng Rehabil 1: 5.
8. Kahn L, Zygman M, et al. (2006) Robot-assisted reaching exercise promotes arm movement recovery in chronic hemiparetic stroke: a randomized controlled pilot study. J Neuroeng Rehabil 3: 12.
9. Leonard EK, Peter SL, Zev WR, David JR (2006) Robot-assisted movement training for the stroke-impaired arm: Does it matter what the robot does? J Rehabil Res Dev 43:619-630.
10. Lum P, Burgar C, et al. (1999) Quantification of force abnormalities during passive and active-assisted upper-limb reaching movements in post-stroke hemiparesis. IEEE Trans Biomed Eng 46: 652-662.
11. Charles GB, Peter SL,Peggy CS,Machiel Van der Loos HF(2000) Development of robots for rehabilitation therapy: the Palo Alto VA/Stanford experience. J Rehabil Res Dev 37: 663-673.
12. Lum P, Burgar C, Machiel VL , Peggy CS, Matra M, et al. (2006) MIME robotic device for upper-limb neurorehabilitation in subacute stroke subjects: A follow-up study. J Rehabil Res Dev 43: 631-642.
13. Loureiro R, Amirabdollahian F, et al.(2003) Upper Limb Robot Mediated Stroke Therapyâ GENTLE/s Approach Autonomous Robots 15: 35-51.
14. Volpe BT, Krebs HI, Hogan N, Edelstein L, Diels C, Aisen M (2000) A novel approach to stroke rehabilitation: robot-aided sensorimotor stimulation. Neurology 54: 1938-1944.
15. Lum PS, Charles GB, Peggy CS, Matra M, Machiel V (2002) Robot-assisted movement training compared with conventional therapy techniques for the rehabilitation of upper-limb motor function after stroke. Arch Phys Med Rehabil 83: 952-959.
16. Hesse S, Schulte-Tigges G, Konrad M, Bardeleben A, Werner C (2003) Robot-assisted arm trainer for the passive and active practice of bilateral forearm and wrist movements in hemiparetic subjects. Arch Phys Med Rehabil 84: 915-920.
17. Zygman ML, Rymer WZ, Reinkensmeyer DJ, Kahn LE (2001) Effect of robot-assisted and unassisted exercise on functional reaching in chronic hemiparesis. Engineering in Medicine and Biology Society 2: 1344-1347.
18. Susan EF, Hermano IK,Joel S,Walter RF, Neville H (2003) Effects of robotic therapy on motor impairment and recovery in chronic stroke. Arch Phys Med Rehabil 84: 477-482.
19. David JR, Craig DT , Wojciech KT, Andrea NR, Leonard EK (2001) Design of robot assistance for arm movement therapy following stroke. Advanced Robotics 14: 625-637.
20. Bambi RB, Roberta K, Yoky M (2004) Effects of visual feedback distortion for the elderly and the motor-impaired in a robotic rehabilitation environment. Robotics and Automation 2: 2080-2085.
21. Wei Y, Bajaj P, Scheidt R, Patton J (2005) Visual error augmentation for enhancing motor learning and rehabilitative relearning. 9th International Conference on Rehabilitation Robotics.
22. Patton JL, Mussa-Ivaldi FA, Rymer WZ (2001) Altering movement patterns in healthy and brain-injured subjects via custom designed robotic forces in Engineering in Medicine and Biology Society 2: 1356-1359.
23. Brewer B, Klatzky R, Matsuoka Y (2008) Visual feedback distortion in a robotic environment for hand rehabilitation. Brain Res Bull75: 804-813.
24. Lum SP, Lehman SL, Reinkensmeyer DJ (1995)The bimanual lifting rehabilitator: an adaptive machine for therapy of stroke patients. Rehabilitation Engineering 3:166-174.
25. Lum SP, Lehman SL, Reinkensmeyer DJ (1993) Robotic assist devices for bimanual physical therapy: preliminary experiments. Rehabilitation Engineering, IEEE Transactions 1:185-191.
26. Taub E, Uswatte G, Pidikiti R (1999) Constraint-Induced Movement Therapy: A New family of techniques with broad application to physical rehabilitation – a clinical review. J. of Rehab. Research and Development 36: 237-251.
27. David JR, Jeremy LE, Steven CC (2004) Robotics, motor learning, and neurologic recovery. Annu Rev Biomed Eng 6:497-525.
28. Robert AS, David JR, Michael AC, Zev WR, Mussa-Ivaldi FA (2000) Persistence of motor adaptation during constrained, multi-joint, arm movements J. Neurophysiol 84:853–862.
29. Rumelhart DE, Hinton GE, Williams RJ(1986) Learning representations by back-propagating errors. Nature 323: 533-536.
30. M Kawato (1990) Feedback-error-learning neural network for supervised learning. Advanced neural computers
31. Wolpert DM, Ghahramani Z, Jordan MI (1995) An internal model for sensorimotor integration Science 269: 1880-1882.
32. Scharver C, Patton J, Kenyon R, Kersten E (2005) Comparing adaptation of constrained and unconstrained movements in three dimensions. IEEE 9th International Conference on Rehabilitation Robotics.
33. Rand MK, Shimansky Y, Stelmach GE, Bloedel JR (2004) Adaptation of reach-to-grasp movement in response to force perturbation Experimental Brain Research 154:50-65.
34. Corke PI, Armstrong-Helouvry B (1994) A search for consensus among model parameters reported for the PUMA 560 robot. Robotics and Automation 2:1608-1613
35. Sciavicco L, Siciliano B (1996) Modeling and Control of Robot Manipulators.
36. Erol D, Sarkar N (2007) Design and Implementation of an Assistive Controller for Rehabilitation Robotic Systems. International Journal of Advanced Robotic Systems 4: 3.
37. Furui W, Barkana ED, Sarkar N (2010) Impact of Visual Error Augmentation When Integrated With Assist-as-Needed Training Method in Robot-Assisted Rehabilitation. IEEE Trans Neural Syst Rehabili Eng 18:571-579.

6

Self-Sufficient Energy Harvesting in Robots using Nanotechnology

Basma El Zein*

Dar Al Hekma University, Saudi Arabia

Abstract

Latest research and technology development have been focusing on improving the performance and the efficiency of intelligent and automated machines / systems used in human services, factories , transportation means, space exploration and may others applications. As the tech world is shrinking rapidly, industries are seeking to miniaturize the devices to the level of nano-machines /nanorobots without compromising their efficiency. Scientists have made significant progress in nanoroboics research field but have not officially released their new products. On the other hand, they opened a new world of discoveries, possibilities and applications in different fields such as military defense, medicine, industry, space exploration, energy. In this article, we will present the role of nanotechnology and nanostructures in energy harvesting systems used in robots to be able to sense and adapt heat, light, sounds, surface texture and chemicals from the environment, as well as to move, communicate and perform complex calculations.

Keywords: Nano robots; Automated machines; Pneumatic sensors

Introduction

Having revolutionary impact on industrial production route, nanotechnology, the engineering of functional system at the atomic scale or nano-scale (10^{-9}m) = 10 Å, is expected to bring many benefits in the automotive industry and robotics. The advances in nanotechnology are very promising to improve our quality of life in all fields ranging from healthcare to electronics. Nanostructures and nano material will be extensively studied and then employed in the design of nano-devices to reduce the material and fabrication cost and improves the efficiency. Being an electromechanical device, a robot is capable of interacting with its environment and is controlled and commanded by a computer program and electronic circuits. Generally a robot is composed of a movable body, actuator with an electric motor, hydraulic or pneumatic sensors to communicate with the programmable brain, through electric and electronic circuits. Nanorobotics, one of the merging fields of robotics at the nanoscale, has a major objective of shrinking device size to nano scale dimensions. A Nanorobot, the controllable machine at the nanometer, is composed of nano-scale component with a smart structure that can actuate, sense, signal, and process information, all at Nano scale. Having so many features such as durability, efficiency, effectiveness and low cost, nano-robots, find wide applications in the field of medicine and space technology. Nano robotic systems deal with vast variety of sciences, ranging from quantum molecular dynamics, to kinematic analysis. Therefore, It depends on the type of nano material and nanostructure that have been employed.

Self Sufficient Energy Harvesting In Nano Robots

As human intervention is not needed, it is important to use the ambient energy (present in the environment) to convert it to electrical energy to drive small electrical and electronic systems making them self-sufficient. For example : One of the applications of nano-robots in the medical field is its inside of the human body to cure many diseases such as diabetes, tumors, blood clots and many others, with no harm to the body. These nano robots are very sensitive to human body temperature, acoustic signals, blood flow, body movement, respiration, body heat…; therefore they can get the energy from the body itself. Power can be supplied when the source is available (battery less), but the most important is to supply electricity when it is needed to match the demand with the supply. Our environment contains a variety of sources available for harvesting presented in different forms not limited to mechanical, thermal and light or solar (Figure 1).

Nanotechnology Role

New innovative and improved products are expect5ed to enter the market due to the promising benefits of Nanotechnology. This later has showed many benefits in different applications in aerospace, agriculture, construction, cosmetics, defense electronics, environment, food, textile and energy. Due to the progress of Nanotechnology, nanostructures, semiconductors materials have facilitated the cost reduction and improvement in energy conversion efficiency. In this section we will focus on the use of nano wire in energy harvesting to be employed in nano-robots. Nano-wires or one dimensional nanostructure have unique optical and electrical properties which makes them more applicable in energy harvesting devices. These elongated solid nanostructure, have high aspect ratio, and offer great possibilities for further development of many optoelectronic devices and sensors, with numerous possibilities for studying exciting physical

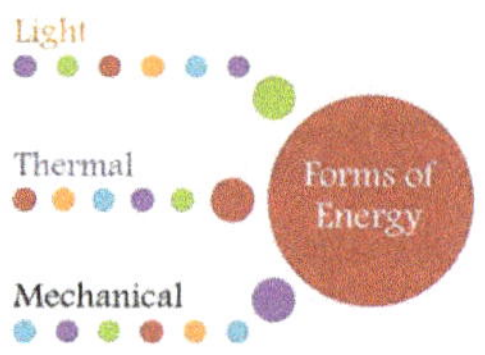

Figure 1: different forms of energy.

***Corresponding author:** Basma El Zein, Dar Al Hekma University, Saudi Arabia, E-mail: basma@ieee.org phenomenon arising from carrier confinement to large surface to volume ratio [1].

Nanostructures in energy harvesting

New nano material's and nanostructures are expected to make a difference in reducing CO_2 generation , using renewable /clean energy , reducing manufacturing cost , improving performance and efficiency of energy harvesting devices such as light emitting diodes(LED), batteries, solar cells, fuel cells and many others . Recently, an increased motivation was noticed that employ nanostructures in energy harvesting and energy storage devices. These include: (a) bulk nano structured materials [3D]; (b) quantum wells [2D]; (c) nano wires [1D]; and (d) quantum dots/nano particles [0D]. Nano wires (NW), elongated solid nanostructure, are semiconducting or metallic structures having a high aspect ratio. Nano wires have been extensively studied in the last decade; and will be presented in the next paragraphs with their application in different forms of energy harvesting.

Light and solar energy harvesting

Recently, solar cell and nanowire researches are considered as hot topics within science and engineering [2]. The need for higher solar cell efficiencies at lower cost has become crucial, and the control of nanostructures has affected the performance of electronic and optical devices [3]. The unique geometry of Nanowires array provides low optical reflection and enhances light trapping and absorption (Figure 2). The free standing NWs array allows the incident light to scatter within its open interiors, which is different from the easy reflection of incident light on the surface of nano-crystallines film. The scattering improves the efficiency of light absorption by increasing the photon path length and diminishing the reflection of incident light. The photo-generated electron –hole pairs in bulk thin film tend to recombine at grain boundaries before they arrive the back contact, while free standing NWs array grown on conducting substrate avoids this shortcoming because the photo-generated charges can be transported to the substrate along individual nanowires which will enhance the photocurrent efficiency. The Effective use of Nanowires (NWs) require the ability to control and tailor their dimensions and morphology (height, diameter, spacing and planar density.

ZnO NWs based Dye Sensitized Solar Cell (DSSC)

Among the excitonic solar cells, the dye sensitized solar cell (DSSC), is considered a low cost solar cell. In a DSSC, dye molecules are used to sensitize wide-band gap semiconductors, such as ZnO and TiO_2, which assist in separating electrons from photo-excited dye molecules. Gerishcer [4] and Memming [5] initiated in 1960 the sensitization of wide-band gap semiconductors by adsorbed monolayers of dye molecules. Thus 1-D wide-band gap semiconductor nanostructures were introduced to improve the charge collection (Figure 3a). The limitation of the DSSC is the poor absorption of low energy photons using available dyes. The redox electrolyte's instability is one of the current challenges, where it degrades over time under UV radiation and may leak if the cell is not perfectly sealed rendering the device inefficient for long term use [6].

Zinc oxide (ZnO) Nanowires Based Quantum Dots Sensitized Solar Cell (QDSSC)

ZnO based DSSCs have still lower conversion efficiencies, caused by :(1) instability of ZnO in acidic dyes by the formation of excessive Zn^{2+} /dye agglomerates; (2) decrease in the kinetics of electron-injection from dyes to ZnO caused by ultrafast electron injection that are opposed by molecular relaxation [7]. To overcome these drawbacks, semiconductor QDs have been used as photo-sensitizers to replace dyes [8]. Theoretically, the conversion efficiency of QDSSCs is expected to reach 44% that is considerably higher than that of DSSCs [9]. The co-sensitization of different sized QuatumDots [10] (Figure 4) will increase the adsorption coverage of solar spectrum and thus increase light harvesting efficiency, (Figure 3b). Many techniques are employed using smaller band-gap materials to expand the absorbance, which in turn increase the light harvesting and overall energy-conversion efficiency of the solar cell.

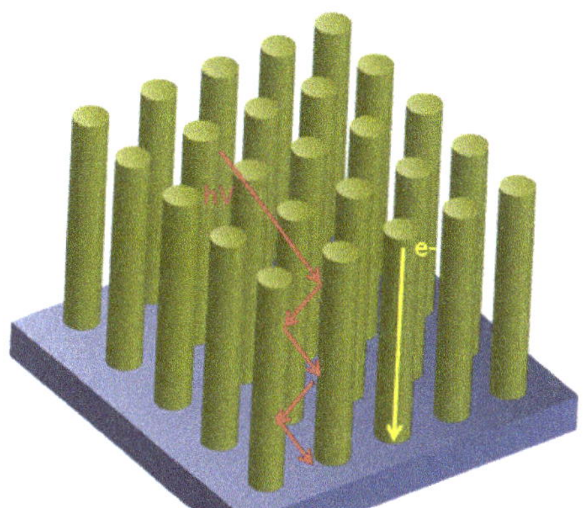

Figure 2: Schematic illustration for Nanowires in light trapping and electron transport to electrodes.

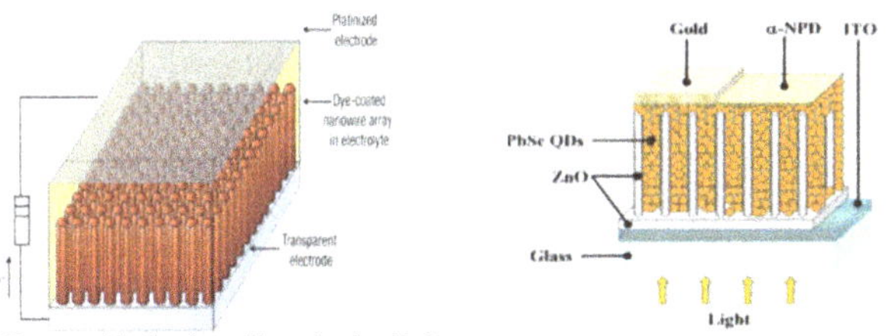

a- Dye sensitized solar cell employing ZnO nanowires [6]

b- PbSe QD/ZnO nanowire solar cell]

Figure 3: Schematic of ZnO Nanowires based a) DSSC , b) QDSSC.

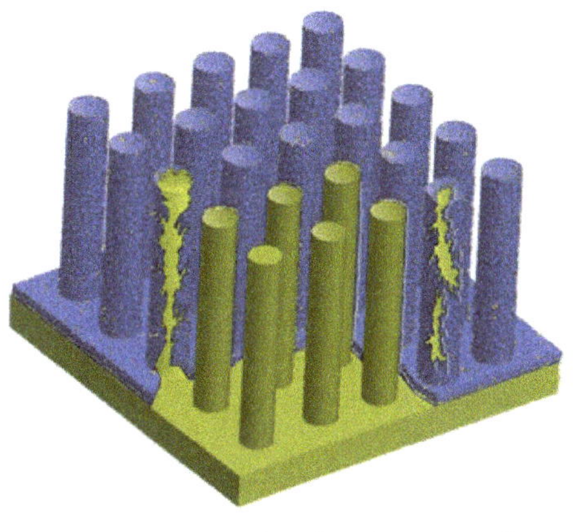

Figure 4: Quantum Dots Sensitized solar cells.

Other Applications Of Nanowires

Uv photodiodes & optical switches

Highly sensitive nanowire switches were demonstrated [11] by exploring the photo-conducting properties of individual semiconductor nanowires. The conductivity of the ZnO NWs is sensitive to ultraviolet light exposure. The increased light-induced conductivity allows reversibly switching the nanowires between "OFF" and "ON" states, an optical gating phenomenon analogous.

Horizontal zno nanowires waveguides

The wave guiding properties [12,13] of single ZnO Nanowires have been investigated by coupling external light into the wire, it is noticed that the light is guided to both ends of the ZnO NW from where it is emitted. While significant scattering is observed in the coupling region, no additional wave guiding losses along the ZnO Nanowires are observed (Figure 5,6).

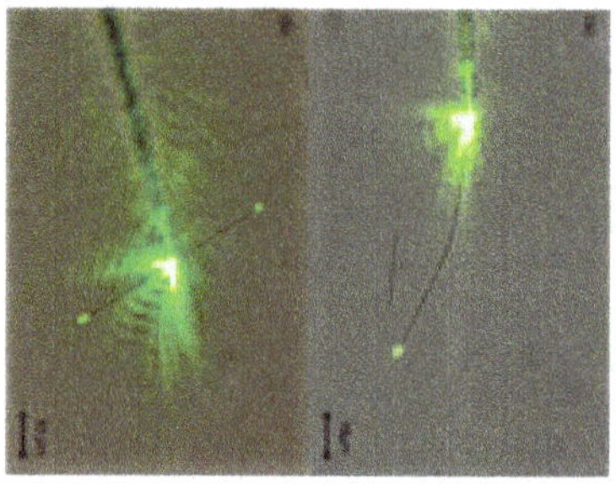

Figure 5: Coupling of green laser light from a silica-tapered fiber into a ZnO NW (a) from one end and (b) at the middle. Image source: Image reproduced from0.

Zno nanowires lasers

One of the most important and widely investigated applications of ZnO NWs is optically pumped ultraviolet lasing. Since the first report by Huang et al [14], many research groups have studied the mechanism of optically excited lasing in ZnO micro- and nanostructures including micropillars, NWs, and nanorods [15,16]. High-quality single-crystalline ZnO NWs constitute ideal lasing nanocavities, which provide both a gain medium and a resonant cavity due to reflection at the planar end-facets.

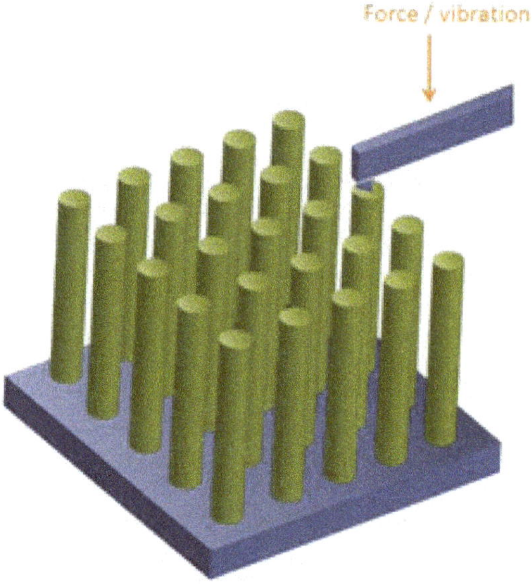

Figure 6: Nanowires arrays with piezoelectric property.

Nanotechnology-enabled piezoelectric mechanical-energy harvesting

Vibration-based mechanical energy is abundant in the environment and more accessible than solar and thermal energy [17,18] Several methods have been established for the conversion of mechanical energy into electricity through the use of piezoelectric materials [19]. Piezoelectric materials have received enormous attention owing to the ability of these materials to convert mechanical energy into electricity directly.Lead zirconatetitanate(PZT), and ZnO nanowires have been demonstrated as the most materials used for mechanicalenergy harvesting [20,21] Piezoelectric NWs excel their bulk counterparts in terms of their enhanced piezoelectric effect, superior mechanical properties, and extreme sensitivity to vibrations of ultra-small magnitude[22,23].

Thermoelectric energy harvesting

An increased interest was given lately to thermoelectric (TE) materials for electric power generation through direct energy harvesting from natural heat source or heat waste. TE effects will be used to generate electricity, to measure temperature, and to cool or heat the devices. Such phenomena include See beck effect (conversion of temperature differences directly into electricity), Peltier effect (production of temperature difference by electric current), and Thomson effect (heating or cooling of a current-carrying conductor with a temperature gradient).An important property of the TE materials is that they are good electrical conductors but bad heat conductors. However, in an environment that the temperature is uniform, the pyroelectric effect has to be used to convert the heat into electricity [24] ZnO (Figure 7) with its pyroelectric and semiconducting properties creates a polarized electric field and charge separation along the nanowires based on time-dependent change in temperature.

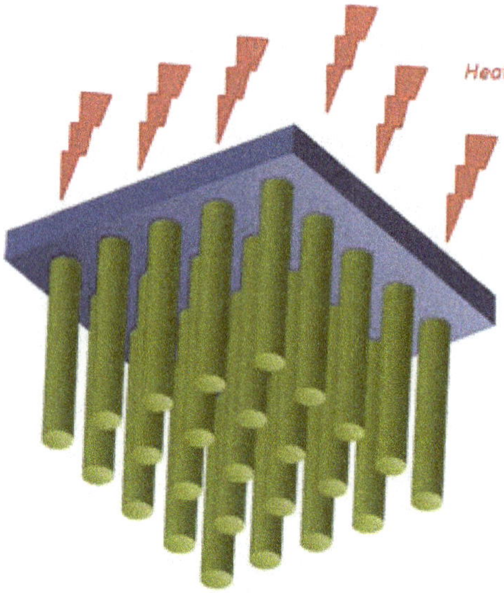

Figure 7: Thermoelectric ZnO NWs Arrays.

Conclusion

Electronic devices used in different sectors such as environment, industry, healthcare and telecommunications are empowered by rechargeable batteries. Using these batteries may result in many challenges in controlling environment pollution, extending the lifetime, recycling and replacing them, In order to effectively solve these challenges, researchers are putting lots of effort on the development of technology using nanostructures and nano-materials to harvest energy from the environment in which those electronics devices are used.

In this article we presented the importance of Nanotechnology in energy harvesting applied in nano-robots. It is expected that this subject create a new revolution world and miracles in the current science and technology.Future engineers, researchers and the scientists have opened the door for more research in the field of nano-robot to improve the device efficiency and improve the different types of energy harvesting devices using new nano-materials and nanostructures.

References

1. B.ElZein, E.Dogheche , Ch2, Lampert academic publishing ISBN: 978-3-659-44003-8
2. Fan Z, Ruebusch DZ, Rathore AA, Kapadia R, Ergen E, et al. (2009) Challenges and Prospects of Nanopillar Based Solar Cells. Nano Research 2:829-843.
3. Ford AC, Ho JC, Chueh YL, Tseng YC, Fan Z et al.(2009) Diameter-Dependent Electron Mobility of In As Nanowires. Nano Letters 9: 360-365.
4. Gerischer H (1972) Electrochemical Techniques for the Study of Photosensitization. Photochem Photobiol 16: 243–260.
5. Memming R (1972) Photochemical and Electrochemical Processes of Excited Dyes at Semiconductor and Metal Electrodes. Photochem Photobiol 16: 325–333.
6. Michael Grätzel (2003) Dye-sensitized solar cells. Jou Photochemistry and Photobiology C: Photochemistry Reviews 4:145-153.
7. Qifeng Z, Christopher SD, Xiaoyuan Z, Guozhong C (2009) ZnO Nanostructures for Dye-Sensitized Solar Cells. Advanced Materials 21:4087–4108.
8. Chen J, Song JL, Sun XW, Deng WQ, Jiang CY, et al. (2009) An oleic acid-capped CdSe quantum-dot sensitized solar cell. Appl Phys Lett.
9. Feifei G,Yuan W,Jing Z,Dong S, Mingkui W, et al. (2008) A new heteroleptic ruthenium sensitizer enhances the absorptivity of mesoporous titania film for a high efficiency dye-sensitized solar cell.Chemical Communications 23: 2635–2637.
10. Chih HC, Shoou JC, Sheng PC, Meng JL, Cherng IC et al.(2010) Fabrication of a White-Light-Emitting Diode by Doping Gallium into ZnO Nanowire on a p-GaN Substrate. J Phys Chem C 114: 12422–12426.
11. H.Kind,H.Yan,B.Messer,M.Law,P.Yang,Adv.Mater 2002,14, No2
12. Robert H, Heinz K (2006) Guided modes in ZnO nanorods. Appl Phys Lett.
13. T. Voss, Advances in Solid State Physics, vol. 48. Berlin, , p. 57, 2009.
14. Michael HH ,Samuel M, Henning F, Haoquan Y, Yiying W, et al. (2001) Room-Temperature Ultraviolet Nanowire Nanolasers. Science 292:1897-1899
15. J. Fallert et al.,Opt. Express, vol. 16, no. 2, pp. 1125–1131, Jan. 2008.
16. M. A. Zimmler, J. Bao, F. Capasso, S. M¨uller, and C. Ronning, Appl. Phys. Lett., vol. 93, no. 5, pp. 051101-1–051101-3, 2008
17. J. A. Paradiso, T. Starner, IEEE Pervas. Comput. 2005, 4, 18– 27
18. Cian ÓM,Terence OD,Rafael VMC,James R,Brendan OF (2008) Energy scavenging for long-term deployable wireless sensor networks.Talanta 75:613-623.
19. Bouendeu E, Greiner A, Smith PJ, Korvink JG (2011) Design Synthesis of Electromagnetic Vibration-Driven Energy Generators Using a Variational Formulation. Microelectromechanical Systems Jou 20:466-475.
20. H. Chen, C. Jia, C. Zhang, Z. Wang, C. Liu in 2007 IEEE International Symposium on Circuits and Systems, Vols. 1 – 11, 2007, pp. 557 – 560.
21. Youfan Hu, Yan Zhang, Chen Xu, Guang Zhu, Zhong LW (2010) High-Output Nanogenerator by Rational Unipolar Assembly of Conical Nanowires and Its Application for Driving a Small Liquid Crystal Display. Nano Lett 10:5025-5031.
22. Xudong Wang (2012) Piezoelectric nanogenerators Harvesting ambient mechanical energy at the nanometer scale. Nano Energy 1:13-24.
23. Zhong Lin Wang (2009) ZnO nanowire and nanobelt platform for nanotechnology.Materials Science and Engineering: R: Reports 64:33-71.
24. Ya Yang,Wenxi Guo,Ken CP,Guang Z,Yusheng Z, et al.(2012) Pyroelectric Nanogenerators for Harvesting Thermoelectric Energy.Nano Lett.12: 2833–2838.

Medical Robot with Electronic Health Record System

Pruthviraj RD[1*], Subhash KC[2], Namratha K[2] and Sushma KR[2]

[1]*Department of Engineering Chemistry, Amruta institute of Engineering and Management Sciences, Bidadi, Bangalore, India*

[2]*Department of Electronics and Communication Engineering, Amruta institute of Engineering and Management Sciences, Bidadi, Bangalore, India*

[*]**Corresponding author:** Pruthviraj RD, Department of Engineering Chemistry, Amruta institute of Engineering and Management Sciences, Bidadi, Bangalore, India
E-mail: namrathakantharaj@gmail.com

Abstract

The healthcare systems require constant monitoring of patients and their records for speedy recovery and also to change the course of tablets at different stages. Most of the healthcare systems are found inefficient in their performance due to human errors in the monitoring of patients who are in the critical care units where human intervention is limited. The obvious reason would also be that the records i.e. the history of the patient may not be available before hand and the doctor will be misguided to prescribe wrong medicines and wrong treatment. Hence, we have come up with the concept of Medical Robot with Electronic health record system-a bridge between the man and the machine, to reduce these errors for effective performance.

Introduction

The concept of creating machines that can operate autonomously dates back to classical times, but research into the functionality and potential uses of robots did not grow substantially until the 20th century [1]. Throughout history, robotics has been often seen to mimic human behaviour, and often manage tasks in a similar fashion. Today, robotics is a rapidly growing field, as technological advances continue; research, design, and building new robots serve various practical purposes, whether domestically, commercially, or militarily. Many robots do jobs that are hazardous to people such as defusing bombs and mines [2].

Today, robotic devices are used to replace missing limbs, perform delicate surgical procedures, and deliver neuro rehabilitation therapy to stroke patients, teach children with learning disabilities, and perform a growing number of other health related tasks [3].

Here is one such concepts which provides the best possible solution for the challenges faced by humans in the field of medicine as shown in the section 1

We have divided the paper into 3 main sections.

Section 1: Problem Statement.

Section 2: Solution – The concept of Medical robot with EHRS.

Section 3: Conclusion.

Problem Statement

We can see huge number of hospitals using the robots for surgical purposes which are giving very good results. But, most of the people do suffer a lot even before the surgery without proper medication, improper method of maintaining patients' record and few human errors [4].

These were the observations from the hospitals which are in rural parts of the country.

Patients suffer due to lack of proper medication leading to death in many cases

Medicine is not "one size fits all." Some medications may trigger allergies, while others may react negatively with previously prescribed drugs. If such risks are ignored, patients can experience severe allergic reactions, major health problems or death.

Identification/information of the patient may not be sufficient/accurate

Patient identification and the matching of a patient to an intended treatment is an activity that should be performed routinely in all care settings. Risks to patient safety occur when there is a mismatch between a given patient and components of their care, whether these components are diagnostic, therapeutic or supportive [5]. Throughout health care, the failure to correctly identify patients and match that information to an intended clinical intervention continues to result in wrong person, wrong site procedures, medication errors, transfusion errors and diagnostic testing errors.

Records may be abused by the third party for illegal reasons

Chances of abusing the medical records of patients for illegal purposes are very high if it is not secured properly.

Illiteracy leads individuals to report non-accurate identity

Illiterate people face many problems in understanding what is written on the records. Most of them faces problem in using ID cards provided by hospitals.

Lack of automation in hospitals

For Hospitals, it is essential to keep constant touch with the staff, patients, utility service provider and suppliers to maintain best standards of the healthcare services [6].

All the above mentioned factors lead to serious consequences on the entire Human Community.

Solution – The Concept of Medical Robot with EHRS

We have come up with a solution to all the above mentioned problems. Burden on the humanity community can be reduced to a very large extent [7]. A robot which can serve patients on time which can be programmed so as to store the entire data of a patient like: disease, name of the tablets, timings at which the tablets should be provided, X-ray report, CT scan report and other types of reports which describes the flaw in the body of human [8].

The robot can also traverse to the other rooms of the hospital in order to serve different patients suffering from various diseases.

EHRS (Electronic Health Record System)

EHRS is software which is also termed as electronic patient record (EPR) or computerized patient record. It is an evolving software concept defined as a systematic collection of electronic health information about individual patients or populations. It is a record in digital format that is capable of being shared across different health care settings, by being embedded in network-connected enterprise-wide information systems [9]. Its purpose can be understood as a complete record of patient encounters that allows the automation and streamlining of the workflow in health care settings and increases safety through evidence-based decision support, quality management, and outcomes reporting.

Design

The front end of EHRS is completely designed using Microsoft Visual basic 6 and is backed up with the MYSQL Database. The EHRS comes in 2 versions.

- Standalone Electronic Health Record System (S-EHRS)
- This version of EHRS does not require any type of network connections such as LAN, WAN, etc.as it has built in database to store and retrieve the data. These types are useful when there is a network error.
- Consolidate Electronic Health Record System (C-EHRS)
- C-EHRS finds its application on the Medical robot while communicating with the server.

Brief working

- The software will be initiated when there is input from RFID Reader to the serial port.
- Now it compares weather the detected RFID tag is valid or not with the data that are included in Database.
- If the Detected RFID tag is a valid one, the software shows the complete details about the RFID tag owner (i.e. Patient). Else, some other action will be taken.
- If the prescription is to dispatch the tablets, then the software goes on checking the Column-"No. of Tablets" and sends a certain no. of bits (it depends on the value given in that column and No. of bits assigned to that Value) from the Serial port which is connected to a Microcontroller Board [Arduino board (ATMEGA8)]. Then the further actions are carried out by Microcontroller.
- The C-EHRS enables robot to retrieve the database from other computer (server) via Wi-Fi network. All the Databases are Password Protected and can be accessed only by robot and the respective doctors.
- The software enables telemedicine facility using Skype protocols when necessary.

Medical robot

Robots are perfect machinery devices which are used to reduce the burden and risks. The robot here is completely embedded with the EHRS and works according to the signals received from the EHRS [10]. This robot is capable of locomotion from one patient to another patient in order to serve the patients as prescribed by the doctors. It is also capable of storing, retrieving the patients data in presence of doctors.

Control system

The control system of Medical robot includes microcontrollers and EHRS software. A microcontroller is a small computer on a single integrated circuit containing a processor core, memory, and programmable input/output peripherals [11]. There are varieties of microcontrollers; in this project, we are using ATMEGA8 to control the locomotion of the robot and ATMEGA328 to work with the EHRS. They were programmed using Arduino IDE.

External devices

The devices that we are using in this project are DC Motors.

They are used to loco mote the robot, and to perform some other tasks that are given by the micro controllers [12].

Amplifiers

In this project, we are using ULN2003AN (7-Darlington emitter follower) IC as a Current amplifier. It amplifies the input current that are produced by the Line Sensors and feed the amplified output to ARDUINO board (Figure 1).

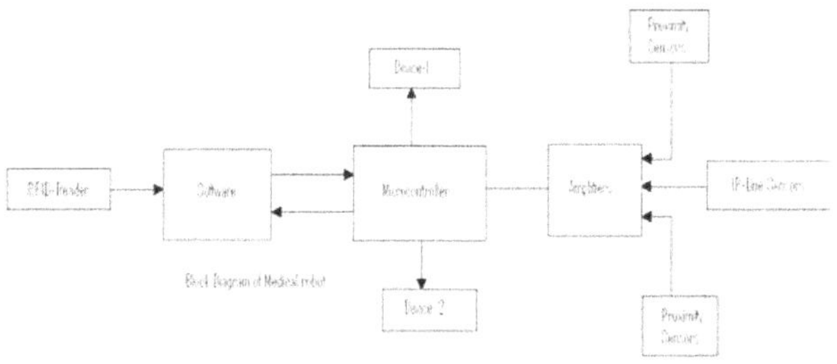

Figure 1: 7-Darlington emitter follower

Sensors

A sensor is a device that measures a physical quantity and converts it into a signal which can be read by an observer or by an instrument [13]. Sensors like Proximity sensors, Line tracking sensors, RFID Readers are used in this project

- A *proximity sensor* is a sensor able to detect the presence of nearby objects without any physical contact. Here, we are using IR Proximity sensors; these sensors help the robot to detect any object / person in front of the robot and takes decision according to the situation.
- An *RFID reader* is a device that is used to interrogate an RFID tag. These are the radio frequency identifying sensors which are used for authentication of patients, doctors, or any other users. The RFID tag talks to the interrogator using what is called the air-interface. This is a specification for how they talk to each other and includes the frequency of the carrier, the bit data rate, the method of encoding. ISO 18000 is the standard for the air interface for item management.
- Line Tracking Sensors are mainly used to track the marked lines. It may be either Black line on a bright background or White line on a Dark background. Usually Line tracking sensors are made up of Photodiodes and IR LED (Figure 2).

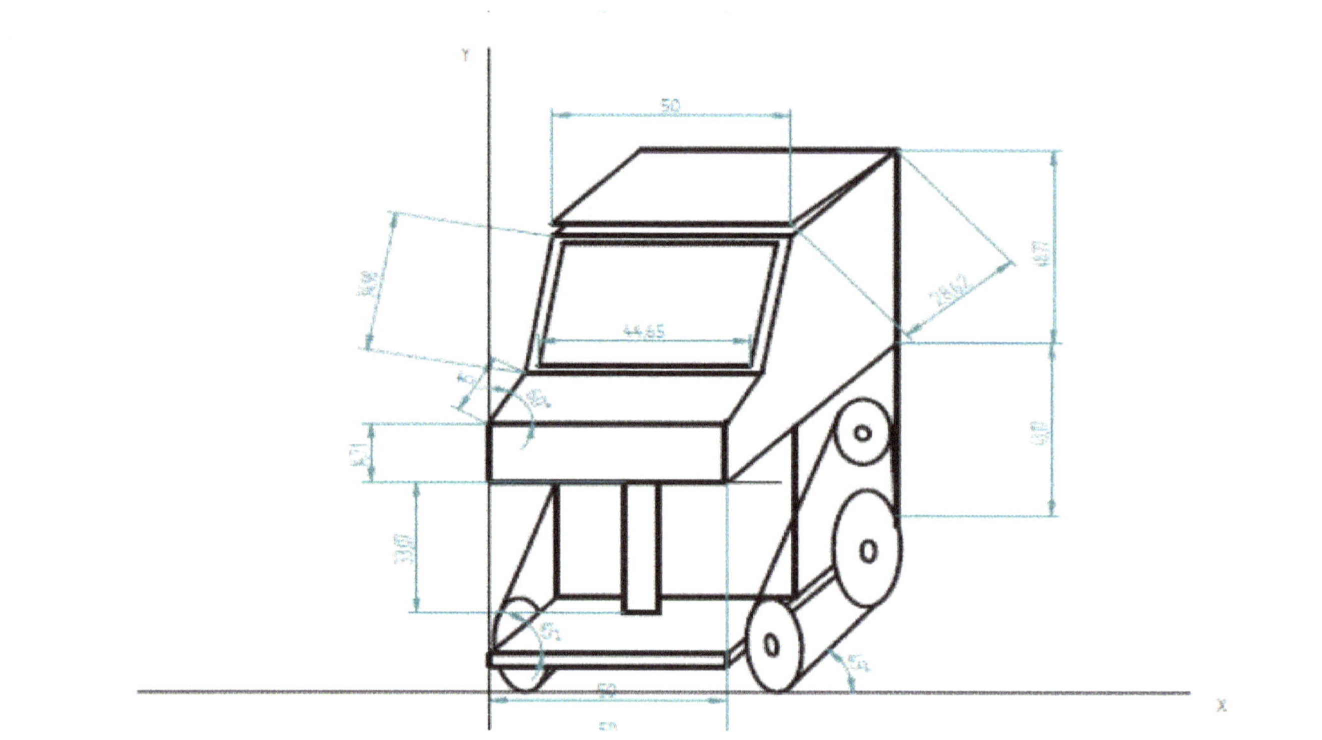

Figure 2: CAED Diagram of the Medical Robot

The Robot's structure

The robot's body is designed in such a way as to maintain its stability even when there is more weight. The locomotion is entirely controlled by the DC Motors having the torque of 78.4532 newton centimeter at the speed 300RPM [14]. The locomotion of the robot is gripped with the rubber track belts to maintain very good friction even on the soft surface.

Conclusion

The EHRS software enables the designed robot to supply the prescribed tablets and treatment synopsis to the respective patients regardless of the liking of the patient and it effectively carries out the orders given by the doctor. It also allows wireless communication between the doctor and the patient saving both time and expenditure.

Special features

- Wireless communication between patients and doctors whenever necessary.
- The data base is secured and can be accessed by doctors only when they are authenticated.
- Telemedicine facility in case of emergency.
- On spot diagnosis option.
- The EHRS can switch between 2 different modes. Standalone mode when there is a proper connection between the robot and the server Consolidate mode when there is a problem in communicating with the server.
- Can be used with any type of electro-medical equipment.

Hence, this concept is proved efficient in utilization of machines for effective performance of the healthcare centres thereby providing solution to one of the global problem.

This project was shortlisted as Top 10 Finalists in the IEEE Region 10's All India Young Engineers' Humanitarian Challenge-2011.

The final design

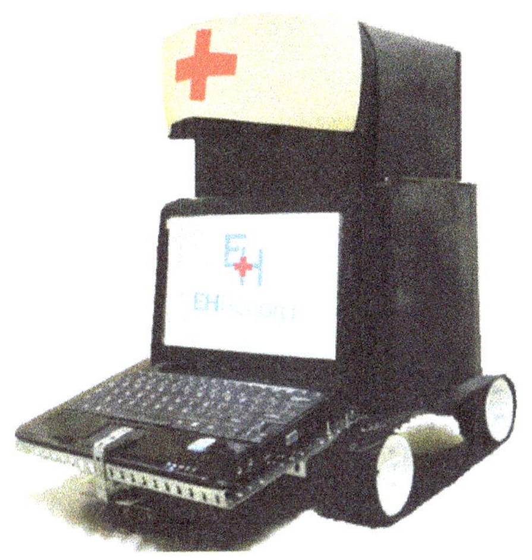

References

1. http://spectrum.ieee.org/robotics/medical-robots
2. www.doc.ic.ac.uk/~nd/surprise_96/journal/vol4/ao2/report.html
3. http://www.ismp.org/Tools/guidelines/ADC_Guidelines_Final.pdf
4. http://www.pppmag.com/documents/V3N6/WDIToT_Pg32.pdf
5. http://en.wikipedia.org/wiki/Arduino
6. http://en.wikipedia.org/wiki/Visual_studio
7. http://en.wikipedia.org/wiki/Multiplexer
8. http://pdf.datasheetcatalog.com/datasheet/fairchild/DM74157.pdf
9. http://www.doyoung.net/video/DATASHEET/PDF/ULN2003.pdf
10. http://users.ece.utexas.edu/~valvano/Datasheets/L293d.pdf
11. http://www.bigresource.com/VB-RFID-programming-wePC2boqvb.html
12. http://www.nejm.org/doi/full/10.1056/NEJMsb1205420#t=article
13. http://jamia.bmj.com/content/16/4/457
14. http://jamia.bmj.com/content/20/1/144.full.pdf+html

Modeling and Simulation of Three Level Piezoelectric Transformer Converters

Prasad Anipireddy[1*], and Challa Babu[1]

[1]*Asst.Proffesor, Department of EEE, SRIST, Nellore, India*

Abstract

This paper proposes a modeling and simulation of three level piezoelectric transformer converters. The increasing utilization of piezoelectric transformers (PT) in power electronics requires fundamental analysis and design of PT power converters. In addition, the possibility of overcoming the essential limitations of PT power ability by a combined operation of PTs needs to be examined. These studies could provide wide opportunities for PT exploitation in power electronic circuits.

The electrical circuit and model of piezoelectric transformers (PTs) ELS-60 is developed. Rosen-type PT and many other non-isolated conversion circuits have the inherit problem of the common neutral between the input and output that make it difficult to connect the output in series like the magnetic transformer. A new method based on the bootstrap method is proposed to solve the aforementioned problem. A new circuit for output summation of voltage level in the output side of each PT is developed. The impact of load and output filter capacitor on the conversion ratio and resonant frequency is discussed. The paper finally proposes a multilevel concept for many PT connected together for voltage summation. The method proposed can be applied to other multiple conversion circuits, which are based on common neutral between the input and output. The MATLAB/SIMULINK software used to design the required circuit diagram.

Keywords: Boost strap method; Converters; Piezoelectricity; Piezoelectric transformer; Rosen type PT

Introduction

Power electronic circuits have traditionally been based on magnetic technology, and until recently, have not been part of the tide of miniaturization and integration advances from which signal-processing integrated circuits have benefited [1]. The transformers and inductors in conventional converters are usually bulky compared to other elements. In an effort to miniaturize power components a piezoelectric, rather than a magnetic transformer could be employed in the power supplies [2]. Utilization of piezoelectric transformers in power electronics became possible owing to new piezoelectric materials that have recently been developed. These materials exhibit improved piezoelectric ceramic characteristics. In the past few decades piezoelectric transformers used widely in many applications, such as DC/DC converters, adapters, and electronic ballasts for fluorescent lamps [3].

Piezoelectric transformers (PTs) are electrical energy transmission devices that combine piezoelectric actuators as the primary side, with piezoelectric transducers as the secondary side. It transforms electrical energy into mechanical energy and back into electrical energy, i.e., unlike a conventional magnetic core transformer in which the magnetic field coupling is used between the primary and secondary windings [4]. Many PTs have been proposed and developed since the first PT was invented by Rosen in 1956, and it has been widely used and rapid developed, as compared with the traditional electromagnetic transformers. PT has big step-up ratio with small size, simple structure, lightweight, high conversion efficiency, good output waveform, no electromagnetic interference, etc [5]. Therefore, the convertors based on PT are very suitable for applications, which need high voltage source but low power. Currently there are three major types of PTs: Rosen, thickness vibration mode, and radial vibration mode [6]. This paper focuses on Rosen-type PT convertor and will use the single-layer Rosen-type model ELS-60 as an example. The model of ELS-60 with approximate parameters will be built in this paper. Because of the good performance of PT such that it can output perfect sine waveform even with square wave input [7].

Modeling of Equivalent Circuits

A PT can be modeled by an equivalent circuit like the one shown in Figure 1. C_{in} is the primary capacitance and C_o is the output capacitance. Some basic electrical specification of ESL- 60 PT is provided by the manufacturer. But other detailed parameters, such as R_m, L_m, C_m, have not been included, which shall be measured prior to any further researches [8]. By using an HP-4194A impedance gain phase analyzer, the electromechanical resonance modes can be measured from the impedance and phase versus frequency spectra.

It is well known that the parameters of PT depend on the load variations, temperature changes, components tolerances, etc. The equivalent circuit for typical DC-DC convertor based on PT for both voltage and current inputs are shown in Figure 2 and 3 [9].

When the output terminal of PT is shortened, its equivalent circuit is shown in Figure 4(a). When the input terminal of PT is shortened, its equivalent circuit is shown in Figure 4(b). Therefore, the parameters of the PT model measured by short-circuit test just are approximate values [10]. In order to obtain the maximum power output or stable

***Corresponding author:** Prasad Anipireddy, Asst.Proffesor, Department of EEE, SRIST, Nellore, India, E-mail: prasad.263@gmail.com

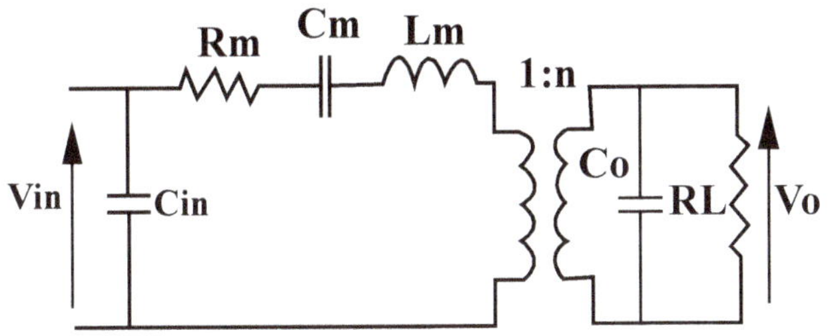

Figure 1: Equivalent circuit PT model

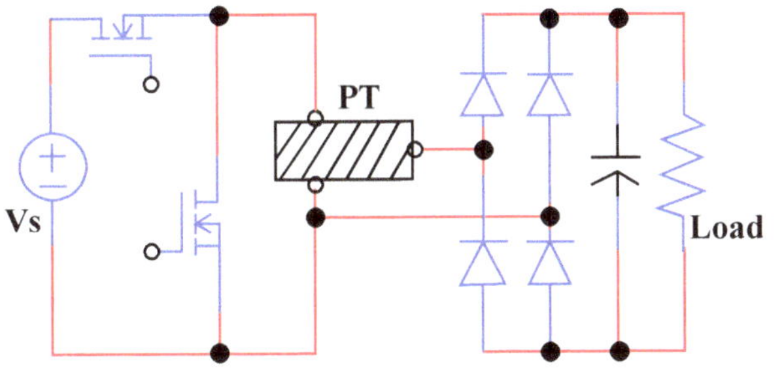

Figure 2: Equivalent circuit for typical DC-DC convertor based on PT: Voltage input

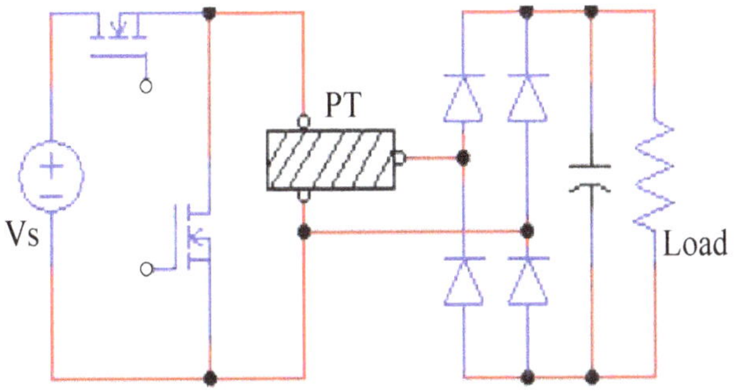

Figure 3: Equivalent circuit for typical DC-DC convertor based on PT: Current input

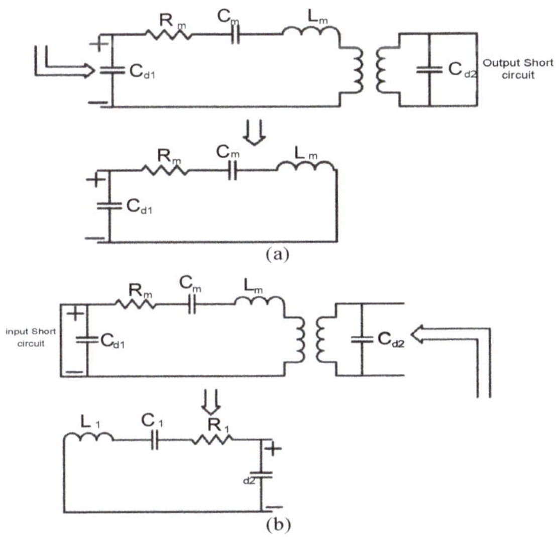

Figure 4: Equivalent circuit model of PT when output shorted and input shorted. (a) Output shorted. (b) Input shorted.

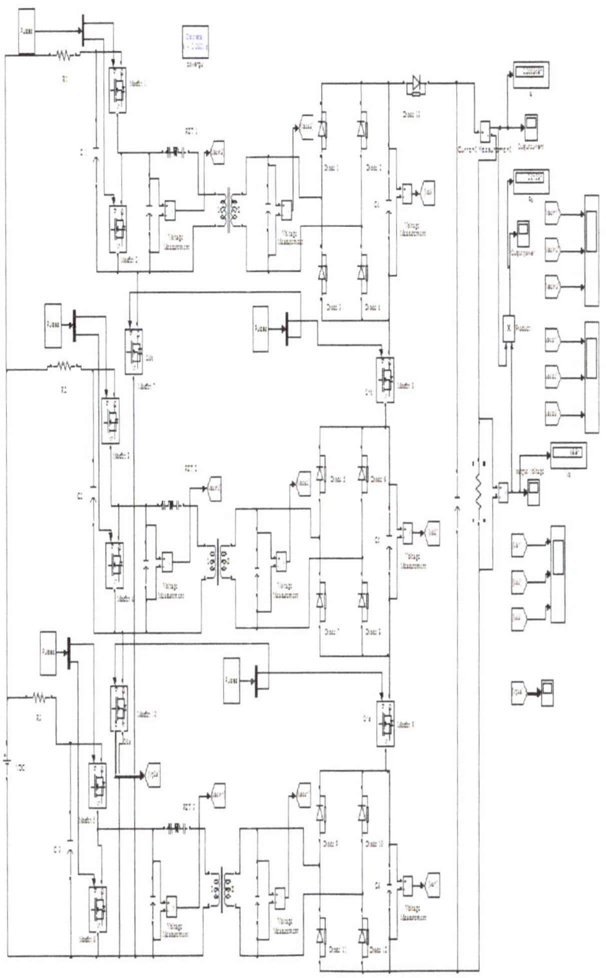

Figure 5: Simulation circuit diagram for three level piezoelectric transformer converters by using Rosen type.

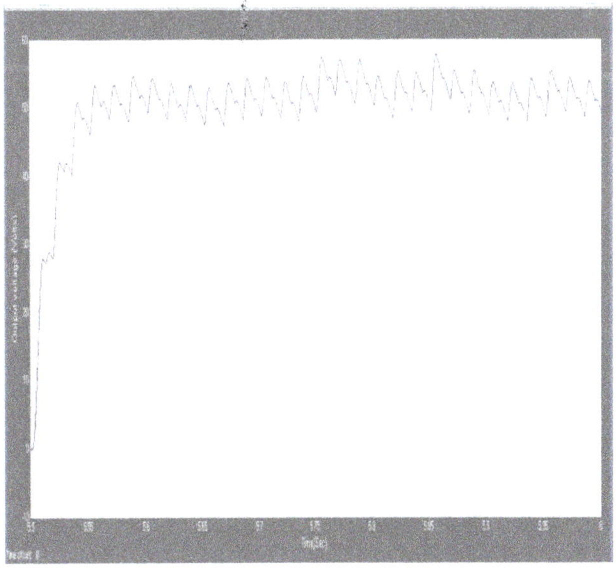

Figure 6: Output voltage wave form

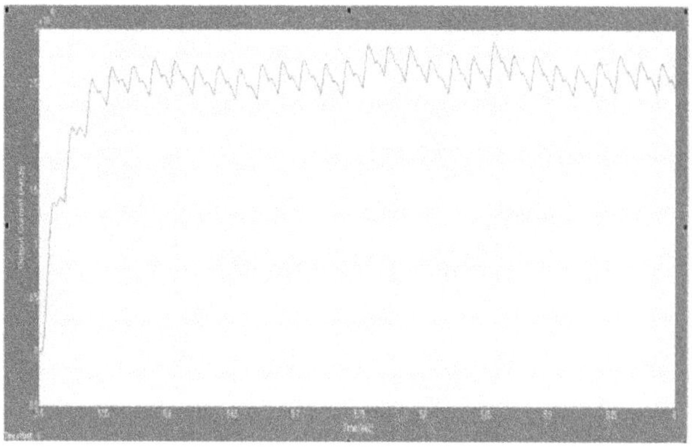

Figure 7: Output Current wave form

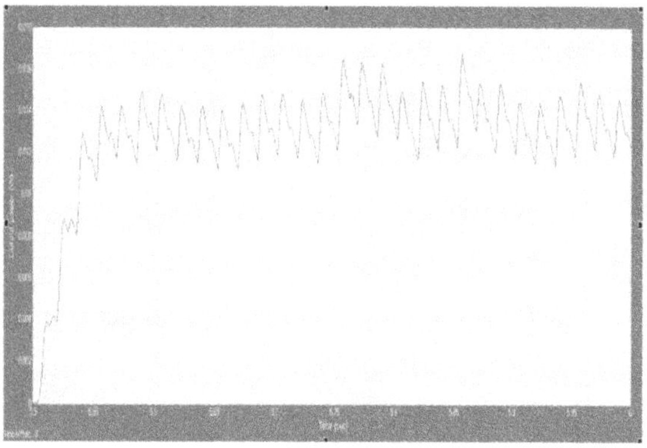

Figure 8: Output Power wave form

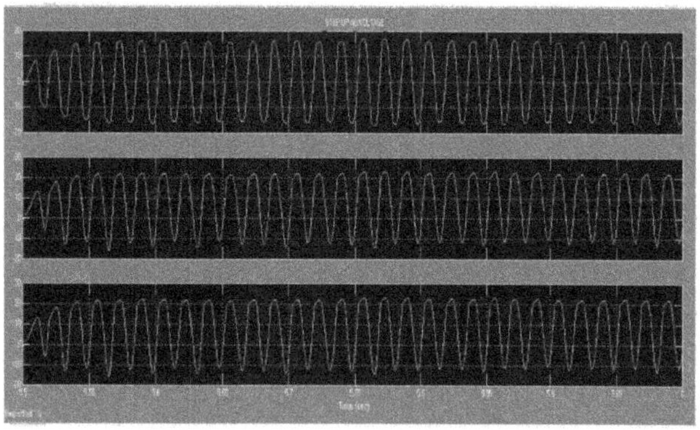

Figure 9: Step-up AC voltage wave form

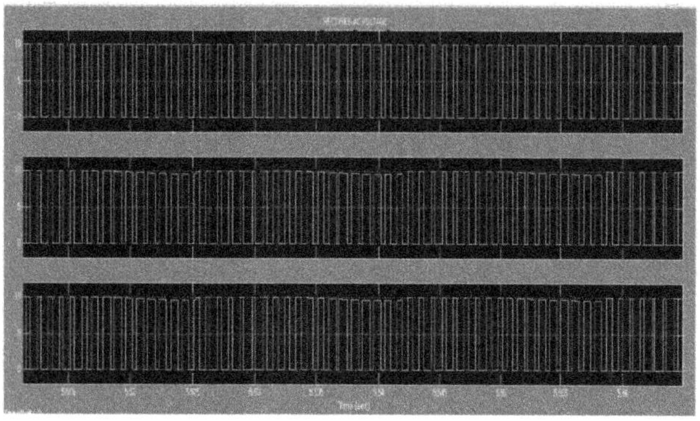

Figure 10: Rectified AC voltage wave form

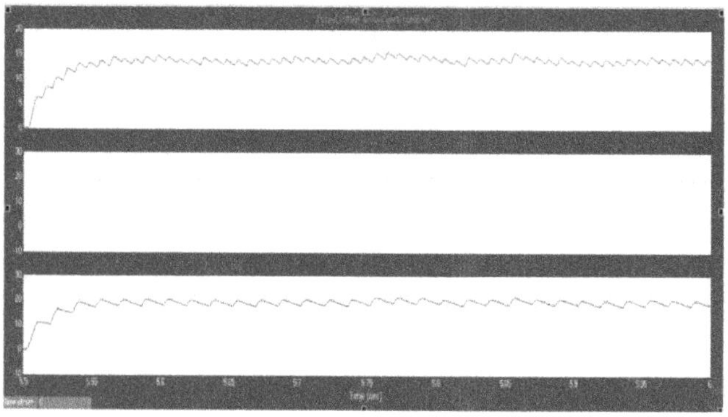

Figure 11: Output voltage wave form across each converter

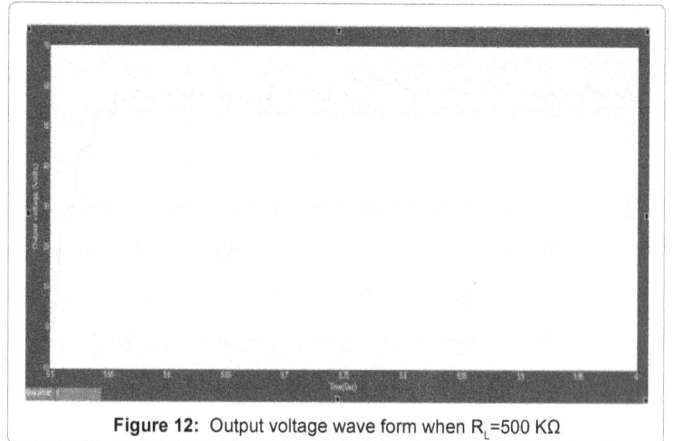

Figure 12: Output voltage wave form when R_L=500 KΩ

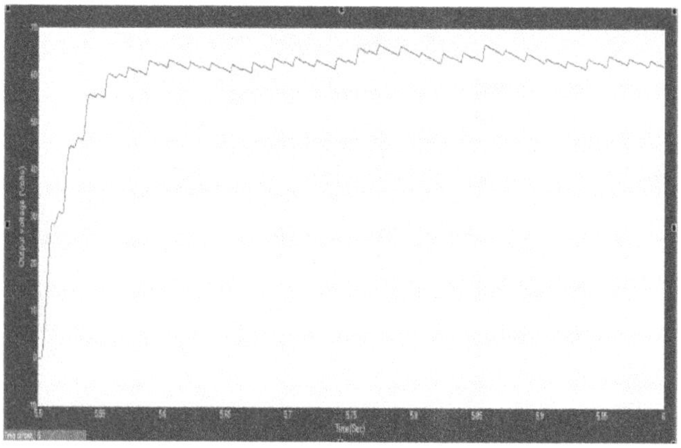

Figure 13: Output voltage wave form when R_L=1000 KΩ

S. No	Parameters	Value
1	V_{in}	10 V
2	R_{in}	1Ω
3	C_{in}	100 e-6
4	C_{pzt}	8.5 e-12
5	C_b	1 e-6
6	C_o	0.01 e-6
7	R_O	200KΩ

Table 1: Parameters of the Converter

S.No	Input voltage (Volts)	Load Resistance (KΩ)	Output voltage (Volts)	Output Current (Amps)	Output power (W)
1	10	100	41.39	0.0004139	0.01713
		200	49.61	0.0002481	0.01231
		400	55.98	0.00014	0.007835
		500	57.72	0.0001154	0.0006663
		1000	62.26	6.226*e-5	0.003877

Table 2: Three level piezoelectric transformer converters

output voltage, some frequency tracking mechanisms are necessary. This is because PT can be seen as a resonant conversion circuit.

In different conditions, it has an optimal frequency to derive maximum power or highest voltage output. It is a pleasure that many frequency tracking mechanisms have been presented in some literatures. The parameters of the PT model measured by short-circuit test are necessary to lock the optimal frequency range [11]. It tracks the optimal frequency to overcome parameters variation.

Modeling of Three Level Converter

The modeling of three level piezoelectric transformer converters by using Rosen type is shown in Figure 5.

The parameters of the three level piezoelectric transformer converters represented in Table 1.

Simulation Results

The three level piezoelectric transformer converters output voltage, output current and output power wave forms are shown I Figures 6-8. Where V_{in}=10v, R_L=200kΩ. From Figure 6 we can observe that the output voltage value is given by 49.61V. The three level piezoelectric transformer converters Step up AC voltage and Rectified AC voltage and V_o across each converter wave forms are shown in Figures 9-11 [12].

Similarly the three level piezoelectric transformer converters output voltages for Constant input (V_{in}=10V) and different load variations are shown in Figure 12 and 13. For constant input and different load variations of three level piezoelectric transformer converters V_0, I_o and P_o values are shown in Table 2.

Conclusion

Rosen-type PT and other non-isolated conversion circuits are widely applied to various occasions. But the common path problem between the input and the output connections is that the resulting multiple connections of non isolated convertor are not simple tasks. A method based on a special bootstrap method has been proposed in order to connect a number of outputs of non-isolated conversion circuits together which allows power to be delivered to output from different sub convertors. The objective of this paper is that through the analysis of multiple PTs convertor, to present a new method, which can obtain the summation of output voltage of multiple conversion circuits, which are based on common neutral between the input and output.

References

1. Yuanmao Y, Eric Cheng KW, Kai Ding (2012) A Novel Method for Connecting Multiple Piezoelectric Transformer Converters and its Circuit Application. IEEE transactions on power electronics 2: 1926-1935
2. Lin CY, Lee FC (1993) Development of a piezoelectric transformer converter.
3. Ivensky G, Zafrany I, Ben-Yaakov SS (2002) Generic operational characteristics of piezoelectric transformers. IEEE Trans. Power Electron 17: 1049-1057.
4. Piezoelectric Transformers in Power Electronics, IEEE transactions on power electronics.
5. Ivensky G, Bronstein S, Ben-Yaakov S (2004) A comparison of piezoelectric transformer ac/dc converters with current doubler and voltage doubler rectifiers. IEEE Trans. Power Electron 19: 1446-1453.
6. Alonso JM, Ordiz C, Costa MAD (2008) A novel control method for piezoelectric-transformer based power supplies assuring zero-voltage switching operation. IEEE Tran. Ind. Electron 55: 1085-1089.
7. S´anchez AM, Sanz M, Prieto R, Oliver JA, Alou P, et al. (2008) Design of piezoelectric transformers for power converters by means of analytical and numerical methods. IEEE Trans. Ind. Electron. 55: 79-88.
8. Lo YK, Pai KJ (2007) Feedback design of a piezoelectric transformer based half-bridge resonant CCFL inverter. IEEE Trans. Ind. Electron. 54: 2716-2723.
9. Liu YC, Chen YM (2009) A systematic approach to synthesizing multi input DC–DC converters. IEEE Trans. Power Electron. 24: 116-127.
10. Qian Z, Abdel-Rahman O, Al Atrash H, Batarseh I (2010) Modeling and control of three-port DC/DC converter interface for satellite applications. IEEE Trans. Power Electron. 25: 637-649.
11. Ho ST (2007) Modeling of a disk-type piezoelectric transformer. IEEE Trans. Ultrason., Ferroelect., Freq. Contr. 54: 2110–2119
12. Piezoelectric Transformer Model: ELS-60 Data Sheet, Eleceram Technology Co., Ltd., Taoyuan, Taiwan.

An Overview of Application of Artificial Immune System in Swarm Robotic Systems

Daudi J*

Department of Aerospace Engineering, School of Engineering, University of Glasgow, UK

Abstract

The Artificial Immune System (AIS) is a biologically inspired computation system based on specifically human immune system. AIS applications in last one decade have been developed to address the complex computational and engineering problems related to classification, optimisation and anomaly detection. Many investigations have been conducted to understand the principles of immune system to translate the knowledge into AIS applications. However, a clear understanding of principles and responses of immune system is still required for application of AIS to Swarm Robotics. This article after a review of AIS models and algorithms proposes an integration of AIS and Swarm Robotics by developing a very clear understanding of immune system structures and associated functions.

Keywords: Immune System (IS); Artificial immune systems (AIS); Artificial Immune Algorithms; Neutrophils; Swarm Robotics (SR)

Introduction

Man has survived through millions of year and the credit goes to the natural defense system our bodies are blessed with, the "Immune System (IS)". The immune system provides protection to a living body against number for foreign molecules (referred to as antigens) e.g. viruses, bacteria, fungi and other parasites. The immune system achieve this objective by observing, studying and identifying foreign molecules that enter our bodies, then, it prompts its response against them by creating and releasing antibodies that attack these antigens, thus eliminating them from our bodies and freeing us from infections. To eliminate the threat, IS has to distinguish between foreign molecules and the molecules/tissues that constitute itself to avoid auto-immune responses.

Immune System possesses excellent ability to recognize the foreign molecules, when they are encountered for the very first time, retain their memory and identify them when encountered at a later stage. IS uses genetic characteristics for biological functioning and thus provide the base for computational modelling of adaptive or in-borne responses. Following human central nervous system; IS is the most complex biological system due to wider and variable responses. Based on its complex and multiple behaviours, its understanding for computational adoption is still inadequate.

Types of Immune System

Immune system is sub-divided into two types of systems i.e. adaptive and innate immunity.

Adaptive immunity is directed against specific disease causing foreign agents and is modified by exposure to such organisms or antigens and keeps strong immunological memory. Adaptive immunity targets the specific pathogens, either previously encountered or not and gets modified by exposure to such pathogens. Adaptive immunity system consists of lymphocytes (white blood cells, more specifically B and T type) which function to recognize and destroy specific substances, and are antigen-specific [1]. Adaptive immune system has Immunological memory which gives the system, capability of more effective immune response against an antigen after its first encounter, leaving the body in a better position to be able to resist in the future against same pathogens. On primary response, the immune system identifies an antigen and responds with release of large number of antibodies to fight with infection and to eliminate the antigen from the body.

Innate immunity is aimed to target any invaders or disease causing agents or pathogen in the body and is non-specific. It is not modified by repeated exposure and thus plays an important role in the initiation and regulation of immune responses. Innate immunity involves number of specialized white blood cells involved to recognize and bind to common molecular patterns of disease causing microorganisms. It does not provide complete protection and is primarily static in nature and does not modify [1]. The innate immune system is considered as the first line of defense [2] which comprised of cells and mechanisms that defend the host from infection by disease causing organisms. Unlike the adaptive immune system (which is found only in vertebrates), it does not confer long-lasting or protective immunity to the host [3]. Innate immune system recruits the immune cells to sites of infection, through the production of specialized chemical mediators, called cytokines. This causes the identification and removal of foreign substances infecting organs, tissues, blood and lymph vessels by specialised white blood cells. Generally, innate and adaptive immunity systems have following characteristics indicated as Figure 1.

All white blood cells (WBC) are known as leukocytes. Leukocytes are not associated with a particular organ or tissue; thus, they function like a unicellular organism. Leukocytes are able to move as singular cell or as a population of cells and respond against foreign agents. The innate leukocytes or white blood cells include: Natural killer cells, mast cells, eosinophils, basophils; and the phagocytic cells including macrophages, neutrophils, and dendritic cells.

Artificial Immune System is actually based upon the capabilities

***Corresponding author:** Daudi J, Department of Aerospace Engineering, School of Engineering, University of Glasgow, UK, E-mail: daudij@yahoo.com

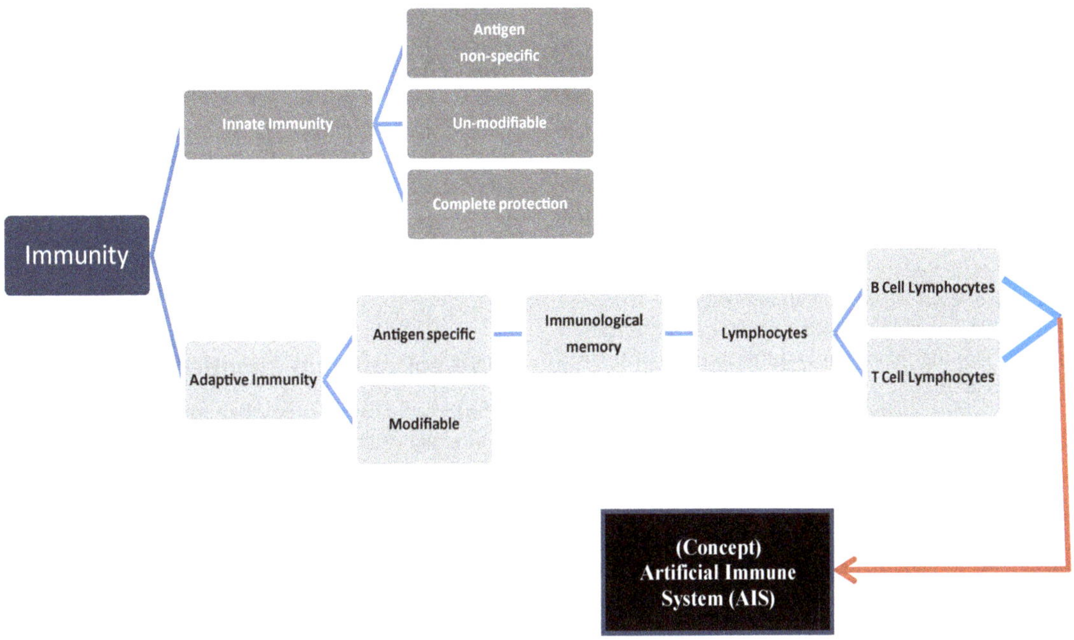

Figure 1: Classification of Human Immune System.

of immune system such as robustness, de-centralization, error tolerance and adaptiveness which enabled the researchers for system computation as "Artificial Immune System (AIS)" [4-6].

Following the patterns and behaviours of IS three basic computational areas are developed such as immune modelling, theoretical AIS and applied AIS. Immune modelling is focused on mathematical models and simulations of natural and artificial immune systems. Theoretical AISs concerned with the theoretical aspects including mathematical modelling of algorithms, convergence analysis, and performance and complexity analysis of such algorithms. Applied AISs includes working on immune-inspired algorithms, building immune-inspired computer systems, to apply AISs to diverse real world applications [7].

AIS follows the IS characteristics for computational application development such as feature extraction, pattern recognition, memory, learning, classification, adaption for utilization in computer security, fraud detection, machine learning, data analysis, optimization algorithms [8].

AIS Algorithms

A critical review of literature has concluded that four major AIS algorithms are focused to develop various AIS applications (Figure 2).

Negative selection based algorithms

Negative Selection Algorithms are based on the principle function that protects the body against self-reactive lymphocytes. IS identifies foreign antigens without reacting with the 'self cells'. Receptors are produced during a pseudo-random genetic re-arrangement process for the generation of T-cells. These receptors undergo negative selection mechanisms in thymus. Those T cells which react against "Self Cells" or against self-proteins are discarded and remaining which do not bind to self-cells are permitted to leave the thymus. These are called as matured T-cells which are allowed free movement throughout the body to contribute in immunological process against foreign antigens [9]. This biological phenomenon has given an inspiration for the developments of most of the existing Artificial Immune systems.

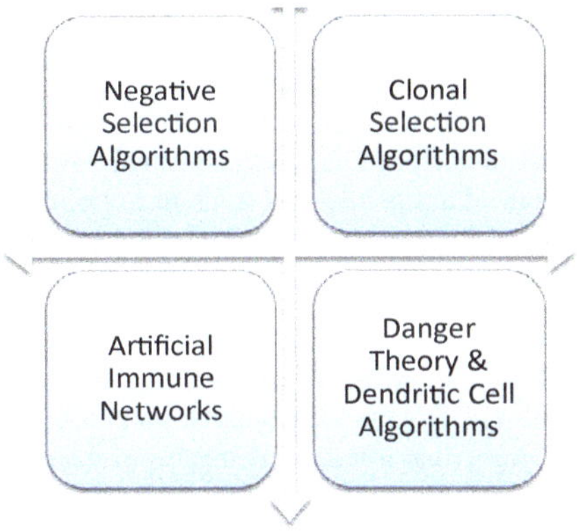

Figure 2: AIS Algorithms.

Clonal selection based algorithms

This Clonal Selection Based Algorithms is formulated on the principle of mechanism of antigen-antibody recognition, binding, cell propagation and separation into memory cell [10]. This is called as Clonal Selection Theory which has resulted in development of several artificial immune algorithms named as clonal selection algorithm (CSA) by Castro and Zuben [11]. Based on clonal selection and affinity maturation principles they named this as CLONALG. CLONALG in 2002.

Artificial immune networks (AINs)

Artificial immune networks (AINs) are other successful models in AISs. Framer's etal proposed their immune network model Farmer et al. which became the fundamental for various AINs algorithm and hence based on this the first immune network algorithm was proposed by Ishida [12]. Later, Timmis et al. re-defined these immune networks, which were formally named as AINE (Artificial Immune Network). According to Knight and Timmis. 2001 AINs uses Artificial Recognition Ball (ARB) to represent identical B-cells [13]. Two B-Cells are joined together, if the affinity between two ARBs is below a network affinity threshold (NAT). A Resource Limited Artificial Immune System (RLAIS) based on AINE is developed in 2001 by Timmis and Neal. This up-gradation of models included the knowledge of the fixed total number of B-cells presented in ARBs with centralized control but having the specific role of each ARB in obtaining resources from the mainstream. Those ARBs not able to obtain resources are removed from the network. The cloning, Mutation and interactions of B-Cells all take place at the ARB level.

Danger theory and dendritic cell algorithms

Dendritic cells within an innate immune system are cells which respond to some specific danger signals. The three main types of dendritic cells such as: 1) Immature Dendritic Cells, which collect parts of the antigen and the signals, 2) Semi-mature Dendritic cells, are immature cells which have decision power against local signals and represent safe and present the antigen to T-cells resulting in tolerance, and 3) Mature Dendritic cells, are capable to react strongly identifying that the local signals represent danger and deliver the antigen to T-cells for reaction against pathogens. The Dendritic Cell Algorithm is biologically inspired development taking inspiration from the Danger Theory of the mammalian immune system with specific function of dendritic cells. Matzinger, first proposed this Danger Theory stating that the roles of the acquired immune system is to respond to signals of danger, rather than discriminating self from non-self [14,15]. The theory states that the helper T-cells activate an alarm signal providing the co-stimulation of antigen-specific cells to respond. The Dendritic Cell Algorithm (DCA) is inspired by the function of the dendritic cells of the human immune system.

An abstract model of dendritic cell (DC) behaviour was also developed and used to form an algorithm, the artificial DCA by Greensmith et al. For this purpose, population based DCA algorithm was applied to numerous intrusion detection problems in computer security including the detection of port scans and botnets, where it has produced impressive results with relatively low rates of false positives [16].

There are several other immunology areas reported in the literature to inspire the development of algorithms and computational tools, for example, humoral immune response, Danger Theory dendritic cell functions, and Pattern Recognition Receptor Model [17-20].

Review of AIS Application Areas

Dasgupta in 1999 under the title "Artificial Immune Systems and Their applications" compiled several useful literature resources including AIS textbooks and application papers. These comprehensive literature resources also addressed the computational models of the immune system and their applications till 1998. Various modeling techniques based on ordinary differential equations, delay differential equations, partial differential equations, agent-based models, stochastic differential equations as well as associated algorithm and simulation frameworks are being used to simulate IS [21].

This diversity of AIS knowledge has helped the researchers to apply AIS to solve several bench-mark problems of the field such as; computer security, numerical function optimization, combinatorial optimization, learning, bioinformatics, image processing, robotics, adaptive control system, data mining, and anomaly or error diagnosis [22]. As a more advance innovation, the applications of AIS in controlling a robotic arm is evident from studies, remote-sensing classification of satellite images, compensating exposure in images with back-lighting, web-mining applications and application in industrial manufacturing process [23-26].

A review of literature has concluded the application of AIS in following three major categories as shown below. However, the newly identified and explored areas may fall beyond these categories Figure 3 [11,23,27-29].

AIS Application in Robotics

AIS can provide a baseline for robots to learn new skills, adapt to new environment throughout its lifetime. With the progress in mechatronics, MEMS (Micro-Electro Mechanical Systems) and nanotechnology the sizes and cost of electronic components (e.g. sensors, actuators and electronic boards) are decreasing, thus robots built from such components if produced at larger size will be very cheap. It would be possible to install/deploy a large number of such small robots and should rightly be called swarm of robots to achieve the desired task. AIS will play its role by bringing intelligence into this swarm of robots to achieve/accomplish the desired task. This idea has been specifically taken from Immune cells of adaptive or innate immunity. In this regard, the behaviour of neutrophils is particularly the interest of author with its potential application in Swarm Robotics.

Neutrophils are the most abundant white blood cells as constitute about 40–70% of the white blood cells in the blood stream [30]. As a first defense line, they respond early to threats against the hosts by detecting changes in the vascular endothelium induced by tissue damage and/or infection [31]. The behaviour of Neutrophils along with other immune cells is quite complex and through the microvascular system initially get access to the source of antigen. This causes changes in the vascular endothelium, giving impetus to neutrophils to exit the microvasculature and to move through the tissue by sensing molecules, or chemokines, produced by damaged or infected tissue [31,32] .

Thus, at the initial phase of inflammation caused by bacterial infection, environmental exposure or by some cancers, the first defensive response is received from the neutrophils which migrate towards the site of inflammation. Neutrophils first migrate through the blood vessels and then through the interstitial tissue, following chemical signals (Chemokines produced by damaged or infected tissues) such as Interleukin-8 (IL-8), C5a, fMLP and Leukotriene B4 in a process called "Chemotaxis" as indicated in Figure 4 [31-33].

This process, termed chemotaxis, plays a significant role in

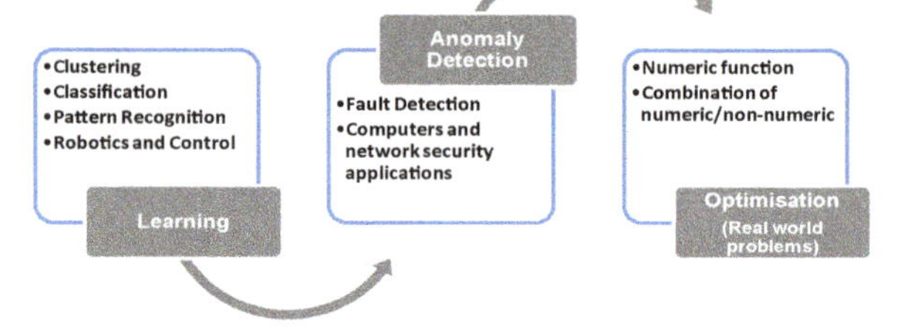

Figure 3: Categorical Areas for AIS Applications.

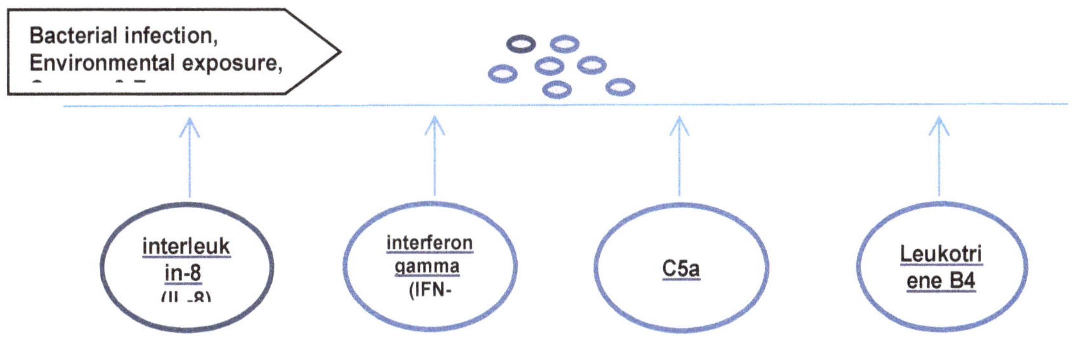

Figure 4: Concept of Chemotaxis.

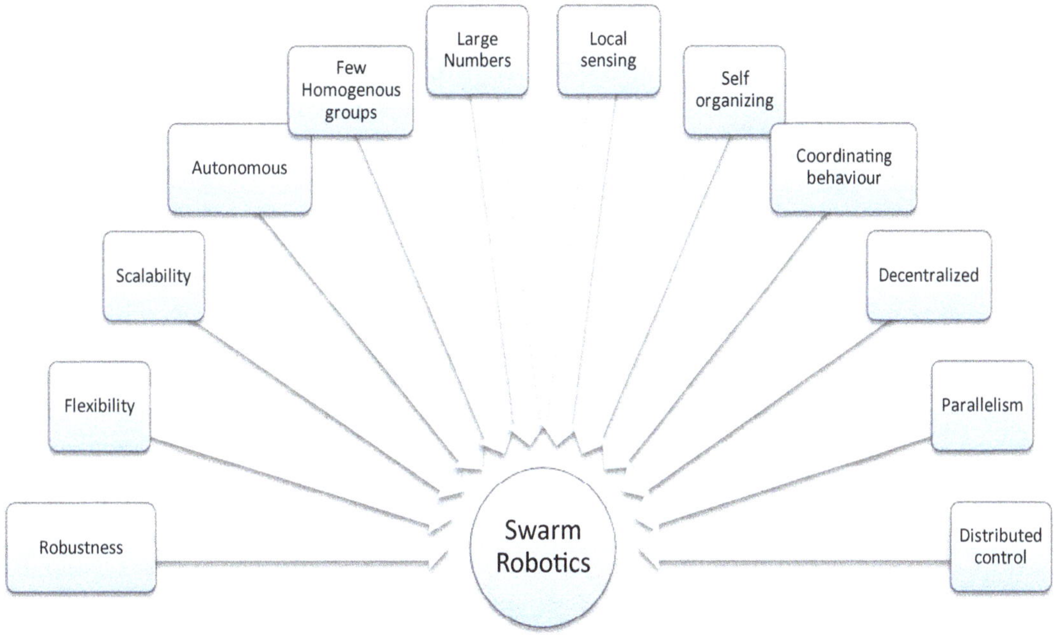

Figure 5: Characteristics of Swarm Robotics.

immune cell motion computation models [34]. These fundamental behaviours of neutrophils are mathematically well explained through the convection-diffusion and reaction-diffusion equations [35,36].

The immune system swarms like neutrophil swarms initially develop from the primary arrest of a small count of neutrophils followed minutes later by massive numbers. This directed migration or dynamic swarming is probably caused by intercellular communication via signaling molecules. These signaling molecules are produced by, and are attracted to, neutrophils. Streaming is another dynamic behavior of neutrophils. It is assumed that neutrophils generate signals to induce swarming and once a swarm reaches a certain size, a large enough signaling center exists to overcome the competing signals of nearby smaller swarms [37].

AIS Applications in Swarm Robotics

Sahin (2005) has defined swarm robotics as "*swarm robotics is the study of how large number of relatively simple physically embodied agents can be designed such that a desired collective behaviour emerges from the local interactions among agents and between the agents and the environment*" [38]. Millonas has also proposed that a swarm system must encapsulate the principles of proximity, quality, diverse response, stability and the adaptability [39].

Swarm robotics is an innovative approach to the coordination of large numbers of robots. Basic motivation comes from the observation of birds/insects as to how these individual entities in itself can cooperate together to carry out complex tasks/goals that cannot be accomplished individually. This sort of coordination capabilities are still beyond the reach of current multi-agent robotic systems. The main characteristics that a swarm robotic system must possess are shown below as Figure 5.

The main feature of swarm behaviour is that each individual (robot) follows simple rules and there is no centralised control dictating their behaviour. Each robot is capable of observing and responding to its environment and directing its activity towards achieving common goal. Robots will collaborate through interaction among themselves exhibiting simple behaviours like self-assembly, self-repair, co-operation, monitoring and responding.

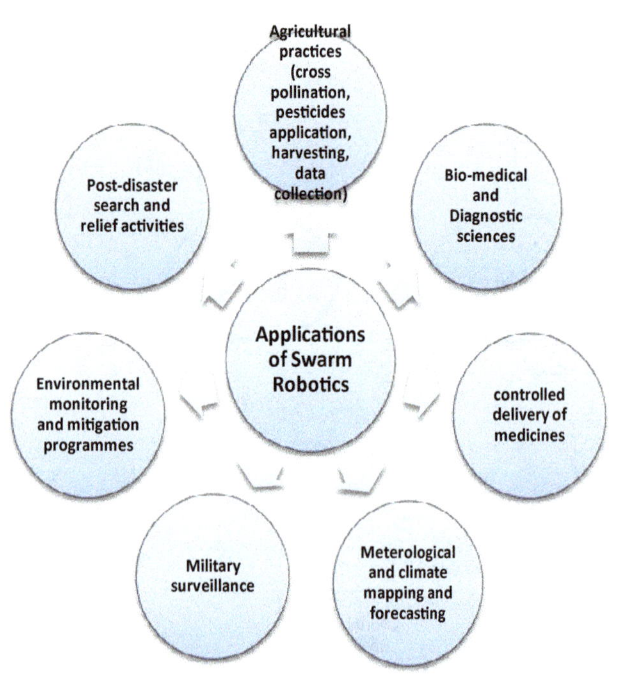

Figure 6: Application of Swarm Robotics in real world.

Following this, it is assumed that mathematical modelling, design and methodology of emergence as well as testing of swarm systems are key tasks to apply neutrophils behaviour in swarm robotics.

Research Needs in AIS and Swarm Robotics

Based on above facts, it is concluded that immune system is a very complex system and still there is a lack of reliable data about many of its constituent cells and molecules, and thus any simulation that intends to model the immune system at full scale will require huge computational resources. Thus, it allows the interconnection of predictive models, defined at multiple scales (molecules, cells, tissues and organs, body-wide systems, the whole organism, and collection of organisms) for conversion into in systemic networks. These networks can address and validate the systemic hypotheses by combining clinical observations, experimentation and predictive modelling. AIS based on neutrophils behavior, will play potential role by bringing intelligence into swarm of robots. An extensive literature review has concluded that Swarm Robotics has high prospects and utilization for a variety of challenging applications as shown in the Figure 6 below.

Application of swarm robotics in wide range of real life fields cannot be ignored for serious consideration for further developments. These applications include biomedical applications for developing micro-nano systems by applying high-resolution monitoring, fast prototyping of micro-environments such as by application of micro-nano robots may be for on-site drug delivery, development of unconventional robots (for DNA devices, nano-particles, synthetic bacteria, bio-bots for advance treatments), synthetic biology applications based on fabrication and manipulation at the micro-nano scale, development of energy-based robot control (magnetic, light-based self-organizing biological and robotic systems) and finally the control of swarm robotic systems with large-scale simulations will be applicable to all such applications. This can be further explored for the development of tools to explore swarm behaviours (bio-inspiration, crowdsourcing, machine learning).

Swarm robots can disperse and perform multiple tasks at difficult and inaccessible sites such as in forests, lakes, hilly areas etc. Swarms of robots because of the robustness of the swarm, can prove highly useful for dangerous tasks including monitoring and mitigating the environmental hazards, like a leakage of a chemical substance, clearing off environment from hazardous wastes.

Environmental pollution is one of the biggest concerns of planet earth in context to survival of all living organisms and thus application of AIS and SR in environmental monitoring and mitigation may prove most beneficial, if applied precisely.

Dramatic incidences of aquatics pollution due to oil spills, organic pollutants and suspended matters in surface water bodies such as oceans, rivers, lakes etc. have highlighted the potential to address the man-made environmental damage. Mitigation of this unwanted situation requires resources in terms of man power, machinery, labour cost and time. Hence, to provide alternative to all such requirements and constraints, research based on application of advance computational approaches is required to address this real world problem of environmental pollution.

Another environmental application may be based on developing in-pipe inspection swarm robots capable to overcome the issues of human factor in labour and time intensive monitoring and also to act in inaccessible environment inside the water or gas pipelines of an area within shortest possible time for long-distance inspection.

Future research is required to investigate deployment possibility of swarm of robots to move on the water's surface autonomously and gather organic pollutants and suspended matters to preserve water resources from quality deterioration. The robots may work together to cover a large area of the water and communicate with one another and with land-bound researchers. This will also help in developing an insight so as to how the problems such as large scale simulation, control of mirco/nano swarm robots can be dealt with by using AIS.

References

1. Castro LNde, Zuben FJV (1999) Artificial Immune Systems: Part I -Basic Theory and Applications. School of Computing and Electrical Engineering, State University of Campinas, Brazil.
2. Grasso P, Gangolli S, Gaunt I (2002) Essentials of Pathology for Toxicologists. (1ST Ed) CRC Press.
3. Alberts B, Alexander J, Julia L, Martin R, Keith R, et al. (2002) Molecular Biology of the Cell. (4th Ed). New York and London, Garland Science.
4. Kephart JO (1994) A biologically inspired immune system for computers artificial life IV. In: Brooks RA, Maes P (Eds), Proceedings of 4th International Workshop on the Synthesis and Simulation of Living Systems, MIT Press.
5. Dasgupta D (1996) Using immunological principles in anomaly detection. In: Proceedings of the Artificial Neural Networks in Engineering (ANNIE96), St Louis, USA.
6. Forrest S, Perelson AS, Allen L, Cherukuri R (1994) Self-nonself discrimination in a computer. In: SP 1994: Proceedings of the 1994 IEEE Symposium on Security and Privacy, IEEE Computer Society, Washington, DC.
7. Dasguptaa D, Yua S, Nino F (2011) Recent Advances in Artificial Immune Systems: Models and Applications. Applied Soft Computing 11: 1574-1587.
8. Dasgupta D (2006) Advances in artificial immune systems. Computational Intelligence Magazine, IEEE.
9. Somayaji A, Hofmeyr S, Forrest S (1997) Principles of a Computer Immune System. In: Proceedings of the Second New Security Paradigms Workshop 75-82.

10. Burnet FM (1959) The Clonal Selection Theory of Acquired Immunity. Cambridge University Press.

11. Castro LND, Zuben FJV (2000) The clonal selection algorithm with engineering applications, Genetic and Evolutionary Computation Conference (GECCO'00) – Workshop Proceedings, Las Vegas, Nevada, USA.

12. Ishida Y (1990) Fully distributed diagnosis by PDP learning algorithm: towards immune network PDP model, IEEE International Joint Conference on Neural Networks, San Diego, USA.

13. Knight T, Timmis J (2001) AINE: an immunological approach to data mining, In: IEEE International Conference on Data Mining , San Jose, CA, USA.

14. Matzinger P (2002) The danger model: are Renewed sense of self. Science 296: 301-305.

15. Matzinger P (1994) Tolerance,danger, and the extended family. Annu.Rev. Immunol 12: 991-1045.

16. Greensmith J, Aickelin J, Cayzer S (2008) Detecting Danger: The Dendritic Cell Algorithm.Copyright Robust Intelligent systems, IGI Publishing.

17. Dasgupta D, Yu S, Majumdar NS (2003) MILA – multilevel immune learning algorithm, in: Genetic and Evolutionary Computation Conference (GECCO 2003), Chicago, IL, USA.

18. Aickelin U, Cayzer S (2002) The danger theory and its application to artificial immune systems. The 1st International Conference on Artificial Immune Systems (ICARIS 2002), Canterbury, England.

19. Greensmith J, Aickelin U, Cayzer S (2005) Introducing dendritic cells as a novel immune-inspired algorithm for anomaly detection. Artificial Immune Systems Lecture Notes in Computer Science 3627: 153-167.

20. Yus S, Dasgupta D (2008) Conserved Self Pattern Recognition Algorithm. Artificial Immune Systems Lecture Notes in Computer Science Volume 5132: 279-290.

21. Kim H (2009) Asymptotic problems for stochastic processes and related differential equations. PhD Dissertation. Department of Mathematics. University of Maryland, USA.

22. Hart E, Timmis J (2008) Application areas of AIS: the past, present and future. Journal of Applied Soft Computing, 8: 191-201.

23. Lee Z J, Lee CY, Su SF (2002) An immunity based ant colony optimization algorithm for solving weapon-target assignment problem. Appl. Soft Comput 2: 39-47.

24. Zhong Y, Zhang L, Huang B, Li P (2006) An unsupervised artificial immune classifier for multi/hyperspectral remote sensing imagery. IEEE Trans. Geosci. Remote Sens 44: 420-431.

25. Nasraoui O, Rojas C, Cardona C (2006) A framework for mining evolving trends in web data streams using dynamic learning and retrospective validation. Comput. Networks 50: 1488-1512.

26. Bailey S (1984) From desktop to plant floor, a CRT is the control operators window on the process. Control Engineering 31: 86-90.

27. Campelo F, Guimarães FG, Igarashi H, Ramírez J, Noguchi S (2006) A modified immune network algorithm for multimodal electromagnetic problems. IEEE Transactions on Magnetics 42.

28. Kalini A, Karaboga N (2005) Artificial immune algorithm for iir filter design. Eng. Appl. Artif.Intell 18: 919-929.

29. Castro LND, Zuben FJV (2002) Learning and optimization using the clonal selection principle. IEEE Transactions on Evolutionary Computation 6: 239-251.

30. Beers MH, Porter RS, Jones TV (2006) The Merck Manual. (18th Ed), Merck and Co., Inc.

31. Janeway CA, Travers P (1997) Immunobiology: The Immune System in Health and Disease. Churchill Livingstone, New York.

32. Ariel A, Serhan NC (2007) Resolvins and protectins in the termination program of acute inflammation, Trends Biochem. Sci. 28: 176-183.

33. Kadirkamanathan V, Anderson SR, Billings SA, Zhang X, Holmes, et al. (2012) The Neutrophil's Eye-View: Inference and Visualisation of the Chemoattractant Field Driving Cell Chemotaxis In Vivo. PLoS One 7: e35182.

34. Nathan C (2006) Neutrophils and immunity: Challenges and opportunities. Nat. Rev. Immunol. 6:173-182.

35. Holmes GR, Dixon G, Anderson SR (2012) Drift-Diffusion Analysis of Neutrophil Migration during Inflammation Resolution in a Zebrafish Model. Advances in Hematology 8.

36. Sua B, Zhoua W, Dormanb KS, Jones DE (2009) Mathematical modelling of immune response in tissues. Computational and Mathematical Methods in Medicine 10: 9-38.

37. Chtanova T, Schaeffer M, Han S, van Dooren G, Nollmann M (2008) Dynamics of Neutrophil Migration in Lymph Nodes during Infection. Immunity 29: 487-496.

38. Şahin E (2005) Swarm robotics: from sources of inspiration to domains of application. Swarm Robotics 3342: 10-20.

39. Millonas MM (1994) Swarms, phase transitions, and collective intelligence. In: Artificial life III. Addison-Wesley, Reading 417-445.

10

Remote Control of Educational Mobile Mini-Robot via Wireless Communication

Manolov OB*

Department of Applied Informatics and Computer Technologies, European Polytechnic University, Pernik

Abstract

This work aims to describe another contemporary manner for interaction between human and mechatronic device by Bluetooth communication with a purpose for implementing wireless remote motion control of an educational mobile robot.

Keywords: Wireless communication; Remote motion control; Educational mobile robot; Bluetooth

Introduction

In recent years, there has been a growing interest in mobile robot motion control. Usually to control the movement of a mobile robot we need to control the speed of rotation of his engines or the rotation direction. This could be done with one of the Wi-Fi or Bluetooth interfaces. The mobile robot has to have the engines connected to a control circuit with one of these interfaces. While controlling the mobile robot with Bluetooth we also have to consider that our application will not only control the robot but also will do other tasks for gathering information about the environment, computing the mobile robot's moving direction and therefore the Bluetooth connection should not block these tasks. Herein the three major topics will be discussed:

1. The educational mobile robot "Audrino".
2. The "Wii Remote Plus" as a console for wireless motion control.
3. The communication between robot and console.

Conceptual Configuration of a Distance Control

The educational mobile platform "Audrino"

Arduino robot is a self-contained platform that allows to be developed an interactive machine to explore an environment. It is the result of the collective effort from an international team in collaboration with Complubot, 4-time world champion in robocup junior robotics soccer, looking at how science can be made fun to learn [1].

For our research and experiments an own mobile robot has assembled, shown on Figure 1. It is based on classical two wheeled platform with reversible DC-servo driven wheels and spherical fulcrum on the rear side. The Motor control module drives the motors, and the Control Board reads sensors and decides how to operate.

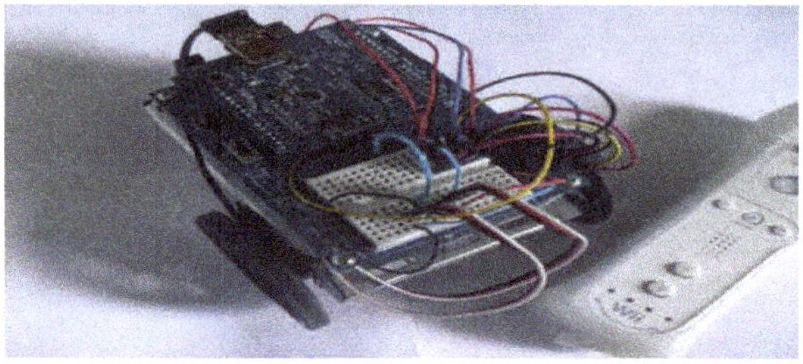

Figure 1: The assembled own "Audrino" type mobile robot.

Each of the boards is a full programmable using the Arduino IDE. The robot's completeness includes the following modules: Arduino Uno Rev3, Parallax Robotics Shield Kit, USB Host Shield 2.0 and Bluetooth communication, which modules are described in detail as follows below. The Arduino Uno Rev3 module, represented on Figure 2, is an open source microcontroller board, based on the Atmel ATmega328 MCU, plus a free software development environment [1]. The module can be used to sense inputs from switches, sensors, and computers, and then to control motors, lights, and other physical outputs.

It has 14 digital input/output pins (of which 6 can be used as PWM outputs), 6 analog inputs, a 16 MHz ceramic resonator, a USB connection, a power jack, an ICSP header, and a reset button. It contains everything needed to support the microcontroller; simply connect it to a computer with a USB cable or power it with a AC-to-DC adapter or battery to get started.

The Arduino Uno Rev3 differs from other preceding boards in that it does not use the FTDI USB-to-serial driver chip. Instead, it features the Atmega16U2 (Atmega8U2 up to version R2) programmed as a USB-to-serial converter. The microcontroller board can be powered via the USB connection or with an external power supply. The power source is selected automatically. The technical specifications of the module Arduino Uno Rev3 are:

- Microcontroller: ATmega328

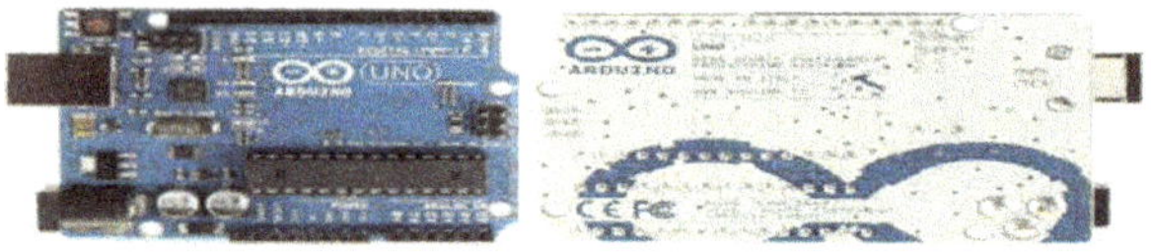

Figure 2: The module Arduino Uno Rev3 (front and back).

***Corresponding author:** Manolov OB, Department of Applied Informatics and Computer Technologies, European Polytechnic University, Pernik
E-mail: ognyan.manolov@epu.bg

- Operating Voltage: 5V
- Input Voltage (recommended): 7-12V
- Input Voltage (limits): 6-20V
- Digital I/O Pins: 14 (of which 6 provide PWM output)
- Analog Input Pins: 6
- DC Current per I/O Pin: 40 mA
- DC Current for 3.3V Pin: 50 mA
- Flash Memory: 32 KB (ATmega328) of which 0.5 KB used by boot-loader
- SRAM: 2 KB (ATmega328)
- EEPROM: 1 KB (ATmega328)
- Clock Speed: 16 MHz

The Arduino free programming software is designed for communication with a computer, another Arduino, or other microcontrollers by UART TTL (5V) serial communication and includes a serial monitor which allows simple textual data to be sent to and from the Arduino board, as it is shown on Figure 3. The RX and TX LEDs on the board will flash when data is being transmitted via the USB-to-serial chip and USB connection to the computer.

Rather than requiring a physical press of the reset button before an upload, the Arduino Uno is designed in a way that allows it to be reset by software running on a connected computer. The Arduino software uses this capability to allow you to upload code by simply pressing the upload button in the Arduino environment. This means that the boot-loader can have a shorter timeout, as the lowering of DTR can be well-coordinated with the start of the upload.

The Parallax Robotics Shield Kit module includes a Board of Education Shield (BOE), represented on Figure 4, which makes it easy to build circuits and connect servo motors to the Arduino Uno Rev3 module [2]. The BOE Shield mounts on metal chassis with motors and wheels.

With this kit and Arduino module is able to activate over 40 hands-on activities in robotics, such as:

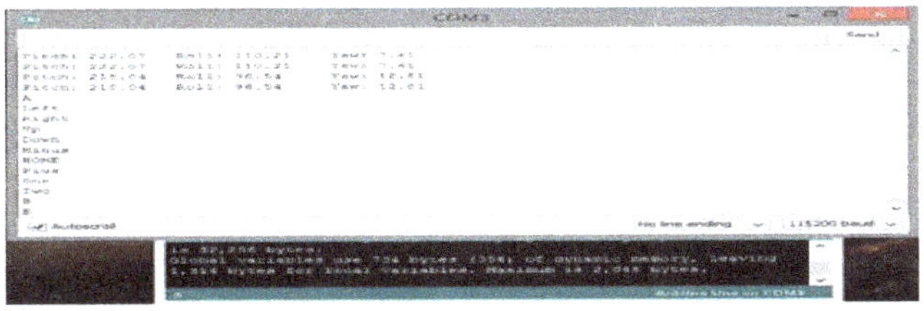

Figure 3: A textual data sent to and from the Arduino board.

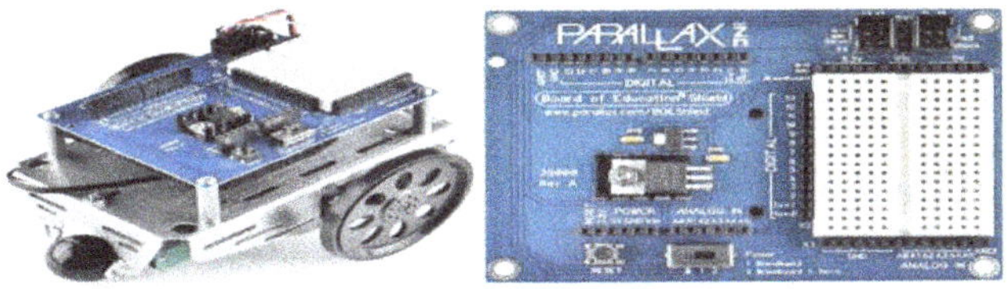

Figure 4: The Parallax Robotics Shield Kit and BOE.

Figure 5: "Wii Remote" one-handed console.

- Learning to program the robot's Arduino Brain
- Calibrating the robot's continuous rotation servo motors
- Using lights and speakers for status indicators
- Assembling the robot
- Preprogrammed navigation
- Using touch-switches to navigate by contact with objects
- Using phototransistors to navigate by light
- Using non-contact infrared sensors to measure distance and avoid or follow objects
- Remote motion control by wireless communication.

The "Wii Remote Plus" as a console for wireless motion control

Motion controllers are used to achieve some desired benefits which can include:

- increased the accuracy of position and speed;
- higher speeds;
- faster time of reaction;
- increased productivity;
- smoother movements;
- integration with other automation;
- integration with other processes;

The "Wii Remote" handle is able to communicate wirelessly with the controller via short-range Bluetooth radio, which permits to operate as far as 10 meters away from the console with up to four controllers. However, to utilize the pointer functionality, the "Wii Remote" must be used within 5 meters [3].

More over, the controller's symmetrical design allows it to be used in either hand and also to use two "Wii Remote" handles in each hand.

The "Wii Remote" handle, shown on Figure 5, represents a one-handed, remote-control based design console, instead of the traditional gamepad controllers.

The handle has the ability to sense acceleration along three axes through the use of an ADXL330 accelerometer (Figure 6). It also features a PixArt optical sensor, allowing it to determine where the console is pointing.

The built into the console BCM2042 microcontroller, shown on Figure 7, includes a large 108 Kb on-chip ROM section for storing firmware. The "Wii Remote" contains a 16 KB EEPROM chip from which a section of 6 kilobytes can be freely read and written by the host.

The "Wii Remote" is a wireless input device, using standard Bluetooth technology and HID protocol to communicate with the host,

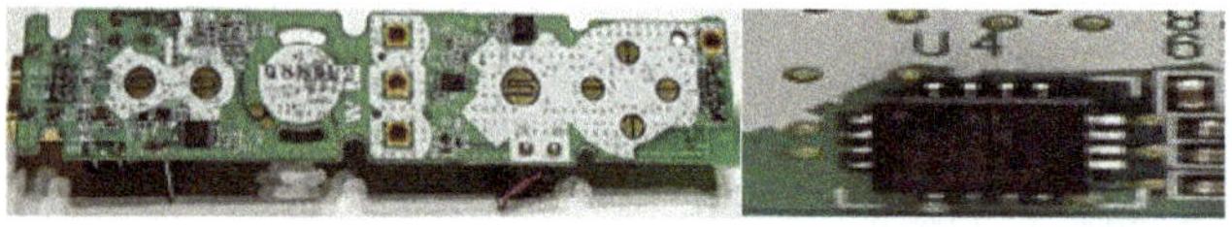

Figure 6: The ADXL330 accelerometer in "Wii Remote".

Figure 7: Wii Remote" with the BCM2042 microcontroller.

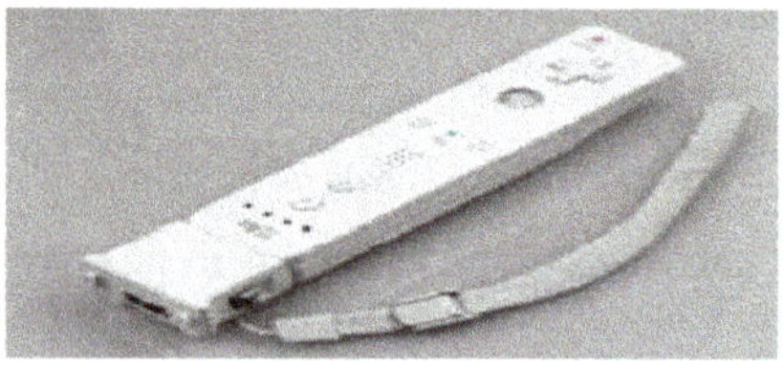

Figure 8: "Wii Remote" with the expansion device "Wii MotionPlus".

Figure 9: The USB Host Shield 2.0 module.

which is directly based upon the USB HID standard.

It is built around a Broadcom BCM2042 Bluetooth System-on-a-chip, and contains multiple peripherals that provide data to it, as well as an expansion port for external add-ons. If the EEPROM chip really contains code for the BCM2042 then this was probably done to make firmware updates possible, so there might be a way of accessing the other parts of the EEPROM via Bluetooth as well.

The "Wii MotionPlus" is an expansion device for the Wii Remote (Figure 8) that allows it to capture complex motion more accurately, as a remote design is fitted perfectly for pointing, and in part to help the console appeal to a broader audience that includes non-gamers.

It incorporates a dual-axis tuning fork gyroscope, and a single-axis gyroscope which can determine rotational motion [4]. The information captured by the angular rate sensor can then be used to distinguish true linear motion from the accelerometer readings.

This allows for the capture of more complex movements than possible with the "Wii Remote" alone. More over, it gives the ability the "Wii Remote" also to be turned horizontally and used like a steering wheel.

The Bluetooth communication between robot and console

The USB Host Shield 2.0 module, shown on Figure 9, is universal connection tool and currently supports the following device classes [5]

- HID devices, such as keyboards, mice, joysticks, etc.
- Game controllers - Sony PS3, Nintendo Wii, Xbox360
- USB to serial converters - FTDI, PL-2303, ACM, as well as certain cell phones and GPS receivers
- ADK-capable Android phones and tables
- Digital cameras - Canon EOS, Powershot, Nikon DSLRs and P&S, as well as generic PTP
- Mass storage devices, such as USB sticks, memory card readers, external hard drives
- Bluetooth dongles.

In order to connect the educational robot Arduino with a wireless controller it was necessary to equip the mobile robot with a wireless communication. Thus, the USB Host Shield 2.0 module with a Bluetooth dongle is used as add-on board for Arduino development platform.

The USB Host Shield 2.0 provides USB Host interface, allowing full and low-speed communication. The board contains Maxim MAX3421E USB host controller, 12MHz crystal, level shifters, resistors, capacitors, Reset button and USB A-type connector. There are also a number of solder pads and jumpers, which are marked with red arrows. The board layout is:

- Power Select
- Power pins
- Analog pins
- GPIN pins
- ICSP connector
- GPOUT pins
- Digital I/O pins 0-7
- Digital I/O pins 8-13
- MAX3421E interface pad
- VBUS power pad

MAX3421E interface pads are used to make shield modifications easier. Pads for SS and INT signals are routed to Arduino pins 10 and 9 via solder jumpers. In case pin is taken by other shield an re-routing is necessary, a trace is cut and corresponding pad is connected with another suitable Arduino I/O ping with a wire. To undo the operation, a wire is removed and jumper is closed. GPX pin is not used and is available on a separate pad to facilitate further expansion. It can be used as a second interrupt pin of MAX3421E.

For activation of USB Host Shield 2.0 module with Bluetooth has to be written communication software and uploaded part of which it is shown on Figure 10. For activation of "Wii Remote" one-handed console with the expansion device "Wii MotionPlus" we need to provide a software initialization procedure where after its fulfillment the "Wii Remote" reached to the return states, represented on Table 1.

For the accelerometer in the expansion device "Wii MotionPlus" we need to provide a software initialization procedure as follows:

```
accXwiiM = ((l2capinbuf[12] << 2) | (l2capinbuf[10] & 0x60 >> 5)) - 500;
accYwiiM = ((l2capinbuf[13] << 2) | (l2capinbuf[11] & 0x20 >> 4)) - 500;
accZwiiM = ((l2capinbuf[14] << 2) | (l2capinbuf[11] & 0x40 >> 5)) - 500;
and for the gyroscopes :
gyroYawRaw = ((l2capinbuf[15] | ((l2capinbuf[18] & 0xFC) << 6)) - gyroYawZero);
gyroRollRaw = ((l2capinbuf[16] | ((l2capinbuf[19] & 0xFC) << 6)) - gyroRollZero);
```

```
.......
} /* fill in setup packet */
setup_pkt.ReqType_u.bmRequestType = bmReqType;
setup_pkt.bRequest = bRequest;
setup_pkt.wVal_u.wValueLo = wValLo;
setup_pkt.wVal_u.wValueHi = wValHi;
setup_pkt.wIndex = wInd;
setup_pkt.wLength = nbytes;
rcode = dispatchPkt(tokSETUP, ep, nak_limit); //dispatch packet
//Serial.println("Setup packet"); //DEBUG
if (rcode) { //return HRSLT if not zero
 Serial.print("Setup packet error: ");
 Serial.print(rcode, HEX);
 return(rcode);
{/* Control transfer with status stage and no data stage */
{ byte rcode;
if (direction)
    { //GET  rcode = dispatchPkt(tokOUTHS, ep, nak_limit);
} else
{  rcode = dispatchPkt(tokINHS, ep, nak_limit);
} return(rcode);
}
/* Control transfer with data Stages 2 and 3 of control transfer. */
{ byte rcode;
if (direction) { //IN transfer
 devtable[addr].epinfo[ep].rcvToggle = bmRCVTOG1;
 rcode = inTransfer(addr, ep, nbytes, dataptr, nak_limit);
 return(rcode);
} else { //OUT transfer
 devtable[addr].epinfo[ep].sndToggle = bmSNDTOG1;
 rcode = outTransfer(addr, ep, nbytes, dataptr, nak_limit);
 return(rcode);
{ .....
```

Figure 10: Example of communication software for activation of USB Host Shield 2.0.

For LEDs	For buttons
0x00, // OFF	0x00008, // UP
0x10, // LED1	0x00002, // RIGHT
0x20, // LED2	0x00004, // DOWN
0x40, // LED3	0x00001, // LEFT
0x80, // LED4	0, // Skip
0x90, // LED5	0x00010, // PLUS
0xA0, // LED6	0x00100, // TWO
0xC0, // LED7	0x00200, // ONE
0xD0, // LED8	0x01000, // MINUS
0xE0, // LED9	0x08000, // HOME
0xF0, // LED10	0x00400, // B
	0x00800, // A

Table 1: Software initialization of "Wii Remote" (return states).

```
gyroPitchRaw = ((l2capinbuf[17] | ((l2capinbuf[20] & 0xFC) << 6)) - gyroPitchZero);
yawGyroSpeed = (double)gyroYawRaw / ((double)gyroYawZero / yawGyroScale);
rollGyroSpeed = -(double)gyroRollRaw / ((double)gyroRollZero / rollGyroScale);
// We invert these values so they will fit the acc values
pitchGyroSpeed = (double)gyroPitchRaw / ((double)gyroPitchZero / pitchGyroScale);
if (!(l2capinbuf[18] & 0x02)) // Check if fast mode is used
yawGyroSpeed *= 4.545;
if (!(l2capinbuf[18] & 0x01)) // Check if fast mode is used
pitchGyroSpeed *= 4.545;
if (!(l2capinbuf[19] & 0x02)) // Check if fast mode is used
rollGyroSpeed *= 4.545;
compPitch = (0.93 * (compPitch + (pitchGyroSpeed * (micros() - timer) / 1000000))) + (0.07 * getWiimotePitch());
```

Conclusion

This working project, nevertheless is still under development and realization, represents the approach and manner for wireless communication by using of Bluetooth connection between human and the educational mobile robot "Audrino" via innovative device - the handle-console "Wii Remote", which enable to provide the so called "master-slave" mode of a distant motion control.

References

1. http://arduino.cc/
2. http://learn.parallax.com/BOEShield
3. http://nintendo.wikia.com/wiki/Wii_Remote
4. http://nintendo.wikia.com/wiki/Wii_MotionPlus
5. http://www.circuitsathome.com

11

A Robotic Path Planner Contender

Kamkarian P[1]* and Hexmoor H[2]

[1]*Department of Electrical and Computer Engineering, Southern Illinois University, Carbondale, USA*
[2]*Department of Computer Science, Southern Illinois University, Carbondale, USA*

Abstract

This article presents a novel offline path planner method to yield a collision-free trajectory among groups of obstacles in a static workspace. It enables a single holonomic point robot acting in a static environment including a fixed initial and goal configurations to achieve its goal toward a collision-free trajectory. In developing our novel path planner, we focused to elevate features that help the planner to route in a wide variety of different situations in regards to lowering the processing time needed for analyzing the workspace and determination of ultimate trajectory. Unlike to some other planners that are able to be applied on some certain obstacle shales such as polygonal, our planner is skillful enough to analyze any types of obstacles, such as circular, spiral, and curved edged obstacles successfully. To increase the performance of our proposed offline path planner, we defined and benefitted from introduction of parameters that help to achieve the best possible results among different scenarios in workspace components arrangements as well as reducing the planner needing to access system resources such as memory or processing unit to analyze the workspace while computing the shortest collision-free path from start point to the goal configuration. The novel planner analyses and transforms the two dimensional representation of the environment into a roadmap consisting a graph of nodes along with all possible routes from the moving robot's initial point into the goal configuration. In order to manage some of the popular problems such as being trapped inside a U-shaped obstacle or routing in narrow pathways among obstacles, that challenge other offline path planners such as Potential Field planner are facing, we used a multi-layer approach in form of different stages to help the planner considering all possible circumstances and hence, compute the best possible route.

Keywords: Path planning; Collision-free trajectory; Rapidly optimizing mapper; Trajectory builder

Introduction

Robotic science is recognized as a powerful tool that the mankind has developed in modern societies to facilitate performing tasks with higher quality and better accuracy rates. Robots are able to perform repetitive tasks tirelessly and without the loss their rate of accuracy because of factors that can cause inaccuracy symptoms contrasted with human performed tasks. Based on constructional and the type of operating tasks, robots are divided into many categories. As a popular category, we cite mobile robots. For a mobile robot, a major activity that has a vital role on operating assigned tasks successfully is the ability to dependably move from the initial to the goal configuration in the workspace. There are many factors involved with the rate of performing a successful movement among obstacles in workspace. For instance, a mobile robot has to have a proper planner to analyze the environment and to build a flawless trajectory in term of safety. The term safety generally refers to an obstacle collision-free path, which has enough distances from every obstacle as well as the workspace boundaries in regards to the accuracy of its sensors and physical actuators to be able to maneuver among obstacles toward goal appropriately. In order to construct a proper path planner, researchers have employed a variety of different methodologies adopted from a wide range of distinctive venues of science. In most cases, developing an ultimate path planner is a long process that is subject to be revised and advanced by different groups of researchers in different research articles. In other terms, when a purely novel path planner has been proposed, its advantages and disadvantages are addressed from the evaluation of applying it on different situations by further research. Our planner is termed rapidly optimizing mapper (ROM). The original contributions of ROM are its efficiency of generating trajectories [1] as well its parametric adjustability [2]. The generated paths are as short as any other planner's output. Our planner can be parametrically adjusted to deal with desired safety and navigation criteria. The addressed issues will be then covered by optimizing or updating the construction of the planner by further research. In the following, we demonstrate a few scenarios of proposed novel path planners along with the processes that further matured them by subsequent research.

Potential Field planner is categorized as a classic and solid path planning method suggested for offline robots. It is constructed based on considering virtual attractive and repulsive forces among components in workspace such as obstacles, start and goal locations along with the workspace boundaries. The fashion that Potential Field algorithm follows is to build trajectory based on deliberating on an attractive influence for the goal configuration as well as repulsive forces from obstacles. In other terms, the goal point has the strongest attraction among other components in workspace which interests robot the most while obstacles based on their shapes, sizes, and locations, have repulsive effects on the robot. Using this strategy helps planner to calculate locus of point in forms of a collision-free trajectory from initial to the goal configuration. The idea of adopting Potential Field for offline robot path planning has promoted by Khatib [3] for the first time. Although adopting the primitive Potential Field method to build trajectories for offline robots has its own advantages, but due to its constraints, it is unable to compute the ultimate trajectory in some certain scenarios. As instances of such conditions we can notice the local minima thatarises

***Corresponding author:** Kamkarian P, Department of Electrical and Computer Engineering, Southern Illinois University, Carbondale, USA
E-mail: pejman@siu.edu

when the robot is located inside a U-Shaped obstacle or is trapped in a closed obstacle. The mentioned cases are addresses by Borenstein and Koren, and Guldner et al. [4,5]. There are several method have proposed to fix the expressed problem by updating the potential field algorithm structure [6,7] as well as adopting other methods such as harmonic functions into the potential field planner [8-10].

As another example, we cite Voronoi path planner. The Voronoi path planner is a method to partition a workspace with the purpose of building a collision-free trajectory and can be applied on environments with both local [11,12] and global knowledge [13] specifications. Trajectory safety is one of the salient key features of employing Voronoi path planner. Considering partitions as the robot pathways crossing from the middle distances of obstacles leads obtaining the maximum path to be used by robot and hence, causing a decent space for robots with poorly detectors or maneuvers skills to reach their goals successfully. Using Voronoi path planner method however, has its own constraints. For instance, in case of building the trajectory in form of a piecewise linear path, the robot, in order to follow the path, has to continuously proceed with the stop and start process and hence spending more time, energy, and effort to reach the goal. Several solutions have been proposed to overcome and improve the mentioned issue such as reconstructing the trajectory by replacing the sharp corners with different types of curves, such as using Bezier curves [11,14-16] or benefiting the technique of connecting vertices with splines [17-19]. As a more advanced strategy to treat the problem of generating sharp corners through the Voronoi method path planning, Grefenstette and Schultz [20] suggested a programming algorithm that dynamically smoothness the piecewise linear trajectories from Voronoi planner.

As the third category of path planners, we consider the randomized path planners. The randomized path planner approaches are generally built based on constructing a set of randomly generated graph of available configurations with the purpose of assessing the workspace in terms of examining different conditions of pathways toward the goal configuration. As the examples of earlier randomized approaches, we can mention probabilistic roadmap method [2,21] and randomized potential field solution [22]. In the probabilistic roadmap path planner method a graph forms from a set of free points randomly distributed in the workspace area at the beginning. The planner then attempts to build the trajectory by considering connecting pairs of nearby configurations from initial to the goal. The Probabilistic Roadmap approach is strong enough to route the trajectory in steerable non-holonomic [23], or holonomic systems, with no problems. It is however, not suitable to be used to build the trajectory in an environment using non-linier controller robots or sophisticated non-holonomic environments. Employing the randomized Potential Field solution is also a challenging task when facing with choosing a proper heuristic potential function in workspaces with several obstacles as well as dynamic and kinematic differential constraints.

Confronted with different variety in workspace objects arrangements such as obstacles and moving robots along with the specifications of each along with different constraints that may cause reduction of the regular path planner's performances, led researchers to consider developing more advanced path planners in form of hybrid path planners. Employing hybrid path planners [24-30] as promising approaches resulted in achievement of better and more reliable trajectories. A hybrid path planner typically benefits from a combination of more than one path planner inherits their advantages in a unique algorithm. Hybrid path planners have usually better performance than each individual path planner. This is because they are typically a collection of other path planners and hence possess more accuracy and possibility to generate the trajectory in a variety of different scenarios successfully.

This paper proposes a novel offline path planner; i.e., ROM, for single point robots. The planner is able to build a collision-free trajectory from initial to the goal configuration. The planner benefits from a multi-layer algorithm which consists of multiple methods with the sole purpose of computing the shortest and the safest possible path for the robot. Since the term safety will vary from a situation to another situation and depends on many factors such as the robot sensors accuracy and its ability to maneuver among obstacles, we considered using many variables that are adjustable based on both point robot physical specifications and the components of the workspace situations, such as the type and locations of obstacles. This paper aims to present our novel path planner in more details. In order to evaluate our planner performances, we will explore it by applying it on special scenarios that usually cause problems on the accuracy of building trajectories for other offline path planners. In addition, this research article has furnished results from applying our planner to more offline path planners. In the next section, we present the mentioned variables along with discussion of the usage and the range of each. The later sections are dedicated to the illustration of the path planner followed by the experimental results from applying it to variety of different workspaces. In order to validate the performance of our novel path planner, in the later section, we have compared it to a few other offline path planners. More detailed reports of these comparisons appear in recent reports [1,31]. Our offline planner can be used by an online planner by repeated applications of our algorithm. However, an online planner is not concerned with the entire route but rather it is seeking best decisions as it navigates. An online planner could be guided by an offline trajectory as generated by our method as a high level plan and can make minor local deviations as its navigation criteria dictates. The latter deviations are similar to local plans in contrast to overall trajectories that are global plans. Although our offline planner is inspired by online planners but it makes no claims about exceeding them.

Definitions and Key Parameters for ROM

In order to fabricate our planner, we considered and defined a few new concepts. These concepts are used in different parts of the planner algorithm construction. In this section we explain each concept in detail. The planner different phases and sections along with the usage of these concepts will be discussed in the next section. Our novel path planner consists of the following key concepts: standoff distance (SD), roadblock obstacle (RO), side edge node, degree of traverse (DT), degree of surface traversal (DST), node visibility, visible pathways, and isolated node. We introduce each of these concepts consequently:

Standoff distance (SD)

The standoff distance is a scalar value that will be adjusted based on the safe boundaries around obstacles that robot is able to maneuver without involving near misses. Moving robots, depending on their usage and the type of environment that they are built to react, are equipped with different types of sensors. The majority tasks for these sensors are to collect and analyze visual data from the robot surroundings. These data will be analyzed with the purpose of recognizing objects around the robot. Robot benefits the processed vision data to make proper decisions to adjust its path to avoid collisions. Moreover, robots, based on the moving equipment, have a variety maneuver skills. In other terms, when faced with an obstacle, robots need to maintain a proper distance from the obstacle to adjust their path and hence, turn around the obstacle successfully. The standoff distance defines the safe width of the pathway around obstacles that robot adjusts its path when reaching it. Depending on the robot maneuvering skills, the SD value can be different. In the other terms, the predetermined standoff distance will be considered a lower value for a robot that has a quicker response and reaction to change its path. The standoff distance, however, needs to be considered with a higher value for robots that depending on their construction need wider safe channel around obstacles. Equation 1 is the mathematical relations for the mentioned concern.

$$\forall (r_1, r_2) \in R, if\ \Delta t_1 < \Delta t_2\ then\ SD_1 < SD_2 \quad (1)$$

In the equation 1, r_1 and r_2 are two arbitrary robots in the robot domain, *R*, and Δt_1 , and Δt_2 indicate the reaction time needed for the first and the second robot respectively. SD_1 and SD_2 are also standoff distances for robot 1, and robot 2 respectively.

Roadblock obstacle

One of the main tasks for the planner to analyze the environment is to recognize roadblock obstacles. The planner benefits from the technique of considering virtual straight lines from the robot starting position to the goal configuration to build trajectory. Obstacles that have at least one collision point with the explained virtual straight line are added to the set of roadblock obstacles. The process of analyzing workspace and recognizing obstacles will be explained in details in the next section. Figure 1 shows a sample workspace along with the determined roadblock obstacles.

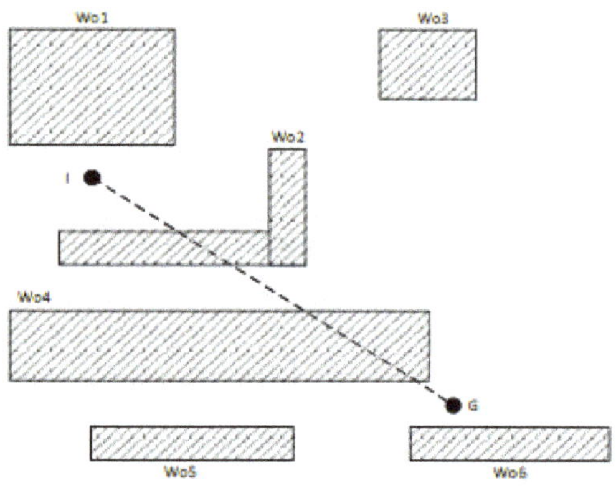

Figure 1: Showcase of roadblock obstacles in a sample workspace.

As it shown in the Figure 1, obstacles 2 and 4 are considered in the set of roadblock obstacles. This is because the virtual straight line from initial point, I, through the goal configuration, G, has intersected with the mentioned obstacles. Equation 2 illustrates the process of determining roadblock obstacles in math formula:

$$\forall w_{o_i} \in W_O, if\ (w_{o_i} \cap IG) \notin \phi\ then\ w_{o_i} \in RO \quad (2)$$

Where O_i represents the ith obstacle and *RO* is the group of roadblock obstacles.

Side edge node

The side edge node is considered as either departure point for roadblock obstacles, or the location that the path builder is able to route a trajectory through the goal configuration. Depending on the workspace constraints as well as roadblock specifications, the location of considering side edge nodes will vary.

Figure 2 shows side edge nodes, n_1 through n_4 for obstacle, w_{o_1} are located in a sample workspace.

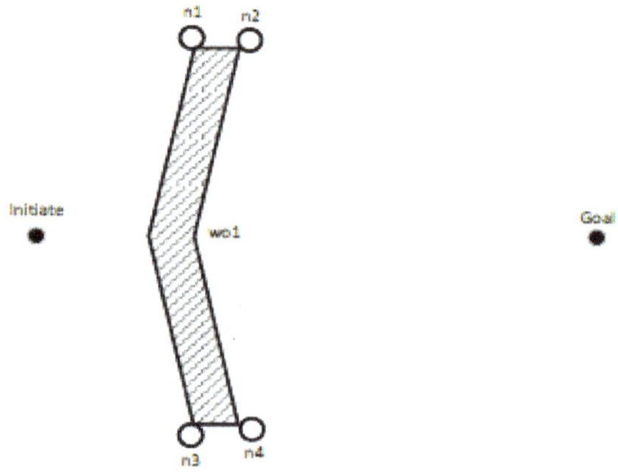

Figure 2: A sample showcase for side edge nodes.

Degree of Traverse (DT)

In order to build the proper collision avoidance trajectory, the path planner analyzes the surface of the roadblock obstacles. The Degree of Traverse is determined based on the sensitivity of the roadblock obstacle surface scanner. The DT value has a direct relation to the final trajectory smoothness. The smaller value for Degree of Traverse, the more smooth and accurate collision-free trajectory. Figure 3 demonstrates a sample DT value for a workspace.

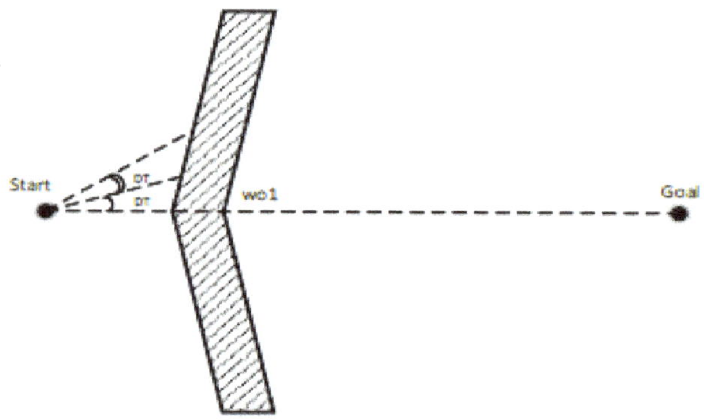

Figure 3: Illustrating DT value for a sample workspace.

Degree of surface traversal (DST)

In certain situations, to reach departure points, a robot needs to perform obstacle surface traverse. This is due to having limitations on either workspace constraints or obstacle specifications (Figure 4).

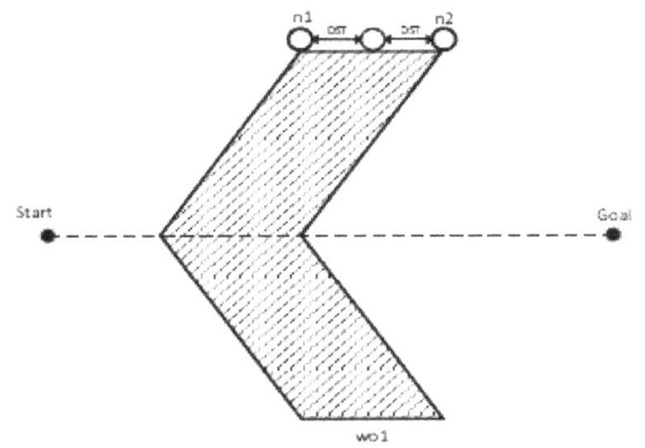

Figure 4: Surface traversing for a sample obstacle in a showcase workspace.

The Degree of Surface Traversal will be determined based on the accuracy of locus for the path crossing from the surface of obstacles. The DST value should be considered based on the robot skills on changing its direction and adjusting its path. A robot with a lower speed and reaction to change its trajectory will have a higher value of DST compared to the robot with a faster processing unit and path adjustment abilities.

Node visibility

Node visibility is a key concept that the planner uses to shorten the trajectory. This feature brings the planner the ability to achieve the best possible result in optimal collision avoidance path, in terms of the trajectory length from the initial to the goal configuration. For a group of nodes that are located on the same obstacle, pairs of nodes that share no intersection with any obstacles through straight rays are considered as visible nodes. In other terms, nodes that are located on the surface of obstacles and are able to see each other without any obstacles in between are categorized as visible nodes. Equation 3 commits the Node visibility concept to a mathematical formula.

$$\begin{aligned}&\forall(w_{o_i} \in W_O)\wedge(\mathrm{n}_j,\mathrm{n}_k \in w_{o_i}),\\&if\left(\left(n_j \cap \textstyle\bigcup_{v=1}^{l} w_{o_v} = \phi\right)\wedge\left(n_k \cap \textstyle\bigcup_{v=1}^{l} w_{o_v} = \phi\right)\right)\\&then(n_j,n_k \in VN)\end{aligned} \tag{3}$$

Where n_j and n_k are nodes located on the obstacle w_{o_i} and VN is the set of visible nodes.

Visible pathway

As one of the most important key features for the planner trajectory optimizer to determine the shortest possible paths toward goal, the Visibility pathways are defined based on the trajectories that connect both ends of a group of visible nodes. Figure 5 demonstrate the Visibility pathway concept.

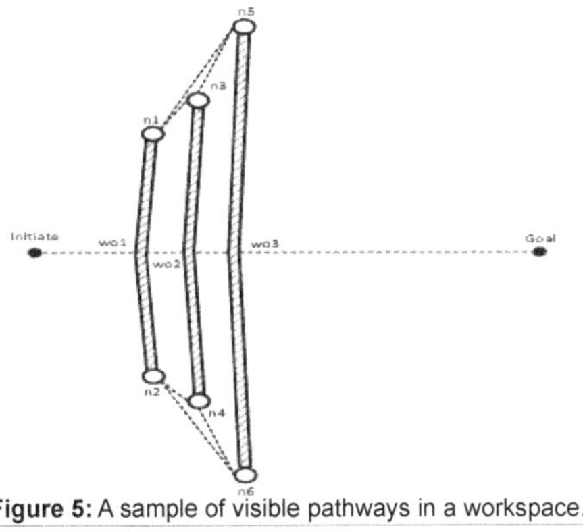

Figure 5: A sample of visible pathways in a workspace.

As it indicated in Figure 5, nodes 3 and 5 are visible from node 1. Nodes 4 and 6 are also visible from node 2. Therefore, connections from node 1 through node 5 and also node 2 through node 6 are considered as visible pathways.

In order to clarify the Visibility pathway concept, it will be illustrated through the following mathematical equation 4.

$$\begin{aligned}&\forall(n_i,n_j,n_k,\in N),\ if\\&\left(\begin{array}{l}\left(\left((\mathrm{n}_i n_j \cap \textstyle\bigcup_{v=1}^{s} w_{o_v}\right)=\phi\right)\\ \wedge\left((n_j,n_k \cap \textstyle\bigcup_{v=1}^{s} w_{o_v})=\phi\right)\\ \wedge\left(((n_i,n_k \cap \textstyle\bigcup_{v=1}^{s} w_{o_v})=\phi\right)\end{array}\right)\\&then(n_i n_k \in VP)\end{aligned} \tag{4}$$

Where n_i, n_j *and* n_k are side edge nodes and VP is the set of visible pathways.

Isolated node

In order to find the optimal trajectory in terms of the shortest collision-free path, our planner considers all possible paths available around each roadblock obstacle. In order to achieve this goal, the planner analyzes each roadblock obstacle side edge nodes to assure there exists a pathway from both sides of roadblock obstacle which connects side edge nodes to one another. Any side edge nodes located on the same roadblock obstacle that is not connected to other nodes from either side of obstacle is considered as isolated node. As an important task for the side edge node analyzer unit of our path planner, it processes any isolated node with connecting them to one another, if possible.

Our algorithm enjoys a rapid computational running time. It examines obstacles once in finding relevant obstacles and then again in graph construction. Graph processing also uses the most computationally efficient algorithm. Therefore, this polynomial time algorithm runs faster than previous planners. Although it handles obstacles that are more complex than polygons, it suffer from computational drawbacks.

The Rom Path Planner Process

Through this section, we introduce our novel path planner construction in detail. The process of optimal trajectory calculation consists of four different and distinct phases as follows: Initial phase, Workspace analyzer, Graph (Roadmap) builder, and shortest path computation unit. The optimal trajectory refers to the shortest collision-free path from start to the goal configuration. Each phase, along with its related sections and steps will be discussed next.

Initial phase

At the beginning of the planning operation, the path planner uses parametric values that are adjusted based on the workspace and robot specifications. The planner uses the locations of the start point and goal configuration, along with the Standoff Distance, Degree of Travers and Degree of Surface Traversal values at the initial phase. These values guide the planner to route the best possible collision avoidance trajectory from start to the goal configuration.

Workspace analyser

This unit of the planner performs two major tasks. Those are analyzing the workspace to determine roadblock obstacles, and roadblock obstacle surface scanning with the purpose of calculating side edge nodes. In order to determine roadblock obstacles, the planner considers a virtual ray in form of a straight line from the initial point into the goal configuration. The initial point will be considered as the start point at the beginning of the workspace analysis. The initial point will be however, considered as the either latest side edge nodes, a hit point, or a leave (i.e., departure) points. The process of determining roadblock obstacles along with determining side edge nodes continues until the planner reaches the goal configuration.

The second step of workspace analyzer unit consists of computing side edge nodes. In order to achieve this goal, the planner starts analyzing the surface of each roadblock obstacles, starting from both sides of each hit point resulting from intersecting the straight ray from the initial point toward goal configuration, for the size of DT. The process of scanningroadblock surface continues until either of the following criteria fulfilled.

a. The analyzer reaches to the leave point;

b. The virtual straight ray intersects any other obstacles.

Graph (Roadmap) builder: In order to find the optimal trajectory, the planner forms a lattice including side edge nodes along with hit and leave points obtained in the workspace analyzer unit. The following steps will be performed to build the graph:

a. All side edge nodes along with the start and goal configurations as well as any hit and leave points belonging to the same roadblock obstacles are joined together;

b. The graph (roadmap) builder removes isolated nodes by connecting them together, if possible;

c. Visible nodes belonging to the same roadblock obstacles will be recognized and visible pathways form based on them;

d. In order to achieve shorter pathways, all nodes located in between of each visible pathways will be removed;

e. Each edge participating to form the graph will be labeled based on Euclidean distances.

Figure 6 is the resultant of applying our path planner on the figure 5 along with the optimal path calculations in terms of the shortest length as well as determining a near miss avoidance trajectory, through the visibility pathways simplifications.

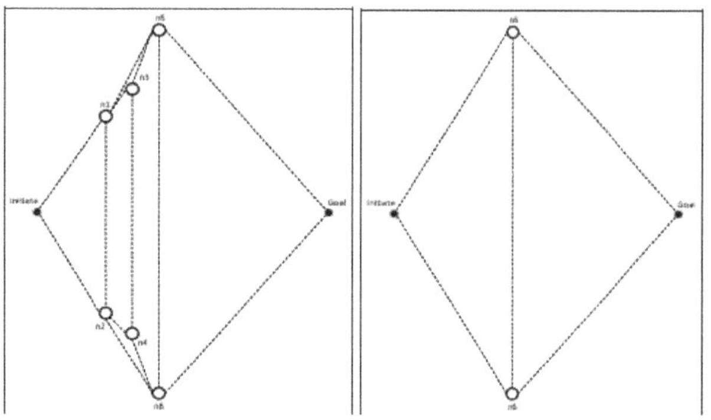

Figure 6: Left: the primitive graph consisting all possible pathways; Right: the optimal graph resulting from removing visible nodes from the graph.

As it shown in the figure 6, (left), the graph (roadmap) builder considers all available paths to form a complete lattice of all possible paths from each point. The graph (roadmap) builder unit will then simplify the graph by removing any visible nodes located among two sides of each visible pathway, as it illustrated in the figure 6, (right).

Shortest path computation unit: The main task of this phase of the planner is to analyze the graph constructed in the graph (roadmap) builder unit with the intention of refining the optimal trajectory from start to the goal configuration. In order to achieve the best possible result, we benefitted using the Dijkstra shortest path algorithm. Having start and goal nodes along with Euclidean distances for each edge of the graph enables Dijkstra algorithm to analyze and hence, determine the shortest trajectory as the optimal output.

Experimental Setup and Results

In order to validate the performance of our planner ROM, we have applied it on many different scenarios. For illustration, we considered five workspaces with having various constrains and limitations for each scenario. Each workspace has the size of 500 by 500 units and a start and goal configurations are considered as small circular points with the equally 10 points in diameters. We aim to evaluate the accuracy and skills of our planner to analyze the workspace and hence, compute the optimal trajectory. These scenarios along with the specification of each as well as the results of applying our path planner on them are categorized and illustrated in the following case studies. This paper focuses on demonstrating our algorithm on workspaces that are challenging for well-established exiting planners. We illustrate that our method overcomes those challenges. Early results show that our method produces the shortest paths [1]. More direct comparisons will be reported in subsequent papers.

Case study 1

For the first case study, we considered a crowded workspace with several obstacles in various shapes, sizes and locations. The aim of employing mentioned configuration is to put the path planner in an intense effort to build the ideal trajectory with the purpose of analyzing its performance along with its accuracy in finding the best possible collision-free route toward goal. We located the start point in the location of 50 by 50. The goal configuration is also considered at the position of 360 by 460 of the workspace Figure 7 shows the workspace arrangement that we considered for the first case study:

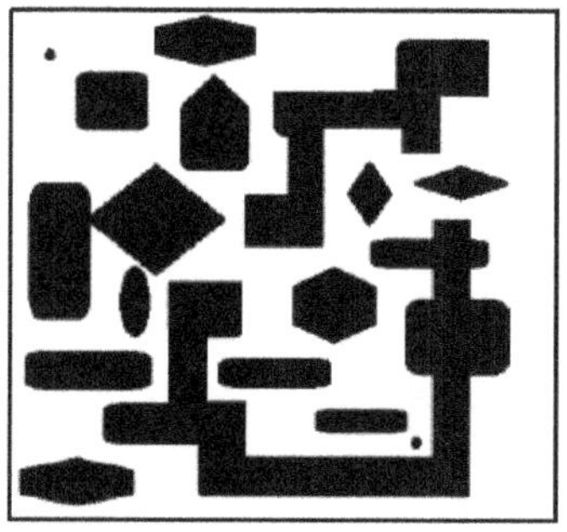

Figure 7: The workspace of the first scenario.

Figure 8 shows the result of building the optimal collision-free trajectory obtained from applying our path planner on the first workspace shown in the Figure 7.

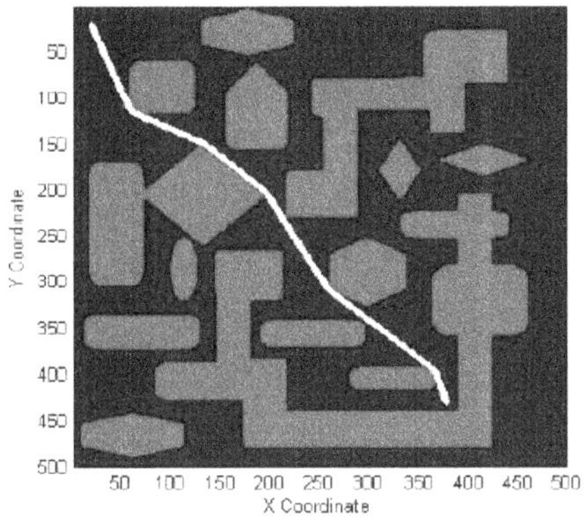

Figure 8: The trajectory resultant from applying our path planner on the first workspace.

As it illustrated in the Figure 8, our path planner was able to build a collision-free trajectory among various obstacles in the first workspace successfully. The distance trajectory from start to the goal configuration is calculated equaled 350 in Euclidean measurement system. We observed many minor and major roadblock obstacles surface scanning along with several trajectory adjustments which is the resultant of operation over a relatively large numbers of obstacles side edge nodes along with the roadblock surface nodes in form a complex lattice of nodes along with all possible joint paths among them.

Case study 2

As it indicated in the earlier sections, our path planner preforms a roadblock obstacle surface scanning when located to the hit point until reaching to a valid leave point. We used the following workspace to evaluate our path planner surface scanning skills. We located the start point at the 330 by 160 and the goal configuration at the 110 by 410 coordinates of the workspace. Generally, a surface scan will be enforced when either start or goal or both of them located inside a spiral or rounded obstacles (Figure 9).

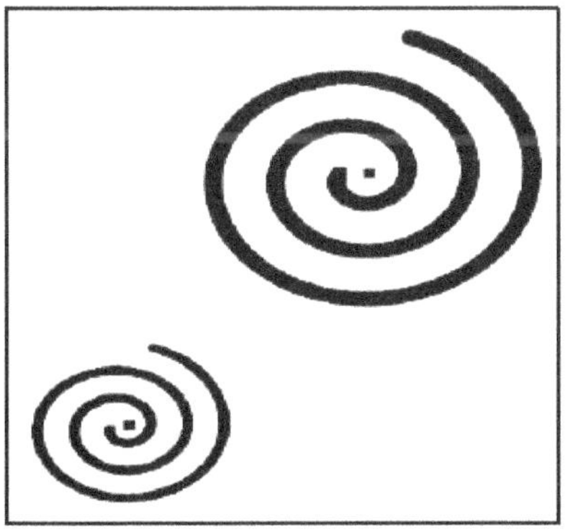

Figure 9: The workspace of the third case study.

As we expected to observe in this scenario, the majority portion of the optimal trajectory is located on the surface of the only roadblock obstacle in the workspace. This is because of the particular shape of the obstacle which prevents the planner to reach to a proper leave point toward the goal and hence, performs a long surface scanning throughout a continuously movement on the surface of the roadblock obstacle, as indicated in Figure 10.

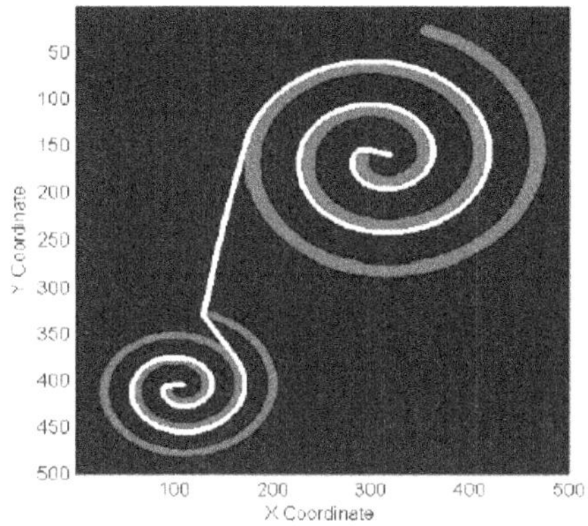

Figure 10: The resultant trajectory obtained from applying the path planner on the third workspace.

The distance of the collision-free trajectory computed by our path planner is equal 1569 units. The majority of the trajectory is located on the surface of the spiral roadblock obstacle except two zones of path from start point to the roadblock obstacle and the only leave point from roadblock obstacle toward goal configuration. This scenario revealed the strength of our path planner skills to compute and build the optimal trajectory in any types of workspaces including spiral obstacles without performance reductions.

Case study 3

In this scenario, we considered the start location inside a U-shaped obstacle while the goal configuration is located in the other side of the obstacle. This arrangement leads the planner evaluation skills in terms of building the shortest collision-free trajectory which is routed partially in opposite direction of the goal configuration. The start location is at the 190 by 230 and the goal configuration is in 450 by 210 coordinates of the workspace. Figure 11 shows the obstacle along with the start and goal points located in the fourth scenario:

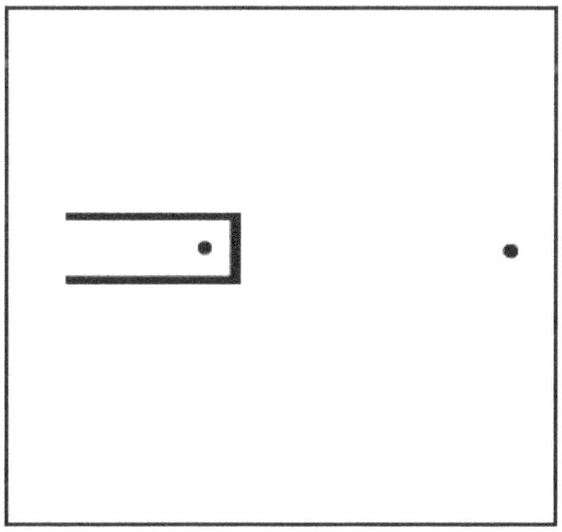

Figure 11: The workspace belongs to the fourth case study.

In this scenario, our path planner calculated the distance of 519 Euclidean for the optimal trajectory as illustrated in Figure 12.

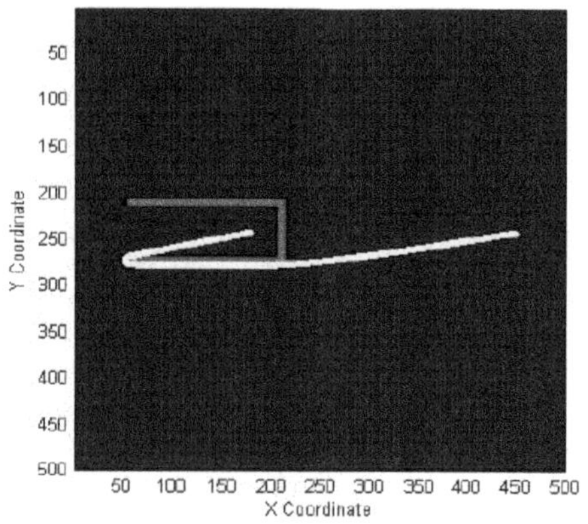

Figure 12: The optimal trajectory obtained from applying our path planner on the forth workspace configuration.

The results concluded from applying our planner on this workspace arrangement reveals that the planner is able to flawlessly handles and hence, draws a collision-free trajectory in situations that either start or goal configuration is located inside a partially closed obstacles along with the presents of various objects that are intensively located close to one another.

Case study 4

Figure 13 shows the workspace that we considered for this case study. This scenario is formed based on a maze shaped arrangement including several edges and walls in different directions. Placing the start point on top of the obstacle and goal configuration inside of the obstacle at the bottom side of it makes the planner to use extensive attempts to continuously analyzing and adjusting the optimal collision-free trajectory. The start and goal points for this scenario is considered to be located at the 150 by 50 and 370 by 470 coordinates of the workspace respectively.

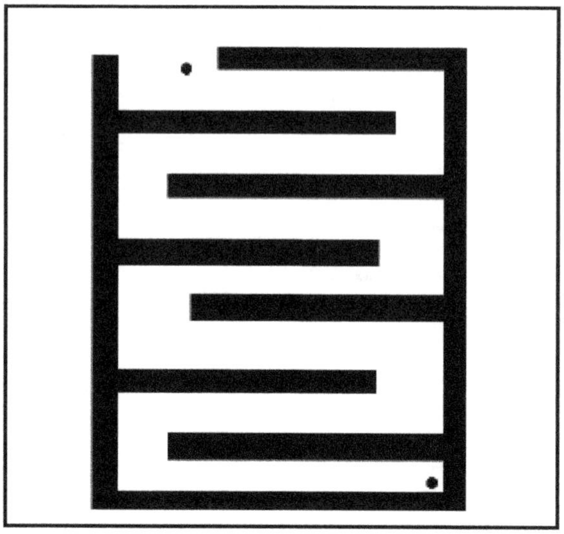

Figure 13: The workspace of the fifth case study.

The resultant of applying our path planner on the last considered scenario is illustrated in Figure 14.

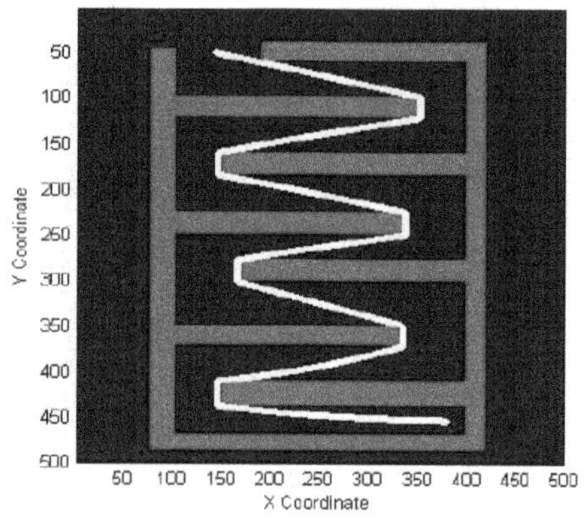

Figure 14: The trajectory resulting from applying our path planner on the fifth case study workspace.

The optimal collision-free trajectory distance that is built by our planner is equal 1568 units. Similar other scenario, our planner was able to successfully analyzing, computing and handling the situation and hence, concluded appropriate solution.

Conclusions

This research article introduced Rapidly Optimizing Mapper (ROM) as a novel approach for single point robots. In order to achieve the best possible result in terms of shortest safe trajectory, the planner algorithm benefits from a combination of many optimization strategies in different layers which enable it to be applied on a variety of different workspaces having different scenarios of start and goal configurations along with obstacle arrangements and shapes. The speed of handling workspaces for generating trajectories is one of our original contributions. The other major contribution is flexibility of handling complex environments using key parameters. These are reported in our most recent papers [1,31]. Despite many other offline path planners that are involved with problems in certain situations and obstacle shapes such as polygonal obstacles, we have observed that ROM is able to plan the trajectory on a more severely limited scenarios, successfully. For example, as it illustrated, it is able to be used to process any types of spiral edged obstacles flawlessly. It is able to detect any possible situations and hence, circumstances on the workspace and determines the best possible route appropriately. As instances of such a situations we can consider the local minima problems that the Potential Field planner is dealing, while ROM is able to successfully overcome to the mentioned situations. ROM analyzes the environment in order to find roadblock side edge nodes, it further transforms the workspace into a graph of roadblock side edge nodes along with recognizing and optimizing isolated nodes and eventually benefits of using Dijkstra's shortest path algorithm to refine the optimal path. The optimal path refers to the shortest path which allows the robot to perform a flawless transition in form of a near miss avoidance movement from initial point to the goal configuration.

References

1. Kamkarian P, Hexmoor H (2015) Efficiency Considerations of an Offline Mobile Robot Path Planner. The 17th International Conference on Artificial Intelligence, (ICAI'15): 35-40.
2. Amato NM, Wu Y (1996) A Randomized Roadmap Method for Path and Manipulation Planning. IEEE International Conference on Robotics and Automation 4: 113-120.

3. Khatib O (1985) Real-Time Obstacle Avoidance for Manipulators and Mobile Robots. IEEE International Conference on Robotics and Automation, St Louis Missouri, pp. 500-505.

4. Borenstein J, Koren Y (1991) The Vector Field Histogram-fast Obstacle Avoidance for Mobile Robots. IEEE Transactions on Robotics and Automation 7: 278-288.

5. Grefenstette J, Schultz AC (1994) An Evolutionary Approach to Learning in Robots. Proceedings of the Machine Learning Workshop on Robot Learning International Conference on Robot Learning. New Brunswick, NJ, pp. 65-72.

6. ArámbulaCosío F, Padilla Castañeda MA (2004) Autonomous Robot Navigation using Adaptive Potential Fields. Mathematical and Computer Modelling 40: 1141-1156.

7. Ge SS, Cui YJ (2000) New Potential Functions for Mobile Robot Path Planning. IEEE Transactions on Robotics and Automation 16: 615-620.

8. Connolly CI, Burns JB, Weiss R (1990) Path Planning using Laplace's Equation. Proceedings of the IEEE Conference On Robotics and Automation, pp. 2102-2106.

9. Utkin VI, Drakunov S, Hashimoto H, Harashima F (1991) Robot Path Obstacle Avoidance Control via Sliding Mode Approach. Proceedings of the IEEE/RSJ International Workshop on Intelligent Robots and Systems, Osaka, Japan, pp. 1287-1290.

10. Guldner J, Utkin V, Hashimoto H (1997) Robot Obstacle Avoidance in N-Dimensional Space using Planar Harmonic Artificial Potential Fields. Journal of Dynamic Systems Measurement and Control 119: 160-166.

11. Guechi E, Lauber J, Dambrine M (2008) On-line Moving-Obstacle Avoidance using Piecewise Bezier Curves with Unknown Obstacle Trajectory. 16th Mediterranean Conference on Control and Automation, pp. 505-510.

12. Mohammadi S, Hazar N (2009) A Voronoi-Based Reactive Approach for Mobile Robot Navigation. Advances in Computer Science and Engineering Springer Berlin Heidelberg 6: 901-904.

13. Bhattacharya P, Grvrilova ML (2007) Voronoi Diagram in Optimal Path Planning. 4th IEEE International Symposium on Voronoi Diagrams in Science and Engineering.

14. Choi J-W, Curry RE, Elkaim GH (2009) Obstacle Avoiding Real-Time Trajectory Generation and Control of Omnidirectional Vehicles. American Control Conference.

15. Hwang J-H, Arkin RC, Kwon D-S (2003) Mobile Robots at Your Fingertip: Bezier Curve On-line Trajectory Generation for Supervisory Control. IEEE International Conference on Intelligent Robots and Systems 2: 1444-1449.

16. Skrjanc I, Klanˇcar G (2007) Cooperative Collision Avoidance between Multiple Robots Based on B´ezier Curves. 29th International Conference on Information Technology Interfaces.

17. Costa TAA, Ferreira AM, Dutra MS (2007) Parametric Trajectory Generation for Mobile Robots. 19th International Congress of Mechanical Engineering 3: 300-307.

18. Eren H, Fung CC, Evans J (1999) Implementation of the Spline Method for Mobile Robot Path Control. 16th IEEE In-strumentation and Measurement Technology Conference 2: 739-744.

19. Magid E, Keren D, Rivlin E, Yavneh I (2006) Spline-Based Robot Navigation. International Conference on Intelilgent Robots and System.

20. Ho Y-J, Liu J-S (2009) Collision-free Curvature-bounded Smooth Path Planning using Composite Bezier Curve Based on Voronoi Diagram. IEEE International Symposium on Computational Intelligence in Robotics and Automation (CIRA).

21. Kavaraki LE, Svestka P, Latombe J-C, Overmars MH (1996) Probabilistic Roadmaps for Path Planning in High Dimensional Configuration Spaces. IEEE Transactions on Robotics and Automation 12: 566-580.

22. Barraquand J, Latombe J-C (1991) Robot Motion Planning: A Distributed Representation Approach. International Journal of Robotics Research 10: 628-649.

23. Laumond JP, Sekhavat S, Lamiraux F (1998) Guid-lines in Nonholonomic Motion Planning for Mobile Robots. In: Laumond (ed.) Robot Motion Planning and Control. Springer-Verlag, Berlin, pp. 1-53.

24. Abinaya S, Hemanth Kumar V, SrinivasaKarthic P, Tamilselvi D, Mercy Shalinie S (2014) Hybrid Genetic Algorithm Approach for Mobile Robot Path Planning. Journal of Advances in Natural and Applied Sciences 8: 41-47.

25. Bashra K, Oleiwi R, Hubert R (2014) Modified Genetic Algorithm Based on A* Algorithm of Multi Objective Optimization for Path Planning. Journal of Automation and Control Engineering 2: 357-362.

26. Chaari I, Koubaa A, Trigui S, Bennaceur H, Ammar A, et al. (2014) SmartPATH: An Efficient Hybrid ACO-GA Algorithm for Solving the Global Path Planning Problem of Mobile Robots. International Jounal of Advanced Robotics Systems: 1-15.

27. Ju M, Wang S, Guo J (2014) Path Planning using a Hybrid Evolutionary Algorithm Based on Tree Structure Encoding. The Scientific World Journal 2014: 1-8.

28. Loo CK, Rajeswari M, Wong EK, Rao MVC (2004) Mobile Robot Path Planning using Hybrid Genetic Algorithm and Traversability Vectors Method. Intelligent Automation and Soft Computing 10: 51-64.

29. McFetridge L, Ibrahim MY (1998) New Technique of Mobile Robot Navigation using a Hybrid Adaptive Fuzzy-Potential Field Approach. Computers and Industrial Engineering 35: 471-474.

30. Yao Z, Ren Z (2014) Path Planning for Coalmine Rescue Robot based on Hybrid Adaptive Artificial Fish Swarm Algorithm. International Journal of Control and Automation 7: 1-12.

31. Kamkarian P, Hexmoor H (2015) A Novel Offline Path Planning Method. The 17th International Conference on Artificial Intelligence (ICAI'15): 10-15.

12

Motion Analysis of Hexapod Robot with Eccentric Wheels

Yujun Wang[1], Can Fang[1] and Qimi Jiang[1,2*]

[1]*School of Computer and Information Science, Southwest University, Chongqing, China*
[2]*Comau Inc, MI, USA*

Abstract

A new concept for developing hexapod robots using eccentric wheels is proposed in this work. Compared with the RHex robot, the proposed hexapod robot can greatly reduce the bumping of the robot body in both smooth ground and rocky terrain. Also, the developed hexapod robot possesses significant advantages over those with common circular wheels in traversing rocky and uneven terrain. Also, the control of the proposed hexapod robot is simple because each eccentric wheel has only one degree of freedom. This work focuses on the kinematics analysis of the proposed hexapod robot. Two types of gaits respectively named as the alternating tripod gait and the hexapod gait are proposed. With the alternating tripod gait, the robot can move continuously. But the hexapod gait is helpful in overcoming the resistant torque caused by the weight on the eccentric wheels. Besides, the effect of the eccentricity on the motion of the robot is analysed. The proposed hexapod robot can be used to detect the gas in underground mines.

Keywords: Hexapod robot; Eccentric wheel; Gas detection; Tripod gait; Hexapod gait; Motion analysis

Introduction

A hexapod robot is a robot that has six legs to walk or move. It is well known that a robot can be statically stable on three or more legs. Since a hexapod robot has several legs, it has a great deal of flexibility in how it can move. If some legs become disabled, the robot may still be able to walk. Furthermore, not all of the robot's legs are needed for stability; other legs are free to reach new foot placements or manipulate a payload. Also, the robot is easily programmed to move around because it can be configured to many types of gaits. There are various designs of hexapod robots with certain functions and advantages. For instance, hexapod robots have been sketched in eight different designs, and every de- sign has its different criteria, specifications, shapes, advantages and disadvantages [1].

Many hexapod robots are biologically inspired by hexapoda locomotion [2-10]. Hexapods may be used to test biological theories about insect locomotion, motor control, and neurobiology. Hexapod designs vary in leg arrangement. Insect inspired robots are typically laterally symmetric. Typically, individual legs range from two to six degrees of freedom. Hexapod feet are typically pointed, but can also be tipped with adhesive material to help climbing walls or wheels so the robot can drive quickly when the ground is flat. The researchers [6] developed a sixlegged walking robot that is capable of basic mobility tasks such as walking forward, backward, rotating in place and raising or lowering the body height. The authors described the first implementation of the neurobiological mechanisms in a physical hexapod robot that is capable of generating adaptive stepping actions with the same underlying control method as an insect [7].

Hexapod robots can be developed to undertake some interesting tasks. For instance, a kind of six-legged robot was developed by Ramesh and Kumar to monitor and perform household tasks independently.

The robot legs movement or method of forward motion using legs is called gait. Most often, a hexapod robot is controlled by gaits, which allow the robot to move forward, turn, and perhaps side-step. One important issue in the development of hexapod robots is to consider the motion and develop proper gaits for the robots [10-22]. For instance, feasible gait patterns were developed by Belter and Skrzypczynski [10] to control a real hexapod walking robot. The use of hybrid genetic gravitational algorithm for generation of the gait for the hexapod robot was described by Seljanko [12].

Gaits for a hexapod robot are often stable, even in slightly rocky and uneven terrain. There are varieties of gaits available. Some of the most common gaits are as follows: (1) alternating tripod gait; (2) quadruped; and (3) crawl. The famous gait used by hexapod robot is the tripod gait. In the tripod gait, there are always three feet of hexapod in contact with the ground. For example, the hexapod robots developed, were used this alternating tripod gait [17-20].

Motion of a hexapod robot may also be nongaited, which means the sequence of leg motions is not fixed, but rather chosen by the computer in response to the sensed environment. This may be most helpful in very rocky terrain, but existing techniques for motion planning are computationally expensive.

As mentioned by Rashid et al. [1], hexapod robots can be designed with certain functions and advantages. Most hexapod robots were designed with only legs. Some hexapod robots were designed with manoeuvrable wheels or combination of legs and wheels [5]. In general, the movement of robot by legs is good for rocky and uneven terrain. But the movement of robot by the wheels is faster than the movement of the robot by legs [1]. The RHex robot could be an exception because the motion of its six C shape legs is similar to the rotation of wheels [2-4].

The RHex robot has a lot of excellent performance such as (1) running on reasonably flat, natural terrain at speeds up to 6 body lengths per second (just over 2.7 m/s); (2) climbing a wide range of stairs and slopes up to 45 degrees; (3) traversing obstacles as high as

***Corresponding author:** Jiang Q, Comau Inc., MI, USA. and Southwest University, Chongqing, China. E-mail: Qimi_J@Yahoo.com

20 cm (about twice RHex's leg clearance) and badly broken terrain with large rocks or obstacles; (4) walking and running upside down; (5) flipping itself over to recover nominal body orientation; (6) leaping across ditches up to 30 cm wide.

However, the RHex robot is not suitable for working in underground mines to detect gas. The reasons are: (1) Its violent bumping and collision of its body with the rocky terrain will be easy to cause sparks leading to explosion in detecting the gas in underground mines. (2) In such an application, the power of every motor is quite limited (only 3 watts). The motors used for the RHex robot cannot be so small. Otherwise, its powerful performance will be greatly compromised. (3) The gas detective robot has a sensor on top of a stick about 1.5 meters high, which stretches up from the robot body. So, walking and running upside down, flipping and leaping are not allowed even if these are good performance of the RHex robot.

In order to develop such a gas detective robot working in underground mines, a new concept for developing hexapod robots using eccentric wheels is proposed in this work. Compared with the RHex robot, the proposed hexapod robot can greatly reduce the bumping of the robot body in both smooth ground and rocky terrain. Also, the developed hexapod robot possesses significant advantages over those with common circular wheels in traversing rocky and uneven terrain. Also, the control of the proposed hexapod robot is simple because each eccentric wheel has only one degree of freedom.

This work focuses on the kinematics analysis of the proposed hexapod robot. In Section 2, the prototype of the proposed hexapot robot with eccentric wheels is briefly introduced. In Section 3, the tripot gait for both straight walking and turning is proposed. In Section 4, the hexapot gait is described. In Section 5, the motion analysis of the proposed hexapot robot is conducted. In Section 6, the advantages of the proposed hexapot robot over both RHex robot and those with common circular wheels are given in several situations. Finally in Section 7, some conclusions and discussions are summarized.

Hexapod Robot with Eccentric Wheels

As shown in Figure 1, the developed hexapod robot with eccentric wheels is traversing an obstacle. This robot is actually symmetric in three directions. First, the robot is symmetric about the central vertical section plane in the moving direction: three eccentric wheels (#1, #2 and #3) at the left- hand side and three eccentric wheels (#4, #5 and #6) at the right-hand side. Second, the robot is symmetric about the vertical section plane passing through the axes of the middle eccentric wheels (#2 and #5). Third, the robot is also symmetric about the horizontal section plane passing through the axes of all six eccentric wheels. In other words, the upper and down parts of the robot are also symmetric about the horizontal section plane passing through the axes of all six eccentric wheels. Hence, the robot is able to walk even if it is overturned. In terms of structure, the developed hexapod is quite similar to the RHex robot in which open curved legs were used.

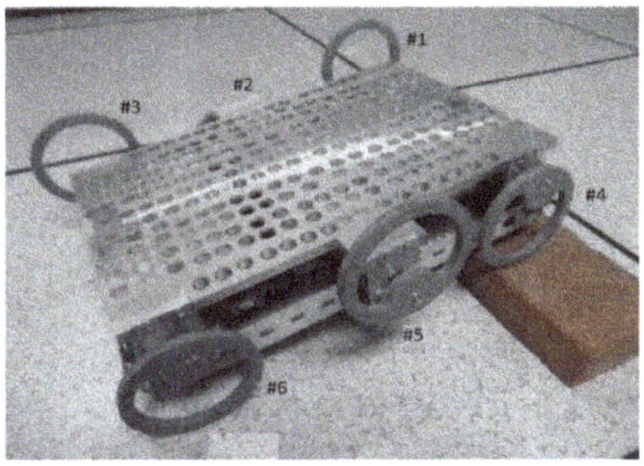

Figure 1: The prototype of hexapod robot with eccentric wheels.

All six eccentric wheels have the same geometric parameters. Every eccentric wheel is driven independently by its own motor. In other words, the axes of wheels #1 and #4 (or wheels #2 and #5, or wheels #3 and #6) are concentric, but not the same. The backward motion of the robot is similar to the forward motion. Every eccentric wheel as shown in Figure 1 does not touch the ground all the time during the walking of the robot. The motion status of the eccentric wheel falls into two phases: (1) supporting phase, and (2) idling phase, as shown in Figure 2, where O is the center of the eccentric wheel; Q is the axis of the eccentric wheel (i.e., the revolute joint); P_1 is the first contact point of the supporting phase during the forward walking; P_2 is the contact point on the diameter passing through the axis Q of the wheel; P_3 is the last contact point of the supporting phase during the forward walking. In other words, the supporting phase starts at the first contact point P_1 and ends at the last contact point P_3. Then, the idling phase starts after the contact point P_3 and ends when the eccentric wheel contacts the ground at P_1 again. Obviously, P_1 and P_3 are two transition contact points.

Tripod Gait

Straight walking

Referring to Figure 1, the six eccentric wheeled legs of the developed hexapod robot can be classified into two groups: Group A consists of wheels (#1, #3 and #5) which form a triangle. Group B consists of wheels (#2, #4 and #6) which form another triangle. The alternating tripod gait for the developed hexapod robot can be shown in Figure 3 and described as follows:

• Status 1 (initial status): All six eccentric wheels contact the ground at their transition points. The three wheels (#1, #3 and #5) of Group A contact the ground at contact point P1. But the three wheels (#2, #4 and #6) of Group B contact the ground at P_3.

• Status 2: The three wheels (#1, #3 and #5) of Group A enter their supporting phase. But the three wheels (#2, #4 and #6) of Group B enter their idling phase. The rotation speed for the supporting phase is ω_s and the rotation speed for the idling phase is ω_i.

• Status 3: This is a transition status. Again, all six eccentric wheels contact the ground at their transition points. The three wheels (#1, #3 and #5) of Group A contact the ground at P3. But the three wheels (#2, #4 and #6) of Group B contact the ground at P1.

• Status 4: The three wheels (#1, #3 and #5) of Group A enter their idling phase. But the three wheels (#2, #4 and #6) of Group B enter their supporting phase.

• Status 5: Again, all six eccentric wheels contact the ground at their transition points. Just as Status 1, the three wheels (#1, #3 and #5) of Group A contact the ground at P1. But the three wheels (#2, #4 and #6) of Group B contact the ground at P_3.

As Status 5 is exactly the same as Status 1, the total number of statuses in one cycle of the tripod gait for straight walking is four: two moving statuses and two transition statuses.

Turning

It is not enough for a robot to just move back and forth along a

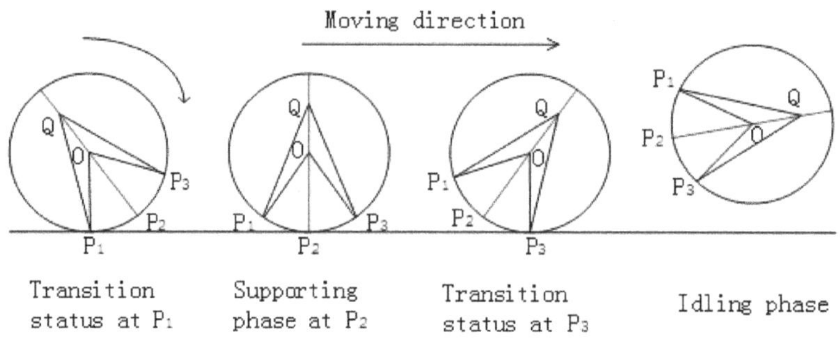

Figure 2: Status of the eccentric wheel during the forward walking of robot.

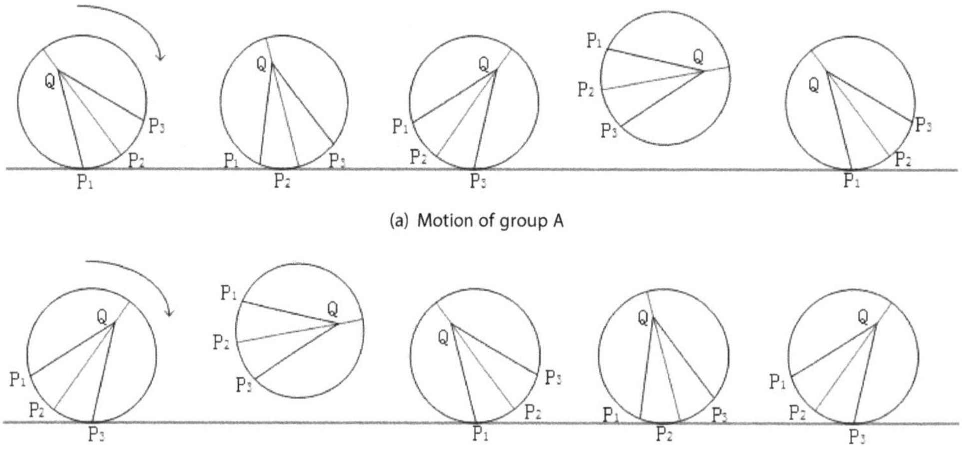

(a) Motion of group A

(b) Motion of group B

Figure 3: The alternating tripod gait for straight walking.

straight line if it is required to undertake a specific task. In other words, the robot should have the ability to change its moving direction. Taking the left turn as an example, the alternating tripod gait can be shown in Figure 4 and described as follows:

- Status 1 (initial status): All six eccentric wheels contact the ground at their transition points. The four end wheels (#1, #3, #4 and #6) contact the ground at P_3. But the two middle wheels (#2 and #5) contact the ground at P_1.

- Status 2: The three wheels (#1, #3 and #5) of Group A enter their supporting phase. The rotation speed of two end wheels (#1 and #3) at the left-hand side is $-\omega_s$. Their contact points start from P_3 and pass the shorter circular arc $\widehat{P_3P_1}$ of the wheels before reaching P_1. As a result, the left-hand side of the robot obtains a backward speed. But the rotation speed of the middle wheel (#5) at the right-hand side is ωs. Its contact point starts from P_1 and passes the shorter circular arc $\widehat{P_1P_3}$ of the wheel before reaching P_3. As a result, the right-hand side of the robot obtains a forward speed. As the left-hand side of the robot obtains a backward speed and the right-hand side of the robot obtains a forward speed, the robot is turning left.

In the meantime, the three wheels (#2, #4 and #6) of Group B enter their idling phase. The rotation speed of the middle wheel (#2) at the left-hand side is $-\omega_i$. It starts from P1 and passes the longer circular arc $\widehat{P_1P_3}$ of the wheel before reaching P_3. But the rotation speed of two end wheels (#4 and #6) at the right-hand side is ω_i. They start from P_3 and pass the longer circular arc $\widehat{P_3P_1}$ of the wheels before reaching P_1.

- Status 3: This is a transition status. Again, all six eccentric wheels contact the ground at their transition points. But now, the four end wheels (#1, #3, #4 and #6) contact the ground at P_1. The two middle wheels (#2 and #5) contact the ground at P_3.

- Status 4: The three wheels (#1, #3, and #5) of Group A enter their idling phase. The rotation speed of two end wheels (#1 and #3) at the left-hand side is $-\omega_i$.

They start from P1 and pass the longer circular arc $\widehat{P_1P_3}$ of the wheels before reaching P_3. But the rotation speed of the middle wheel (#5) at the right-hand side is ω_i. It starts from P3 and passes the longer circular arc $\widehat{P_3P_1}$ of the wheel before reaching P_1.

In the meantime, the three wheels (#2, #4, and #6) of Group B enter their supporting phase. The rotation speed of the middle wheel (#2) at the left-hand side is $-\omega_s$. Its contact point starts from P_3 and passes the shorter circular arc $\widehat{P_3P_1}$ of the wheel before reaching P_1. As a result, the left-hand side of the robot obtains a backward speed. But the rotation speed of two end wheels (#4 and #6) at the right-hand side is ω_s. Their contact points start from P_1 and pass the shorter circular arc $\widehat{P_1P_3}$ of the wheels before reaching P_3. As a result, the right-hand side of the robot obtains a forward speed. As the left-hand side of the robot obtains a backward speed and the right-hand side of the robot obtains a forward speed, the robot is turning left.

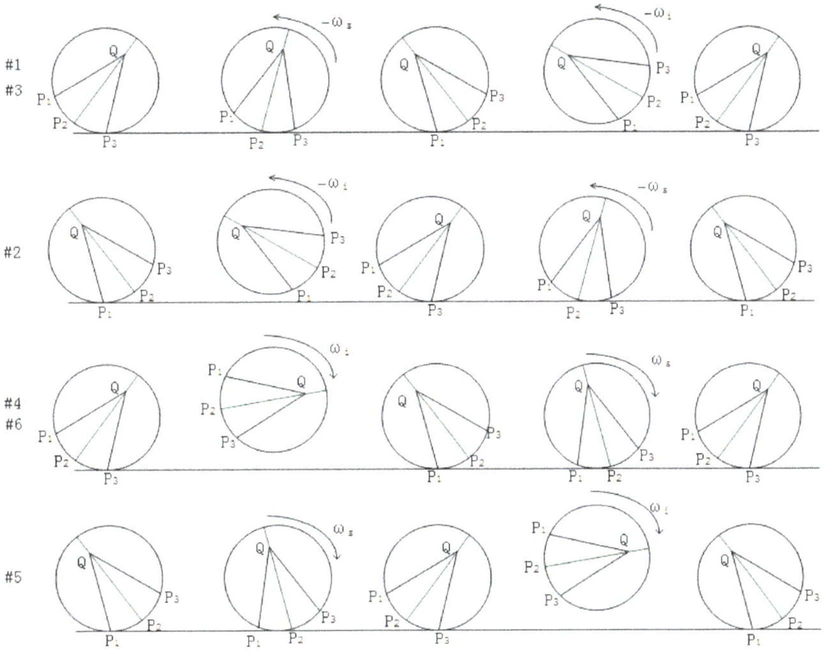

Figure 4: The tripod gait for left turn.

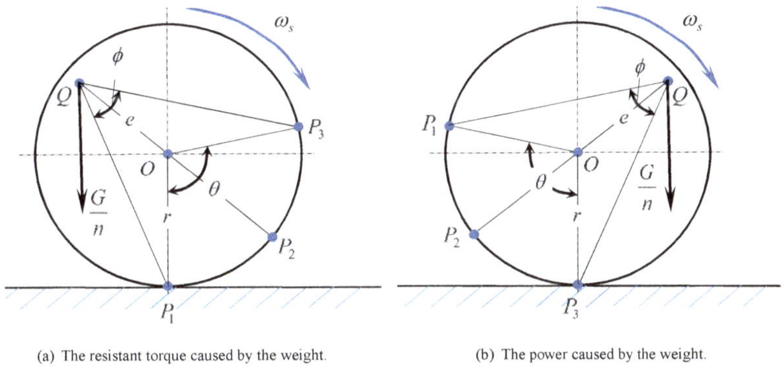

(a) The resistant torque caused by the weight. (b) The power caused by the weight.

Figure 5: The torque caused by the weight.

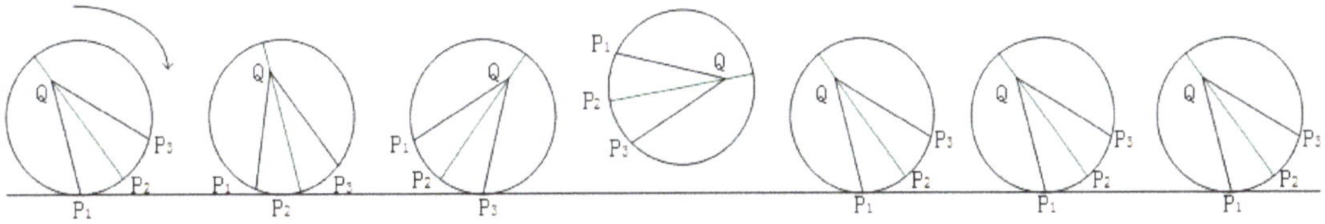

(a) Motion of group A

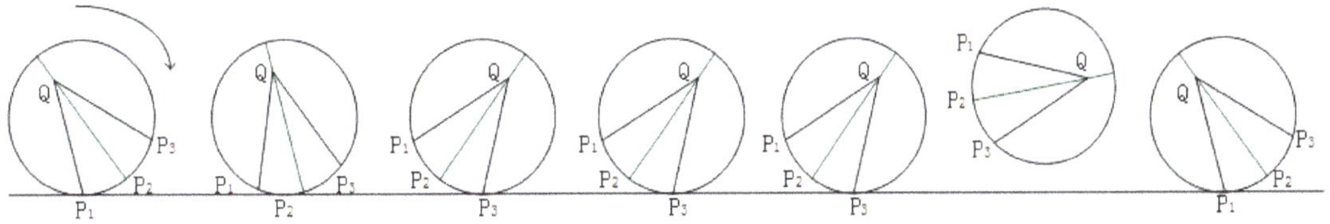

(b) Motion of group B

Figure 6: The hexapod gait for straight walking.

• Status 5: Again, all six eccentric wheels contact the ground at their transition points. Just as Status 1, the four end wheels (#1, #3, #4 and #6) contact the ground at P_3. But the two middle wheels (#2, #5) contact the ground at P_1.

As Status 5 is exactly the same as Status 1, the total number of statuses in one cycle of the tripod gait for turning is also four: two moving statuses and two transition statuses.

It can be seen that for the left turn, no matter in supporting phase or idling phase, the rotation speed of each wheel at the left-hand side of the robot is negative. But the rotation speed of each wheel at the right-hand side of the robot is positive. It can be easily to deduce that for the right turn, the rotation speed of each wheel at the left-hand side of the robot will be positive. But the rotation speed of each wheel at the right-hand side of the robot will be negative.

There are some similarity and differences between the gait for straight walking and the gait for turning. Similarity: In each case, when the three wheels of Group A form a triangle at their supporting phase, the three wheels of Group B also form another triangle at their idling phase. After that, when the three wheels of Group A form a triangle at their idling phase, the three wheels of Group B also form another triangle at their supporting phase. This is why both gaits are called as alternating tripod gaits.

Difference: In the gait for straight walking, the three wheels of each group stand at the same transition contact point (P_1 or P_3) at each transition status. The rotation speed of every wheel at the moving statuses is always positive. But in the gait for turning, the three wheels of the same group do not stand at the same transition contact point at the transition statuses. For instance, if the wheels (#1 and #3) at the left-hand side of the robot stand at P_1, the wheel (#5) at the right-hand side stands at P3. Also, the sign of the rotation speed of the wheels at the left-hand side of the robot is different from the wheels at the right-hand side.

Hexapod Gait

As mentioned in the introduction, the proposed hexapod robot can be used as the mining robot for detecting the gas in underground mines. Hence, the power of the motors is quite small. Otherwise, it would be easy to cause sparks that could lead to explosion. On the other hand, the weight of the robot causes a resistant torque on the eccentric wheel as shown in Figure 5a because of the eccentricity. (Of course, when the axis Q lies at the other side of the vertical diameter through the center O, the torque caused by the weight becomes a kind of power source as shown in Figure 5b. The n in Figure 5 is the number of wheels contacting the ground at the same time. For the tripot gait, n=3. In order to reduce this resistant torque caused by the weight, one solution is to increase the number n of the wheels contacting the ground. If we can make all six wheels to contact the ground at the same time (i.e., n=6), the resistant torque caused by the weight on the wheel will decrease by half. To this end, the hexapod gait is proposed in this section.

Straight walking

The hexapod gait for the developed hexapod robot can be shown in Figure 6 and described as follows:

• Status 1 (initial status): All six eccentric wheels contact the ground at their transition points P_1.

• Status 2: All six wheels enter their supporting phase. This is a moving status.

• Status 3: All six eccentric wheels contact the ground at their transition points P_3. This is a transition status.

• Status 4: The three wheels (#1, #3 and #5) of Group A enter their idling phase. But the three wheels (#2, #4 and #6) of Group B keep their poses at their transition points P_3. This is an adjusting status.

• Status 5: The three wheels (#1, #3 and #5) of Group.

A contact the ground at their transition points P_1. But the three wheels (#2, #4 and #6) of Group B still keep their poses at their transition points P_3. This is a transition status.

• Status 6: The three wheels (#1, #3 and #5) of Group A keep their poses at their transition points P_1. But the three wheels (#2, #4 and #6) of Group B enter their idling phase. This is an adjusting status.

• Status 7: All six eccentric wheels contact the ground at their transition points P_1. This is a transition status, which is exactly the same as Status 1.

As Status 7 is exactly the same as Status 1, the total number of statuses in one cycle of the hexapod gait for straight walking is six: one moving status, two adjusting statuses and three transition statuses.

Turning

Similarly, the hexapod gait for turning also falls into several statuses. Taking the left turn as an example, the hexapod gait can be shown in Figure 7 and described as follows:

• Status 1 (initial status): All six eccentric wheels contact the ground at their transition points. The three wheels (#1, #2 and #3) at the left-hand side contact the ground at P_3. But the three wheels (#4, #5 and #6) at the right-hand side contact the ground at P_1.

• Status 2: All six wheels enter their supporting phase. The rotation speed for the three wheels (#1, #2 and #3) at the left-hand side is $-\omega_s$. So, the left-hand side of the robot obtains a backward speed. But the rotation speed for the three wheels (#4, #5 and #6) at the right-hand side is ω_s. So, the right-hand side of the robot obtains a forward speed. As the left-hand side obtains a backward speed and the right-hand side obtains a forward speed, the robot turns to the left. Hence, this status is a turning status.

• Status 3: All six eccentric wheels contact the ground at their transition points. The three wheels (#1, #2 and #3) at the left-hand side contact the ground at P_1. But the three wheels (#4, #5 and #6) at the right-hand side contact the ground at P_3. This is a transition status.

• Status 4: The three wheels (#1, #3 and #5) of Group.

A enter their idling phase. The rotation speed of the two wheels (#1 and #3) at the left-hand side is $-\omega_i$.

But the rotation speed of the wheel (#5) at the right- hand side is ω_i. The three wheels (#2, #4 and #6) of Group B keep their poses. This is an adjusting status.

• Status 5: All six eccentric wheels contact the ground at their transition points. The four end wheels (#1, #3, #4 and #6) contact the ground at P_3. But the two middle wheels (#2 and #5) contact the ground at P_1. This is a transition status.

• Status 6: The three wheels (#2, #4 and #6) of Group.

B enter their idling phase. The rotation speed of the wheel (#2) at

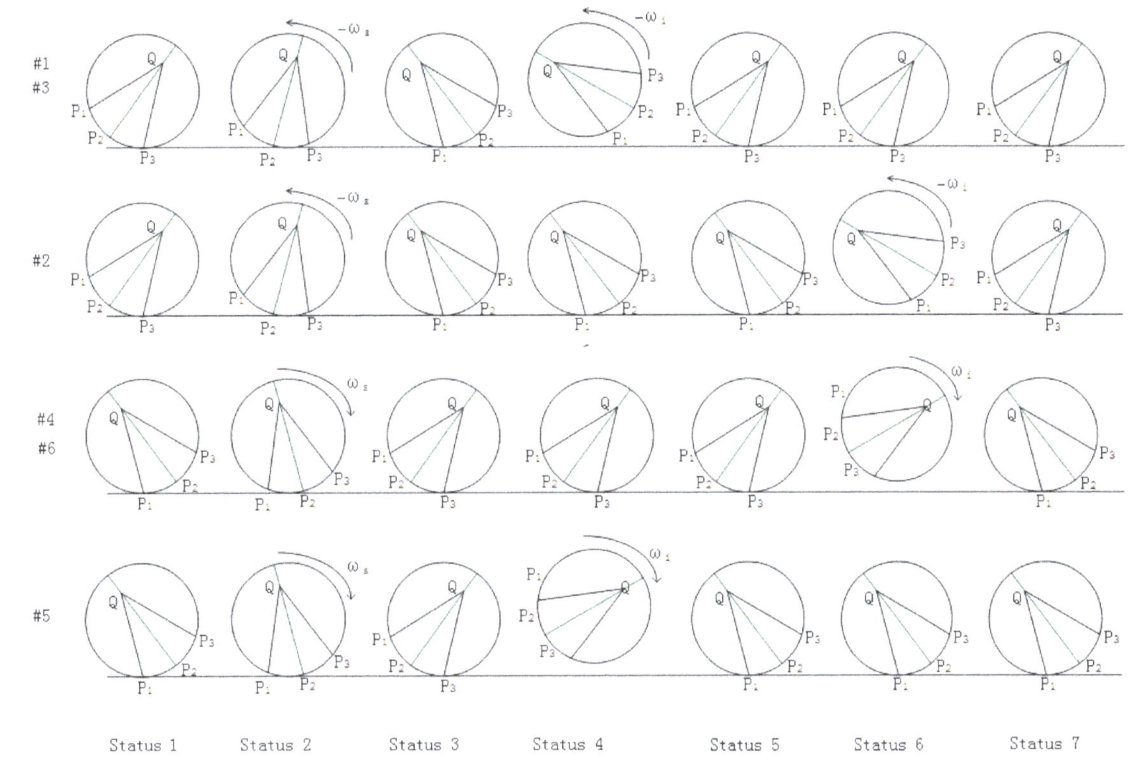

Figure 7: The hexapod gait for left turn.

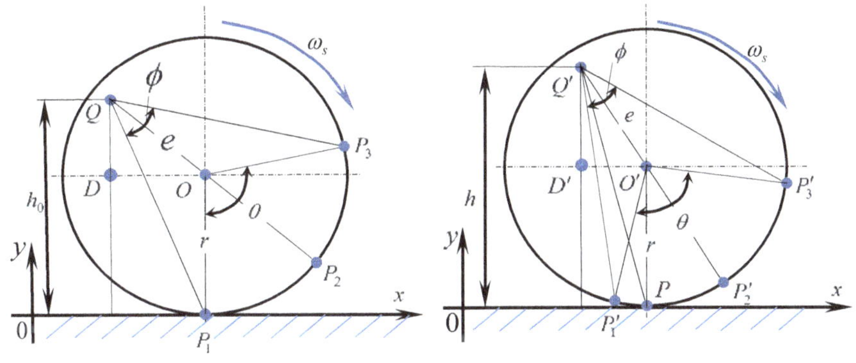

Figure 8: Motion analysis.

the left-hand side is $-\omega_t$. But the rotation speed of the wheels (#4 and #6) at the right-hand side is ω_t. The three wheels (#1, #3 and #5) of Group A keep their poses. This is an adjusting status.

• Status 7: All six eccentric wheels contact the ground at their transition points. The three wheels (#1, #2 and #3) at the left-hand side contact the ground at P_3. But the three wheels (#4, #5 and #6) at the right-hand side contact the ground at P_1. This is a transition status, which is exactly the same as Status 1.

Motion Analysis

Determination of geometric parameters

Referring to Figure 8, the central angle for the supporting phase is θ, the angle between P_1Q and P_3Q is φ, i.e., P1QP3= φ. Hence, the displacement of the rotation axis Q of the eccentric wheel can be given as φ. In the above sections, we mentioned two transition contact points P_1 and P_3. But how to determine these two transition contact points? To do this, we need to determine some geometric parameters. First, one basic requirement for the robot is that the robot should keep the minimum height so that the bottom of its body cannot touch the ground. In other words, the minimum height h0 of the rotation axis Q of the eccentric wheel should be given at the design stage. Referring to Figure 8a, we get

$$|OD| = \sqrt{|OQ|^2 - |QD|^2} = \sqrt{e^2 - (h_0 - r)^2} \tag{1}$$

where r and e are respectively the radius and eccentricity of the eccentric wheel. So,

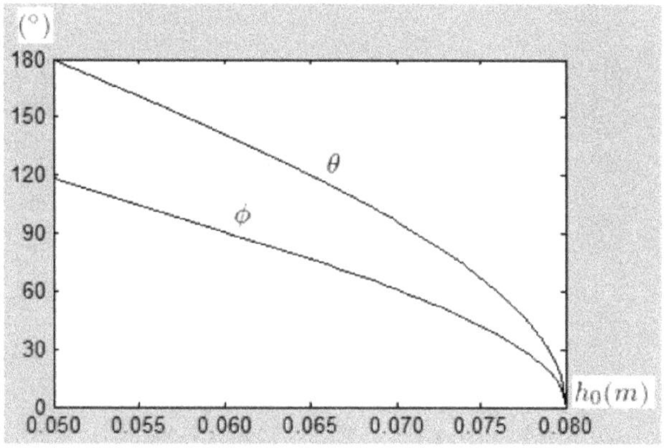

Figure 9: The variation of **φ** and **θ** with respect to h_0 for given e = 0:030 m.

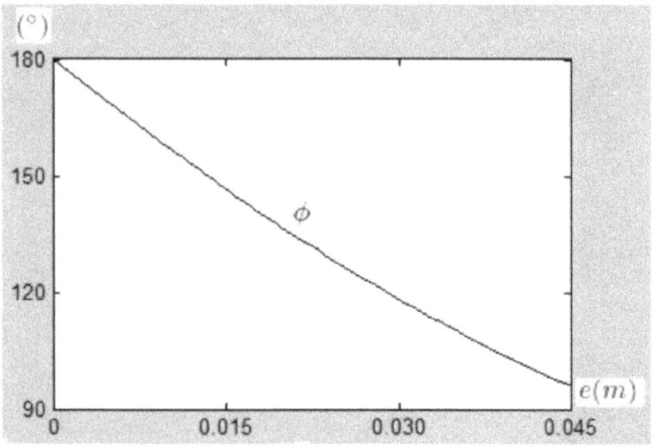

Figure 10: The variation of with respect to e for h_0 = r.

$$\angle OQD=\tan^{-1}\frac{|OD|}{|QD|}=\tan^{-1}\frac{\sqrt{e^2-(h_0-r)^2}}{h_0-r}. \tag{2}$$

$$\angle P_1QD=\tan^{-1}\frac{|OD|}{h_0}=\tan^{-1}\frac{\sqrt{e^2-(h_0-r)^2}}{h_0-r}. \tag{3}$$

From Equation (2) and (3), we get

$$\begin{aligned}\phi &= \angle P_1QP_3 = 2\angle P_1QP_2 = 2(\angle OQD-\angle P_1QD)\\ &=2\left(\tan^{-1}\frac{\sqrt{e^2-(h_0-r)^2}}{h_0-r}-\tan^{-1}\frac{\sqrt{e^2-(h_0-r)^2}}{h_0}\right).\end{aligned} \tag{4}$$

$$\theta=\angle P_1OP_3=2\angle P_1OP_2=2\tan^{-1}\frac{\sqrt{e^2-(h_0-r)^2}}{h_0-r}. \tag{5}$$

Motion status of Q

Suppose that the eccentric wheel contacts the ground at P_1 at the initial moment (t=0) as shown in Figure 8a. It will contact the ground at point p at the moment t as shown in Figure 8b from which we get

$$\begin{aligned}|O'D'| &= |O'Q'|\sin\angle O'Q'D' = e\sin\angle P_2'O'P\\ &= e\sin(\theta/2-\omega_s t).\end{aligned} \tag{6}$$

$$\begin{aligned}|Q'D'| &= |O'Q'|\cos\angle O'Q'D' = e\cos\angle P_2'O'P\\ &= e\cos(\theta/2-\omega_s t).\end{aligned} \tag{7}$$

The displacement of the center O of the eccentric wheel is

$$x_O = r\angle P_1'O'P = r\omega_s t,\ y_O = r, \tag{8}$$

Hence, the displacement of the rotation axis Q of the eccentric wheel can be given as

$$\begin{aligned}x &= x_O + (|OD|-|O'D'|)\\ &= r\omega_s t+\sqrt{e^2-(h_0-r)^2}-e\sin(\theta/2-\omega_s t),\\ y &= y_O+|Q'D'| = r+e\cos(\theta/2-\omega_s t)\end{aligned} \tag{9}$$

Differentiating Eq.(9) with respect to time t, the velocity of Q can be given as

$$\begin{aligned}v_x &= \dot{x} = r\omega_s + e\omega_s\cos(\theta/2-\omega_s t)\\ v_y &= \dot{y} = e\omega_s\sin(\theta/2-\omega_s t).\end{aligned} \tag{10}$$

Differentiating Eq.(10) with respect to time t, the acceleration of Q can be given as

$$\begin{aligned}a_x &= \ddot{x} = e\omega_s^2\sin(\theta/2-\omega_s t)\\ a_y &= \ddot{y} = -e\omega_s^2\cos(\theta/2-\omega_s t).\end{aligned} \tag{11}$$

Motion status of the robot

For straight walking, no matter which gait (the tripot gait or the hexapod gait) is used, all wheels in the supporting status have the same motion status. Hence, the motion status of the robot body is exactly the same as the motion of the rotation axis Q given by Equation (9) - (11).

For turning, the displacement, velocity and acceleration of the robot body in the vertical direction are respectively given as yQ, $\dot{y}$Q, $\ddot{y}$Q which are the same as the straight walking. But in the horizontal direction, considering the symmetric structure of the robots, the turning (rotation in the horizontal plane) speed and acceleration can be given as

$$\begin{aligned}\omega &= 2v_x/W_r\\ \alpha &= \dot{\omega} = 2a_x/W_r\end{aligned} \tag{12}$$

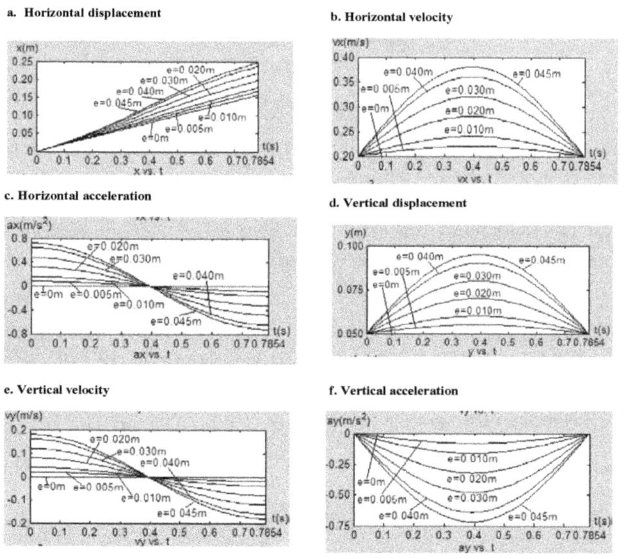

Figure 11: Displacement, velocity and acceleration vs time t for different e values.

Where W_r is the width of the robot.

Numerical results

Referring to the developed prototype as shown in Figure 1, the dimension of the robot body is 0.362 m × 0.232 m × 0.071 m. The overall weight is about 58.9 N. The radius and eccentricity of the eccentric wheels are respectively 0.05 m and 0.03 m. The rotation velocity at the supporting phase is $\omega_s = 4.0$ rad/s.

The effect of the minimal height h0 on θ and φ: In general, the axis Q of the eccentric wheel should not lower than the center O of the wheel for the supporting phase, i.e., $h_0 \geq r$. This is because the wheel cannot be designed too large for the limited structure of the robot. However, for different h_0, the two angles (θ and φ) are different. Figure 9 shows the variation of θ and φ with respect to h0 for the given eccentricity e=0.03 m. It can be seen that for h0=r=0.05 m, θ=180° and φ=118.073°. When h0=r + e=0.08 m, both θ and φ are 0, because Q and O are on the same vertical diameter in this case.

The effect of the eccentricity e on φ: As shown in Figures 2-8, the central angle corresponding to the supporting phase is less than 180° (i.e., θ<180°). For the tripot gait, the rotation speed for supporting phase is much lower than the rotation speed for the idling phase (i.e. $w_s<w_i$) because the time (T_s) used for the supporting phase is the same as the time (T_i) used for the idling phase (i.e. $T_s=T_i$). For the hexapod gait, as there are two adjusting statuses in one cycle, even if $\omega_i>\omega_s$, the time interval between two moving statuses is long. In order to simplify the control of the tripot gait, we can make $\omega_s=\omega_i$. This requirement leads to θ=180° and h_0=r. When θ=180°, for the hexapod gait, if we make $\omega_i>\omega_s$ (say $\omega_i=2\omega_s$), the time interval between two moving statuses can be reduced. When $h_{0=}$r, θ(= 180°) does not change with e. But φ will change with respect to the eccentricity e as shown in Figure 10.

The effect of the eccentricity *e* on the motion of the robot: As mentioned in the above subsections, when h_0=r, θ=180°. As the rotation velocity at the supporting phase for the developed prototype is ω_s=4.0 rad/s, we can calculate the time used for the supporting phase as $T_{s=}$0.785 s. In order to investigate the effect of the the eccentricity e on the motion of the robot, several e values (0 m, 0.005 m, 0.010 m, 0.020 m, 0.030 m, 0.040 m, and 0.045 m) are chosen for comparison. Among these e values, e=0.030 m is the eccentricity of the wheels adopted in the developed prototype robot. Substituting the above parameters into Equation (9) - (11), we can calculate the motion of the robot for straight walking. The results of one supporting phase T_s(= 0.785 s) can be given in Figure 11.

Figure 11a shows the horizontal displacement of the robot. Figure 11b shows the horizontal velocity of the robot. Figure 11c shows the horizontal acceleration of the robot. Figure 11d shows the vertical displacement of the robot.

Figure 11e shows the vertical velocity of the robot. Figure 11f shows the vertical acceleration of the robot. From Figure 11, we can see that when e=0, the horizontal displacement of the robot is a slant straight line. But the vertical displacement, the velocity and acceleration are all 0. This result is obvious and logical.

When e>0, the variation of the displacement, velocity and acceleration with respect to time for different e values is similar and can be described as follows:

The horizontal displacement of the robot increases with respect to time. In the first half of T_s, the horizontal velocity increases. At the moment $t=T_s/2$=0.393 s, it reaches its maximum. After that, the horizontal velocity decreases to its original value. The horizontal acceleration decreases with respect to time t. At the moment $t=T_s/2$=0.393 s, it reaches 0. After that, the horizontal acceleration becomes negative. Similarly to the horizontal velocity, the vertical displacement increases in the first half of T_s. At the moment $t=T_s/2$=0.393 s, it reaches its maximum. After that, the vertical displacement decreases to its original value. Similarly to the horizontal acceleration, the vertical velocity decreases with respect to time t. At the moment $t=T_s/2$=0.393 s, it reaches 0. After that, the vertical velocity becomes negative. The vertical acceleration of the robot is always negative. In the first half of T_s, it decreases. At the moment $t=T_s/2$=0.393 s, it reaches its minimum. After that, the vertical acceleration increases to its original value.

Referring to Eq.(12), the turning angular velocity is similar to the horizontal velocity as shown Figure 11b and the turning angular acceleration is similar to the horizontal acceleration as shown in Figure 11c.

Advantages of the Eccentric Wheels

As mentioned in the introduction section, the goal for developing the proposed hexapot robot with eccentric wheels is to detect the gas in underground mines. Avoiding potential explosion is the key point. Hence, any factor which could cause sparks is not expected. Although it has a lot of excellent performance, the RHex robot cannot satisfy this specific requirement because it has violent bumping and collision of its body with the rocky terrain. Also, its driving motor cannot be small enough (with a power less than 3 watts) to keep its all excellent performance.

Figures 12-14 can be used to exemplify the advantages of the proposed eccentric wheeled leg over the C shape leg of the RHex robot. Figure 12a shows the eccentric wheeled leg of the proposed hexapot

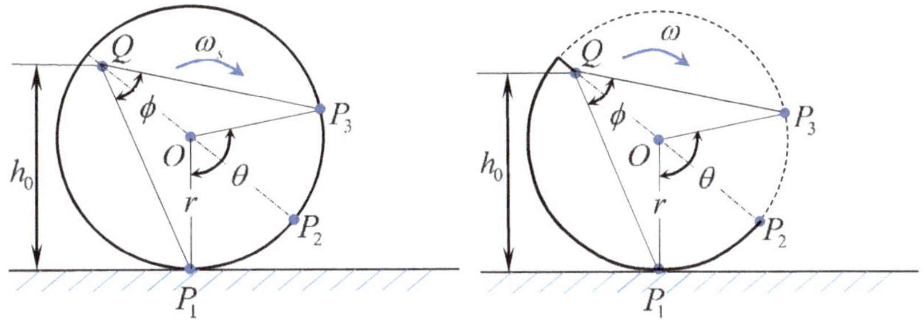

(a) Eccentric wheeled leg (b) RHex leg.

Figure 12: Supporting legs.

Figure 13: The RHex robot with the supporting legs at their tips P2 (Courtesy of UPenn).

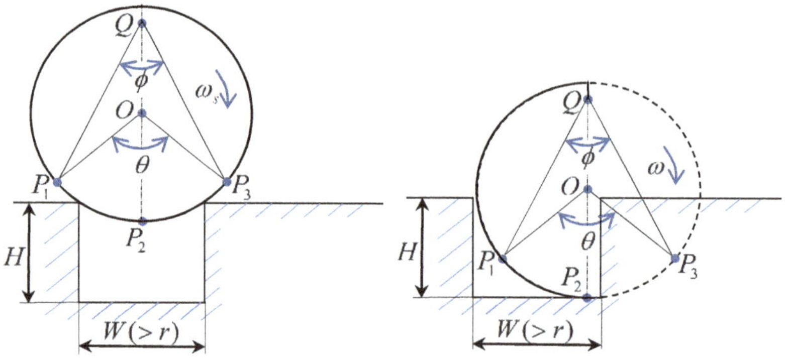

Figure 14: Supporting legs over the ditch.

robot. Figure 12b shows the C shape leg of the RHex robot. To be convenient for comparison, here the C shape leg of the RHex robot is supposed to be half circle with the same eccentricity e. Obviously the supporting arc of the RHex C shape leg is only half of the supporting arc of the eccentric wheeled leg. When the supporting point reaches P_2, both the proposed hexapot robot and the RHex robot reach their highest positions. After that, the proposed hexapot robot will be supported by arc P_2P_3. However, the body of the RHex robot will fall a distance $(\Delta h = QP_2 - h_0)$ before the P_1 points of another group C shape legs touch the ground (Figure 13). The time interval is $\Delta t = \theta / (2\omega)$. In other words, during such a $\Delta t = \theta / (2\omega)$, no legs are supporting the RHex robot body. As a result, its body falls down by $(\Delta h = QP_2 - h_0)$. This definitely causes bumping of the RHex robot. This bumping is inherent for the RHex robot.

As mentioned in the introduction section, in the tripod gait, there are always three feet of hexapod in contact with the ground. Obviously, the RHex robot is an exception.

Another violent bumping and collision of the RHex robot comes from the badly broken terrain. Referring to Figure 14b, as long as the width of a ditch is larger than the radius (i.e., W> r), the RHex C shape leg falls into the ditch. This results in violent bumping of the robot body. If the depth of the ditch is larger than the diameter (i.e., H>2r), the whole C shape leg falls into the ditch and the bottom of the RHex robot body collides the ground. Such violent bumping and collision is not expected in a context of gas detection application in underground mines. Referring to Figure 14a, on the contrary, as long as the width of the ditch is less than the diameter (i.e., W < 2r), the eccentric wheeled leg will not fall down a distance more than the radius, no matter how deep the ditch is. Hence, on one hand, the eccentric wheeled leg greatly reduces the bumping of the robot body. On the other hand, it greatly reduces the possibility for the bottom of the robot body to collide the rocky terrain.

Another advantage of the proposed eccentric wheeled hexapot robot is: Its backward motion is exactly the same as its forward motion. This is because the eccentric wheeled leg is symmetric (Figures 12 and 14). But for the RHex robot, its backward motion is completely different from its forward motion. Actually, the RHex robot uses its backward motion for leaping and flipping itself. However, these violent bumping is not expected in a context of gas detection application in underground mines.

Figure 15 shows the situation for the proposed hexapot robot to traverse an obstacle which is higher than the radius of the wheel. For the common circular wheels with the axis Q coinciding with the center O, it is impossible to traverse such an obstacle. But for the proposed hexapot robot, it is easy to climb and traverse this obstacle. This mainly depends on the motion of wheel #1 as shown in Figure 15a. Of course, wheel #4 as shown in Figure 15b also provides some assistance. Hence, the proposed eccentric wheel possesses significant advantages over the common circular wheel in climbing and traversing an obstacle.

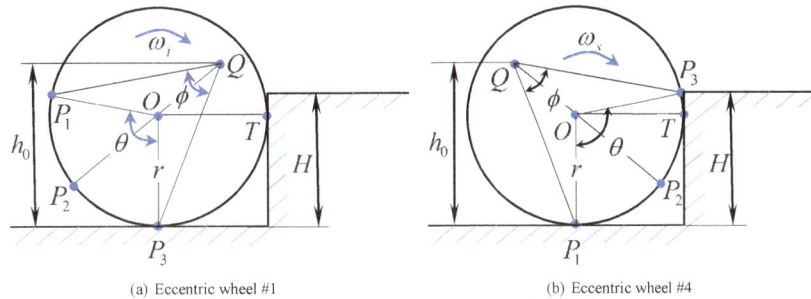

(a) Eccentric wheel #1   (b) Eccentric wheel #4

Figure 15: Passing the obstacle.

Conclusions and Discussions

A new concept for developing hexapod robots using eccentric wheels was proposed in this work. Compared with the RHex robot, the proposed hexapod robot can greatly reduce the bumping of the robot body in both smooth ground and rocky terrain. Also, the developed hexapod robot possesses significant advantages over those with common circular wheels in traversing rocky and uneven terrain. This work focused on the kinematics analysis of the proposed hexapod robot.

• The proposed alternating tripod gait can make the robot to move continuously. If the central angle $\theta=180°$, we can get $\omega_s=\omega_i$. All wheels can move with the same rotation velocity. This makes the control very easy. However, if the eccentricity e is large and the power of the motor is small (For instance, the motors used in the mining robot for detecting gas in underground mines are very small), the weight could cause a big resistant torque on the three driving wheels. This could be a big burden for the robot.

• The hexapod gait benefits in overcoming the resistant torque on the wheels caused by the weight. With six driving wheels, the resistant torque caused by the weight on each wheel can be reduced by half. The disadvantage of the hexapod gait is that there are two adjusting statuses. If $\theta=180°$ and $\omega_s=\omega_j$, the stopping time is twice of the moving time. Hence, for the hexapod gait, we have to make $\omega_i >> \omega_s$ depending on different applications of the robot.

• Based on the parameters used in the prototype, the motion analysis of the developed robot is conducted. The results for the cases with e>0 show that even if the rotation velocity ωs of the wheels is constant during the supporting phase, the velocities of the robot in both horizontal and vertical directions cannot be constant.

The future work will focus on

• Dynamic modeling and analysis.

• Vision system development and remote control.

• Further experimental studies.

Acknowledgements

The authors gratefully acknowledge the support from China State Key Laboratory of Silkworm Genome Biology.

References

1. Rashid MZA, Aras MSM, Radzak AA, Kassim AM, Jamali A (2012) Development of Hexapod Robot with Manoeuvrable Wheel. International Journal of Advanced Science and Technology 49: 119-135.

2. Saranli U, Buehler M, Koditschek DE (2001) RHex: A Simple and Highly Mobile Hexa-pod Robot. The International Journal of Robotics Research 20: 616-631.

3. Galloway KC, Haynes GC, Ilhan BD, Johnson AM, Knopf R, et al. (2010) X-RHex: A Highly Mobile Hexapedal Robot for Sensorimotor Tasks, Technical Report.

4. Altendorfer R, Moore N, Komsuoglu H, Buehler M, Brown (2001) RHex: A biologically inspired hexapod runner. Autonomous Robots 11: 207-213.

5. Chen SC, Huang KJ, Chen WH, Shen SY, Li CH, et al. (2013) Quattroped: A Leg–Wheel Transformable Robot. IEEE Transactions on Mechatronics 19: 730-742.

6. Aparna K, Geeta S (2013) Insect Inspired Hexapod Robot for Terrain Navigation. International Journal of Research in Engineering and Technology 2: 63-69.

7. Lewinger WA, Quinn RD (2010) A hexapod walks over irregular terrain using a controller adapted from an insects nervous system. Proceeding of the IEEE/RSJ international conference on intelligent robots and systems(IROS), Taiwan.

8. William A, Lewinger H, Reekie M (2011) A hexapod robot modeled on the stick insect, carausius morosus. Proceedings of the 15th international conference on advanced robotics,Tallinn, Estonia.

9. Lewinger WA, Branicky MS, Quinn RD (2005) Insect Inspired, Actively Compliant Hexapod Capable of Object Manipulation. Proceedings CLAWAR 2005 8th International Conference on Climbing and Walking Robots.

10. Belter D, Skrzypczynski P (2010) A Biologically Inspired Approach to Feasible Gait Learning For A Hexapod Robot. Int J Appl Math Comput Sci 20: 69-84.

11. Ramesh AP, Kumar CSK (2010) Autonomous Home Automated Hexapod Robot. International Journal on Computer Science and Engineering 2: 3016-3020.

12. Seljanko F (2011) Hexapod walking Robot gait generation using genetic gravitational hybrid algorithm. proceeding of the 15th International Conference on Advanced Robotics, Tallinn, Estonia.

13. Daud M, Nonami K (2012) Autonomous walking over obstacles by means of LRF for hexapod robot COMET-IV. Journal of Robotics and Mechatronics 2455-2463.

14. Mehdigholi H, Akbarnejad S (2012) Optimization of watts six bar linkage to generate straight and parallel leg motion. International Journal of Advanced Robotic Systems 9: 1-6.

15. Cal M, Fatuzzo G, Oliveri SM, Sequenzia G (2007) Dynamical Modeling And Design Optimization of A Cockroach-Inspired Hexapod. Proceedings of IMAC-XXV: Conference & Exposition on Structural Dynamics.

16. Burkus E, Odry P (2007) Autonomous Hexapod Walker Robot. Proceedings of 5th International Symposium on Intelligent Systems and Informatics.

17. Duan X, Chen W, Yu S, Liu J (2009) Tripod Gaits Planning and Kinematics Analysis of a Hexapod Robot. Proceedings of IEEE International Conference on Control and Automation.

18. Polulu Coorporation (2010) Sample project: Simple Hexapod Project. Pololu Corporation.

19. Billah MM, Ahmed M, Farhana S (2008) Walking Hexapod Robot in Disaster Recovery: Developing Algorithm for Terrain Negotiation and Navi-gation. World Academy of Science, Engineering and Technology 18: 328-333.

20. Birkmeyer P, Peterson K, Fearing RS (2009) DASH: A Dynamic 16g Hexapedal Robot. Proceedings of IEEE/RSJ International Conference on Intelligent Robots and Systems.

21. Tan X, Wang Y, He X (2011) The gait of a hexapod robot and its obstacle-surmounting capability. Proceedings of IEEE 9th World Congress on Intelligent Control and Automation (WCICA), Taipei, Taiwan.

22. Li J, Wang Y, Wan T (2012) Design of A Hexapod Robot. Proceedings of 2nd International Conference on Consumer Electronics, Communications and Networks, China.

Optimal Strategies for Virus Propagation

Soumya B[1-4*]

[1]*Department of Computer Science, University of New Mexico, USA*

[2]*Ronin Institute, Montclair, USA*

[3]*Complex Biological Systems Alliance, USA*

[4]*Broad Institute of MIT and Harvard, USA*

[*]**Corresponding author:** Soumya B, Department of Computer Science, University of New Mexico, USA
E-mail: soumya.banerjee@roninstitute.org

Abstract

This paper explores a number of questions regarding optimal strategies evolved by viruses upon entry into a vertebrate host. The infected cell life cycle consists of a non-productively infected stage in which it is producing virions but not releasing them and of a productively infected stage in which it is just releasing virions. The study explores why the infected cell cycle should be so delineated, something which is akin to a classic "bang-bang control" or all-or-none principle. The times spent in each of these stages represent a viral strategy to optimize peak viral load. Increasing the time spent in the non-productively infected phase (τ_1) would lead to a concomitant increase in peak viremia. However increasing this time would also invite a more vigorous response from Cytotoxic T-Lymphocytes (CTLs). Simultaneously, if there is a vigorous antibody response, then we might expect τ_1 to be high, in order that the virus builds up its population and conversely if there is a weak antibody response, τ_1 might be small. These tradeoffs are explored using a mathematical model of virus propagation using Ordinary Differential Equations (ODEs). The study raises questions about whether common viruses have actually settled into an optimum, the role for reliability and whether experimental infections of hosts with non-endemic strains could help elicit answers about viral progression.

Keywords: Viral dynamics; Ordinary differential equations; Optimization; Bang-bang control; Viral strategies; Optimal control theory

Introduction

A normal cell upon infection goes through a life cycle characterized by 2 phases: a stage in which it is producing virions but not releasing it (non-productively infected stage) and a stage in which it is releasing virions into the outside environment (productively infected stage). Hence there is a delay between infection and release of virions. In whatever follows, we denote the time spent in the non-productively infected stage as τ_1 and the time spent in the productively infected stage as τ_2. The time in τ_1 is spent in viral penetration, uncoating of viral core, transcription and assembly.

The number of virions produced over the entire infected cell life cycle is directly proportional to $\tau_1 + \tau_2$. It is asked whether the virus might be trying to maximize this quantity in order to optimize "virulence" (a quantity which shall be concretized shortly). The question of why there need be 2 distinct phases and not just one where virion production and release occur simultaneously, also cries out for explanation. Such forms of delineation are called "bang-bang control" or the all-or-none principle and are characterized by a phase of proliferation and then terminal differentiation, and are frequently encountered in optimal biological systems [1].

If the total length of the infected cell lifetime is a measure of "virulence", we can then set a theoretical upper bound on it and then compare it with its actual value from field measurements. This would give us a qualitative understanding of "how far" the virus can still go in optimizing itself e.g. it can be used to determine if the avian-influenza virus is already as virulent as it can be or is it still sub-optimal.

The rest of the paper is organized as follows: Section 2 discusses arguments for optimization in biological systems and Section 3 introduces the principle of "bang-bang control". The hypotheses and questions are posed in Section 4 and Section 5 outlines the mathematical model. Section 6 contains the results and discussions and concluding remarks are presented in Section 7.

Optimization in a Biological System

Before commencing with the mathematical analysis we state what our modeling philosophy will be and give some justification for employing such an approach. First of all, there is certainly no a priori reason why virus propagation or any other biological system should operate in an "optimal" fashion. Indeed there is a substantive issue as to whether the notion of "optimality" can be given an operational meaning for many biological systems. Typically, an organism or a virus is forced to cope with a number of competing influences so that an improvement in one direction involves a sacrifice in another. Thus optimality must be interpreted in a broader sense as a "best compromise" solution. Beyond this consideration, however, there are at least two major reasons why a particular biological system might not be performing its function in the most expeditious fashion. First, despite the fact that one tends to think of natural selection as an inherently optimizing process, improvements on existing mechanisms generally proceed by small modifications of existing structures. Thus there is ample opportunity for the system to become trapped in "local" maxima; there may be "nearby" structures with higher fitness but to reach them may require a temporary, but fatal, decrease in overall fitness. Second, while the system may be constantly improving, evolution is a slow and erratic process so that any system we examine may not have had time to optimize under existing selective pressures. Both of these objections may be partially circumvented by restricting attention to systems which appear to have been evolutionarily static for a long time. The mammalian immune system and viruses surely fulfill this criterion.

A virus typically only wants to proliferate in a host only so much as to ensure transmission to another host (an exception is the Ebola virus which kills its host so fast as to prevent propagation to another host). Hence it is trying to optimize the basic reproductive ratio R0 in epidemic models. In vector-borne pathogens, the peak viremia in blood serum is a very good determinant of R0 [2]. Hence, we assume that the virus is trying to optimize the peak viral load in blood serum (P_v).

Bang-Bang Control

In a seminal paper Perelson et al. [1], examined the mammalian immune system and looked at optimal strategies for B-cell proliferation and differentiation. They used control theoretic principles to analyze the minimum time taken by the immune system to eliminate a fixed amount of antigen in the shortest span of time. The problem briefly stated is as follows: given an initial population of B-cells (which secrete antibodies at a modest rate, proliferate into B - cells or differentiate into plasma cells) and plasma cells (which secrete antibodies at a very high rate but do not proliferate), how do you apportion the total population between B-cells and plasma cells? Does the optimal strategy involve proliferation of B-cells followed by differentiation into plasma cells? Or does it involve simultaneous B-cell proliferation and differentiation? The authors showed using optimal control theory that the optimal strategy for B-cells is to go through a stage of proliferation (to build up their population) and then differentiate into plasma cells. Such a control is called "bang-bang" or all-or-none. It is not immediately evident or intuitive that a strategy of simultaneous B-cell proliferation and differentiation is not optimal.

A parallel is drawn between that work and the problem at hand here, where the infected cell also goes through a phase of production of virions followed by a phase of virion release. The reasons for "bang-bang control" in the infected cell system and its implications are explored in the following sections.

Hypotheses and Questions

This section explores some of the hypotheses proposed and frames some questions. There are 2 hypotheses about the non-productively infected stage of the infected cell:

Hypothesis 1

The virus is not trying to optimize the duration of the non-productive infected stage (τ_1). Hence this time is exactly equal to the time required for viral penetration, uncoating of viral core, transcription and assembly. The interpretation is that as soon as the first complete virions is assembled, the infected cell immediately proceeds to release the virion i.e. it switches to the next phase of productive infection. The obvious disadvantage of this strategy is that the amount of virions produced would be reduced, compared to an approach in which τ_1 is increased. Clearly this strategy is sub-optimal and we do not explore it further.

Hypothesis 2

The virus is trying to optimize peak viral load and hence viral production. However it cannot increase the duration of the productive infected stage (τ_2). This is so because there are physiological limits imposed by the area and strength of the cell wall, which will constrain the duration of virion release. After a threshold, the cell wall will simply fall apart. It can only increase the duration of the non-productive infected stage (τ_1).

- Having a high τ_1 would imply an increased virion release count. However, this would come at the cost of increased susceptibility to lysis by Cytotoxic T-Lymphocytes (CTLs). A lower τ_1 would reduce the susceptibility to CTL mediated lysis at the expense of a reduced virion count.

- The virus might optimize itself such that it bursts early in the face of a weak antibody response. Conversely, it could burst later (after building up a pool of virions) when confronted with a vigorous antibody response.

Question 1: Why cannot τ_1 increase indefinitely?

Question 2: Why is the optimal control "bang-bang"?

Mathematical Formulation

A standard mathematical model of virus propagation adapted from Baccam et al. [3] is constructed to test the hypothesis. Ordinary Differential Equations (ODEs) are used to represent populations of virus, infected cells and normal cells. The equations are shown below:

$$\begin{cases} \frac{dT}{dt} = -\beta TV \\ \frac{dI_1}{dt} = \beta TV - kI_1 - \omega_{CTL}I_1 \\ \frac{dI_2}{dt} = kI_1 - \delta I_2 \\ \frac{dV}{dt} = pI_2 - cV \end{cases}$$

where T = target cell population,

I_1 = non-productively infected cell population

I_2 = productively infected cell population

V = virus population

β = rate constant of infection

k = rate of death of non-productively infected cells

δ = rate of death of productively infected cells

ω_{CTL} = rate of CTL-mediated lysis of non-productively infected cells

p = number of virions produced per productively infected cell per time step c = rate of clearance of free virus particles

We also get $\tau_1 = 1/k$ and $\tau_2 = 1/\delta$

In this simple ODE model, the population of target cells (normal and uninfected cells) are represented by the variable T. They are also lost due to infection, which is represented by the term $-\beta TV$. The non-productively infected cells (I_1) are supplied by the loss from the target-cell pool and die at a rate proportional to their number density with constant of proportionality k. They are also lysed by (Cytotoxic T-Lymphocytes) CTLs at a rate proportional to their density and with a constant of proportionality of ω_{CTL}. Productively infected cells (I_2) are replenished from the non-productive pool and die at a rate proportional to their density and with a constant of proportionality of δ. New virions (V) are produced by infected cells at the rate p_{I2} and virions are lost at a rate proportional to the virus concentration with constant of proportionality c (representing antibody -mediated virion clearance).

The variation in ω_{CTL} has been modelled in a time-dependent fashion. Namely it is made to mimic the clonal expansion of a pool of effector CTLs after day 4.

$$\omega_{CTL}(t) = \begin{cases} 0, t < 4 \\ \Omega \times e^{\theta \times (t-4)}, t \geq 4 \end{cases}$$

The model was parameterized from a study of experimental infection of Influenza A virus in humans [3]. The model was implemented in the Berkeley Madonna package [4] and the code is freely available for download [5]. The model parameters are shown below in Table 1.

Parameters	β [(TCID50 /ml)$^{-1}$ × day-1)]	δ (day^{-1})	p [(TCID50 /ml)$^{-1}$ × day^{-1})]	k (day^{-1})	c (day^{-1})	T_0	V_0 (TCID50 /ml)
Value	4.9×10^{-5}	4.2	2.8×10^{-5}	3.9	4.3	4×108	4.3×10^{-2}

Table 1: Estimated parameter values from Baccam et al. [3]

Results and Discussion

The model as outlined in the previous section, thus parameterized, was used to test the hypothesis.

Test of Hypothesis 2a

Restating, having a high τ_1 would imply an increased virion release count with cost of an increased susceptibility to lysis by Cytotoxic T-Lymphocytes (CTLs). A lower τ_1 would reduce the susceptibility to CTL mediated lysis at the expense of a reduced virion count.

Result

We observed that the optimal strategy was to increase τ_1 till a threshold (in this particular case it was found to be just less than 4 days). Incidentally, day 4 is also the time at which CTL action is initiated. Hence, the optimal strategy for the virus is to continue the non-productively infected phase till just before CTL initiation. Till CTL action is initiated, the virus will continue to build its population. Increasing τ_1 beyond 4 days would lead to loss of produced virions due to CTL-mediated infected cell lysis. Any decrease below 4 days would reduce the total virus production and hence peak viremia. Hence the optimal control is "bang-bang". Bang-bang control strategies have also been known to be optimal in other biological systems like differentiation of B-cells and production of plant seeds [1]. Note that due to the use of a continuous ODE system (which mimics biology more closely) as opposed to a delay-differential equation, some infected cells do burst earlier than day 4 (Figure 1).

Test of Hypothesis 2b

Restating, the virus might optimize itself such that it bursts early in the face of a weak antibody response. Conversely, it could burst later (after building up a pool of virions) when confronted with a vigorous antibody response.

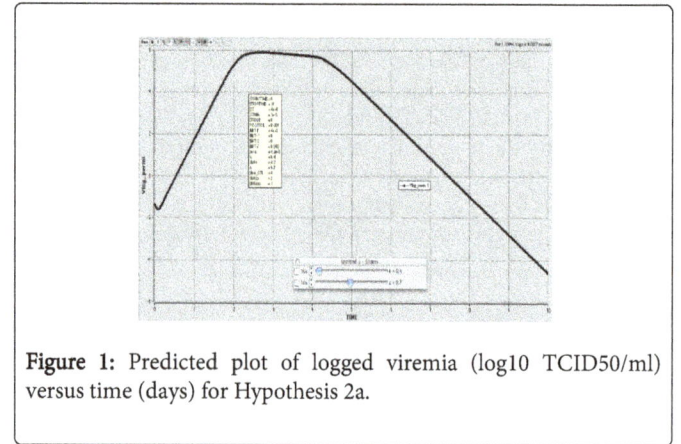

Figure 1: Predicted plot of logged viremia (log10 TCID50/ml) versus time (days) for Hypothesis 2a.

Result

The antibody response was varied by manipulation of the virion clearance term c in the ODE system. It was found that the optimal strategy remained conserved under variations in the antibody response i.e. the optimal strategy for the virus was always to burst at $\tau_1 = 4$ days. We can reason about this in the following manner: increasing τ_1 beyond 4 days would lead to loss of produced virions due to CTL-mediated infected cell lysis and any decrease below 4 days would reduce the total virus production and hence peak viremia. Hence antibody response has no effect on τ_1 - a fact that is perhaps not intuitively obvious (Figure 2).

Question 1: Why cannot τ_1 increase indefinitely?

From the preceding discussion, it becomes evident that if τ_1 were to increase indefinitely beyond the time to CTL initiation, then there would be a concomitant decrease in virion output due to CTL-mediated infected cell lysis. Hence the time to CTL initiation sets an upper bound on τ_1.

Question 2: Why is the optimal control "bang-bang"?

Due to physiological limits on cell wall integrity, the time spent in the productively infected phase (τ_1) must be limited. Any attempt to increase it beyond a threshold would merely cause the whole cell wall to break down. Hence, in order to increase virus production, the only "recourse" the virus has is to increase the time spent in the non-productively infected phase and build up the virus population until onset of CTLs. This naturally gives rise to 2 delineated phases ("bang-bang control"). Any intermediate graded response i.e. virion production occurring simultaneously with release is essentially equivalent to the productively infected phase and since the time that can be spent in it is severely limited, we see that it is a sub-optimal strategy. Such strategies are also optimal in diverse biological systems ranging from differentiation of B-cells in the immune system to allocation of energy to seeds in plants [1].

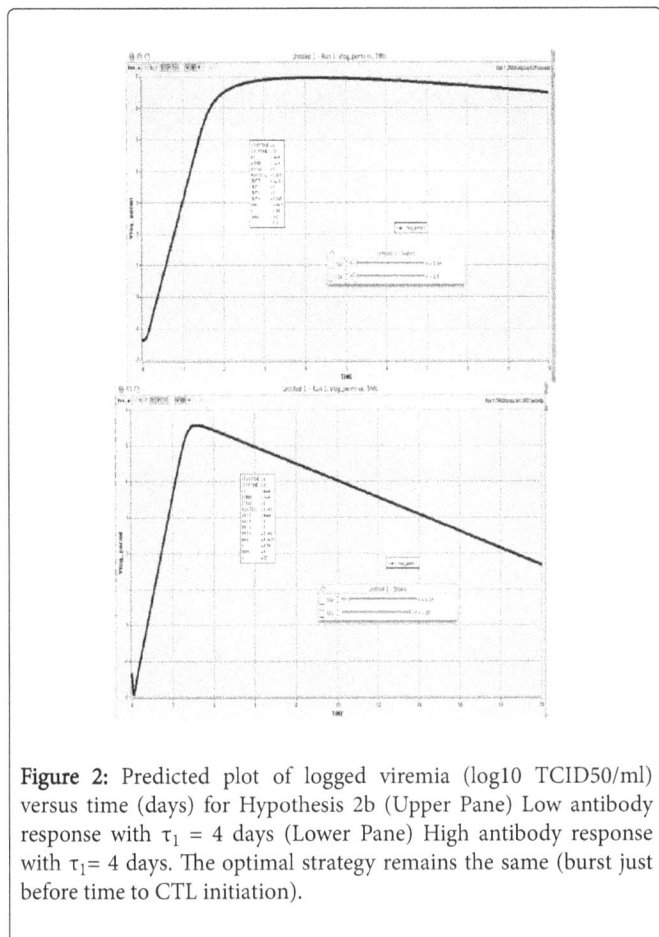

Figure 2: Predicted plot of logged viremia (log10 TCID50/ml) versus time (days) for Hypothesis 2b (Upper Pane) Low antibody response with τ_1 = 4 days (Lower Pane) High antibody response with τ_1= 4 days. The optimal strategy remains the same (burst just before time to CTL initiation).

Conclusions

This work visits virus proliferation from an optimization viewpoint. A few basic assumptions are made: a) the time spent in the productively infected phase is constant and cannot be subjected to optimization beyond a threshold, and b) the virus is trying to optimize virion production and hence peak viremia. Starting from these assumptions, it is posited that the optimal strategy for virus proliferation is to delay burst till onset of Cytotoxic T-Lymphocytes. This so called "bang-bang control" or all-or-none principle is exhibited in many other biological systems like ant colonies and annual plants [1]. However, optimization may not be the only principle at work. In fact, considerations of reliability may be invoked to explain the presence of long-lived latently infected cells (e.g. HIV). These long-lived cells evade detection by CTLs and ensure a prolonged viremia in hosts.

Another conclusion, which is not intuitively obvious, is the fact that the optimal strategy of allocating the maximum time in the non-productively infected phase remains invariant even in the face of a varying antibody response. This strategy is insensitive to the humoral response and depends only on the time to CTL initiation.

The total length of the infected cell lifetime is a measure of "virulence", and a theoretical upper bound has been set on it. Comparing this value to the actual value from field measurements would give us a qualitative understanding of "how far" the virus can still go in optimizing itself e.g. it can be used to determine if the avian-

influenza virus is already as virulent as it can be or is it still sub-optimal. In the case of the Influenza A virus from the Baccam et al. study [3], the theoretical upper bound on $\tau 1$ is around 4 days, whereas the observed is around 12 hours, suggesting that the virus is still operating sub-optimally and still has scope to improve by mutating itself. Insights like these could be crucial for bio-surveillance efforts and help inform strategies to cope with future pandemics caused by virus mutations.

Lastly, it is instructive to note that experimental infections of hosts with non-endemic strains (viral strains that have not co-evolved with the host and hence are not operating in an optimal manner) could affect experiment outcome. This would elicit a lower than normal viral response, since the viral strategy would now be characterized by Hypothesis 1 i.e. the time spent in the productively infected phase (τ_1) would just constitute the time required for viral penetration, uncoating of viral core, transcription and assembly and no more.

Clearly more work needs to be done to verify these arguments and an extensive analytical treatment of these arguments coupled with more experimental work will be the subject of future work. The current work highlights the significance of simple mathematical and dynamical models that reveal insights into biological processes as has been done previously in immunology and cell biology [6-15].

Acknowledgments

The author wishes to acknowledge fruitful discussions in the 2008 Santa Fe Institute Complex Systems Summer School.

References

1. Perelson AS, Mirmirani M, Oster GF (1976) Optimal Strategies in Immunology: B-Cell Differentiation and Proliferation. Journal of Mathematical Biology 3: 325-367.

2. Komar N, Langevin S, Hinten S, Nemeth N, Edwards E, et al. (2003) Experimental infection of North American birds with the New York 1999 strain of West Nile virus. Emerg Infect Dis 9: 311-322.

3. Baccam P, Beauchemin C, Macken CA, Hayden FG, Perelson AS (2006) Kinetics of Influenza A Virus Infection in Humans. Journal of Virology 80: 7590-7599.

4. Macey RI, Oster G, Zahnley T (2001) Berkeley Madonna, version 8.0. Technical report. University of California, Berkeley, California.

5. Soumya B (2015) Optimal Strategies for Virus Propagation. Model file for Berkeley Madonna for simulations.

6. Soumya B, Melanie M (2010) Scale invariance of immune system response rates and times: perspectives on immune system architecture and implications for artificial immune systems. Swarm Intelligence 4: 301-318.

7. Soumya B (2013) Scaling in the Immune System. PhD Thesis, University of New Mexico, USA.

8. Soumya B, Hentenryck P, Cebrian M (2015) Competitive dynamics between criminals and law enforcement explains the super-linear scaling of crime in cities. Palgrave Communications.

9. Peng L, Calderon A, Georgios K, Soumya B (2014) A bioorthogonal small-molecule switch system for controlling protein function in cells. Angewandte Chemie.

10. Graessl M, Koch J, Calderon A, Banerjee S, Mazel T, et al. (2014) A mechano-sensitive excitable system causes local Rho activity oscillations.

11. Banerjee S, Levin D, Moses M, Koster F, Forrest S (2011) The value of inflammatory signals in adaptive immune responses. 10th International Conference pp: 1-14.

12. Soumya B, Melanie M (2009) A hybrid agent based and differential equation model of body size effects on pathogen replication and immune system response. 8th International Conference, Germany.

13. Soumya B (2015) Analysis of a Planetary Scale Scientific Collaboration Dataset Reveals Novel Patterns. Cornell university library.

14. Soumya B (2009) An Immune System Inspired Approach to Automated Program Verification. Cornell university library.

15. Soumya B, Melanie M (2010) Modular RADAR: An immune system inspired search and response strategy for distributed systems. Artificial Immune Systems, 9th International Conference, Germany.

14

Cooperation of Two Robot Tractors to Improve Work Efficiency

Zhang C and Noguchi N*

Laboratory of Vehicle Robotics, Graduate School of Agriculture, Hokkaido University, Kita-9, Nishi-9, Kita-ku, Sapporo 060-8589, Japan

Abstract

A system for cooperation of two robot system was developed to solve the problem of shortage of labor and to improve efficiency of field work. A safety model of a robot tractor was proposed for coordination and cooperation of two robots. Each robot has its own pre-determined path, and tracking the path is independent of the other robot. Thus, by controlling the velocity of each robot, the robots can keep a certain shape when working due to radio communication between the two robots. As for headland turn, the two robots turn together as long as they are in a safe condition. Computer simulation was conducted to confirm the effectiveness of this system. The results showed that it is possible for two robots to work safely together applying the developed safety model. Compared with work by a single robot, this system using two robots can improve work efficiency by at least 80 percent.

Keywords: Leader-follower pattern; Formation control; Safety model; Robot tractor

Introduction

Many efforts have been made by engineers in advanced countries to reduce labor and work time in agriculture due to the decreasing agricultural labor forces. Global demand for higher productivity in agriculture field has led to the need for more cooperation between agricultural machines. Advances in industrial mechanization and automation have inspired engineers to develop robots that can perform various agriculture field tasks. There are two major categories of research on agricultural robotics: ground sensing systems that use machine vision, odometers, accelerometers, etc. [1-4] and satellite-based system that use GPS for navigation [5-7]. The error in accuracy of a real-time kinematic (RTK) GPS system is now only 1 to 2 cm per 10 km [8]. Generally, the error in accuracy of a guidance system using RTK-GPS is 3 to 5 cm.

As for robot technology, many researchers have developed sweep coverage algorithm for multi-agent robot system [9-11]. The agriculture robot tractor is similar to the sweep coverage robot. Both of them need to cover the whole area using the minimum time. The difference between the robot tractor and sweep coverage robot is that the robot tractor need to cover each place only one time. In many cases, cooperation and coordination of two robots is necessary to improve work efficiency and to reduce work time and work strength. For instance, for a robotic combine harvester harvesting in a field, an on-the-go unloading system with a transport robot that moves the harvested products to collection positions helps improve harvesting efficiency since the harvester does not need to stop. Another example is that a robot tractor doing tillage can be followed by another robot seeding and fertilizing at the same time. In these operations, the two robots need to be controlled carefully to prevent loss of harvested products and collision of the robots. In agriculture, researchers have focused on the master-slave approach for coordination. Noguchi et al. [12] developed a master-slave robot system to conduct farm work. The system mainly includes GOTO and FOLLOW algorithms. The master controls the slave to follow a parallel path at a given distance and angle from the master or to go to a certain point along any path as long as it does not collide with the master. Zhang [13] proposed an intelligent master-slave system that enables a semi-autonomous agricultural vehicle (slave) to follow a master with a given lateral and longitudinal offset. They used a state space dynamic model and a proportional-derivative controller with state feedback and disturbance feed-forward for the tractor. Vougioukas [14] proposed a distributed control framework to coordinate the motions of teams of autonomous agricultural vehicles operating in the same field. The framework includes master-slave and peer-to-peer modes, which are based on nonlinear model predictive tracking controllers that communicate with each other.

However, the use of two robots also has some disadvantages. If it is not properly constructed, the two robots may actually increase the complexity of the system instead of simplifying it. The multiple robot system has to address many issues that do not appear in single robot system, such as communication and interaction with the other robot, formulation and headland turn cooperation. Generally, research on coordination of robots has mainly focused on formation control [15-19], which means the control of a group of robots in a coordinated way to maintain the formation of a certain shape. Formation control was also part of this article. The main aim of this article was to determine how to use two robots in one field and how much work efficiency can be improved by simultaneously using two robots compared with using one robot. Two robots can cooperate in many different patterns. For instance, they can start from different sides of the field and work to the center or they can work from the center to the sides. In addition, the two robots can work in parallel and share the same performance or they can work like a master-slave system with one following the other. The collaboration of two robots differs depending on the field conditions or work operations. In this study, we classified the cooperation patterns into two categories: 1) a leader-follower pattern, which means the two robots keep a constant shape during the operation, and 2) a free pattern, which means the formation shape of two robots is unlimited. It would be easier to understand the leader-follower pattern since the two robots will maintain a certain shape, regardless of whether they are in parallel or front and back form. In this kind of pattern, each robot needs to adjust its velocity to maintain the formation shape. However, the free pattern does not limit the formation shape of two robots, which

***Corresponding author:** Noguchi N, Laboratory of Vehicle Robotics, Graduate School of Agriculture, Hokkaido University, Kita-9, Nishi-9, Kita-ku, Sapporo 060-8589, Japan, E-mail: noguchi@bpe.agr.hokudai.ac.jp means that even though the two robots communicate each other, they will not adjust their velocities as long as they are in a safe condition. In this article, we focus on the leader-follower pattern because it has more applications than the free pattern. For instance, the on-the-go unloading system is a typical application of leader-follower pattern.

Two robot tractors were used in this study, and the desired positions were given to each robot. Unlike a master-slave robot system, in the newly developed system, each robot tracks its own desired path, which means steering control is independent. Each robot adjusts its velocity to formulate a certain shape during the operation. However, during headland turn, the two robots do not need to keep a certain shape considering the best use of the headland. If the two robots continue to keep the shape in the headland turn operation, there would need to be more space in the headland, which would decrease the cultivating area of the field. Thus, during a headland turn, each robot uses its own turn method, and the trajectory of the robots may collide. A safety model for a robot tractor is proposed for cooperation by two robots during headland turn to avoid collision.

The cooperation system by two robots has several advantages. Firstly, two robots are used simultaneously, which increases work efficiency compared with the use of one robot. Secondly, compared with a large robot tractor, the use of two small robot tractors helps for reduction of damage to the crops and severe soil compaction. Finally, each robot's navigation is independent, which means it is easy to use them separately as a single robot. This autonomy of each robot enhances the usefulness of the two robot system.

The rest of this article is structured as follows. The method used for the cooperation system, including the safety model, formation control and turning cooperation are presented in section 2. Results of simulation and experiment of the cooperation system are presented in section 3, and the improved work efficiency of this system is also discussed. Finally, conclusions are given in section 4.

Methods

Communication structure

A robot tractor used in this study had already been developed [20]. In that study, an RTK-GPS and an IMU were used to provide position and posture data for robot's navigation. A computer was used as a controller to communicate with the tractor's ECU through CAN BUS. In this study, the navigation method was the same as Yang's work [20].

ROBOT-1 (abbreviated as RT1) and ROBOT-2 (abbreviated as RT2) were used as unmanned tractors. Both of them are basically commercial tractors. RT1 is wheel-typed tractor, while RT2 is type of half-track crawler, as shown in Figure 1. The length, width and height of RT1 are 3.9 m, 1.75 m and 2.62 m, respectively, and the length, width and height of RT2 are 4.26 m, 1.81 m and 2.68 m, respectively. The weights of RT1 and RT2 are 2840 kg and 3820 kg, respectively. The power of RT1 is 61.0 kW, and that of RT2 is 77.2 kW. The specifications of each tractor include steering control, a switch for forward and backward movements, easy-change transmission, a switch for power take off, hitch functions, engine speed set, engine stop and brake.

Figure 1 shows the structure of the communication method between two robots. The robot tractor's navigator software communicates with the tractor's ECU through a CAN BUS and exchanges information with the robot tractor's Client/Server software through a memory. The robot tractor's Client communicates with the Server through Bluetooth. The tractor's navigator software was developed to follow a predetermined path, and the steering control is independent from other software. The tractor's Client/Server were developed for cooperation work. For instance, the Client was on RT1 and the Server was on RT2. The Client reads information about RT1 and sends it to the Server. The Server obtains information from RT1 and RT2, do calculation and send a command to the Client, and the Server sends a command to RT2.

Tractor model for safety evaluation

If the two robots are far away and can do their work without colliding, we can skip this issue. For example, if the two robots work in separate fields, there is no need to worry about the robots colliding. However, if the two robots work in the same field, as long as the two robots are cooperating, the safety issue has to be considered. In this study, the safety zone of the robot is defined as a circle or rectangle. It can be concluded whether the two robots are safe by judging the relative positional condition of two safety zones. Take a circular zone as example, if the two circles are separated throughout the work task, then we can conclude that it is a safe situation; on the other hand, if the two circles are intersected, which means the two robots have collided, then we can conclude that it is a dangerous situation.

Circle model: Firstly, each robot's safety zone was simplified as a circle, as shown in Figure 2a, where $RT_i.x$, $RT_i.y$ are central coordinates of the circle and $RT_i.r$ is the radius of the circle. Eq. (1) shows the definition of $RT_i.r$ where $RT_i.l_{width}$ is the width of the equipment carried by RT-*i*, $RT_i.l_{front}$ is the distance from the center to front, and $RT_i.l_{rear}$ is the distance from the center to back. The distance between two circles can be calculated by Eq. (2), where $d_{1\ to\ 2}$ is the distance between

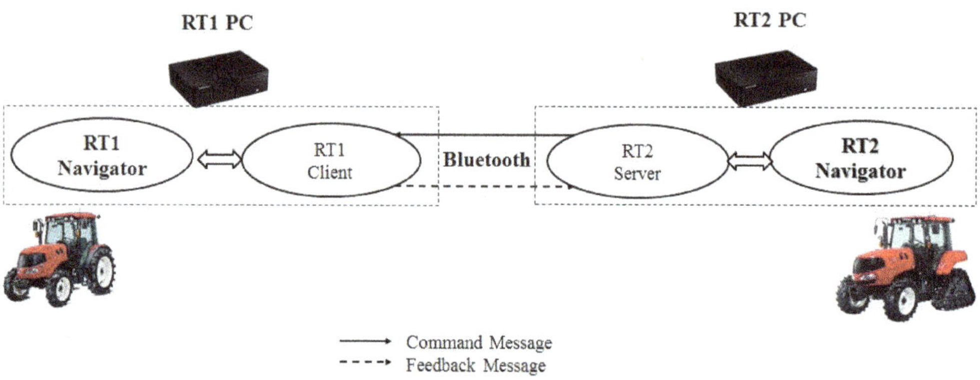

Figure 1: Communication structure of cooperation system.

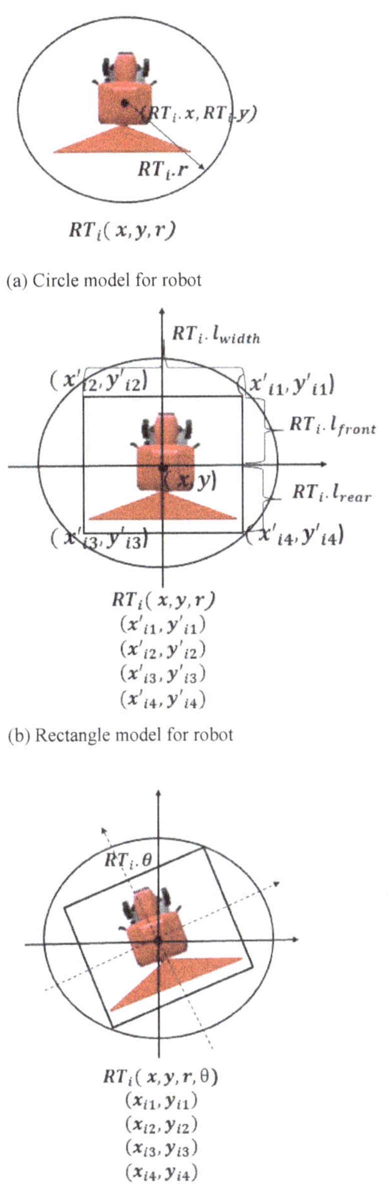

(a) Circle model for robot

(b) Rectangle model for robot

(c) Rotation of a robot

Figure 2: Model of robot tractor.

two safety zones. If $d_{1\ to\ 2}$ is more than zero, the two robots are safe; otherwise, the two robots are in danger.

$$RT_i.r = \sqrt{\left(0.5RT_i.l_{width}\right)^2 + \left(\max\left(RT_i.l_{front}, RT_i.l_{rear}\right)\right)^2} \qquad (1)$$

$$d_{1to2} = \sqrt{\left(RT_1.x - RT_2.x\right)^2 + \left(RT_1.y - RT_2.y\right)^2} - RT_1.r - RT_2.r \qquad (2)$$

Rectangle model: The safety zone of the robot tractor can also be simplified as a rectangle, as shown in Figure 2b. The rectangular zone is more compact than the circular zone, thus it is more effective on space utilization. However, a rectangular zone needs more calculation since it is more complex than a circular zone. The coordinates of the four corners are different depending on the size of the tractor and equipment. The four corners can be simplified as $P'_{i1}(x'_{i1}, y'_{i1})$, $p_{i2}(x'_{i2}, y'_{i2})$, $p_{i3}(x'_{i3}, y'_{i3})$ *and* $p_{i4}(x'_{i4}, y'_{i4})$ where *i* represents the ID of robot.

Eq. (3) shows the equation to calculate the coordinates of the four corners. However, when a robot is used in the field, it always rotates and changes direction, especially in a turning operation, as shown in Figure 2c, and the coordinates of the four corners should thus be transformed according to the rotation. Eq. (4) shows the transformation equation. Finally, for each robot, we have central coordinates $RT_i.x$, $RT_i.y$ $(i\ \{1,2\})$ circle radius $RT_i.r$ $(i \in \{1,2\})$ rotation yaw angle $RT_i.\theta$ $(i \in \{1,2\})$ and four corners' coordinates

$P_{ij}(x_{ij}, y_{ij})$ $(i \in \{1,2\})$, $j \in \{1,2,3,4\})$,

$$\begin{cases} x'_{i1} = RT_i.x + 0.5RT_i.l_{width} \\ y'_{i1} = RT_i.y + RT_i.l_{front} \\ x'_{i2} = RT_i.x - 0.5RT_i.l_{width} \\ y'_{i2} = RT_i.y + RT_i.l_{front} \\ x'_{i3} = RT_i.x - 0.5RT_i.l_{width} \\ y'_{i3} = RT_i.y - RT_i.l_{rear} \\ x'_{i4} = RT_i.x + 0.5RT_i.l_{width} \\ y'_{i4} = RT_i.y - RT_i.l_{rear} \end{cases} \qquad (3)$$

$$\begin{cases} x_{ij} = x'_{ij}\cos RT_i.\theta + y'_{ij}\sin RT_i.\theta \\ y_{ij} = y'_{ij}\cos RT_i.\theta - x'_{ij}\sin RT_i.\theta \end{cases} \qquad (4)$$

The two rectangles are checked to see whether they are intersected. If they are intersected, it means the two robots have already collided. However, in the field, the rotation angles of the two robots are different, which means we cannot judge whether the two robots are intersected by absolute coordinates. The following methods are used to check whether the two rectangular zones are intersected.

Each rectangle can be simplified as four segments, which means we can judge whether the 8 segments are intersected. Suppose there are segments L_{1k} (p_{1i}, p_{1j}), $(i,j \in \{1,2,3,4\})$ from RT1 and L_{2k} (p_{2i}, p_{2j}), $(i,j \in \{1,2,3,4\})$ from RT2, where $p_{1i}, p_{1j}, p_{2i}, p_{2j}$ are start points and end points of these segments. Firstly, suppose two rectangles T_1 and T_2 are in an absolute coordinate system and their sides are parallel with the *x-axis* or *y-axis*. The diagonal line of T_1 is L_{1k} and the diagonal line of T_2 is L_{2k}. If the two rectangles T_1 and T_2 are not intersected, then the two segments are not intersected, as shown in Eq. (5). If the two rectangles are intersected, further judgment is needed. Secondly, if the vectors $v_{1i}(p_{1i}, p_{1j})$, $v_{2i}(p_{2i}, p_{2j})$, $s_1(p_{1i}, p_{2i})$, $s_2(p_{1j}, p_{2i})$, $s_3(p_{2i}, p_{1i})$ and $s_4(p_{2j}, p_{1i})$ satisfy Eq. (6), the two segments are separated; otherwise, they are intersected.

$$x_{1i} > x_{2j} \| x_{1j} < x_{2i} \| \max(y_{1i}, y_{1j}) < \min(y_{2i}, y_{2j}) \| \min(y_{1i}, y_{1j}) < \max(y_{2i}, y_{2j}) \qquad (5)$$

$$\begin{cases} s_1 \times v_{2i} * s_2 \times v_{2i} > 0 \\ s_3 \times v_{1i} * s_4 \times v_{1i} > 0 \end{cases} \qquad (6)$$

The minimum distance between two rectangular zones, $d_{1\ to\ 2}$ is also needed for safety evaluation. Normally, the distance between two robots means the distance between two GPS receivers. The minimum distance between two rectangular zones indicates the space between two robots. As mentioned before, each rectangle can be simplified as four segments, and thus the minimum distance between two rectangular zones can be represented by the minimum distance between two sets of segments. Firstly, suppose there are segments L_{1k} (p_{1i}, p_{1j}) from RT1 and L_{2k} (p_{2i}, p_{2j}) from RT2. Suppose point M(X,Y) is on $L_{1,k}$ and point N(U,V) is on $L_{2,k}$. The coordinates of *M* and *N* can be expressed as Eq. (7), where s,t $\in$ (0,1). Thus, the length of *MN* can be expressed as Eq. (8). Secondly, suppose *f (s,t)* is equal to the square of $\overline{MN}$, as shown in Eq. (9). To find the minimum value of *f (s,t)* derivative *f (s,t)* on the *s* and t partial derivatives and let them equal zero, as shown in Eq. (10). Thus, we can get Eq. (11) to obtain the values of s and t. If the calculated values of s and t belong to 0 to 1. These values can be taken back to Eq. (7) and Eq. (8) to calculate $\overline{MN}$. Thirdly, if the values of s and t do not belong to 0 to 1, $\overline{MN}$ should be the minimum value of each point of RT1 to the segments of RT2, as shown in Eq. (12).

$$\begin{cases} X = x_{1i} + s\left(x_{1j} - x_{1i}\right) \\ Y = y_{1i} + s\left(y_{1j} - y_{1i}\right) \\ U = x_{2i} + t\left(x_{2j} - x_{2i}\right) \\ V = y_{2i} + t\left(y_{2j} - y_{2i}\right) \end{cases} \quad (7)$$

$$\overline{MN} = \sqrt{(X-U)^2 + (Y-V)^2} \quad (8)$$

$$f(s,t) = \overline{MN}^2 = (X-U)^2 + (Y-V)^2 \quad (9)$$

$$\begin{cases} \dfrac{\partial f(s,t)}{\partial s} = 0 \\ \dfrac{\partial f(s,t)}{\partial t} = 0 \end{cases} \quad (10)$$

$$\begin{aligned} &\left[\left(x_{1j} - x_{1i}\right)^2 + \left(y_{1j} - y_{1i}\right)^2\right]s \\ &-\left[\left(x_{1j} - x_{1i}\right)\left(x_{2j} - x_{2i}\right) + \left(y_{1j} - y_{1i}\right)\left(y_{2j} - y_{2i}\right)\right]t \\ &= \left[\left(x_{1i} - x_{1j}\right)\left(x_{1i} - x_{2i}\right) + \left(y_{1i} - y_{1j}\right)\left(y_{1i} - y_{2i}\right)\right] \\ &-\left[\left(x_{1j} - x_{1i}\right)\left(x_{2j} - x_{2i}\right) + \left(y_{1j} - y_{1i}\right)\left(y_{2j} - y_{2i}\right)\right]s \\ &+\left[\left(x_{2j} - x_{2i}\right)^2 + \left(y_{2j} - y_{2i}\right)^2\right]t \\ &= \left(x_{1i} - x_{2i}\right)\left(x_{2j} - x_{2i}\right) + \left(y_{1i} - y_{2i}\right)\left(y_{2j} - y_{2i}\right) \end{aligned} \quad (11)$$

$$\overline{MN} = minp_{1i} \rightarrow L_{2k}; p_{1j} \rightarrow L_{2k} \;\; \mathrm{i}, j, k \in \{1,2,3,4\} \quad (12)$$

The minimum distance between two rectangular zones is used to judge whether it is safe for robots to continue work. For instance, by setting a limited value, e.g., 0.5 m, if the distance is less than 0.5 m, RT2 needs to stop work and wait until RT1 goes away.

Formation control in a work operation

Formation control is mainly used in the leader-follower pattern. Each robot tractor's path is already known, and the lateral distance between two robots is limited to the desired path. The longitudinal distance is used for formation control. The velocities of the robots are changed according to the longitudinal distance between the two robots. Figure 3 shows the control parameters used in velocity control. (x_0, y_0) and $(x_0^{'}, y_0^{'})$ are the start and end points of the current path of RT1, respectively. RT_1.x, RT_1.y and RT_2.x, RT_2.y are the positions of the robot tractors. l is the lateral displacement between the two robots, d is longitudinal displacement between the two robots, L_1 is the distance from RT1 to the current path, L_2 is the distance from RT2 to the current path, d_1 is RT1's distance to the end point of the current path, d_2 is RT2's distance to the end point of the current path, and d_3 is the length of the current path. L, d_1, d_2 *and* d are used for velocity control. L, d, d_1, d_2, d_3, L_1, L_2 can be calculated by Eq. (13).

$$L_1 = \frac{\left|\left(y_0^{'} - y_0\right)RT_1.x + \left(x_0 - x_0^{'}\right)RT_1.y + \left(x_0^{'}y_0 - x_0y_0^{'}\right)\right|}{\sqrt{\left(x_0^{'} - x_0\right)^2 + \left(y_0^{'} - y_0\right)^2}}$$

$$L_2 = \frac{\left|\left(y_0^{'} - y_0\right)RT_2.x + \left(x_0 - x_0^{'}\right)RT_2.y + \left(x_0^{'}y_0 - x_0y_0^{'}\right)\right|}{\sqrt{\left(x_0^{'} - x_0\right)^2 + \left(y_0^{'} - y_0\right)^2}}$$

$$l = L_1 - L_2$$

$$d = \sqrt{\left(RT_1.x - RT_2.x\right)^2 + \left(RT_1.y - RT_2.y\right)^2 - l^2}$$

$$d_1 = \sqrt{\left(RT_1.x - x_0^{'}\right)^2 + \left(RT_1.y - y_0^{'}\right)^{2-} L_1^2}$$

$$d_2 = \sqrt{\left(RT_2.x - x_0^{'}\right)^2 + \left(RT_2.y - y_0^{'}\right)^{2-} L_2^2}$$

$$d_3 = \sqrt{\left(x_0^{'} - x_0\right)^2 + \left(y_0^{'} - y_0\right)^2} \quad (13)$$

Eq. (14) shows the equations of the velocity controller, where k, a and b are control gains. In this system, the change value, *temp*, is departed by two parts, for one robot is speed up, for the other one is speed down. Also, maximum velocity is set to ensure safety. In some conditions, the velocity of the tractor cannot increase much since the power of the tractor is not sufficient for the work. In that case, the maximum velocity is used to limit the velocity of the tractor.

$$\begin{cases} temp = kd + a\int dd_t + b\dfrac{d_d}{d_t} \\ V_1 = V_{set} + 0.5temp \\ V_2 = V_{set} - 0.5temp \\ V_1 \in \left[0, v_{max}\right], V_2 \in \left[0, v_{max}\right] \end{cases} \quad (14)$$

Path planning and skip path method

A U-turning method has been proposed for headland turn. Figure 4 shows the U-turning method. A flag is added to record the current turning status of the robot, simplified as T_F, and T_F is used in the turning cooperation. The turning procedure is composed of nine steps. The steps of the turning are as follows.

Step 1: The robot went straight forward from A to B (T_F=1) and then turned at the maximum steering angle to point C (T_F=2).

Step 2: The robot calculated the distance between the current path and the next path, which is w, and then decided the distance between point C and point D, which is w-$2r$, where r is the minimum turning radius of the robot tractor. If w was less than $2r$, the robot would go

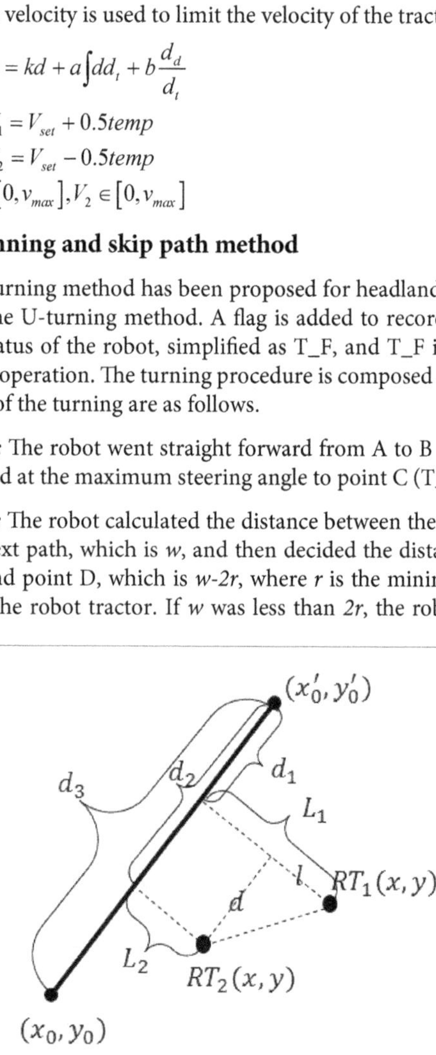

Figure 3: Control parameters of velocity control.

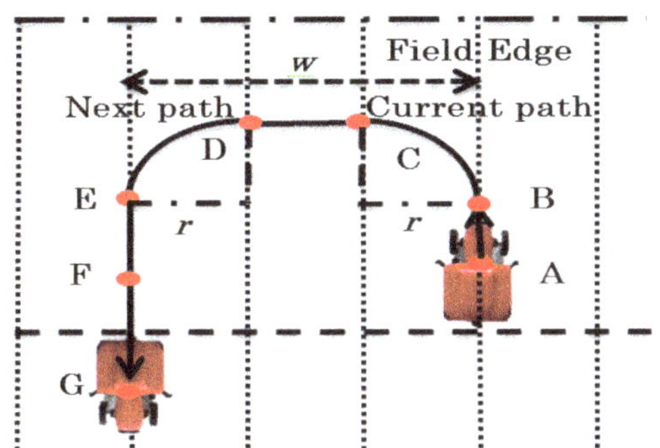

Figure 4: U-turning method.

backward to ensure a turning radius (T_F=3); otherwise, the robot went forward from point C to point D (T_F=4).

Step 3: The tractor turned to the next path from point D (T_F=5) and went straight forward to point F (T_F=6) and then went back to point E (T_F=7).

Step 4: The tractor restarted initialization from point E to point G (T_F=8), and the turning was completed at point G (T_F=9).

In general, similar to the human driver's usual practice, the robot turns to the next path adjacent to the current path. However, in this case, the robot goes backward during point C to point D since the turn radius of robot is larger than the path width.

When the robot goes backward, whatever condition it is in, it is a danger to the other robot. If the robot always goes forward to enter the next path, the width between the current path and next path should be more than 8 m, which means the robot should skip at least one path from the current path to next path. Then the path number should satisfy Eq. (15). Each set includes m paths, and Eq. (16) shows the sequence of paths. Taking (m=5, n=2) as an example, the total number of paths should be 11. Suppose the path order starts from path 1, according to Eq. (16), the path sequence should be 1->4->2->5->3->6->9->7->10->8->11.

$$\text{Path number} = mn + 1; \quad n \in N,\ m \in \{m \mid m = 2x + 1, x > 1 \, and \, x \in N\} \quad (15)$$

$$a_{mn+i} = \begin{cases} a_{mn+i-1} + \left[\frac{m}{2} \right], i \in even \, And \, i \in [2, m+1] \\ a_{mn+i-1} - \left[\frac{m}{2} \right], i \in odd \, And \, i \in [2, m+1] \end{cases} \quad (16)$$

Turning cooperation

The turning cooperation of RT1 and RT2 can be divided into four steps.

Step 1: RT1 starts turning and RT2 continues working if it is safe. If RT2 reaches a given position, which means d_2 is less than a given limited value, and RT1 is still turning from the current path, which means T_F is less than 4, RT2 will stop and wait.

Step 2: RT1 continues turning and begins to turn to the next path. RT2 continues working and starts turning. RT1 and RT2 will turn together. If it is in a dangerous situation, RT2 will stop and wait until it is safe. If RT2 has already stopped, and T_F of RT1 equals 7 and the distance between two safety zones is less than the limited value, then RT1 will stop the current operation and skip to the next operation, which means RT1 will stop going backward and will go forward to continue the turning operation.

Step 3: RT1 finishes turning and RT2 continues turning. RT1 will stop and wait at a given position, which means d_3-d_1 is larger than a given limited value.

Step 4: RT2 finishes turning, and RT1 and RT2 will work together. Formation control will be used to ensure the two robots keep a certain shape.

Step 1 and 2 are also used to avoid deadlock. In general, the deadlock may happen when T_F of RT1 equals 3 or 7. According to step 1, RT2 stops at a given position until RT1 finishes status 3. Thus it is safe for both robots. As mentioned in step 2, when it comes to a dangerous situation and RT2 already stopped, RT1 will stop current operation (T_F=7) and move to next operation (T_F=8) to avoid deadlock. In addition, before the field test, simulation is needed to check whether it is a safe situation.

Simulation Results

Simulation was needed to help us check the cooperation status of the two robots and also to determine appropriate control gains. For the robot tractors used in this study, $RT_1.l_{width}$, $RT_1.l_{front}$ and $RT_1.l_{rear}$ are 2.3 m, 3 m and 2.7 m, respectively, and the $RT_2.l_{width}$, $RT_2.l_{front}$ and $RT_2.l_{rear}$ are 2.3 m, 3.3 m and 2.4 m, respectively. According to practical experience, the velocity of a robot tractor is 3.0 km/h when conducting rotary tillage. As mentioned in formation control, the maximum velocity of a robot tractor is 3.5 km/h. The longitudinal distance between the two robots is 12 m, which is calculated by Eq. (17).

$$d_{RT1betRT2} = RT1.l_{rear} + RT2.l_{front} + \frac{1}{2} * v_{set} * t_{Em_stop} \quad (17)$$

Where, $d_{RT1betRT2}$ is the set longitudinal distance between two robots, and t_{EM_stop} is the stopping time of the robots. In this study, v_{set} was replaced by v_{max} of tractor (not v_{max} of robot), which is 15 km/h, t_{EM_stop} was 2.5 s. That is how 12 m was calculated. In addition, the minimum longitudinal distance between two robots was 8 m (2.7m + 3.3m + 0.5*1m/s*2.5s). Thus, the longitudinal distance between two robots should be more than 8 m.

As for the turning cooperation, the limited value of RT1 is the same as the longitudinal distance, and the limited value of RT2 is 2 m. For simulation, taking into account overlap, the path width is 2.2 m and the path length is 100 m.

Each robot starts from the related start point, and changes velocity according to the longitudinal distance between two robots. During headland turn operation, the simulation software simulates the trajectory of the robot based on the distance between current path and next path (in this case, the distance is 4.4 m) and the robot followed the trajectory. If the distance between two robots or the distance between two safety zones was less than a given limited value, the RT2 would stop and wait until it was safe.

Turning to an adjacent path

A navigation map with six paths was made for the simulation, and the path order was 1→2→3→4→5→6. A circular model was first used in the simulation.

Figure 5 shows the routine of the two robots. The two robots turn to adjacent path. The red line indicates RT1 and blue line indicates RT2. The first figure of Figure 6a shows the work status of two robots.

0 means headland turn and 1 means working. The minimum distance between two robots is 6.6 m. The longitudinal distance between the two robots is zero and the lateral distance between the two robots is 6.6 m, which means that the two robots are parallel to the path direction but moving in different directions, as shown in Figure 6b.

Figure 7 shows the velocity data of two robots. The waiting time of RT1 is 40.8 s and that of RT2 is 38.3 s during each headland turn.

To reduce the waiting time of two robots, rectangular model is used to retake the simulation. Figure 8a shows the distance between two robots and the distance between two rectangular zones. The minimum distance between two robots is 5.18 m, and the distance between two rectangular zones is 1.24 m. The minimum distance between two rectangular zones is 0.76 m, and the distance between two robots is 5.2 m, as shown in Figure 8b.

Figure 9 shows the velocity data of the two robots. RT1 stops and waits 25.5 s, and RT2 stops and waits 24.4 s during each headland turn.

The minimum distance between the two robots when using a rectangular model is 5.18 m, 1.42 m less than that when using a circular model. Also, compared with using a circular model, the waiting time of RT1 is reduced by 15.3 s by using a rectangular model. The advantages of a circular model are that it is easy to realize and has less calculation, but a circular model is less effective than a rectangular model for space utilization, which increased the waiting time of the two robots and decreased the work efficiency.

Skip path turn simulation

The comparison of the circle model and the rectangle model was discussed, and the rectangle model was effective on space utilization, thus the rectangle model was used in the following simulation. Skip path turn method is used to retake the simulation, as shown in Figure 10. The robot always turns left and skips 1 or 2 paths to enter the next path. Thus, the robot does not go backward to enter the next path during headland turn.

Figure 11a shows the distance between two robots and the distance between two rectangular zones when the robots skip 2 paths. The minimum distance between the two robots is 7.8 m, and the distance between the two rectangular zones is 1.88 m. The minimum distance between the two rectangular zones is 1.62 m, and the distance between

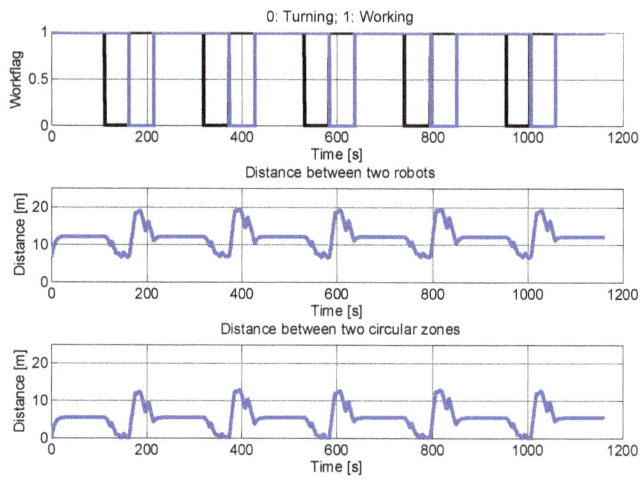

(a) Distance between two robots and two circular zones

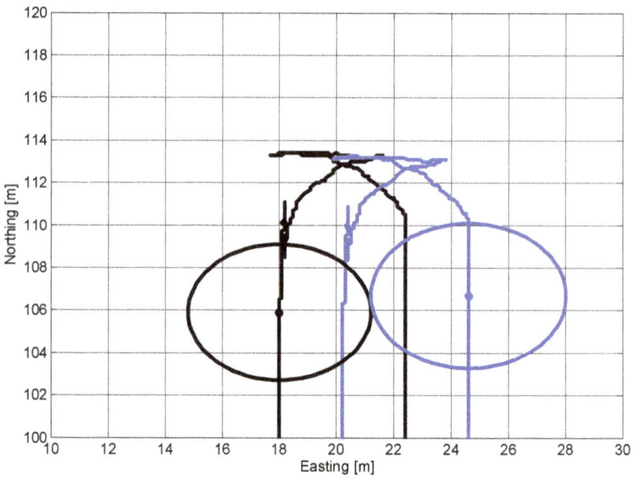

(b) Robots' status at the minimum distance between two robots

Figure 6: Performance of distance using circular model.

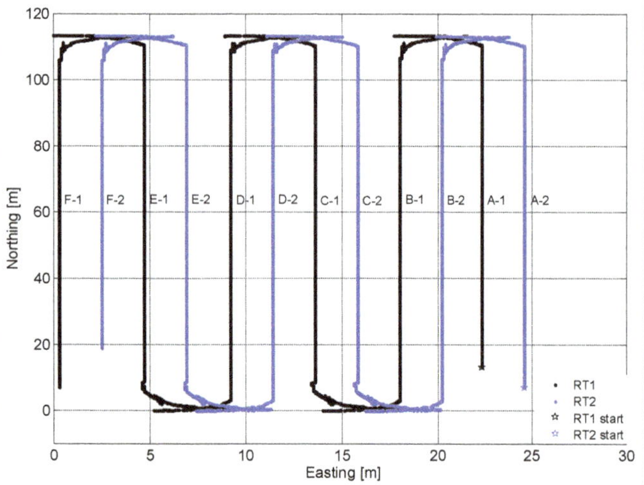

Figure 5: Simulation of routine of two robots when they turn to adjacent path.

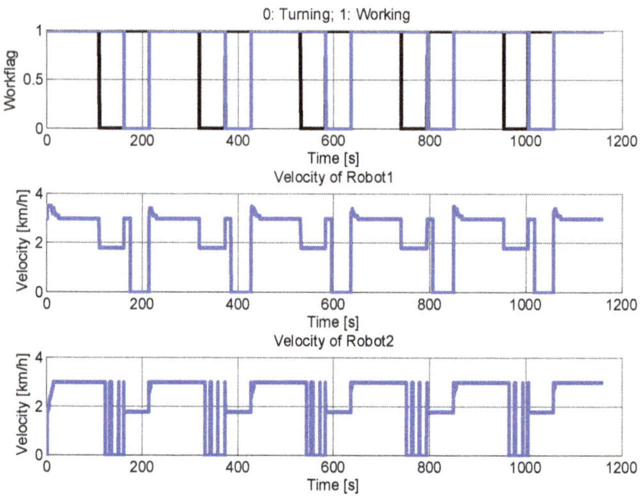

Figure 7: Velocity data of two robots using circular model.

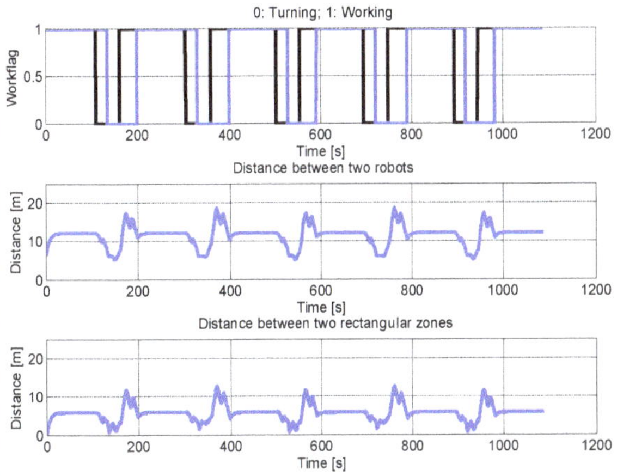

(a) Distance between two robots and two rectangular zones

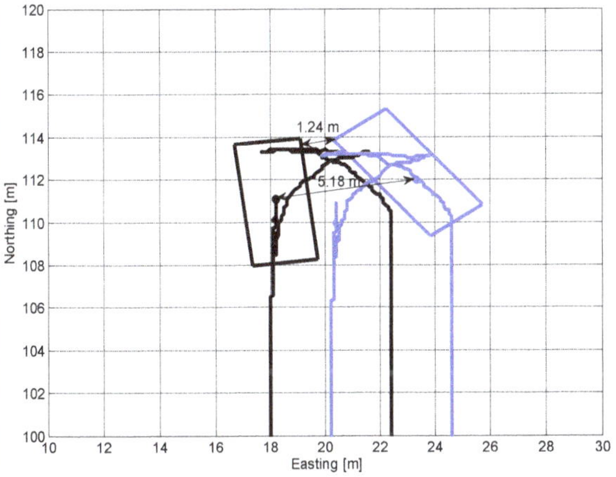

(b) Robots' status at the minimum distance between two rectangular zones

Figure 8: Performance of distance using rectangular model.

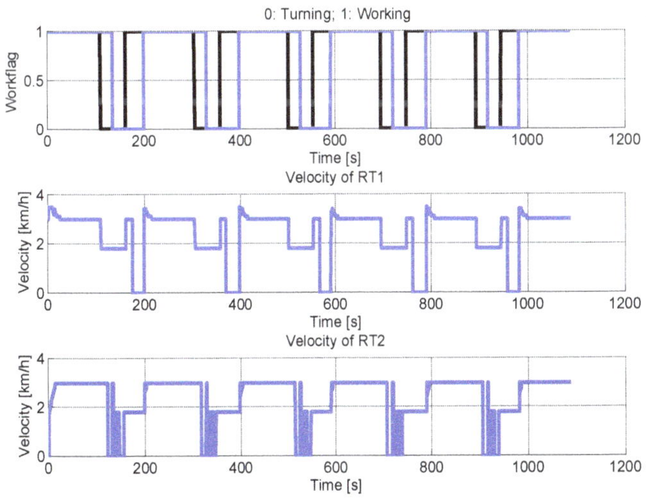

Figure 9: Velocity data of two robots using rectangular model.

the two robots is 7.9 m, as shown in Figure 11b.

Figure 12 shows the distance between two robots and the distance between two rectangular zones when the robots skip 1 path to enter the next path. The minimum distance between the two robots is 5.2 m, and the distance between two rectangular zones is 0.76 m, as shown in Figure 12b. This is a slightly closer situation than that when 2 paths are

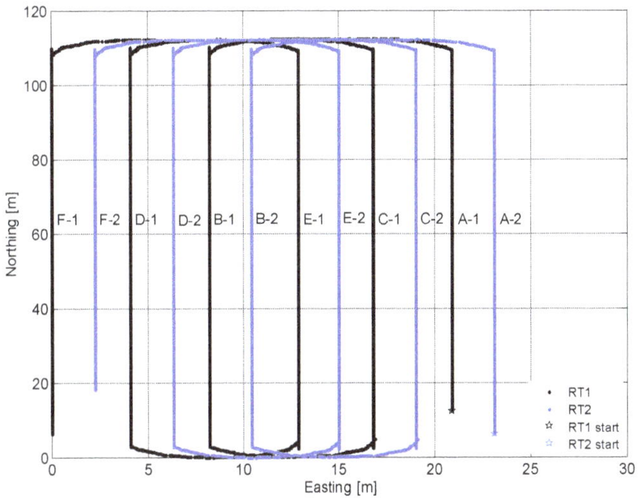

Figure 10: Simulation of routine of two robots using skip path turn method.

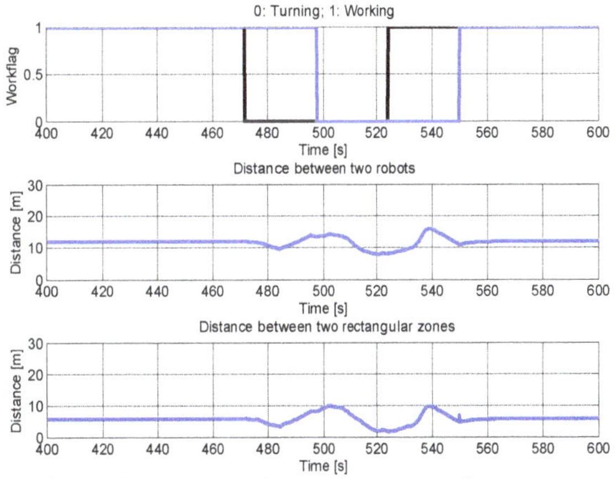

(a) Distance between two robots and two rectangular zones

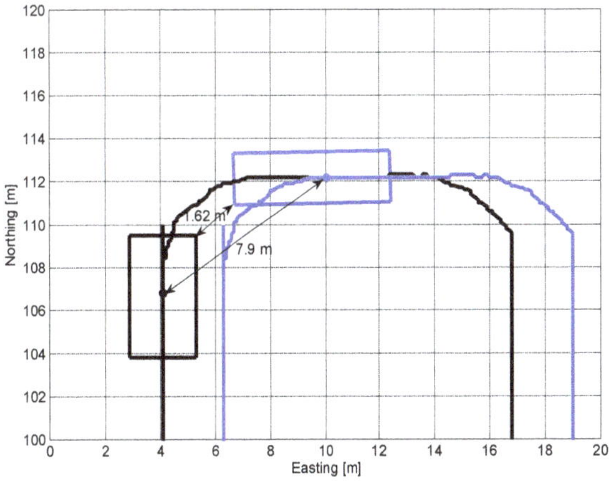

(b) Robots' status at the minimum distance between two rectangular zones

Figure 11: Performance of distance when robots skip 2 paths.

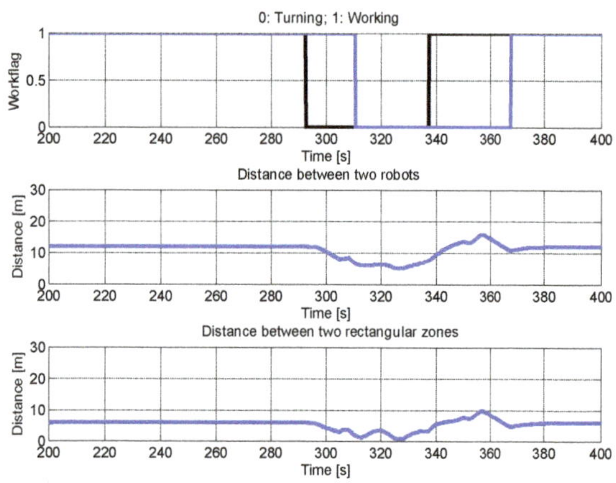

(a) Distance between two robots and two rectangular zones

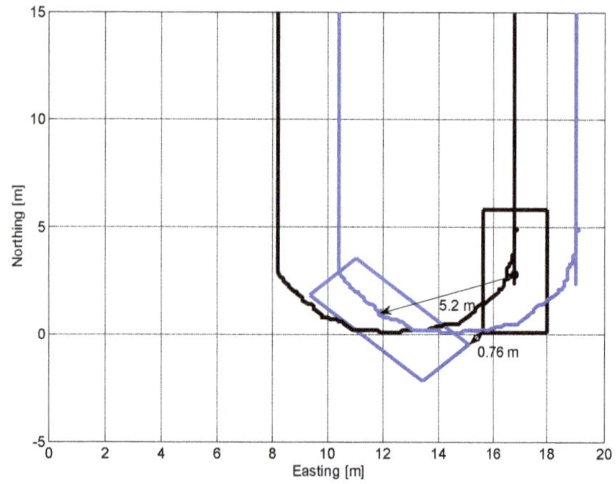

(b) Robots' status at the minimum distance between two rectangular zones

Figure 12: Performance of distance when robots skip 1 path.

skipped, but it is also an acceptable value.

Figure 13a and 13b show the velocity data of two robots when they skip 2 paths and 1 path to enter the next path, respectively. If the robots skip 2 paths, the waiting time of RT1 is 12.4 s and waiting time of RT2 is 11.2 s. If the robots skip 1 path, the waiting time of RT1 is 17.6 s and waiting time of RT2 is 16.2 s.

Work efficiency

The waiting times of RT1 and RT2 using different methods are shown in Table 1. By skipping 2 paths, the waiting time of RT1 was reduced by 7.9 s compared with that when turning to the adjacent path and was reduced by 5.4 s compared with that when skipping 1 path. In conclusion, the rectangle model is more effective than the circle model on waiting time, and skipping a path is more effective than turning to the adjacent method.

Take turning to an adjacent path as example. The setting longitudinal distance between the two robots is 12 m and turning time of RT1 is 52.3 s. It takes two robots 19.8 min to finish the work and it takes one robot 34.5 min to finish the same work, which is an improvement in efficiency of 74.2 percent. To improve the work efficiency, we need to reduce the waiting time of RT1. Increase the longitudinal distance between two robots can reduce the waiting time. Take the longitudinal distance as 34 m as an example and do the simulation. The velocity data is shown in Figure 14 Both of the robots continue working without stopping during the whole operation. In this simulation, it will take two robots 17.23 min to finish the work, an improvement in efficiency of almost 100 percent compared with that using one robot. Generally, the distance between two robots has a positive effect on waiting time of the two robots, and waiting time has a negative effect on work efficiency.

Results of field test

Figure 15 shows the trajectory of two robots of the experiment. The experiment was taken in a farm in Hokkaido University. The length of the path was 100.8 m, and the width of the each robot's path was 2.2 m. The velocity of two robots was 3.0 km/h and the maximum velocity of two robots was 3.5 km/h. The longitudinal distance between two robots

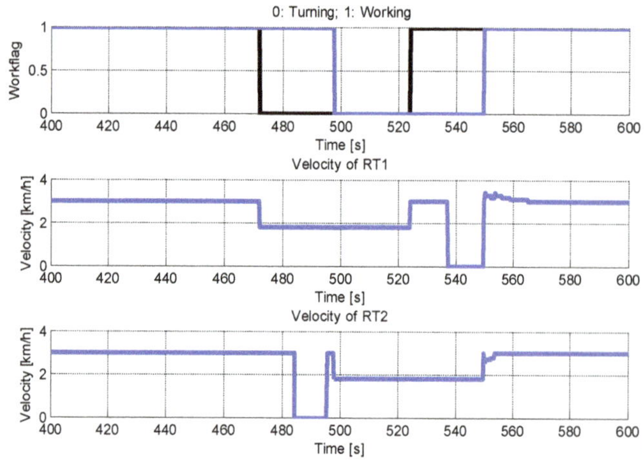

(a) Velocity data of two robots when skip 2 paths

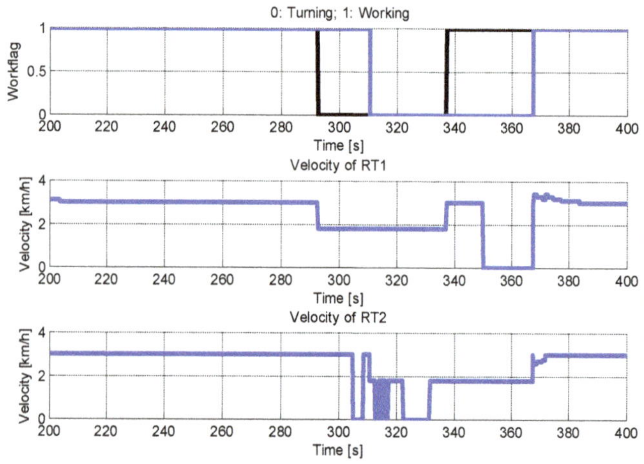

(b) Velocity data of two robots when skip 1 path

Figure 13: Velocity data of two robots using skip path turn method.

	Waiting time of RT1	Waiting time of RT2	Total waiting time (each turn)
Skip 0 path	25.5 s	24.4 s	25.5 s
Skip 1 path	17.6 s	16.2 s	17.6 s
Skip 2 path	12.4 s	11.2 s	12.4 s
Skip 0 path (Circle model)	40.8 s	38.3 s	40.8 s

Table 1: Waiting time of two robots.

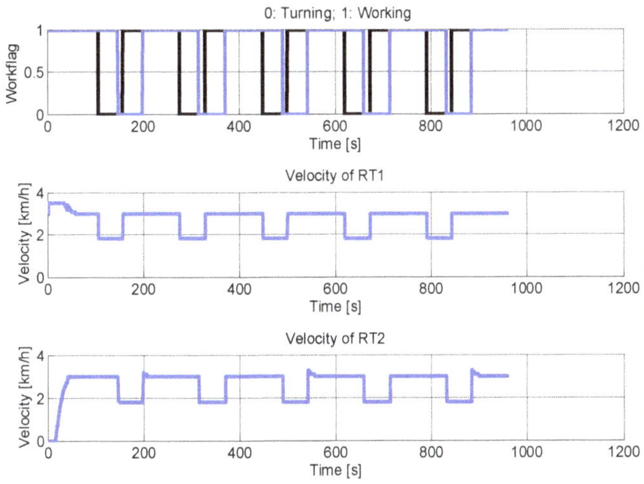

Figure 14: Velocity data of two robots when waiting time of RT1 is zero.

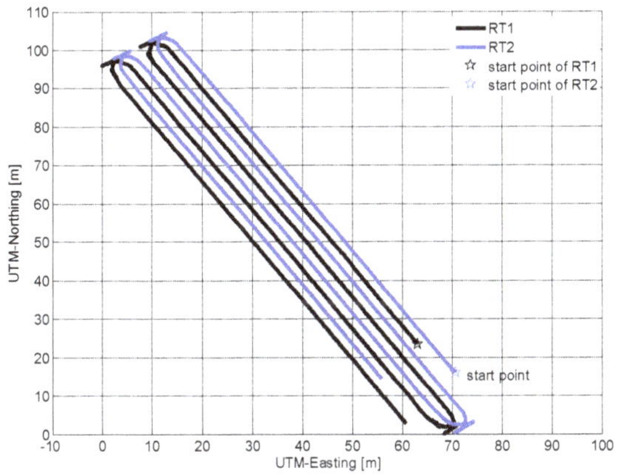

(a) Trajectory

(b) Photo of two robots

Figure 15: Trajectory of two robots.

was 12 m.

Figure 16 shows the performance of distance of two robots. The minimum distance between two robots was 5.12 m, and the distance between two rectangles was 1.02 m. This distance is 0.39 m larger than that of simulation. The minimum distance between two rectangles was 0.85 m, which is safe for two robots to work. The average error of distance between two robots when they were working on the path is 0.19 m, and the RMS of the distance is 0.22 m. The average error of distance between two rectangles when two robots were working on the path is 0.13 m, and the RMS of this distance is 0.34 m.

Figure 17 shows the performance of velocity of two robots. According to the experiment results, RT1 stopped and waited 18 s during each turning, 7.5 s shorter than simulation. RT2 stopped and waited 18.4 s during each turning, 6 s shorter than simulation. The main reason of this is that the real velocity of the robot tractor is always changing, and delay exists between command and response. The delay is about 1.8 s.

Figure 18 shows the accuracy of each robot. Lateral error is used to evaluate the robot's performance. Lateral error is the distance error between robot's position and pre-determined path. The average lateral error of RT1 is -0.03 m, and RMS of lateral error of RT1 is 0.05 m. The average lateral error of RT2 is -0.02 m, 0.01 m less than RT1. And RMS of lateral error of RT2 is 0.03 m.

The total work time of the experiment is 11.3 min, improved 79.6% work efficiency compared with using one robot, which takes 20.3 min to finish the work.

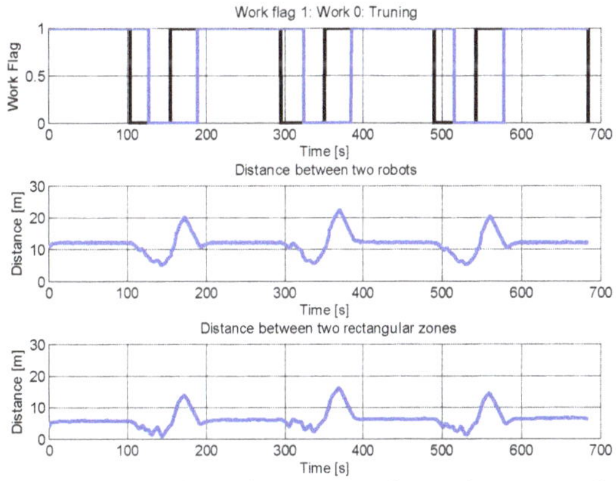

(a) Experiment of distance between two robots and two rectangular zones

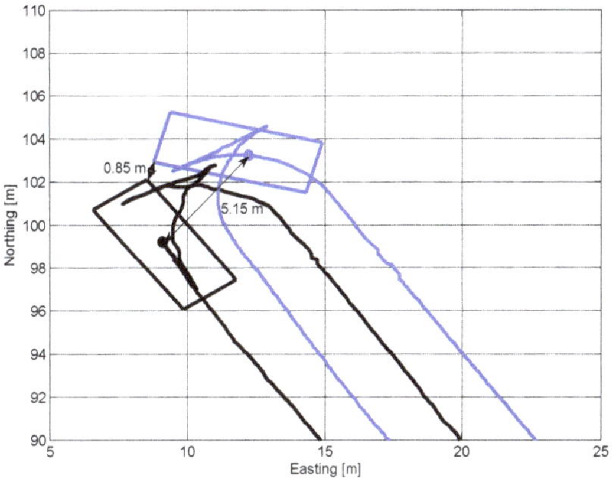

(b) Experiment of robots' status at the minimum distance between two robots

Figure 16: Performance of distance between two robots.

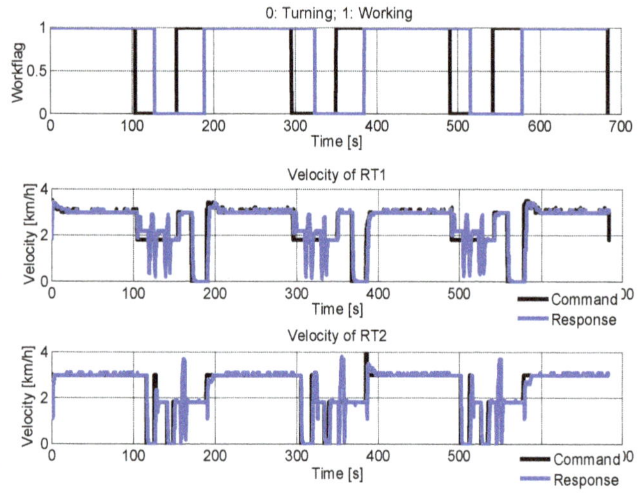

Figure 17: Experiment of velocity data of two robots.

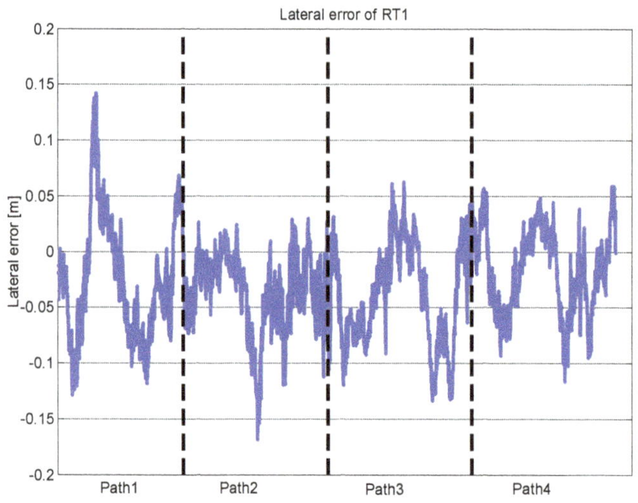

(a) Lateral error of RT1

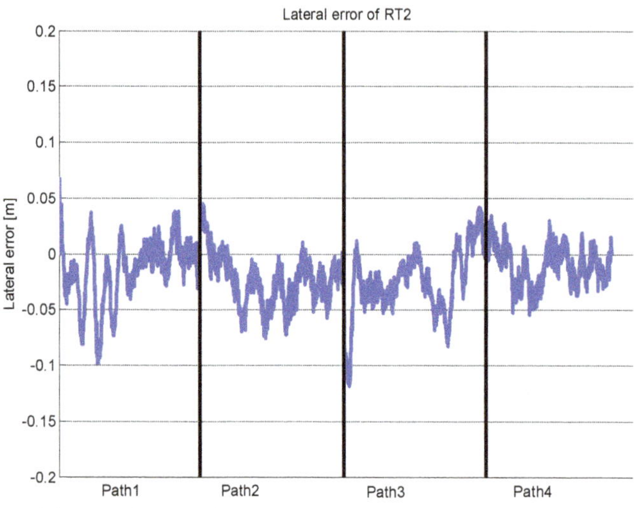

(b) Lateral error of RT2

Figure 18: Lateral error of two robots.

Conclusion

Cooperation of two robot tractors was proposed in this article. Each robot individually tracks its desired path and maintains a certain shape during the work operation. To ensure the safety of robot tractors, a rectangle model and a circle model were proposed. Path planning and turning cooperation were used to improve the efficiency of this system.

The results of simulation showed that the rectangle model reduced waiting time by 15.3 s compared with that using the circle model when the robots turn to adjacent paths. By skipping 2 paths to turn to the next path, the waiting times of RT1 and RT2 were reduced by 5.2 s and 5 s, respectively, compared with that when skipping 1 path. If the robots turn to adjacent paths, the waiting times of RT1 and RT2 would be 25.5 s and 24.4 s, respectively. The work efficiency was limited by the length of field, set velocity of robots and waiting time of RT1. If the length is 100 m, set velocity is 3.0 km/h, and set distance between the two robots is 12 m, the efficiency will be improved by 74.2 percent compared with that using one robot. However, if the length of the field is increased to 500 m, the efficiency improves by 92.5 percent. If the set distance is increased to 34 m, the efficiency improves almost 100 percent compared to that using one robot. In addition, the improved work efficiency have a positive effect on the energy consumption ($E_{con}=2/(1+E_w)$, where, E_{con} is the efficiency of energy consumption, E_w is the improved work efficiency. If E_w increases, then E_{con} decreases; the smallest E_{con} possible is optimal. For instance, if the E_w increase from 60 percent to 100 percent, the E_{con} will decrease from 1.25 to 1.

In conclusion, cooperation of two robot tractors can reduce work time and work strength. Work efficiency is based on the setting parameters and the system can be improved close to 100 percent efficiency compared with using one robot.

Acknowledgement

This study was supported by the Cross-ministerial Strategic Innovation Promotion Program (SIP) managed by Cabinet Office, Japan.

References

1. Billingsley J, Schoenfisch M (1997) The successful development of a vision guidance system for agriculture. Computers and Electronics in Agriculture 16: 147-163.
2. Hague T, Marchant JA, Tillett ND (2000) Ground based sensing systems for autonomous agricultural vehicles. Computers and Electronics in Agriculture 25: 11-18.
3. Kaizu Y, Imou K (2008) A dual-spectral camera system for paddy rice seedling row detection. Computers and Electronics in Agriculture 63: 49-56.
4. Kise M, Zhang Q (2008) Development of a stereovision sensing system for 3D crop row structure mapping and tractor guidance. Biosystems Engineering 101: 191-198.
5. Stombaugh TS, Benson ER, Hummel JW (1998) Automatic guidance of agricultural vehicles at high field speeds. ASAE paper.
6. Bell T (1999) Automatic tractor guidance using carrier-phase differential. G.P.S. Computers and electronics in agriculture: Special Issue Navigating Agricultural Field Machinery.
7. Noguchi N, Reid JF, Zhang Q, Will JD, Ishii K (2001) Development of Robot Tractor based on RTK-GPS and Gyroscope. ASAE Paper.
8. Berber M, Arslan N (2013) Network RTK: A case study in Florida. Measurement 46: 2798-2806.
9. Cheng TM, Savkin AV, Javed F (2011) Decentralized control of a group of mobile robots for deployment in sweep coverage. Robotics and Autonomous Systems 59: 497-507.
10. Ni W, Wang X, Xiong C (2013) Concensus controllability, observability and

robust design for leader-following linear multi-agent systems. Automatica 49: 2199-2205.

11. Zhai C, Hong Y (2013) Decentralized sweep coverage algorithm for multi-agent systems with workload uncertainties. Automatica 49: 2154-2159.
12. Noguchi N, Will J, Reid J, Zhang Q (2004) Development of a master-slave robot system for farm operations. Computers and Electronics in Agriculture 44: 1-19.
13. Zhang X, Geimer M, Noack PO, Grandl L (2010) Development of an intelligent mastereslave system between agricultural vehicles. In Proceedings of the 2010 IEEE intelligent vehicles symposium pp: 250-255.
14. Vougioukas SG (2012) A distributed control framework for motion coordination of teams of autonomous agricultural vehicles. Biosystems Engineering 2: 284-297.
15. Cheng TM, Savkin AV (2011) Decentralized control of multi-agent systems for swarming with a given geometric pattern. Computers and Mathematics with Applications 61: 731-744.
16. Shen D, Sun W, Sun Z (2014) Adaptive PID formation control of nonholonomic robots without leader's velocity information. ISA Transactions 53: 474-480.
17. Abbaspour A, Alipour K, Jafari HZ, Moosavian SAA (2015) Optimal formation and control of cooperative wheeled mobile robots. Comptes Rendus Mecanique 343: 307-321.
18. Agarwal M, Agrawal N, Sharma S, Vig L, Kumar N (2015) Parallel multi-objective multi-robot coalition formation. Expert Systems with Applications 42: 7797-7811.
19. Dai Y, Kim Y, Wee S, Lee D, Lee S (2015) A switching formation strategy for obstacle avoidance of a multi-robot system based on robot priority model. ISA Transactions 56: 123-134.
20. Yang L, Noguchi N (2014) Development of a Wheel-Type Robot Tractor and its Utilization. In Proceedings of the 19th World Congress of the IFAC, 11571-11576. Cape Town, South Africa.

15

The Generalized Architecture of the Spherical Serial Manipulator

González-Palacios MA*, Ortega-Alvarez CJ, Sandoval-Castillo JG, Cuevas-Ledesma SM and Mendoza-Patiño FJ

Division of Engineering Campus Irapuato-Salamanca, University of Guanajuato, Salamanca Carr - V of Santiago, Community Palo Blanco, Salamanca, Mexico

Abstract

It is well known that the inverse kinematics problem for the spherical serial manipulator has been solved in the past by diverse methods; but this problem for a generalized architecture based on the Denavit-Hartenberg parameters has not been treated. Therefore, this paper considers such treatment in a geometric analysis to derive a closed-form solution of the inverse kinematics problem, whose algorithm is validated by simulating pick and place operations. With the code implementation of a novel linear tracking algorithm introduced here, this application is accomplished in real time with the aid of a development software devoted to simulate robotic applications in real time, allowing the visualization of the performance of all possible architectures including the eight types defined in this paper; It is also shown that the Stanford arm is comprised within this classification. In order to demonstrate the great potential offered by combining the algorithms released here in simulations for industrial applications requiring a quick response, a case study is presented by taking as examples, the Stanford manipulator and a spherical manipulator with generalized architecture.

Keywords: Inverse kinematics problem; Industrial serial manipulators; Hartenberg-Denavit notation; Real time simulation; Linear path tracking; Robot architecture

Introduction

Some architectures of serial manipulators like the RRP have been studied in different ways, for instance, in [1] a RRP manipulator is studied to solve the inverse kinematics problem, IKP for short, near singularities applying the least-square method. Furthermore, by Saeidpourazar [2,3] the forward and inverse kinematics problems are analyzed in a RRP nanomanipulator called MM3A, which is able to perform specific tasks. It was also developed a controller which evaluates the vision and strength of the MM3A.

A numerical method that solves the IKP of six-degree-of-freedom manipulators with combinations of revolute, prismatic or cylindrical joints, is proposed in [4]. A similar study in [5] shows an algorithm for the solution of the inverse kinematics problem of six- or fewer degree-of-freedom manipulators, where the PUMA robot is analyzed as a case study, and being implemented in the KINEM package, which is a software that solves the inverse kinematics problem of several serial manipulators.

DYNAMAN [6], is another simulation software package developed in FORTRAN, where a method to formulate the dynamic equations is applied on serial robots with multiple links, with either revolute or prismatic joints. This software does not have an interface to modify the DH parameters in a simple way. Besides, the model generated lacks of details.

SYMORO [7] and OpenSYMORO [8] is an open source software package developed fundamentally in Pythom dedicated to the symbolic modeling of robots. This package provides support to robot models with flexible joints and to mobile robots. It offers support to serial robots with open and closed chains. A visualization tool to perceive the robot structure is also provided. Another software Package, RobSim [9], whose platform is MATLAB, is able tosimulate five manipulator types: spherical, cartesian, cylindrical, and two architectures of the articulated robot (KAU) PYR and RPR, where link dimensions and robot poses can be specified. It also displays the workspace in 2D.

Some references discussed the IKP for different architectures in the same configuration as in [10], which is obtained for a 5R manipulator using Clifford's exponential algebra, it is also presented in [11] a unified solution in closed form of a serial robot with six-degree-of-freedom by Pieper's geometry or the Duffy's geometry. On the other hand, Pieper [12], studies rigid body motions onsix-degree-of-freedom manipulators, with revolute or prismatic joints. FORTRAN is applied for the position analysis and trajectory generation. By means of using velocity methods and Newton-Rapshson technics, numeric solutions of the IKP of the general case are presented. Although Pieper's work is focused in the methodology required to solve any type of manipulator with six-degree-of-freedom, only presents the study case of the Stanford manipulator, where the equations to obtain the joint variables are shown.

Paul and Shimano [13], provide rules to assign coordinate frames to manipulator links, including simple cases and prismatic joints. In this work, the Stanford manipulator is presented as study case to prove their algorithm to solve position and orientation.

Khalil and Murareci [14] establish the possibility of considering special values of the geometric parameters of manipulators to take the characteristic polynomial to the simplest form using a minimum number of equations. This polynomial, for special architectures, can be obtained numerically. The RRP architecture is studied only when $\alpha_1 = n\pi$ and $\alpha_2 = n\pi$ (the first three axes are parallel), besides, an equation for each joint variable is not presented explicitly.

Much of the kinematic analysis presented in the literature provides numerical solution approaches, as in the case of [15], where a finite element method is applied to solve the forward kinematics

***Corresponding author:** González-Palacios MA, Division of Engineering Campus Irapuato-Salamanca, University of Guanajuato, Salamanca Carr - V of Santiago, Community Palo Blanco, Salamanca, Mexico, E-mail: maxg@ugto.mx problem of the Stanford manipulator using the software Maple V. A probabilistic approach, specifically the Monte Carlo method, is applied in [16] to model the kinematic and dynamic random errors of various manipulator parameters, such as those related to the Stanford manipulator. In [17] the equations of motion are developed for any manipulator with two - or three-degree-of-freedom, where the PUMA and the Stanford robots are presented as case studies.

An important subject is the validation of the results obtained in mathematical models with the application of computer packages for simulation. For example, in [18] are implemented in LabVIEW the forward kinematic and the dynamic analysis of a SCARA manipulator, whereas in [19], MATLAB is applied to solve the forward kinematics problem of a spherical manipulator by implementing a FPGA.

It was found, in general, that the specific solution of the IKP of the spherical manipulator is based on either the basic architecture or the architecture defining the Stanford arm, but to the best of our knowledge, the solution to a generalized architecture has not been reported explicitly in the literature. This generalized solution involves the formulation that embrace eight different possible architectures that the spherical manipulator can hold depending on the values of the first two twist angles according to the Denavit-Hartenberg notation [20].

The generalized closed-form solution of RRP serial manipulators presented in this paper, is an important contribution to the research and development field, since the resulting algorithm, represents a powerful tool to study in real time, the motion of all possible architecture combinations within the same framework.

In this research work, a section is dedicated to establish the guidelines of what we call the generalized architecture of the spherical manipulator. Then, we define the eight types of the generalized spherical manipulator, followed by the formulation of the inverse kinematics solution, which in most cases is represented as a closed-form solution. This formulation includes the pseudo-code that was applied as template to implement the code in ADEFID's platform [21]. We have dedicated a section to present an algorithm that generates intermediate poses by providing a single parameter, once the initial and final poses of a linear trajectory have been defined.

In the last section, we present a case study related to a simulation of pick and place operations. We integrate in the simulation, the algorithm for the inverse kinematics solution and the algorithm for the linear trajectory planning. In order to provide a closer taste of the performance, sets of frames are sequentially presented. The simulation is running in real time within the software developed for this application.

The Generalized Spherical Manipulator

The manipulator designed with spherical architecture for the positioning of the wrist center point, is the well-known Stanford arm, whose pioneer model is shown in Figure 1a, while Figure 1b shows a more up-to-date version.

The full architecture of the manipulators described above is identified as RRPRRR. Since the last three joints are connected with the architecture of a spherical wrist (RRR), we will refer to these type of manipulators by the first three joints only, namely, the RRP spherical manipulator.

Our study is focused in the derivation of the inverse kinematics solution for the positioning of the wrist center point of what we call the generalized architecture of the spherical manipulator, in other words, according to the definitions of the links parameters following the Denavit-Hartenberg's notation [20], DH notation for short, all link offsets and link lengths are not zero.

The skeleton representation of the basic architecture of the spherical manipulator along with the corresponding DH parameters, is shown in Figure 2, based on the values given in Table 1, where all link lengths a_i, as well as b_2 are zero. The architecture of the Stanford manipulator, shown in Figure 3, is obtained from the basic architecture of the spherical manipulator by setting b_2, the offset of the second link, not equal to zero, as shown in Table 2. Now, the generalized architecture, with the parameters given in Table 3, is plotted in Figure 4.

On the Classification

Applying the same concept introduced in [22] regarding the manipulators' architecture, we define here eight different types depending on α_1 and α_2 values. Thus, α_1 can be either 90° or 270°, if the manipulator is right shoulder (R) or left shoulder (L), respectively. Whereas α_2, which defines the direction of motion of the wrist center point while the manipulator is set at the zero configuration with $b_3 \neq 0$, can hold 0, 90°, 180° , and 270°. Table 4 shows the nomenclature defined for this classification, and Figures 5 and 6, the skeleton representation of right shoulder and left shoulder.

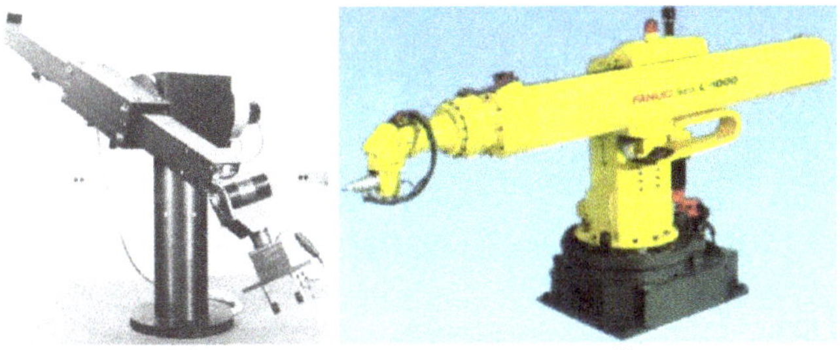

(a) (b)

Figure 1: Stanford manipulator.

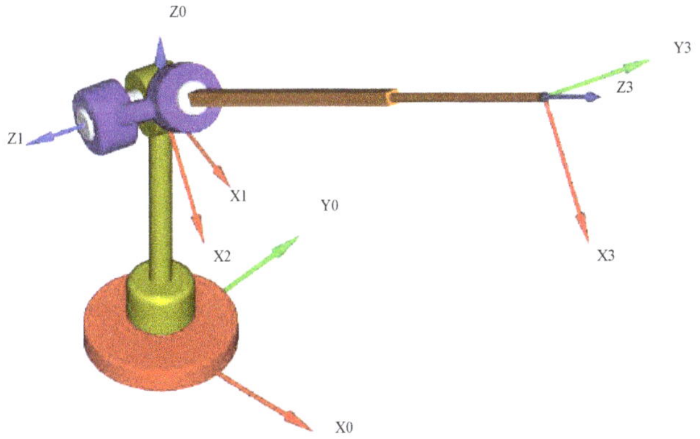

Figure 2: Skeleton representation of the spherical architecture.

Link	θ_i	b_i	a_i	α_i
1	θ_1^*	b_1	0	90°
2	θ_2^*	0	0	-90°
3	0	b_3^*	0	0

*Joint variables

Table 1: DH parameters of the spherical architecture.

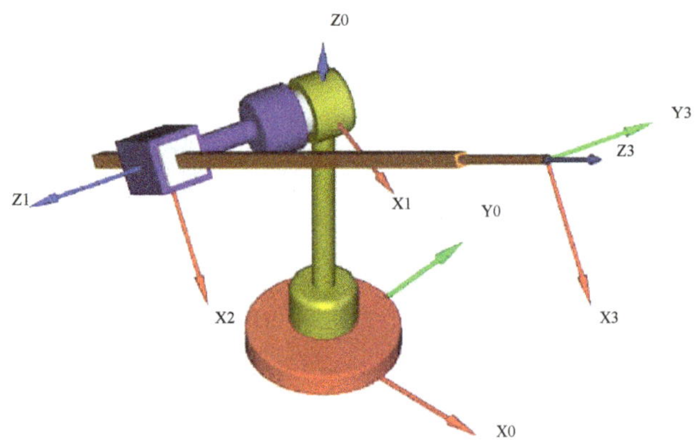

Figure 3: Skeleton representation of the Stanford architecture.

Link	θ_i	b_i	a_i	α_i
1	θ_1^*	b_1	0	90°
2	θ_2^*	b_2	0	-90°
3	0	b_3^*	0	0

*Joint variables

Table 2: DH parameters of the stanford architecture.

Link	θ_i	b_i	a_i	α_i
1	θ_1^*	b_1	a_1	α_1
2	θ_2^*	b_2	a_2	α_2
3	θ_3	b_3^*	a_3	α_3

*Joint variables
α_1=90°, 270°
α_2=0,90°, 180°, 270°

Table 3: DH parameters of the generalized spherical architecture.

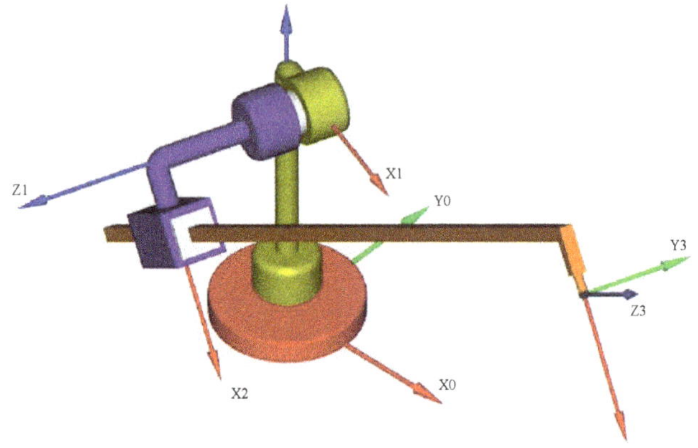

Figure 4: Skeleton representation of the generalized spherical architecture.

The Generalized Geometric Approach

In this section, we establish an algorithm that describes the inverse kinematics solution of the generalized spherical architecture RRP, with spherical wrist. The approach presented here, holds the eight architectures described in Section 3, nevertheless, we select the RD type to show the variables and to state their geometric relationship that will be applied along the formulation. Therefore, we depart from the fact that except for the joint variables, we know all the values given in Table 5. We also assume that the position of the wrist center point $p=(p_x, p_y, p_z)^T$, is known.

To obtain θ_1, the joint variable that orientates x_1 with respect to x_0, we use Fig. 7 as reference, where all the linear dimensions shown, lie in parallel planes to the x_0-y_0 plane. Thus, θ_1 is obtained as

$$\theta_1 = \phi + \beta \quad (1)$$

From eq. (1), ϕ is related with $r = \sqrt{p_x^2 + p_y^2}$ by,

$$\phi = \arctan(p_y/p_x) \quad (2)$$

while β is associated with h and r as,

$$\beta = \arctan(h/\sqrt{r^2 - h^2}) \quad (3)$$

with

$$h = b_2 + \lambda_2 b_3 \quad (4)$$

In eq. (4), $\lambda_2 = \cos\alpha_2$. Further down we will take into account that $\lambda_i \equiv \cos\alpha_i$ and $\mu_i \equiv \sin\alpha_i$ for i=1,2,3. It is important to note that the equations derived for the inverse kinematic solution are obtained for an arbitrary value of α_2, and the four values for the architectures defined here become particular cases Figure 7.

Now, to obtain θ_2, the joint variable that orientates x_2 with respect to x_1, we use Figure 8, in this case, all the linear dimensions shown, lie in parallel planes to the x_1 - z_0 plane. Thus, θ_2 is obtained as

Architecture	Description		α_1	α_2
RD	Right Shoulder	Down Wrist	90°	90°
RU		Up Wrist		270°
RE		External Wrist		0°
RI		Internal Wrist		180°
LD	Left Shoulder	Down Wrist	270°	270°
LU		Up Wrist		90°
LE		External Wrist		180°
LI		Internal Wrist		0°

Table 4: Classification according to the twist angles.

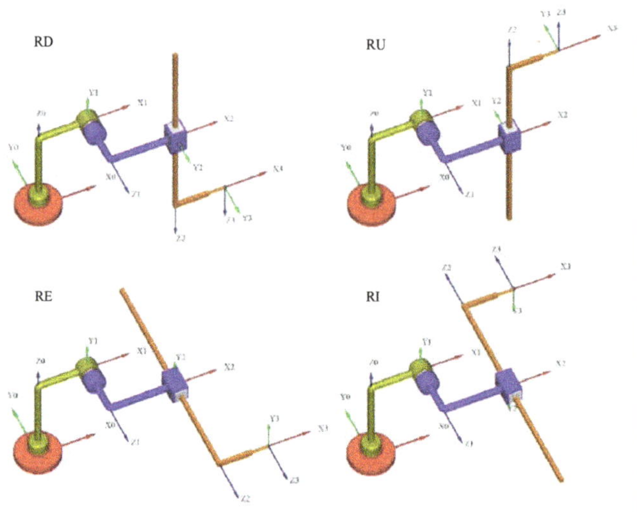

Figure 5: Right shoulder architectures.

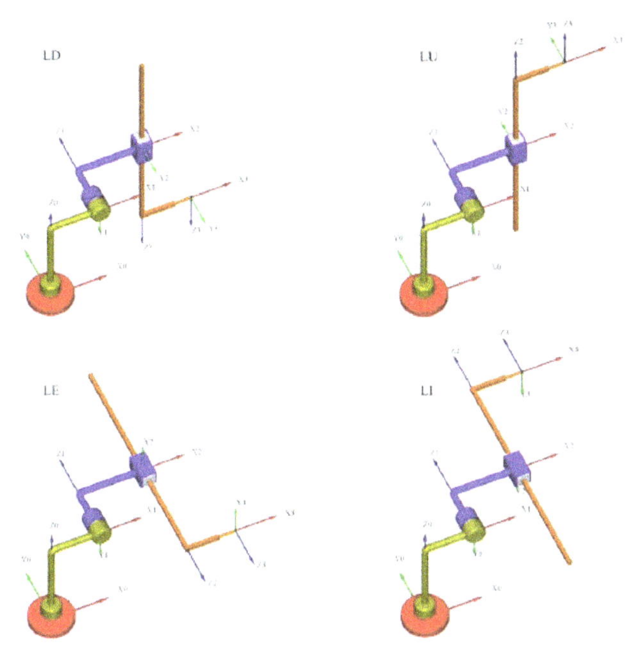

Figure 6: Left shoulder architectures.

θ_i	b_i	a_i	α_i
θ_1^*	b_1	a_1	90°
θ_2^*	b_2	a_2	α_2
θ_3	b_3^*	a_3	α_3

*Joint variables

Table 5: DH table for the right shoulder architecture.

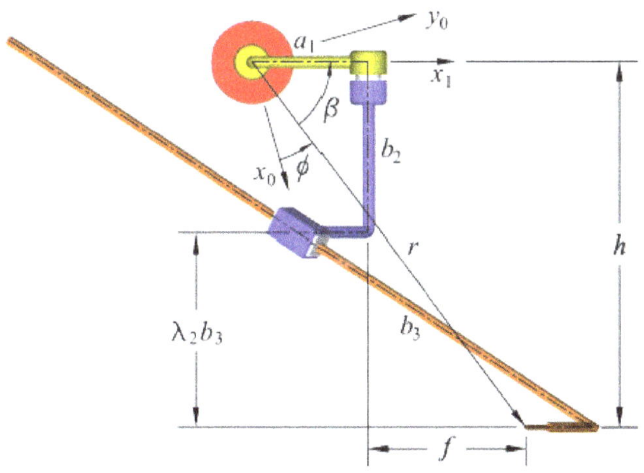

Figure 7: Plant view of a RD architecture.

$$\theta_2 = \gamma + \psi \tag{5}$$

In eq. (5), γ is associated with a_2, a_3 and a projection of b_3 as,

$$\gamma = \arctan(\mu_2 b_3/(a_2+a_3)) \tag{6}$$

While ψ is related with d and f by,

$$\psi = \arctan(d/f) \tag{7}$$

where,

$$d = p_z - b_1 \tag{8}$$

$$f = \sqrt{r^2 - h^2} - a_1 \tag{9}$$

Now, a projection of $b3$ is related to $a2$, $a3$ and e, namely,

$$\mu_2^2 b_3^2 = e^2 - (a_2 + a_3)^2 \tag{10}$$

with

$$e = \sqrt{d^2 + f^2} \tag{11}$$

To solve θ_1 and θ_2, from eqs. (1) and (5), respectively, we need ϕ, β, γ, and ψ. From eq. (2) we see that ϕ depends on the position of the wrist center point, which is given, but the other three angles, are functions of b_3, therefore, we must solve eq. (10) first, which for arbitrary values of α_2, must be evaluated numerically. However, if $a_1 = 0$, then b_3 is obtained in closed-form as,

$$b_3 = \sqrt{d^2 + r^2 - (\mu_2 b_2)^2 - (a_2 + a_3)^2} - \lambda_2 b_2 \tag{12}$$

A similar analysis can be performed to solve the inverse kinematics of the left shoulder manipulator. Now, for the eight types defined in Section 3, a closed-form solution for b_3 is also obtained. Taking into account that $b_3 > 0$, there are two possible solutions for a given architecture, and they depend on the position of the manipulator's body, which is represented by the first joint. The first set of solutions

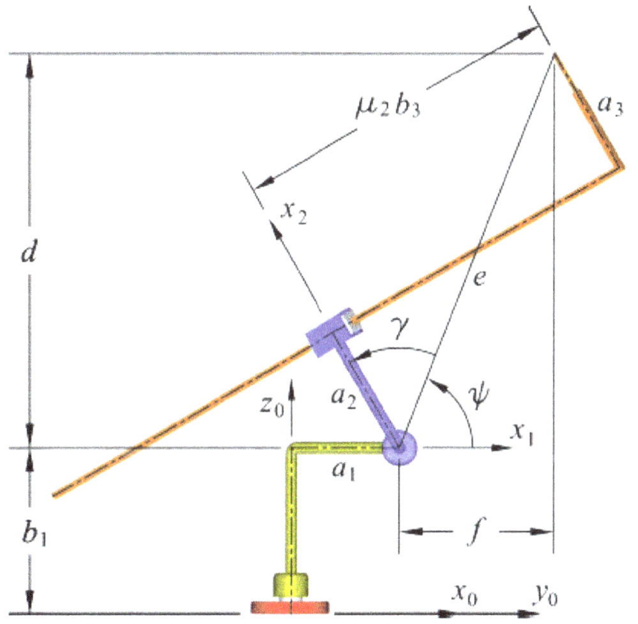

Figure 8: Side view of a RD architecture.

Arch.	b_3	θ_1	θ_2
RD	$\sqrt{d^2 + ([r^2 - b_2^2]^{1/2} - a_1)^2 - (a_2 + a_3)^2}$	$\phi+\beta$	$\gamma+\psi$
RU			
LD			
LU			
RE	$\sqrt{r^2 - ([(a_2 + a_3)^2 - d^2]^{1/2} + a_1)^2} - b_2$		
LE			
RI	$-\sqrt{r^2 - ([(a_2 + a_3)^2 - d^2]^{1/2} + a_1)^2} + b_2$		
LI			

Table 6: Joint variables according to the eight architectures. First body solution.

<table>
<tr><th>Arch.</th><th>b_3</th><th>θ_1</th><th>θ_2</th></tr>
<tr><td>RD</td><td rowspan="4">$\sqrt{d^2+([r^2-b_2^2]^{1/2}+a_1)^2-(a_2+a_3)^2}$</td><td rowspan="8">$(\phi-\beta)+\pi$</td><td rowspan="8">$(\gamma-\psi)-\pi$</td></tr>
<tr><td>RU</td></tr>
<tr><td>LD</td></tr>
<tr><td>LU</td></tr>
<tr><td>RE</td><td rowspan="2">$\sqrt{r^2-([(a_2+a_3)^2-d^2]^{1/2}-a_1)^2}-b_2$</td></tr>
<tr><td>LE</td></tr>
<tr><td>RI</td><td rowspan="2">$-\sqrt{r^2-([(a_2+a_3)^2-d^2]^{1/2}-a_1)^2}+b_2$</td></tr>
<tr><td>LI</td></tr>
</table>

Table 7: Joint variables according to the eight architectures. Second body solution.

Link	θ_i (deg)	b_i (mm)	a_i (mm)	α_i (deg)
1	30*, 170.35**	300	100	90
2	110*, -137.31**	150	100	90
3	0	400*, 571.5**	75	0
*Body solution 1, **Body solution 2				

Table 8: DH parameters for the solutions of the RD architecture.

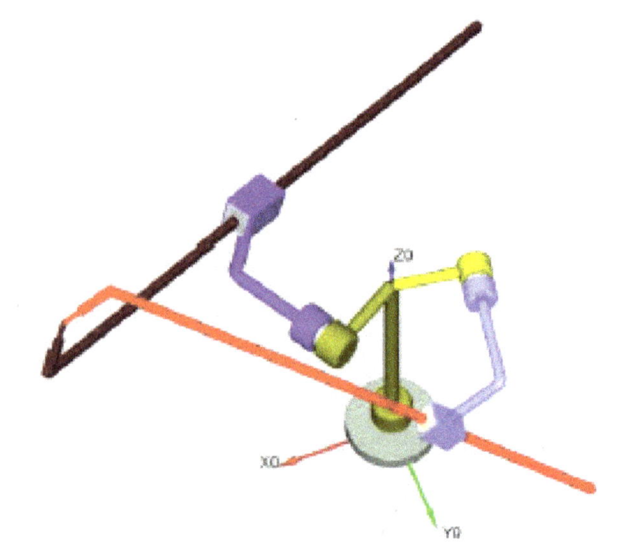

Figure 9: Representation of the two solutions of the RD architecture with p = (435,78,601)^T.

is listed in Table 6, while the second set, in Table 7. To the best of our knowledge, these sets have not been reported in the literature.

To illustrate the two solutions to reach a given wrist center point, an example of a RD manipulator is presented here. The coordinates of p are (435,78,601) mm, and the two solutions together with the design parameters are shown in Table 8. The simulation is achieved in ADEFID's platform [21], and Figure 9 illustrates the two poses of the manipulator.

With the closed-form solution for the inverse kinematics positioning approach, above introduced, we have derived a pseudo-code that consider all possible solutions in which b_3 can be evaluated with no need of a numerical solution. Furthermore, α_1 and α_2 are symbolically expressed so that not only the eight architecures are contemplated but also those architectures with an arbitrary value of α_2 if $a_1 = 0$. Such pseudo-code is listed next:

```
beginfunSetInversePosition(DH,p)
    r² ← px² + py²
    d ← (pz − b1)μ1
    ϕ ← arctan(py/px)
    ifa1 == 0
        κ ← d² + r² − μ2²b2² − (a2 + a3)²
        ifκ<0  return FALSE
        b3 ← √κ − λ2b2
    else  if ABS(μ2) < 0                    //If  α2 = 0° or 180°
        if IsBodySol1 == TRUE
            κ ← r² − (√((a2 + a3)² − d²) + a1)²
        else
            κ ← r² − (√((a2 + a3)² − d²) − a1)²
        ifκ<0  return FALSE
        b3 ← (√κ − b2)λ2
    else if ABS(μ2) == 1                 //If α2 = 90° or −90°
        if IsBodySol1 == TRUE
            κ ← d² + (√(r² − b2²) − a1)² − (a2 + a3)²
        else
            κ ← d² + (√(r² − b2²) + a1)² − (a2 + a3)²
        ifκ<0  return FALSE
        b3 ← √κ
    h ← (b2 + λ2b3)μ1
    β ← arctan(h/√(r² − h²))
    ifIsBodySol1 == TRUE
        f ← √(r² − h²) − a1
    else
        f ← √(r² − h²) + a1
    ψ ← arctan(d/f)
    γ ← arctan(μ2b3/(a2 + a3))
    ifIsBodySol1 == TRUE
        θ1 ← ϕ + β
        θ2 ← γ + ψ
    else
        θ1 ← (ϕ − β) + π
        θ2 ← (γ − ψ) − π
    return TRUE
endfun
```

Note that two arguments should be passed when Set Inverse Position ( ) is called, namely,

- The DH parameters of the manipulator through a structure defined as DH, which stores the values θ_i, b_i, a_i, and αi, with I=1,2,...,6.
- The desired position vector p of the wrist center point.
- Now, the pseudo-code for the full inverse kinematics solution is readily defined as:

```
beginfunSetFullInverse(DH,Q,d)
    p ← d − Q[a6, 0, b6]^T
    θ1, θ2, b3 ←SetInversePosition(DH,p)
    θ4, θ5, θ6 ←See [15] for the inverse
    orientationsolution
endfun
```

In this case, besides the argument DH, there are other two arguments to pass when calling Set Full Inverse ( ):

- The desired orientation of the end effector given by Q.
- The desired position vector of the end-effector reference point d.

Linear Path Tracking

In order to better understand the case study presented in the following section, we report here an algorithm devoted to generate discreet poses given by a single parameter and departing from the knowledge of the initial and final poses of a linear trajectory. Any

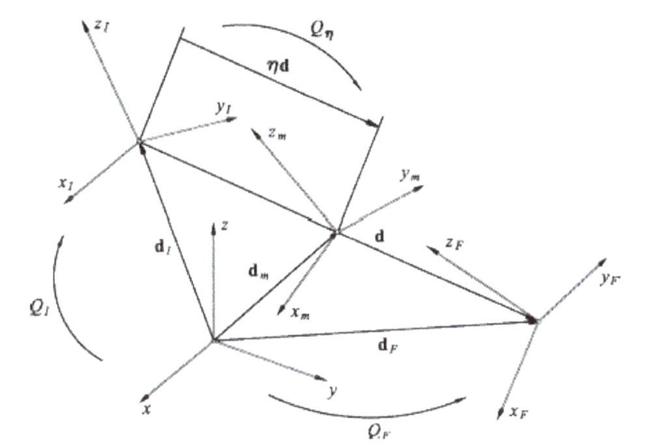

Figure 10: Linear path tracking by defining initial and final poses.

desired pose of the end effector is represented by its coordinate frame associated with a homogeneous matrix T containing the matrix Q and the vector d, which respectively represent the frame rotation and the position vector with respect to a given reference coordinate frame, as indicated in eq.(13).

$$T = \begin{bmatrix} Q & d \\ 0 & 1 \end{bmatrix} \tag{13}$$

Once the initial and final frames, S_I and S_F, of the corresponding poses are defined by T_I and T_F, respectively, intermediate poses in a line path, represented by T_m, can be readily obtained with the aid of a single parameter η, with $0 \le \eta \le 1$. To achieve this, it is necessary to define some invariants, such as vector d, defining the position vector of the origin of frame S_F with respect to frame S_I, and the roll, pitch and yaw angles, which we denote as φ_x, φ_y and φ_z. These angles define the orientation of frame S_F with respect to frame S_I by performing basic rotations about x_I, y_I and z_I, represented as $Q_x(\varphi_x)$, $Q_y(\varphi_y)$, and $Q_z(\varphi_z)$, respectively. In Figure 10 we can observe the relationship between the mentioned poses. The function that evaluates the invariants is named Set Invariants ( ), and the pseudo-code is presented below:

beginfun SetInvariants(T_I, T_F)
 $Q_I \leftarrow$ **GetRotMatFromPose**(T_I)
 $Q_F \leftarrow$ **GetRotMatFromPose**(T_F)
 $\mathbf{d}_I \leftarrow$ **GetVecFromPose**(T_I)
 $\mathbf{d}_F \leftarrow$ **GetVecFromPose**(T_F)
 $\varphi_x, \varphi_y, \varphi_z \leftarrow$ **GetRPYAngles**$(Q_I^T Q_F)$
 $\mathbf{d} \leftarrow \mathbf{d}_F - \mathbf{d}_I$
endfun

Now, with the invariants defined with Set Invariants ( ), it is possible to obtain any intermediate pose by calling the function named Get Pose ( ), whose pseudo-code is given as:

beginfun GetPose(η)
 $Q_m \leftarrow Q_I Q_z(\eta\varphi_z) Q_y(\eta\varphi_y) Q_x(\eta\varphi_x)$
 $\mathbf{d}_m \leftarrow \mathbf{d}_I + \eta\mathbf{d}$
 $T_m \leftarrow \begin{bmatrix} Q_m & \mathbf{d}_m \\ \mathbf{0} & 1 \end{bmatrix}$
 returnT_m
endfun

Functions Get Rot Mat From Pose ( ) and Get Vec From Pose ( ) take care of retrieving respectively, Q and **d** from a given T matrix. Whereas Get RPY Angles ( ), returns roll, pitch and yaw angles by solving the following equation:

$$Q_I^T Q_F = Q_Z(\varphi_Z) Q_y(\varphi_y) Q_x(\varphi_x) \tag{14}$$

Case Study

The results obtained in this paper were validated through an on-line simulation developed in ADRS (Architecture Design and Robot Simulation) which is an application developed in ADEFID platform. The simulation consists in the execution of a pick and place operation. The full cycle of this task is composed on a set of control poses along the planned trajectory. In this case it takes four poses besides the home pose to complete the task.

In order to have a better picture of the great advantage of applying the linear tracking approach introduced here in combination with the closed-form solution of the IKP, we describe the algorithm that takes the robot smoothly from the initial pose to final pose of the linear trajectory.

To begin with, we distinguish two main control levels in the simulation program. The first level consisting in the invariants definition of a desired trajectory, and the second level, in the execution of the process to move from one control point to another.

Within a complete cycle, the robot is set in different states. For each state the robot moves from the beginning to the end of the linear trajectory. When the robot reaches the end pose of a given state, it is set as SUSPEND state, meaning that the robot waits for the next control step, and a control step might be triggered by an input signal, either virtual or real.

Let us consider that the action to move from pose A to pose B is defined by the state STATE1, and the input signal is the Boolean variable named Task from A to B. Then the pseudo-code to be performed in the first level, which is part of an infinite loop, is defined as follows:

```
...
if SUSPEND ==RobotState&& TRUE == TaskFromAtoB
       SetInvariants(TA,TB)
       Timer.Init(Period_AB)
       SetState(STATE1)
...
```

Note that at this control level, the statements are executed only once. Now, for the second level, a switch is applied to consider all the

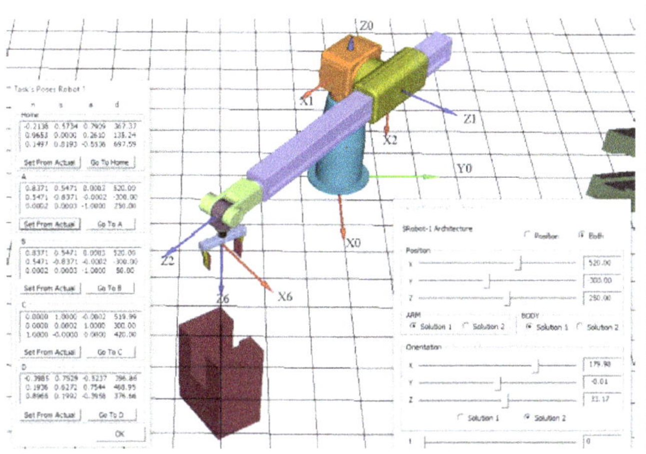

Figure 11: Setting the poses of the pick and place simulation.

θ_i (deg)	b_i (mm)	a_i (mm)	α_i (deg)
θ_1	400	0	-90
θ_2	120	0	90
0	b_3	0	0

Table 9: DH table, stanford manipulator (LU-manipulator).

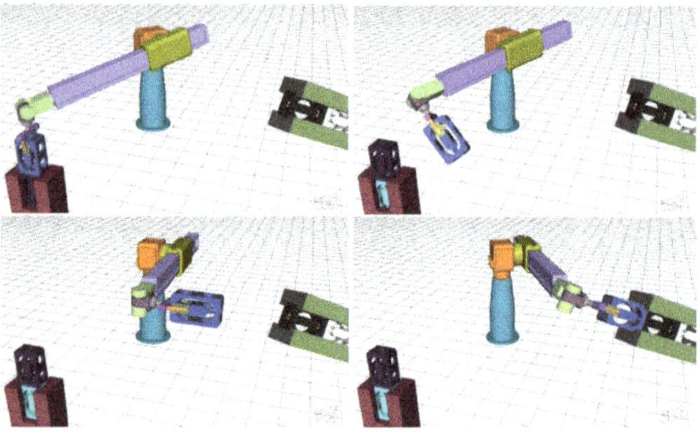

Figure 12: Stanford manipulator performing a pick and place task. First body solution.

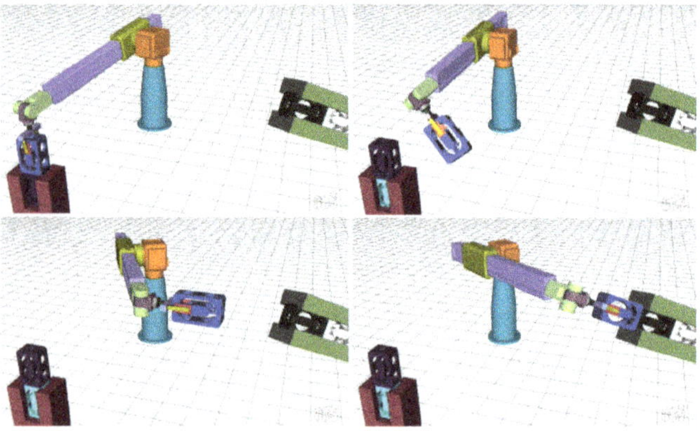

Figure 13: Stanford manipulator performing a pick and place task. Second body solution.

cases established for the simulation. In the pseudo-code presented below, the case STATE1 will be accessed several times until the state is changed to SUSPEND:

```
...
beginswitch(nState)
        ...
        begincase STATE1
                if FALSE ==Timer.bIsExpired()
                        if TRUE == GetFrameParameter(Timer,&η)
                                Q_m,d_m ← T_m ← GetPose(η)
                                SetFullInverse(DH,Q_m,d_m)
                else
                        SetSate(SUSPEND)
        endcase
        ...
endswitch
...
```

Setting the control poses that the manipulator must follow during the simulation execution becomes a simple task due to the visualization aids the program provides to the user, in other words, a dialog with tools like sliding bars helps to evaluate in-line the inverse kinematics. While interacting with the sliding bars the manipulator moves accordingly. Now with the aid of another dialog, the user can set a given position as one of the five control poses mentioned above. Figure 11, displays an image in which the two dialogs are opened together with a sample of a Stanford robot. Note that in the image, Pose A was set by just pressing the button labeled Set From Actual.

Now, we applied the algorithm to a simple case, i.e., the Stanford manipulator, which is set as a LU -manipulator. On the ADEFID platform we simulate it with the parameters shown in Table 9.

We have already indicated that for a given architecture there are two possible solutions with $b_3 > 0$, then we present still images of the simulation while performing the task. In Figure 12, four frames show the sequence of the simulation, starting from the top left corner, where the block is picked from the feeding conveyor, and ending to the right bottom corner where the block is delivered. The same simulation is performed in Figure 13, but in this case, with the second solution. These solutions are chosen by setting the variable IsBodySol1 true or false in the Inverse Position ( ) function.

The capacity to import STL files of models created in any CAD software having the feature to export this type of files, is an ADEFID's advantage that allows to create more realistic simulations. Such task,

θ_i (deg)	b_i (mm)	a_i (mm)	α_i (deg)
θ_1	270	153	90
θ_2	300	182	90
0	b_3	80	0

Table 10: DH table of a RD manipulator with generalized architecture.

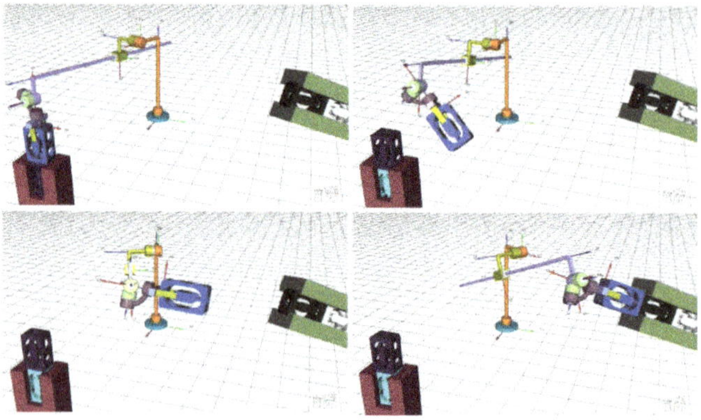

Figure 14: A generalized spherical architecture. First body solution.

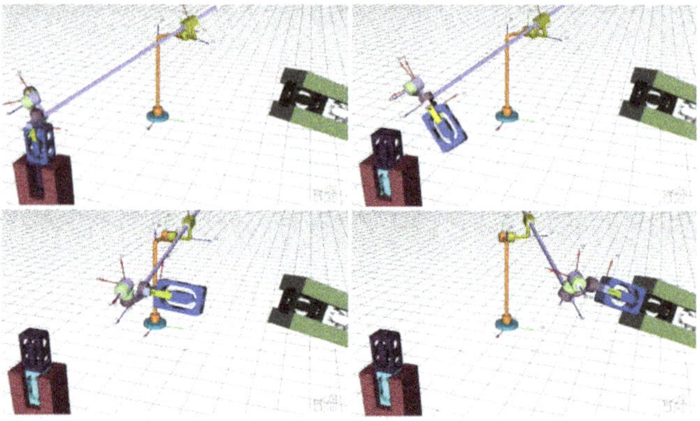

Figure 15: A generalized spherical architecture. Second body solution.

is achieved through the application of the DrawSTL ( ) function pertaining to the CSTL Draw class. With this function, an OpenGL list for each link of a robot is generated, and all lists are handled to achieve simulations as those shown in Figures 11-13.

As a second example we test the algorithm with the generalized spherical architecture. The DH design parameters applied in this case are given in Table 10. The first and second solutions are shown in Figures 14 and 15, respectively. Considering that it is a generalized architecture, and there is no a specific design of the links, the manipulator is shown in the skeleton form.

Conclusions

The unified orthogonal architecture concept presented in [22] for serial 6R manipulators, motivated the idea to extend the concept on spherical serial RRPRRR manipulators, and we were able to derive a formulation to solve the inverse kinematic problem for a generalized architecture, in other words, for cases in which all the Denavit-Hartenberg parameters involved for the positioning problem are not zero.

With the twist angles combination of the two first joints of the generalized architecture of the spherical serial manipulator, eight different types were defined, and it was shown that the Stanford arm belongs to one of these. It was also found that for each type, two possible solutions exist for the inverse kinematics problem on the positioning of the wrist center point.

Although the general solution reported here for the inverse kinematics problem is numerical, we obtained a closed-form solution for the case in which a_1=0 with an arbitrary value of α_2. Besides, closed-form solutions were also obtained for the eight architectures defined in this paper.

We were able to handle all closed-form solutions with the approach of a single algorithm. An original feature of this algorithm is that the user can control the solutions allowing to choose the one that best suits the application requirements. A novel algorithm to follow a linear path of the end effector was also introduced in this paper, knowing its initial and final poses, by controlling the position and orientation along the path with a single parameter. The corresponding pseudo-codes of both algorithms were developed, implemented, and validated within a case study focused in pick and place operations, with the aid of ADEFID through ADRS (Architecture Design and Robot Simulation), which is an open source application dedicated to design and simulate in real time, serial manipulators for industrial and research purposes.

Acknowledgement

The first author acknowledges the support from SNI, (Sistema Nacional de Investigadores), México. The second to the fifth authors acknowledge the support from CONACYT (Consejo Nacional de Ciencia y Tecnología), México.

References

1. Kircanski MV (1993) Inverse kinematic problem near singularities for simple manipulators: symbolical damped least-squares solution. Robotics and Automation Proceedings.
2. Saeidpourazar R, Jalili N (2008) Nano-robotic manipulation using a RRP nanomanipulator: Part A – Mathematical modeling and development of a robust adaptive driving mechanism. Applied Mathematics and Computation 206: 618-627.
3. Saeidpourazar R, Jalili N (2008) Nano-robotic manipulation using a RRP nanomanipulator: Part B – Robust control of manipulator's tip using fused visual servoing and force sensor feedbacks. Applied Mathematics and Computation 206: 628-642.
4. Raghavan M, Roth B (1993) Inverse kinematics of the general 6R manipulator and related linkages. ASME 115: 502-508.
5. Manocha D, Zhu Y (1994) A fast algorithm and system for inverse kinematics of general serial manipulators. Proceedings of IEEE Conference on Robotics and Automation 4: 3348-3353.
6. Sreenath N, Krishnaprasad PS (1986) DYNAMAN: A tool for manipulator design and analysis. Robotics and automation proceedings.
7. Khalil W, Creusot D (1997) SYMORO+: A system for the symbolic modelling of robots. Robotica 15: 153-161.
8. Khalil W, Vijayalingam A, Khomutenko B, Mukhanov I, Lemoine P, et al. (2014) Open SYMORO: An open-source software package for symbolic modelling of robots. International Conference on Advanced Intelligent Mechatronics.
9. Balamesh AS, Almatrafi TD, Aljawi AN, Akyurt M (2002) RobSim - A simulator for robotic motion. The 6th Saudi Engineering Conference, KFUPM, Dhahran.
10. Gracia PA, McCarthy JM (2006) Kinematic synthesis of spatial serial chains using clifford algebra exponentials. Journal of Mechanical Engineering Science 220: 1-16.
11. Xijian H, Yiwei L, Li J, Hong L (2015) VRM: A unified framework for closed-form solutions of a special class of serial manipulators. Int. J. Advanced Robotic Systems 12: 1-16.
12. Peiper DL (1968) The kinematics of manipulators under computer control. Stanford Artificial Inteligence Project.
13. Paul RP, Shimano B (1978) Kinematic control equations for simple manipulators. Decision and Control including the 17th Symposium on Adaptive Processes, IEEE Conference.
14. Khalil W, Murareci D (1994) On the general solution of the inverse kinematics of six-degrees-of-freedom manipulators. Advances in Robot Kinematics and Computationed Geometry pp: 309-318.
15. Torby BJ, Kimura II (1999) Dynamic modeling of a flexible manipulator with prismatic links. J Dyn Sys Meas Control 121: 691-696.
16. Rao SS, Bhatti PK (2001) Probabilistic approach to manipulator kinematics and dynamics. Reliability Engineering & System Safety 72: 47-58.
17. Ravishankar AS, Ashitava G (1999) Nonlinear dynamics and chaotic motions in feedback-controlled two- and three-degree-of-freedom robots. Indian Inst of Science pp: 1-16.
18. Kaleli A, Dumlu A, Corapsız MF, Erentürk K (2013) Detailed analysis of SCARA-type serial manipulator on a moving base with LabVIEW. Int J Advanced Robotic Systems 10: 1-10.
19. Sanchez DF, Muñoz DM, Llanos CH, Motta JM (2009) FPGA implementation for direct kinematics of a spherical robot manipulator. Reconfigurable Computing and FPGAs.
20. Denavit J, Hartenberg RS (1955) A kinematic notation for lower pair mechanisms based on matrices. Journal of Mechanical Design 77: 215-21.
21. González-Palacios MA (2012) Advanced engineering platform for industrial development. The Journal of Applied Research and Technology 10: 309-326.
22. González-Palacios MA (2013) The unified orthogonal architecture of industrial serial manipulators. J Robotics and Computer-Integrated Manufacturing 29: 257-271.

Use of Internet of Things in a Humanoid Robot

Jalamkar D and Selvakumar AA*

School of Mechanical and Building sciences, VIT University, Chennai, India

Abstract

Internet of things is becoming the most growing technology in recent days. Main idea behind the IoT is to extract the various values from various sensors which are attached to various objects by connecting them to the network and automating the actions performed by the object or a system. In this review paper, a study is made to understand the importance of use of IoT in humanoid robots. In order to appreciate this growing technology, this technology should be implemented to as many objects as possible so that everything will get connected in future. Along with other hardware developments, 5G internet technology is also getting developed rapidly. Use of IoT in humanoid will make many things easier to monitor and control in many ways which is discussed further in this paper. Case studies are also involved in this study to get the brief idea about how the hardware development is being done in this area. Comparison of various hardwares available to make the system work is done considering the factors like cost, ease of development, simplicity, etc.

Keywords: Humanoid; IoT; Internet of things; Control and automation; 5G internet

Introduction

Internet of Things is all about connecting various devices by using internet and letting them communicate with each other and the remote server. Large amount of data transfer takes place between various things and the server [1]. So it also demands the continuous high speed internet connectivity. "With 5G technology, getting and staying connected will get easier", said Aicha Evans, Intel's Corporate Vice President and General Manager of the Communication and Devices Group [2]. After introduction of 5G, around 50 billion things will be connected and IoT will become more efficient, faster and effective [2]. Even today with 3G and 4G internet, IoT is working because the connected devices are very less. According to IDC, worldwide market of internet of things is going to grow up around $1.7 trillion up to 2020. Other areas in which IoT will play a very important role is industrial plants and healthcare industry. IoT is being used to monitor the various parameters of the machineries in the industries which will help to predict the breakdowns. On other hand in healthcare industry, wearable tech devices are already here to keep database of body functionality [3]. Robotics is also the very much important area to consider for implementation of the IoT. Connecting various robots in the industry, remote places, homes, etc. to internet is also important. It is possible to monitor and control all the robots at different locations centrally. Current research is leading the robotics field to use the internet thus giving birth to the new term "Internet of Robotics". Role of IoT in the area of robotics is to ease the control and communication of human and robot. This can be classified into two types of robots such as Industrial Robots and Domestic robot like a humanoid robot [4]. This paper especially concentrates on using IoT in humanoid robots.

Motivation

There are several important reasons for developing a humanoid robot with IoT capability. Humanoid robots are mostly being developed as domestic robots which will stay in house, act like humans in house, do household tasks, etc. Implementation of IoT in humanoid will make this job a lot easier for the robot as well as for humans. Home security is the most important task that needs to be performed by the robot in absence of the house owner. Multiple tasks in this area like intruder detection, avoiding disastrous situations like fire, attending the visitor at door, etc. All these tasks can be monitored by the owner from anywhere in the world. Even when in home, the owner will be updated with all the surrounding parameters and situations with help of various sensors on the robot. With IoT implemented to the robot, one can control the robot and have access to all data on the mobile phone anytime anywhere. This makes it even easier with a dedicated mobile application which may be on android, ios, or any other operating system. IoT can be seen as revolutionary technology of future which have potential to make every job easy and almost everything accessible for human beings.

Available IoT Technologies

Smart phones and tablets are the most important and common tools that are required to communicate in the field of IoT.

Currently many hardware developments are taking place in this area. Some of them are Wi-Fi shields for various microprocessor or microcontrollers, RFID tags [5], NFC enabled devices, Ethernet Shields, etc. RFID are wireless microchips which are used to tag any object and to make it readable [6]. RFID reader is located centrally and objects are tagged by using RFID tags. Every tag contains some information which is read by the reader. Most of the readers are smartphones (Figure 1) or handheld devices which are connected to the internet for Iot applications [7]. Since RFID is a short range communication technology, it may not be suitable for all the applications like a humanoid application which is being discussed in this paper.

Ethernet shield (Figure 2) tacked on arduino uno microcontroller is also an example of hardware related to the IoT. Arduino is connected to the internet by using ethernet shield and other sensors and actuators

***Corresponding author:** Selvakumar AA, School of Mechanical and Building sciences, VIT University, Chennai, India
E-mail: arockia.selvakumar@vit.ac.in

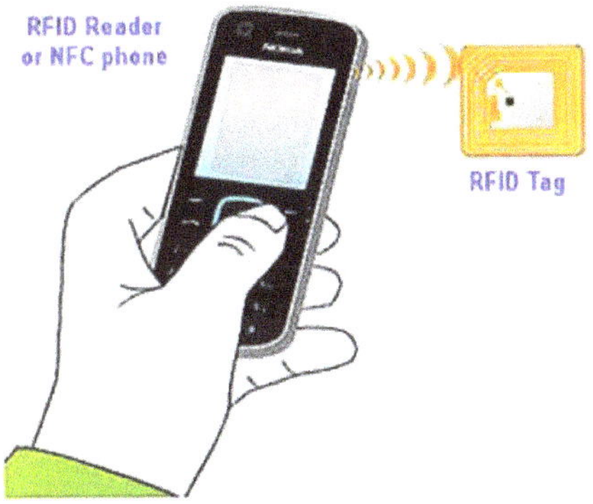

Figure 1: RFID reader and RFID tag.

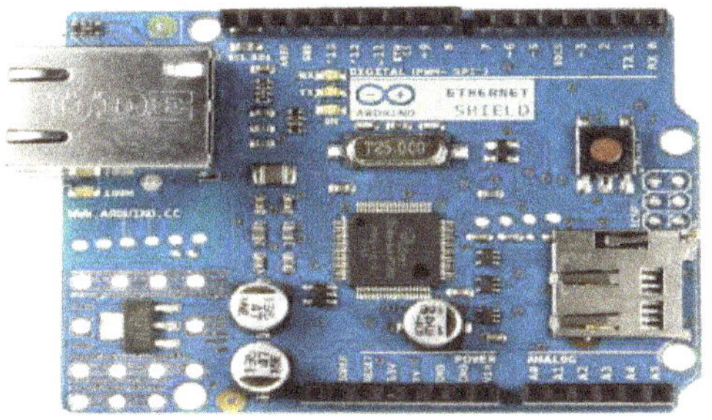
Figure 2: Ethernet shield.

connected to the arduino [8]. But disadvantage of this system is that it is not truly wireless. Ethernet shield needs to be connected by a cable to provide internet connection.

Most of the hardwares are compatible with arduino microcontroller. The Arduino is a very good and versatile microcontroller and development environment that can be used to control a variety of devices very effectively, it is also able to read many kinds of data from variety of sensors available today [9]. It is simple, cheap, easy to use and program and it has many software libraries and supporting extension hardware available ready to use. It is the open hardware platform for working and experimenting with many projects related to the Internet of Things.

ZigBee is a networking technology in wireless platform for data transmission. Zigbee module (Figure 3) is used along with the microcontroller for internet connectivity. It has advantages such as low power consumption, small and compact in size, moderate cost. A zigbee network which consists of cluster mesh and star is ideal for application of IoT [10]. Suggested structure of zigbee network can form a complex network which has further more advantages like self-healing.

Use of GSM shields (Figure 4) is also an option in parallel with above mentioned methods [11]. But with the use of GSM shield with microcontroller, there comes a problem with continuous network connectivity. If SIM card loses the signal which is very frequent and common problem, the system may stop working. As most of the GSM shields work with GPRS, bandwidth problem may occur where high bandwidth is required for specific applications. Another problem with GSM is costly internet plans with less validity.

This disadvantage of GSM can be avoided by making use of wired broadband internet connection to a wi-fi router and creating a hotspot for Wi-Fi module.

Wi-Fi module like esp. 8266 is recently developed hardware which is updated from version 1 to the latest version 14. This hardware is the most promising in terms of connecting anything to the internet. Various versions of esp. 8266 comes in different configurations for different applications. They vary in terms of number of GPIOs, cost, size, processing speed, memory, etc. [12]. This is very cheap module with system on chip. Esp8266 can work independently without connecting to any microcontroller or microprocessor. ESP 01 (Figure 5) is the most widely used module for small applications where less GPIOs are needed. This is cheapest version available today.

Figure 3: Zigbee module.

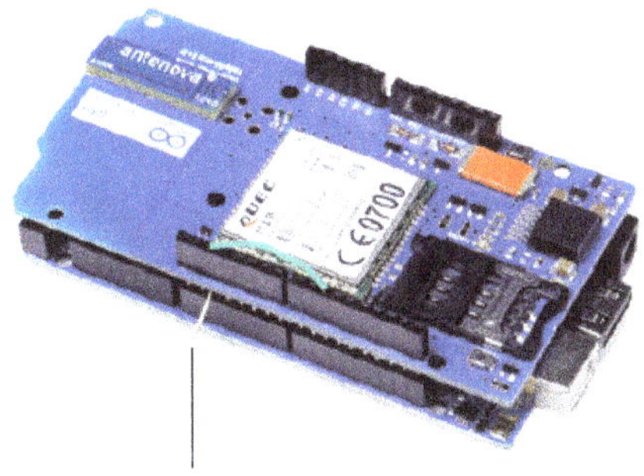
Figure 4: GSM shield.

Figure 5: ESP8266 01.

ESP 12 is widely used where more GPIOs are needed. Developing IoT applications with ESP 12 is much simpler than ESP 01 because of many design improvements that are made lately (Figure 6).

Best advantage with this module is that multiple modules can be connected to each other in order to increase the number of GPIOs which is the need of a humanoid robot application. This way it is possible to connect large number of sensors and actuators for large applications. Among all the discussed hardware above, ESP8266 module family seems the most promising and cheapest option for IoT applications.

Table 1 shows the comparison between various versions of ESP8266 module. Selection of any version can be made as per the various requirements of GPIOs, cost, size, memory, speed, etc. [12].

Protocols for IoT

Standard protocols are required for the operation if IoT. Like internet protocols IoT also needs some protocols. MQTT and CoAP are two most promising protocols for IoT. Both protocols can be used for long range applications and both are open standard [13].

MQTT protocol

MQTT is a messaging protocol which stands for Message Queuing Telemetry Transport. This protocol is very good for remote location messaging. MQTT has the client/ server architecture (Figure 7) in which every sensor and actuator is connected to the server or broker. There are many brokers available for IoT applications with various feature sets. Messaging is done in the form of chunk data to the server.

Figure 6: ESP8266 12.

Board ID	pins	pitch	LEDs	Antenna	Dimensions mm
ESP-01	8	0.1"	Yes	Etched-on PCB	14.3 × 24.8
ESP-02	8	0.1"	No?	None	14.2 × 14.2
ESP-03	14	2 mm	No	Ceramic	17.3 × 12.1
ESP-04	14	2 mm	No?	None	14.7 × 12.1
ESP-05	5	0.1"	No	None	14.2 × 14.2
ESP-06	12+GND	misc	No	None	?
ESP-07	16	2 mm	Yes	Ceramic	20.0 × 16.0
ESP-08	14	2 mm	No	None	17.0 × 16.0
ESP-09	12+GND	misc	No	None	10.0 × 10.0
ESP-10	5	2 mm?	No	None	14.2 × 10.0
ESP-11	8	1.27 mm	No?	Ceramic	17.3 × 12.1
ESP-12	16	2 mm	Yes	Etched-on PCB	24.0 × 16.0
ESP-12-E	22	2 mm	Yes	Etched-on PCB	24.0 × 16.0
ESP-13	18	1.5 mm	?	Etched-on PCB	? × ?
ESP-14	22	2 mm	1	Etched-on PCB	24.3 × 16.2

Table 1: ESP8266 module family [12].

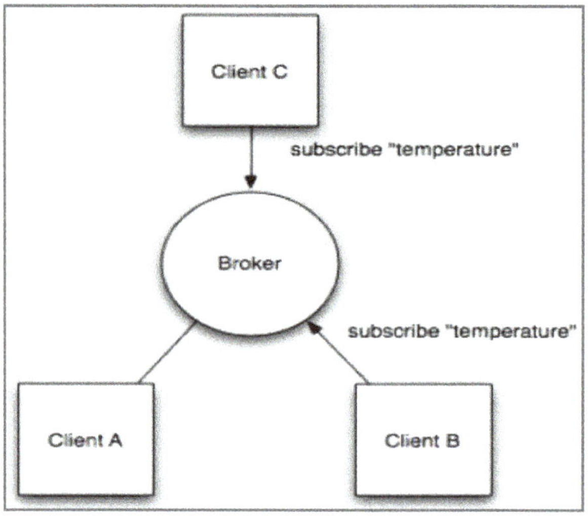

Figure 7: MQTT Architecture [13].

Server then publishes this data to the connected devices and clients [13].

It is mostly used for wireless sensor networks specifically for embedded devices [14].

There are 5 ways in which MQTT protocol works [15].

- **Connect:** Connects to the server wirelessly.
- **Disconnect:** Disconnects after finishing the task.
- **Subscribe:** Waits for subscribing the client to the broker.
- **Unsubscribe:** Client can be unsubscribed from the topics or connected things.
- **Publish:** Returns to the application thread after finishing the task and displaying value to the client.

MQTT also takes care of the security by asking username and password to connect to the broker.

CoAP protocol

CoAp stands for The Constrained Application Protocol. CoAp protocol is from "CoRE (Constrained Resource Environments) IETF group". This protocol is specifically designed for machine to machine interaction which is also a part of IoT [16]. CoAP is based on UDP. UDP enables the hardware to quickly wake-up. It transmits smaller packets of data with low overhead. This makes possible for a device to stay in a sleep mode for long time, to save the battery [17].

CoAp is a document transfer protocol similar to HTTP.

CoAP also follows client/server architecture. Client makes requests and servers send the responses back in the form of data.

But for the application of IoT, MQTT is more stable, reliable and lightweight protocol than the other protocols. MQTT is easy to get up in working network.

Case Studies

Case study on ESP8266 Version 01

Experiment: A study was done on ESP8266 01 Wi-Fi module for a small IoT application. An experiment was performed to turn on and off the LEDs connected to the module.

Connections were made as per the requirement for flashing the firmware into the chip. Firmware was custom made as per the requirement. After flashing the firmware, the module was taken out from the flash mode by changing the wiring and the code was uploaded to it.

Thingspeak.com was used as a remote server. Wi-Fi module was connected to the Thingspeak.com by using the API key and channel id provided by the server. On the other end, a smart phone with an android application was also connected to the thingspeak.com by using same API key and channel id using any internet connection.

Result: From the smart phone, the LEDs were able to turn on and off successfully. But a problem was faced regarding the response time of total operation. The response was not quick, a delay of about 8 to 20 seconds was observed between each switching and actual operation of LED. Figure 8 shows the HI / LOW status of the LEDs updated on the thingspeak.com account.

Also, this has a disadvantage of limited GPIOs. Only two LEDs were able to connect at a time. There problems are not desirable in order to use this module to implement IOT in a humanoid robot.

Case study on ESP8266 version 12

Experiment: A similar experiment was carried out with ESP8266 version 12 wi-fi module. This module is available in the form of a development board with onboard LEDs for testing, which is much easier to use than the former versions.

In this experiment, cloudmqtt.com was used as a server. Module was connected in a similar way to the server by using the key provided by the server. This was also operated by a smart phone with an android application.

Result: In this experiment the delay between the switching was reduced to almost less than a second which is desirable.

This experiment was also carried out to actuate multiple servo motors along with the LEDs as shown in Figure 9. A PWM signal was generated from the android application by using a slider interface. Values from 0 degree to 180 degrees were set to the PWM signal in the arduino microcontroller. By sending the signals from smart phone to the server then to the wi-fi module and then to the arduino microcontroller and finally to the servo motors, multiple servos were successfully actuated from multiple sliders in the android application.

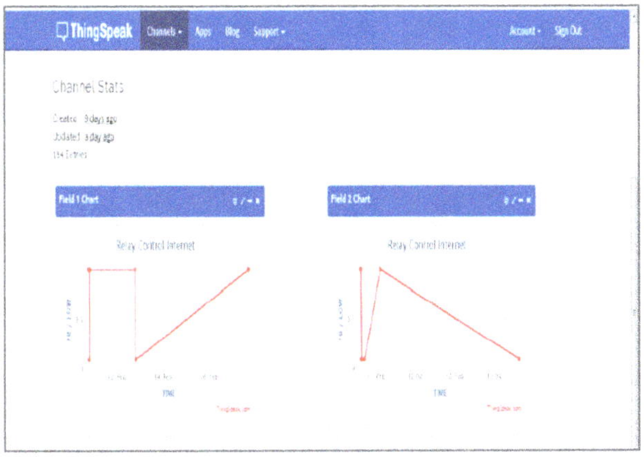

Figure 8: HI / LOW status of LED.

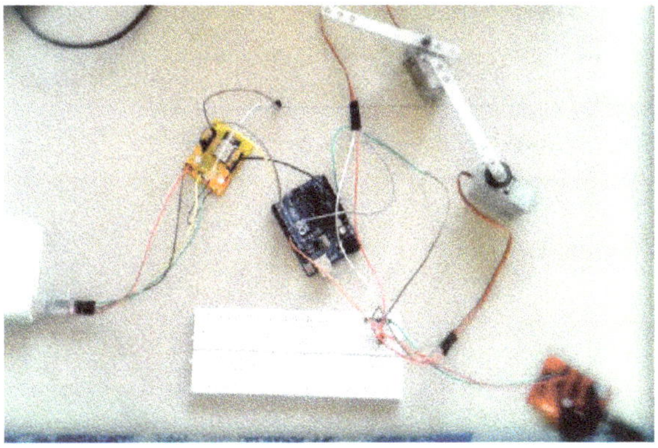

Figure 9: Actuation of servo motors and LEDs.

Conclusion

In order to implement Internet of Things in the Humanoid application, it is necessary to identify the tasks that are to be performed by the humanoid robot. Various methods can be used to implement IoT depending on the tasks. Humanoid robot mainly consists of servo motors as the main actuators.

As the number of actuators increases the need of more GPIOs also increases. Humanoid robots are very sophisticated in terms of number of sensors and actuators used. Selection of the proper IoT hardware is very important so that it should withstand the processing speeds and communication bandwidths. ESP 8266 version 12 wi-fi module is a very good hardware with many advantages as discussed earlier, for moderate level projects. Sensors attached on the humanoid robot such as temperature sensor, accelerometer sensor, PIR sensor, gyroscope, strain gauges, etc. can be successfully connected to the ESP8266 wi-fi module. Single servo at a time can also be actuated with this method. This method makes use of MQTT protocol for communication which is the most accepted and stable protocol for internet of things. Pros and cons of various methods, hardware, architectures, protocols for IoT should be considered in order to make the proper selection depending upon the need of application.

References

1. Sarma S (2016) IoT is not hype, but it's also not some magic technology. Leslie D'Monte.
2. Landau DM (2016) How 5G will Power the Future Internet of Things. Iq Intel.
3. Roy U (2016) Internet of Things: Let devices do the talking. Tech News Technology.
4. Grieco LA, Rizzo A, Colucci S, Sicari S, Piro G, et al. (2014) IoT-aided robotics applications: Technological implications, target domains and open issues. Computer Communications 54: 32-47.
5. Buschmann T, Sebastian L, Heinz U (2009) Humanoid robot lola: Design and walking control. Journal of physiology-Paris 3: 141-148.
6. Dmitri S (2016) RFID technology and internet of things. Slide share.
7. Ray Floyd (2014) RFID and the Internet of Things. Engineering.com.
8. Kadir W, Muhamad HW, Reza ES, Ibrahim BSK (2012) Internet Controlled Robotic Arm. Procedia Engineering 41: 1065-1071.
9. Doukas C (2012) Building Internet of Things with the Arduino. Create space.

10. Liguo Q, Huang Y, Tang C, Han T (2012) Node design of internet of things based on ZigBee multichannel. Procedia Engineering 29: 1516-1520.

11. https://www.arduino.cc/en/Guide/ArduinoGSMShield

12. http://www.esp8266.com/wiki/doku.php?id=esp8266-module-family

13. Toby J (2014) MQTT and CoAP, IoT Protocols. Eclipse.

14. http://mqtt.org/documentation

15. http://docs.oasis-open.org/mqtt/mqtt/v3.1.1/mqtt-v3.1.1.html

16. http://coap.technology/

17. James S (2015) MQTT and CoAP: Underlying Protocols for the IoT. Electronic design.

17

Contactless Medium Scale Industrial Robot Collaboration

Asif S* **and Webb P**

School of Aerospace, Transport and Manufacturing, Cranfield University, Cranfield, MK43 0AL, UK

***Corresponding author:** Seemal Asif, School of Aerospace, Transport and Manufacturing, Cranfield University, Cranfield, MK43 0AL, UK
E-mail: s.asif@cranfield.ac.uk

Abstract

A new age of industrial robotics is here which includes safety rated human friendly robots. These types of robots are more suitable in high volume and small scale industries. The operations in these areas can be fully automated. There are other industries which have specialty for low volume and high variability. High variety means there are less chances of fully automation as one product will be varying from other and in this complete automation is not viable option. The industrial robots which deal with high volume and can adopt variability through programming. There are plenty of manual operations which can be semi-automated. To address this human robot collaboration is a complete fit. The experiment has been undertaken to make one of these medium scale robot to collaborate with human to perform a task together. The contactless leap motion device has been used for controlling robot motions.

Keywords: Axis industrial robot; Contactless device, Human robot collaboration; Leap motion; Comau NM45

Introduction

The growing cost of High-Value/Mix and Low Volume (HMLV) industries like Aerospace is heavily based on industrial robots and manual operations done by operators [1]. Robots are excellent in repeatability by HMLV industries need changes with every single product. On the other hand human workforce is good at variability and intelligence but cost a lot as production rate is not comparable to robots and machines. There are flexible systems which have been specifically introduced for this type of industry FLEXA is one of them. But still there is need of collaboration between human and robot to get the flexible and cost effective solution [2]. A comprehensive survey has been conducted specifically on the issue of Human Robot collaboration [3] which laid out many advantages of this approach includes flexibility, cost-effectiveness and use of robot as intelligent assistant. There are several attempts have been made for Human Robot Collaboration for HMLV industry and Chen et al. attempt is one of them [4].

The conventional 6 axis arm robot is being used in several industries like automotive, aerospace, etc. Problem with these types of robot is they are not safety rated hence they have to be operated behind the cage and they have to be pre-programmed. The attempt of Human robot collaboration along with safety precautions has been made [5] in which they used 3D model and multiple depth images to monitor the robot cell. Other attempts to address safety has also been made [6] this case study used Pilz SafetyEYE and a SICK Safety Laser Scanner integrated via a Pilz Safety Relay. This safety system not only monitored the zone from top of the robot and assembly item but also under the robot arm and assembly item. Along with all these attempts to address safety of Human Robot Collaboration safety regulations has also been updated which allows Human Robot Collaboration in certain situations [7].

The research has been carried out in Cranfield University which shows the successful human robot collaboration using hand gesture control. The research used the Kinect Camera to carried out the task and monitor the zone. A case study for collision detection and avoidance has been presented which used 3D model of robot and depth images from Kinect camera [8]. The result shows that industrial robot can be used for human robot collaboration if safety systems are there to monitor [5].

The safety light curtains were being used in Boeing subsidiary of Australia where they used it for the safety of operator where operation needs manual intervention. The successful deployment of the system prove the little manual intervention of human for manual operation is safe with use of appropriate technology [9].

Human Robot interaction traditionally been done through teach pendants but that required trained operator. Other methods includes wearables like gloves. But these types of wearable technologies are not flexible enough in working environment as it needs frequently wearing and removing [10]. This can cause damage to the wearable and can potentially increase the amount of maintenance.

This paper presents a human robot collaborative system and their contactless interaction. A system in which operator can guide movements of robot using a contactless device. The leap motion is used as contactless device which does not need extensive training. A user with basic training can operate the system [11]. The use of light curtains enables to remove the cage around the robot to create feel of collaboration and enhances safety of the operator. The robot will stop movement immediately as the light curtains detect presence of operator on the boundary of cell.

The focus of our research is on human robot collaboration using contactless device like leap motion. The System used human hand guidance to move robot in certain direction.

Medium Scale Collaborative (MSC) Cell

System description

The system consists of medium scale industrial robot with 45 Kg payload. The robot used for this setup is Comau NM45 with C4G controller. The system also includes light curtains setup to remove cage and the Leap Motion device to move robot arm and control gripper. Figure 1 shows the system setup which includes following:

- Safety Light Curtains: For the safety of
- Leap Motion: To guide robot movement and control Gripper (end effector)
- Robotiq Gripper: It is three fingers gripper to grab items.
- Computer with Central Control Software System
- Comau NM45 Robot Arm with C4G Controller

Figure 1: MSC Setup.

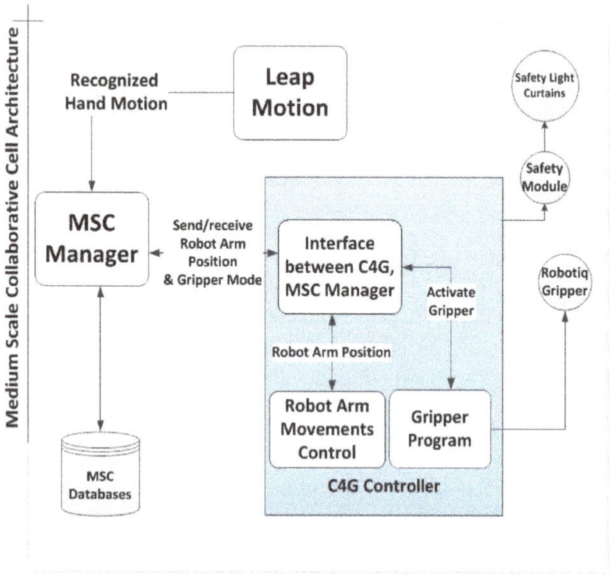

Figure 2: MSC System Architecture.

Architectural design

The system designed on modular architecture. It consists of number of sub systems/modules. Figure 2 shows MSC System Architecture which includes following main sub systems:

- **MSC Manager:** Which includes Control and coordination of all modules.
- **Hand Movement Detection:** Detects the movement of hands; Detects Gripping and un-Gripping.
- **Interface between C4G and MSC Manager**: Sen/Receives Robot Arm position and Gripping Data.
- **Robot Arm Movements Control:** Translates the received position into robot arm coordinates and sends the translated position to the robot; Sends the information back to the interface once robot arm been moved to requested position.
- **Gripper Program:** Controls the movements of gripper with specified force to grab/release the item.
- **MSC Database:** Stores System's data like points and Logs.
- **Safety Module:** Attached with the C4G controller it stops robot arm movement as soon as safety light curtains detects any obstruction between curtains.

Human robot contactless interaction

The system used two both hands to interact with robot arm using contactless Leap Motion device. The left hand has been used to direct robot movement towards Left, Right, Up and Down positions. The right hand has been used to control the gripper movement either open/close. Figure 3 shows the control movements with Left hand. Figure 4 shows the control of Gripper with right hand.

For each gesture except Close (for closing gripper fingers) users need to have fingers apart in order to recognize by leap motion.

The system starts with hand detection trigger. It determines the hand type whether it is left or right and then follows the procedure according to the detected hand type. If the hand is right hand then user can control gripper movements and can close and open the gripper depending upon the current arm position. The gripper movements can only be controlled if robot arm is in down position.

The left hand can control movements of robot arm and can use signals illustrated in Figure 3 to move the robot arm to the desired position. The flow of the system described in Figure 5 below. The system completely stops once light curtains detect any movement / obstruction between lights.

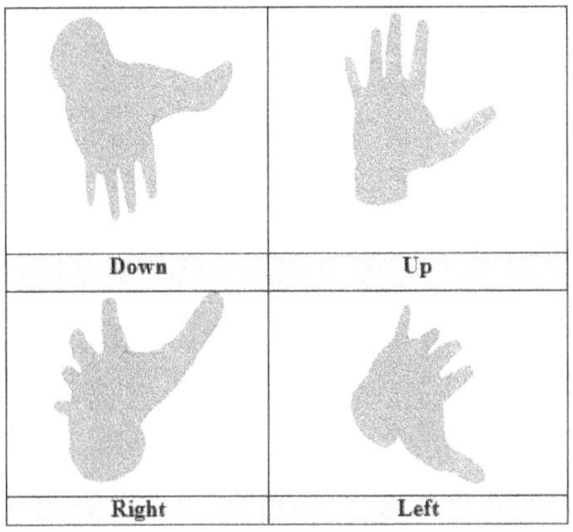

Figure 3: Left Hand Direction to Control Robot Arm Movements.

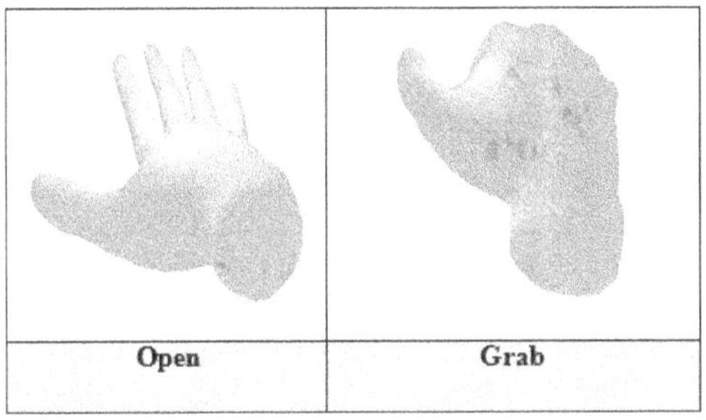

Figure 4: Right Hand to Control Gripper.

System Evaluation

The experiment has been setup to evaluate the system. A simple task was devised to fulfil the requirement. The user was asked to pick and place pipes from one position to another. Table 1 shows the Robot action according to user hand movements.

The pipe was originally placed on position A. The user was given task to move pipes between positions as illustrated in Table 2. The users were also given time to practice gestures before starting the actual task.

There were 6 participants from mixed professional background to perform the test which excludes the pilot test. Figure 6 shows the MSC setup on which participants performed tests. All participants were given same task as described in Table 2 above. Participants were also given time for training and practicing gestures.

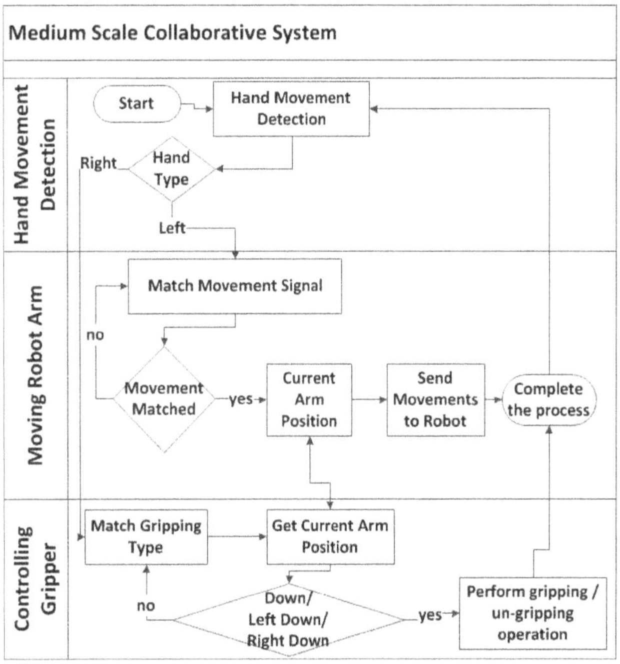

Figure 5: MSC System Architecture Data Flow.

Hand type	Hand position/movement	Action
Left	Hand flat above leap motion	Robot stops
Left	Tilt hand left	Robot moves to your left
Left	Tilt hand right	Robot moves to your right
Left	Tilt hand upwards	Robot moves up
Left	Tilt hand downwards	Robot moves down
Right	Grab fingers and thumb together	Robot moves gripper
Right	Pull fingers and thumb apart	Robot moves gripper

Table 1: Hand movements and robot actions.

Results and Discussion

System was recording participants' time to perform each gesture. A separate program was written to get participants' time and perform evaluation on data.

Figure 7 shows chart plot for the time taken to complete a single gestures. First 21 gestures data were taken for each participant to do the comparison. The mean gestures completion time was between 2 to 5 seconds. The Figure 8 shows the mean time to complete one gesture.

Total time to complete the task ranges between 1 min and 13 sec to 4 minutes 10 seconds. The mean time for all of the participants to complete the task was 2 minutes and 23 seconds. Figure 9 shows the total time to complete the test by participants.

Step	Action
1	Pick up pipe A
2	Move pipe A to position C
3	Drop pipe at position C
4	Move to position B (Up)
5	Move to position C
6	Pick up pipe C
7	Move the pipe C to position B
8	Drop pipe at position B

Table 2: Test task description.

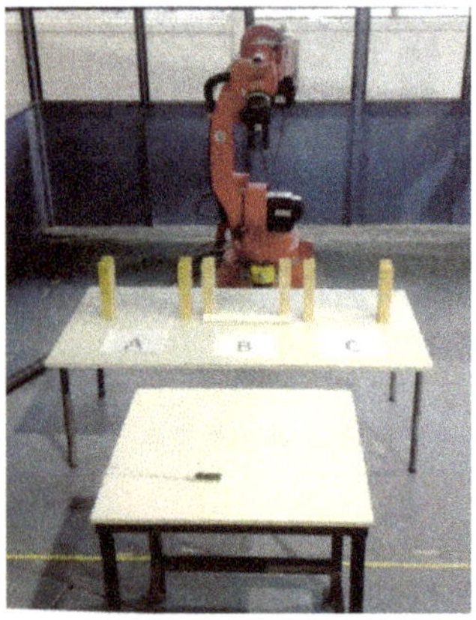

Figure 6: MSC Test Setup.

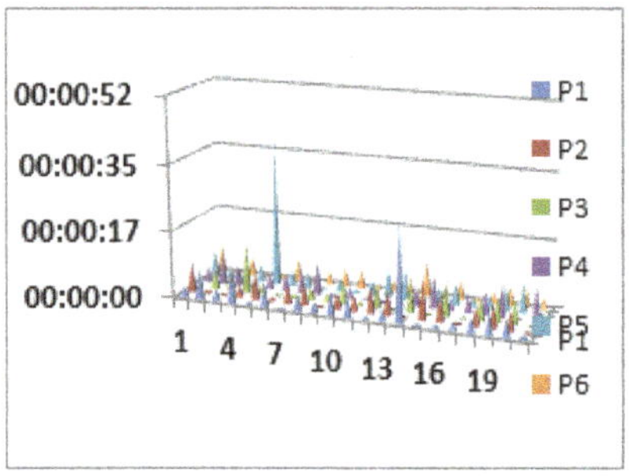

Figure 7: Time between Gestures – x-axis: gesture number, y-axis: time to complete gesture (sec).

The maximum time to finish same task was spent by participant 3 which is also reflective in Figure 10 as number of duplicate gestures were detected in higher number for that particular participant.

Total number of hand movements was range from 22 to 38 except for participant 3 whose hand movements recorded 88 in number as shown in Figure 11. This also proves the data shown in Figures 9 and 10.

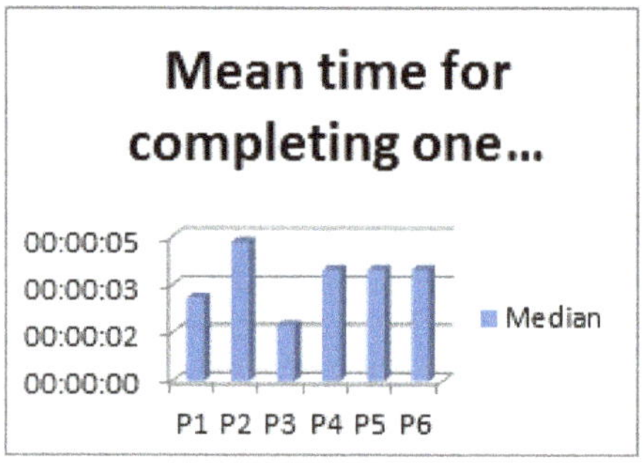

Figure 8: Mean time for completing one Gesture– x-axis: participants, y-axis: mean time to complete one gesture (sec).

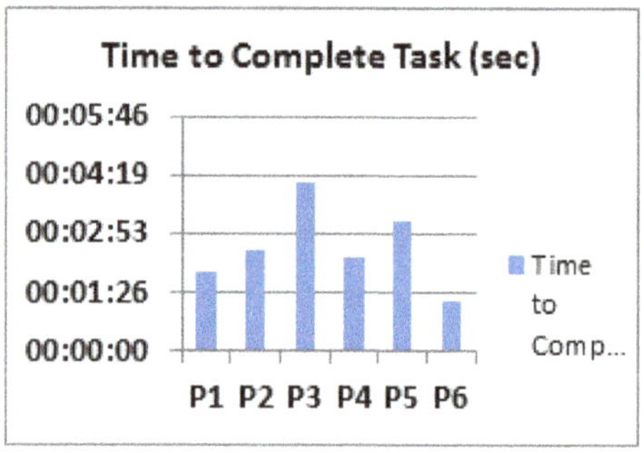

Figure 9: Time to Complete Task– x-axis: participant number, y-axis: time to complete task (sec).

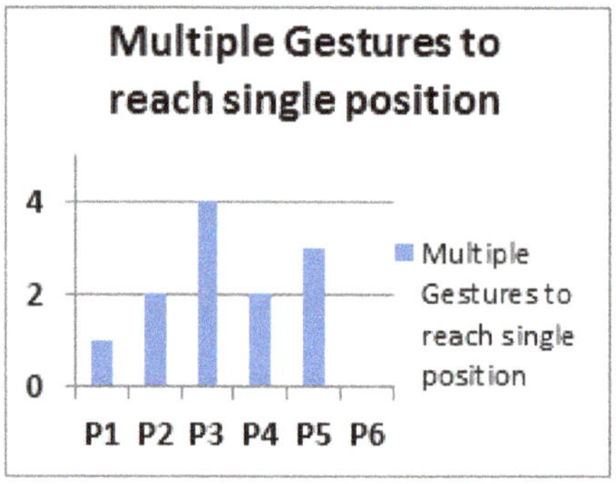

Figure 10: Multiple Gestures to reach single position – x-axis: participant number, y-axis: number of multiple gestures/duplicate gestures to reach single position.

The feedback taken from participants also reflects that for some participants hand gestures were not recognized merely on titling sometimes they have to move hand completely to get right and left movement. For other gestures this problem was not present. This shows room of improvement in generalizing the hand detection for different types of hands. There is also another impact on the test which was un-controlled light as we had bright sunny spells during the test. This also impact the efficiency of leap motion to detect hands and hence system to recognize gesture type. The lighting conditions were beyond control but one certain improvement can be made which is to generalize parameters to recognize left and right gesture for different hand types.

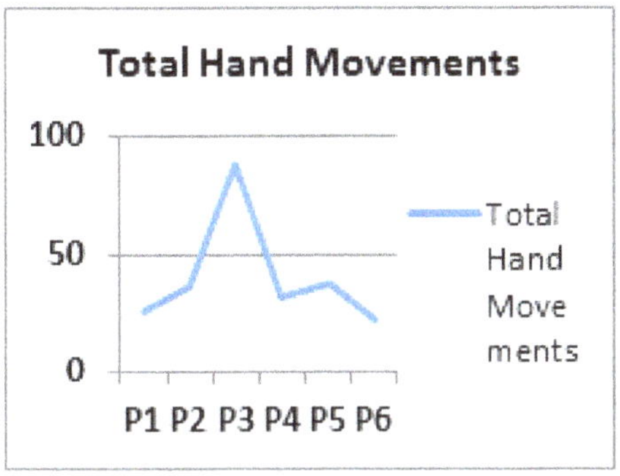

Figure 11: Total Hand Movements– x-axis: participant number, y-axis: total number of hand movements to complete task.

Summary

The experiment shows positive results for the contactless control of robot. Mean time to complete task for all participant was 2 minutes and 23 seconds which also shows the productiveness of system. However there is still room for improvement especially to generalize the hand's parameters for left and right gesture. This kind of case-study can be used in HMLV industry in which there is still need of manual intervention from operators where robot can act as assistant to the operator.

Acknowledgments

This research is supported by the EPSRC Centre for Innovative Manufacturing in Intelligent Automation. Particular thanks to to John Thrower, Senior Technical Officer in Cranfield University, for his technical input and Support.

References

1. Pycraft M, Singh H, Phihlela K (1997) The Volume-Variety Effect on Design. Operations Management, Pearson Education, Limited.
2. Webb P, Asif S (2015) Enhanced Cell Controller for Aerospace Manufacturing. Aircr Eng Aerosp Technol.
3. Krüger J, Lien TK, Verl A (2009) Cooperation of human and machines in assembly lines. CIRP Ann Manuf Technol 58: 628-646.
4. Chen F, Sekiyama K, Zasaki H, Huang J, Sun B, et al. (2011) Assembly Strategy Modeling and Selection for Human and Robot Coordinated Cell Assembly. International Conference on Intelligent Robots and Systems Pp: 4670-4675.
5. Schmidta B, Wanga L (2013) Contact-less and Programming-less Human-Robot Collaboration. Forty Sixth CIRP Conference on Manufacturing Systems pp: 545-550.
6. Hamilton A, Webb P (2015) Assessing the Safety Risk of Collaborative Automation within the UK Aerospace Manufacturing Industry. Safety-Critical Systems Symposium, Bristol, UK.
7. (2011) BS ISO 10218-2: robots and robotic devices – safety requirements for industrial robots Part 2: robot systems and integration. BS Institution.
8. Tang G, Asif S, Webb P (2015) The integration of contactless static pose recognition and dynamic hand motion tracking control system for industrial human and robot collaboration. Ind. Robot An Int. J 42: 416-428.
9. Atkinson J, John H, Jones S, Gleeson P (2007) Robotic Drilling System for 737 Aileron. SAE Int.
10. Neto P, Pires N, Paulo M (2013) High-level programming and control for industrial robotics: using a hand-held accelerometer-based input device for gesture and posture recognition. Ind Robot An Int J 37: 137-147.
11. Marin G, Dominio, F and Zanutiggh P (2014) Hand gesture recognition with leap motion and kinect devices. EEE international conference on image processing (ICIP).

Development of Modular Robotic Design Concepts for Hot Cell Applications

Ahmed Sherif El-Gizawy[1,2*], Abdulraheem Kinsara[2], Ghassan Mousa[2] and Andrew Gunn[1]

[1]*Industrial and Technological Development Center Mechanical and Aerospace Engineering, University of Missouri-ColumbiaColumbia, Missouri-65211, USA*
[2]*Center of Excellence for Industrial Design and Manufacturing Research (CEIDM) Faculty of Engineering, King Abdulaziz University Jeddah, Saudi Arabia*

Abstract

The production of isotopes for diagnosis and treatment of cancer patients involves handling and processing of irradiated materials. This process is performed inside heavily shielded workstations termed Hot Cells. A modular robotic design for handling irradiated materials inside hot cells is introduced. The new robotic system is reconfigurable in order to enhance versatility of applications and precision of its tasks. The reliability of the introduced robot control system is assessed using Failure Trees (FT) Methodology. The technology developed in the present work allows for improving productivity and cost effectiveness for production of medical isotopes.

Keywords: Modular robotic design; Kinematics and control strategy; Hot cell; Medical isotopes

Introduction

The production of medical isotopes, involves handling and processing of irradiated materials. It is performed inside heavily shielded workstations termed Hot Cells. Most of the handling tasks inside hot cells are conducted with the aid of master-slave mechanical manipulators in order to reduce radiation exposure to operators. Figure 1 displays a pair of such manipulators that are currently used in the production of isotopes inside hot cells. These manipulators transfer the kinematic motion of user-operated controls located outside the hot cell, to end effector claws inside the hot cell [1]. Nevertheless, they are bulky structures, slow in motion and suffer from not capable of transferring rotational motion with precision, and very expensive (exceeding $200,000 per unit). An effective alternative for handling radioactive materials inside hot cells is the use of cost effective semiautonomous robots [1,2]. The main objective of the present research is to develop and support modular robotic design concepts to improve task versatility, reduce personnel radiation exposure, and improve productivity and cost effectiveness in production of radioactive isotopes.

The design requirements for the required robots should follow the reported conditions [1-5]. The platform should cope with different types of applications inside the hot cell to ensure mobility and accuracy in positioning. The platform has to have rigid structure to be able to carry sensor and tools payload without excessive deformation. These requirements, in addition to the nature of the radioactive environment, impose important constraints on the motor power and the power supply.

Figure 1: Operators control manipulators that mimic the human movement through pneumatic servos and hydraulics, used for handling tasks inside hot cells.

Hence, all evolved design specifications should ensure reliable deployability of the introduced modular robots for hot cell environment.

The engineering design process

The design process road map [2] is used in the present work for developing the required devices. It illustrates the flow of the design process from preconception, to conceptualization, to configuration, and lastly, to design finalization and validation.

Quality function deployment

The Quality Function Deployment (QFD) is a preliminary planning mechanism that is used to aid in the design process. The QFD relates the customer needs to specific engineering characteristics of the potential design to better understand which characteristics have greater priority over others. The Quality Function Deployment for the Single-Arm Robot can be seen in Figure 2. Three classes of requirements are included in the reported QFD: Functions; Ergonomics; and Constraints (cost, size, environment). The most important engineering characteristics are shown to be: remote robust control of motion, high degree of freedom, and the ability to perform in a high risk environment (radioactive tolerance).

Kinematics and control strategy

Kinematics analysis is applied to the design concepts of modular robots. It refers to the determination of required joint values and end effector values, as done in path planning. The kinematic equations of the manipulator can be used to handle redundancies (different combinations of joint angles that produce the same end effector orientation and position), collision avoidance, and avoidance of singularity solutions [3]. Once the kinematic equations have been

***Corresponding author:** El-Gizawy AS, University of Missouri-Columbia, USA, E-mail: sherifelg@yahoo.com developed, they can be used to create a dynamic model of the robot, whereupon the effects of forces upon movements are analyzed. The results of this analysis are used to improve the control algorithms of the robot. The control system and electronics for robots are analyzed for robustness. Evaluation of power sources (batteries... etc.) to be used is conducted for optimum battery configuration. The introduced control system for modular robots is wireless using Bluetooth wireless capabilities.

Modular robot structural design requirements

In order to reduce cost of materials and manufacturing of the required devices, thermoplastic with high strength are used in construction of most of the structure components. Most polymers and elastomers are not significantly affected by radiation doses below 10,000 Gy (1 Mrad). Teflon is a unique exception and parts that include Teflon insulation must be replaced by parts using less affected insulation. For example, hookup wire with cross-linked polyolefin insulation is tolerant to a radiation dose of 1 MGy. Structural analysis for strength and failure are conducted using the FEM Simulation of SOLIDWORKS computer aided design and engineering platform.

Design for manufacturing and assembly

Upon realizing the initial concepts of modular robot, all of them and their components are reviewed and modified for ease of assembly and manufacturing by Direct Digital Manufacturing (DDM) technique. DDM describes a class of relatively new manufacturing process in which parts are fabricated by additive manufacturing. In this process, parts are fabricated layer-by-layer in thin cross sections from geometries supplied by a computer aided (CAD) model of the part. One of the key benefits of this type of manufacturing is the ability to achieve complex and intricate geometries that could not be done with conventional methods (e.g. ducting with internal features). This, in combination with the fact that no expensive tools are required to produce a part dramatically, reduces lead times that would otherwise be required to produce functional products [6]. Additionally, DDM processes use much less material than conventional methods because fabrication is a result of adding material instead of removing it, thus reducing waste. At the time the present project was undergoing, the research teams have access to DDM technologies at both MU and KAU, specifically Fusion Deposition Modeling (FDM) and Selective Laser Sintering (SLS). FDM process is displayed in Figure 3, while SLS is displayed on Figure 4. Both technologies were use in the early prototypes. Because FDM uses larger variety of thermoplastic materials than SLS, the final concepts were built using FDM process.

DDM (3D Printing) by fused deposition modeling (FDM) is a manufacturing process that creates parts by depositing polymer filaments in successive layers (Figure 3). The polymer filament passes through the extrusion nozzle, which heats the filament to a semi-molten state and deposits it on the build platform [6]. The platen build platform then descends within the work build chamber to allow for the deposition of the next layer upon the previous one. The variety of materials available, ease of producing complex geometries, and extremely fast fabrication times make FDM a very attractive manufacturing option for several industries.

Modular robotic design and testing

Following the engineering design process presented earlier, a modular robot was developed to be reconfigurable in order to perform

9 = Strong
3 = Moderate
1 = Weak

	Programmable Motion Software	Joint-Linkage Connections	High Number of D.O.F	Motor Torque	Material Strength	Solid Mode Design	Manufacturing Cost	Temperature Tolerance	Radioactive Tolerance	Customer Importance	Existing Products	Planned Product	Improvement Ratio	Sales Points	Improvement Ratio	Relative Weight
Functions																
1. Repeatable Motion	9	1								5	5	5	1.0	1.0	5.0	0.10
2. Range of Motion		3	9							4	3	3	1.3	1.5	8.0	0.15
3. Solid Structure					9					4	4	4	1.0	1.0	4.0	0.08
4. Maximum Load				9	9					2	3	3	0.7	1.0	1.3	0.03
Ergonomics																
5. Practical Gripper						9				4	3	5	1.3	1.0	5.3	0.10
6. Simplified User Controls	9									3	2	3	1.5	1.5	6.8	0.13
7. Sleek Design (Hidden Components)						9				2	4	2	0.5	1.0	1.0	0.02
Miscellaneous																
8. Cost							9			4	3	3	1.3	1.5	8.0	0.15
9. High Risk Environment Design								3	9	5	3	5	1.7	1.5	12.5	0.24
Absolute Importance	2.04	0.56	1.39	0.23	0.92	1.1	1.39	0.72	2.17	10.5					51.9	1.00
Relative Importance	0.19	0.05	0.13	0.02	0.09	0.1	0.13	0.07	0.21							
Competitive Assessment			3			3	3	2	2							
Technical Difficulty	X	X		X	X											
Target Values			4			4	2	2	4							

Figure 2: QFD for a modular robot.

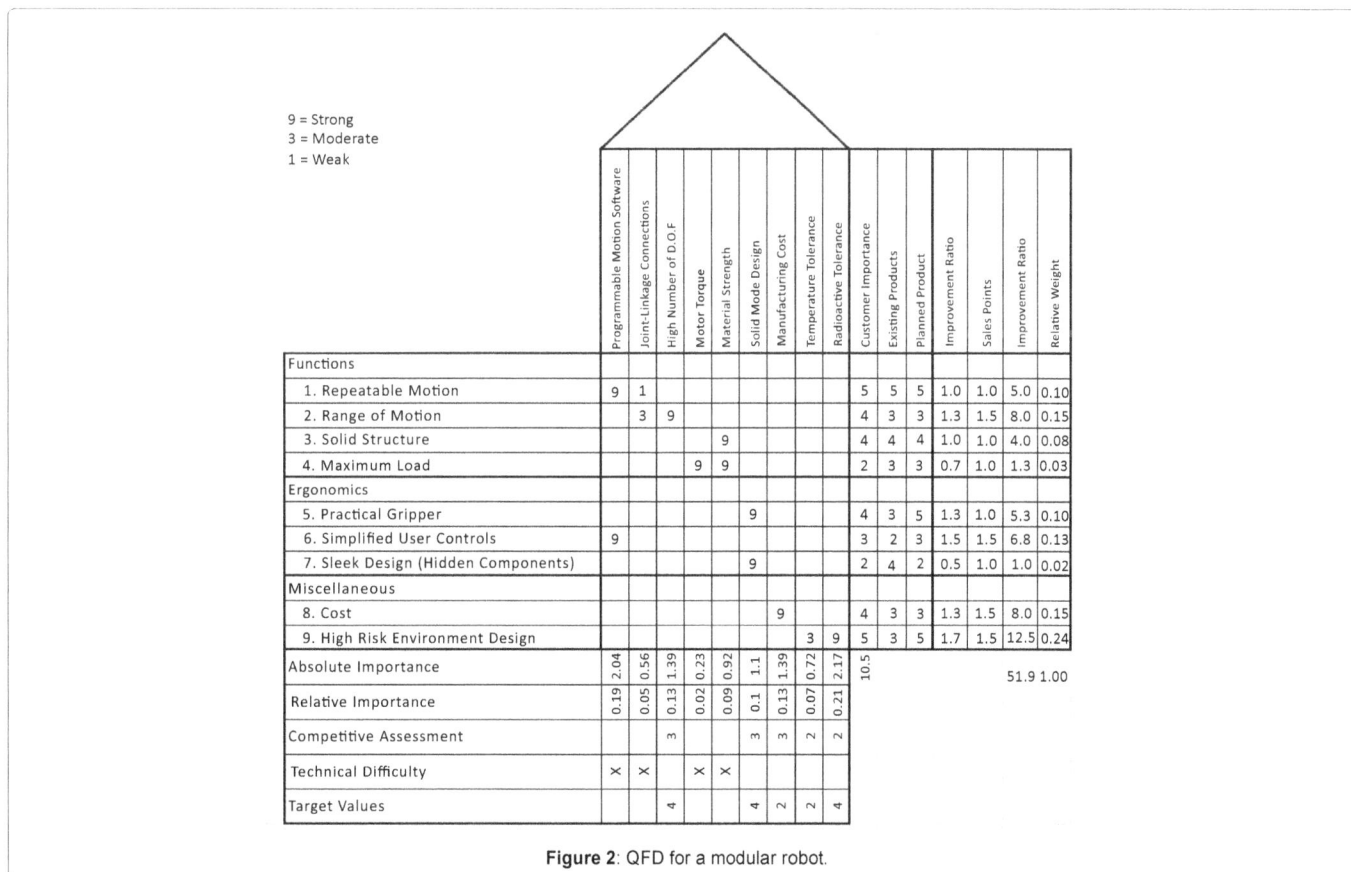

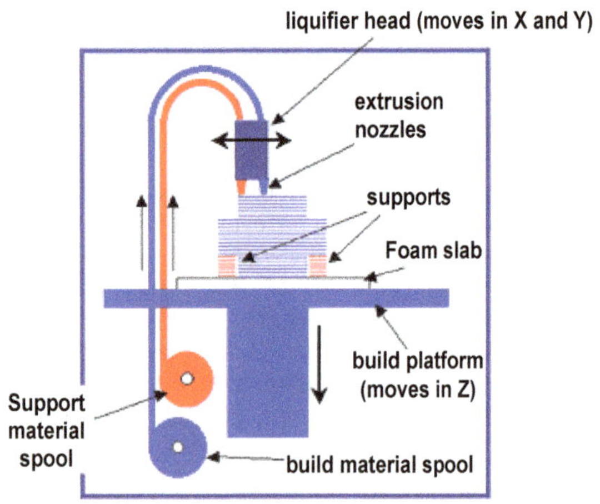

Figure 3: Fusion deposition modeling (FDM) technique for rapid digital manufacturing.

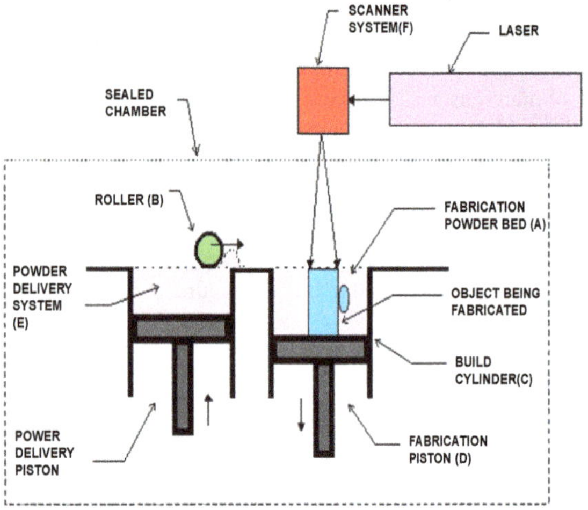

Figure 4: Selective laser sintering (SLS) technique for rapid digital manufacturing.

different materials handling and processing tasks inside hot cells. The first concept selected for evaluation is the two-arm robot moving on tracks (Figure 5).

The goal of the initial kinematic analysis was to determine the "envelope" of the Robot. At its resting position, the robot measures 5.95 in wide × 12.95 in tall × 10.74 in deep. The robot was animated to achieve its most extreme positions. The analysis indicates that the envelope of the first configuration of the Modular Robot is 19.62 in wide × 16.66 in tall × 15.54 in deep. When redesigning the robot chassis, the battery and control boards must be considered along with additional space for wiring. All of the electrical control components can be inserted and attached to the main chassis of the robot. A top view and cutaway top view of the robot chassis can be seen in Figures 6 and 7. The battery was the largest component and was given space underneath the center of the robot. The battery slides into place and is secured using an end cap that can be inserted into a slot on the front of the robot. In the back of the robot, the motor control board is inserted into a slot at the bottom of the chassis cavity. The voltage regulator and 9V battery are inserted into side channels along the main battery slot. The servo board is then screwed down on the top of the cavity with the servo control board underneath. All of the wiring is fed to the motor control board through a slot in front of it. The servo board is also rotated so that the serial port is directly out of the back of the robot. This orientation will allow for connection of a serial cable to the port without interference with the tracks. The Bluetooth board is secured to the top of the chassis near the servo board via double sided tape. The torso turret was also repositioned to allow easier connection of motor wires to the track motors. Figure 8 displays the redesigned Electrical System for efficient assembly.

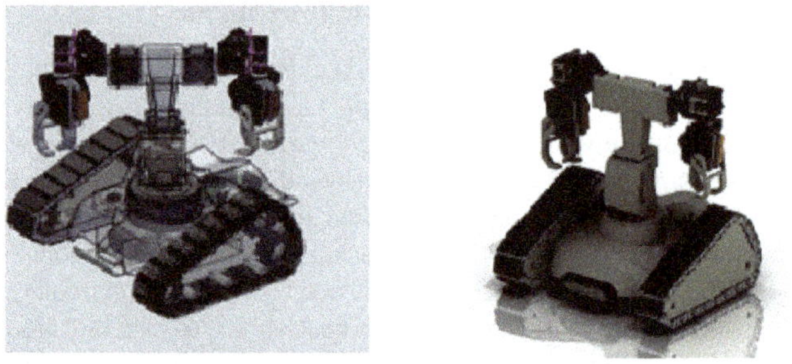

Figure 5: Two-arm modular robot on tracks.

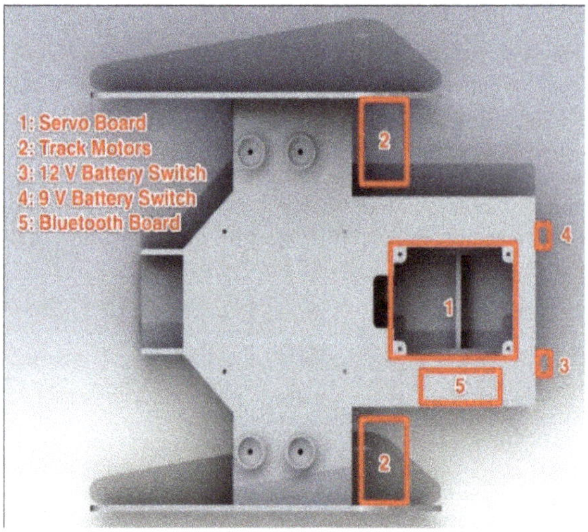

Figure 6: Top view of robot chassis.

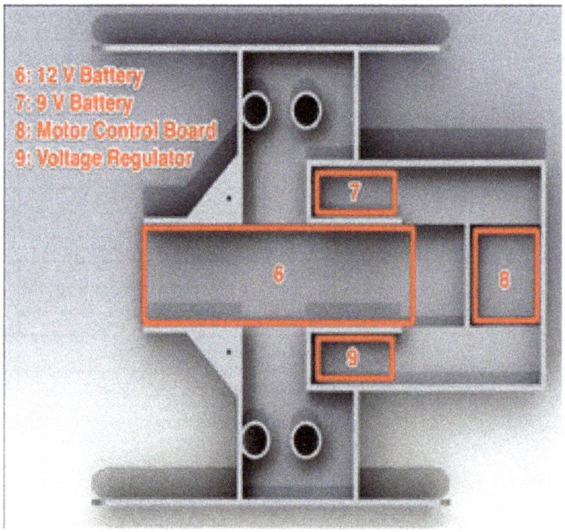

Figure 7: Cutaway top view.

Assembly of final prototype

Figure 9 explain about Exploded view and Final Product of the Redesigned Robot for Efficient Assembly.

Assessing reliability of the developed modular robot

In order to assess the reliability and risk level of different systems in the new Modular Robot, fault tree analysis was conducted on all of them. A fault tree analysis for the electronics and control system of the robot is displayed in Figure 10. The top event of the tree is the failure of the electronic system. This event can occur if four other events happen, which include motor board failure, servo board failure, servo failure or motor failure. If failure occurs in the motor board, the tracks will not be able to move. If the servo board fails, then the servomotors will not be able to control the position of the torso and arms. The board failures can be caused by three different events, electrical shortages, power loss, Bluetooth out of range. The estimated failure rate of the robot control system during operation is 2%, which is considered acceptable for the intended applications.

Experimental verification of the performance of a single arm of the developed modular robot was tested using the experimental setup displayed in Figure 11. Initial testing began with rudimentary motions controlled by computer Servo Position inputs. Basic functions allowed for co-ordinates to be input to the arm and the arm would move to the location. Several objects vary in size were used for evaluating the functions of the developed robotic system. Accuracy analysis in determining object position and orientation was performed using repeated tests. The attached vision system determines the orientation of an object and location in X-Y coordinates. The followings are samples of the repeatability tests for finding object location using Cartesian coordinates where base of robot is [0,0]:

Trial 1: Detected at: X=157.00 and Y=135.00

Trial 2: Detected at: X=158.00 and Y=135.00

Trial 3: Detected at: X=157.00 and Y=136.00

Repeatability determined from the above tests is 99.63% which is considered very well for the intended applications.

Other configurations of the developed modular robot

Other configurations of the presented Modular Robot are displayed in Figures 12 and 13. Both, the two-arm and single arm robots are mounted on a screw-driven, rail mounted platform (linear slider mechanism) to add a seventh degree of freedom to the modular robot system. The two-arm robot is suitable for repair and assembly tasks inside the hot cell while the single arm robot is configured for precision materials handling tasks such as those used for disassembling annular targets for removal of irradiated nuclear

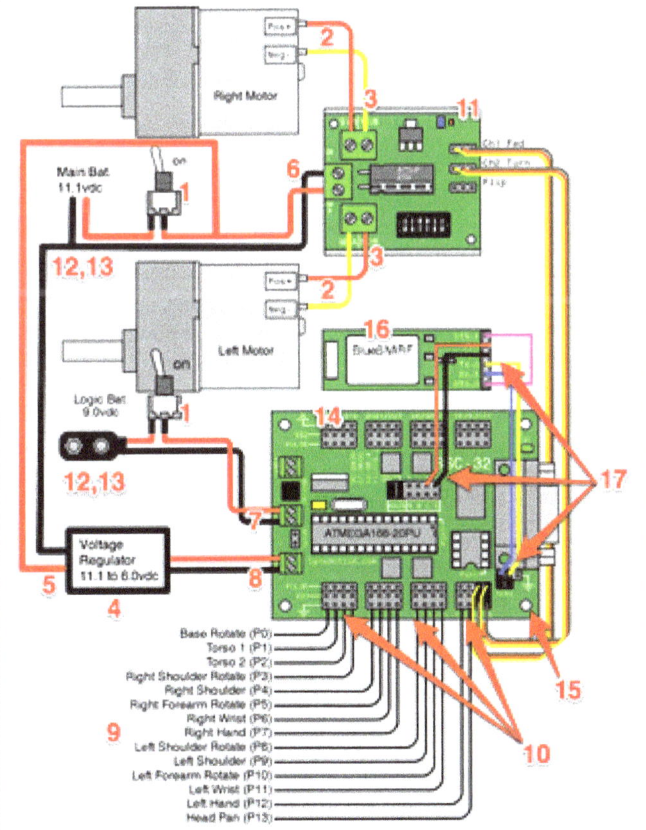

Figure 8: Diagram of redesigned electrical system for efficient assembly.

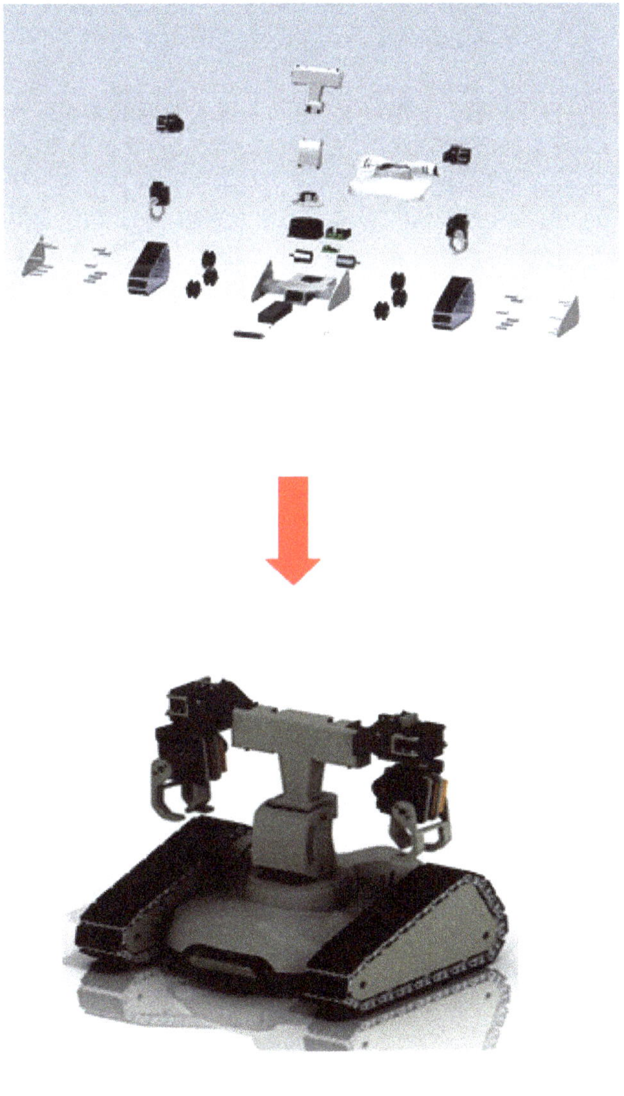

Figure 9: Exploded view and final product of the redesigned robot for efficient assembly.

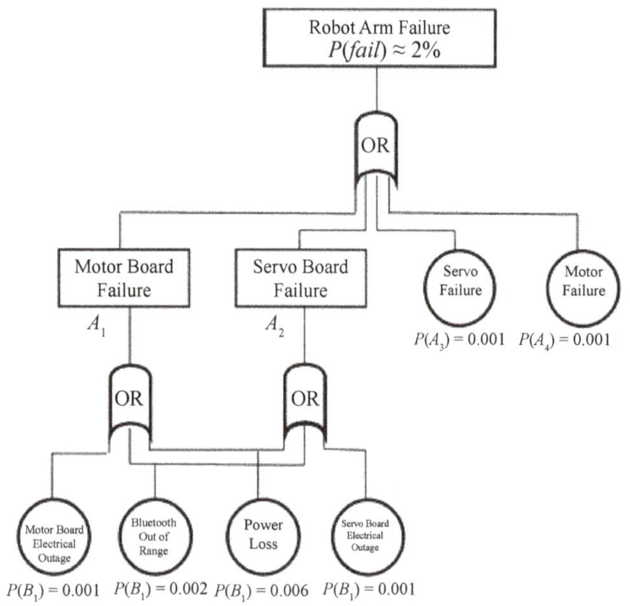

Figure 10: Fault tree for the control system.

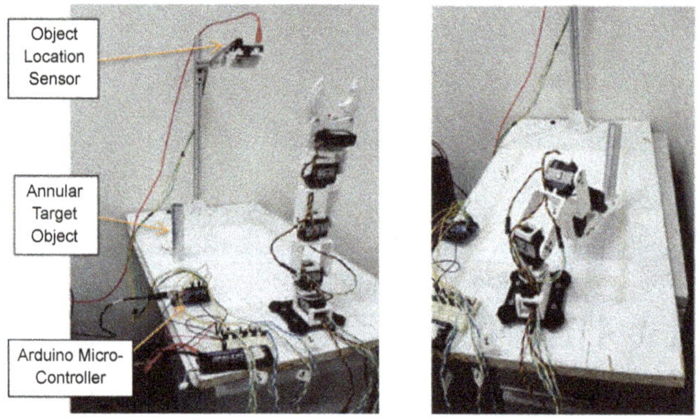

Figure 11: Setup for evaluating target reorientation and placing.

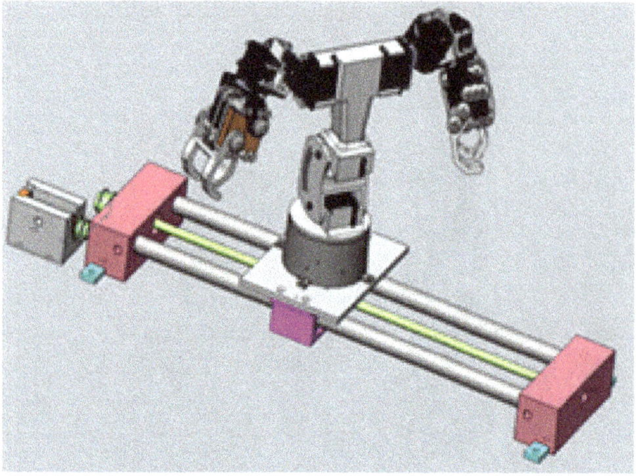

Figure 12: Two-arm modular robot on slider mechanism.

foil during the production of Mo-99 medical isotopes. It should be mentioned that all configurations of the modular robot are sized to have small footprints and to fit available hot cells. Any of the

Figure 13: Single arm modular robot with seven degrees of freedom.

mentioned configurations would cost \$5,000-\$10,000 compared with an average cost of \$150,000-\$300,000 for a master-slave mechanical manipulator or \$75,000 for commercially available 7-degrees-of-freedom robot that does not fit inside the hot cell and has limited capabilities for handling precision tasks.

Conclusions

- A modular robotic design for handling irradiated materials inside hot cells is introduced.
- The new robotic system is reconfigurable in order to enhance versatility of applications and precision of its tasks.
- The robotic control system is developed to be wireless using Bluetooth wireless capabilities.
- The reliability of the introduced robot control system is assessed using Failure Trees (FT) Methodologies.
- The technology developed in the present work helps in supporting modular robotics design for hazardous environments and allows for improving productivity and cost effectiveness for production of medical isotopes.

Acknowledgements

The authors wish to acknowledge the financial supports of King Abdul Aziz University, Saudi Arabia, STRATASYS, Inc. (USA), ITECH D&M, LLC (USA), and Industrial Technology Development Center (University of Missouri), for the present research. Our appreciations are also extended to Engineering Students at University of Missouri and King Abdul Aziz University for their participation in the Robotic Projects and for providing the necessary efforts to make them successful.

References

1. Fogle RF (1992) The use of Teleoperators in Hostile Environment Applications. Proc. Int. Conference. On Robotics and Automation (ICRA '92) Nice, France. pp: 61-66.
2. Ferner E, Hoyer A, El-Gizawy AS (2011) Remote Disassembly System for Annular Targets during Production of Mo-99. 2011 ANS Winter Meeting and Nuclear Technology Expo, Technical Conference Proceedings: The Status of Global Nuclear Deployment, Washington DC, American Nuclear Society (CD-ROM). p: 3

3. Doroftei D, Matos A, de Cubber G (2014) Designing Search and Rescue Robots towards Realistic User Requirements, Applied Mechanics and Materials 658: 612-617.

4. Doroftei D, Cubber GD, Chintamani K (2012) Towards collaborative human and robotic rescue workers, 5th International Workshop on Human-Friendly Robotics (HFR2012).

5. Crane CD, Duffy J (2008) Kinematic Analysis of Robot Manipulators. Cambridge University Press.

6. Gizawy ASE (2011) Process-induced Properties of FDM Products. Proceedings of The ICMET, International Conference on Mechanical Engineering and Technology Congress & Exposition, ICMET 2011, Paris, France. p: 7.

19

Variable Admittance Control for Climbing Stairs in Human-Powered Exoskeleton Systems

Ahmed AIA[1]*, Cheng H[1], Lin X[1], Omer M[2] and Atieno JM[3]

[1]*Center for Robotics, School of Automation, University of Electronic Science and Technology of China, 611731 Chengdu, China*
[2]*School of Automation, University of Electronic Science and Technology of China*
[3]*School of Electronic Engineering, University of Electronic Science and Technology of China, 611731 Chengdu, China*

Abstract

Online gait control in human-powered exoskeleton systems is still rich research field and represents a step towards fully autonomous, safe and intelligent navigation. Admittance Controller performs well on flat terrain walking in human-powered exoskeleton systems for acceleration and slowdown. We are the first who proposed Variable Admittance Controller (VAC) for smooth stair climbing control in Human-Powered Exoskeleton Systems. Trajectory correction technique transforms the interaction forces exerted on the exoskeleton from the pilot to appropriate intended joint flexion angles through dynamic viscoelastic models. We demonstrate the proposed control strategy on one degree-of-freedom (1-DOF) platform first, and then extend to the Human power Augmentation Lower Exoskeleton (HUALEX). The experimental results show that the proposed gait transition control strategy can minimize the interaction dynamics with less interaction force between the pilot and the exoskeleton compared to the traditional admittance controller. Compared to Ordinary Admittance Controller, the proposed VAC significantly improve the normalized Mean Squared Error (nMSE) of trajectory tracking from 2.751° to1.105°.

Keywords: Admittance control; Variable admittance; Gait transition; Interaction force; Coupled human-exoskeleton System

Introduction

A lot of researches on Human-Powered Exoskeleton Systems has been focused on flat terrain walking while stair ascent and vice versa is frequent process on daily life activities. Human-exoskeleton systems designed for constraining human movements to allow people to operate more easily or more efficiently in a variety of situations, required consideration of economical issues. Here we mean these systems must have simple and available control methods so can have further developments. If we take BLEEX control methodology Sensitivity Amplification Controller (SAC), efficient control way to shadow human motion but so expensive and resource consumer (hardware and software) [1].

During the navigation of coupled human-exoskeleton systems stairs ascending or descending is frequent process especially when applied for indoors mission. Several studies were performed to investigate normal human stair ascent and descent [2], investigations of biomechanics and motor coordination in human lower extremity during ascent and descent walking at different inclinations [3-5]. Other investigations such as staircase climbing of patients with knee and hip [5,6].

The physical Human-Robot Interaction pHRI as a result for human mind intention is a cooperative activities between separately human and exoskeleton [7] or close physical interaction [8] are big challenge, but in coupled human-robot system is more challenge because additional issues must considered because coupled human-robot system or exoskeleton robots act directly on the human body .The pressure-sensitive devices are widely used for gait analysis [9], a few researches are conducted to use such technique for measure gait variability [10] and for gait control [11,12]. Around thigh force-sensitive device is used for wearer intention estimation it can autonomously achieve smooth transition between different gates. Around thigh wearable force sensor device is designed to act as a sensory feedback tool to monitor the human exoskeleton interaction as well as to use it for gait transition control. The use of this interaction force as a feedback signals for our control system is efficient way for good results. The use of force-sensitive device in the end-effector of the intended joints can Figure out human intentions in clear and accurate way. For autonomous and smooth transition between different gaits we apply calibrated and efficient signal collector sensors. The wearable force sensor devices comparing to fixed (eneorized treadmills) are the best solution for on-line applications. The extracted feedback signals can monitor the human exoskeleton interaction as well as control the gait transitions.

An admittance controller also called position based impedance controller, uses the end-effector interaction force feedback to estimate the appropriate joint position. The desired joint position and position feedback are used to estimate the appropriate actuators inputs [13]. Since Neville Hogan first introduced impedance controllers [14], they have become well established specially in robotics and coupled human-exoskeleton system. The application of an impedance control strategy to coupled human-exoskeleton system will allow variable deviation from a given joint trajectory to intended maneuvers according to pilot. The Lokomat is a bilateral robotic orthosis that is used in conjunction with a body weight support system to control patient leg movements in the sagittal plane. The Lokomat hip and knee joints are actuated by linear drives, which are integrated in an exoskeletal structure. Admittance control extensively utilized in rehabilitation robotics but also was applied to upper limb power augmentation exoskeletons in

***Corresponding author:** Abusabah IA Ahmed, Center for Robotics, School of Automation, University of Electronic Science and Technology of China, Chengdu, China. E-mail: abusabah22@hotmail.com many researches [15-19], for robot human Quadrocopter [20] and for wheel chair control [21].

We propose variable admittance controller for smooth stairs ascent to handle trajectory changes. The implemented control strategy minimizes interaction forces during stairs climbing (i.e. the exoskeleton and the pilot are in contact in variety of places).Our proposed strategy minimizes the interaction forces by translate them to desired motion and keep interaction force within accepted specified thresholds. In this work, we show how to learn admittance controller parameters online using observations stairs ascent trails. To prove the performance efficiency of proposed we show how to learn admittance controller stiffness parameters on-line using observations stairs ascent trails. To prove the performance efficiency of proposed control strategy a simulation on 1-DOF exoskeleton platform is demonstrated for normal admittance control and modified one. The result comparison shows great minimization on interaction force and in computational effort. We choose admittance controller in this work because our exoskeleton system HUALEX designed with sensitive force sensors which can provide robust feedback signal for control scheme.

The paper is organized as follows: Section 4 shows the description and integration of the human-exoskeleton system, and the ordinary admittance control performance on gait transition. We validate the efficiency of variable admittance control in Section 5 by simulation on 1-DOF platform. Section 6 shows the performance efficiency of proposed strategy on interaction force and tracking error minimization. Finally, conclusions and some perspective on future uses and further development of this technique drawn in section 7.

Human-Exoskeleton System Design

A. Dynamic models

Totally 10-DoFs revolute joints are adopted for the HUALEX design. Among them, two-linkage revolute mechanism regarded as a 2-DOFs multilink pendulum is actuated parallel to the human thigh and shank when the exoskeleton is coupled to the pilots. The remaining DOFs are un-actuated and passively driven with in human joints movement. The appearance diagram of HUALEX is shown in Figure 1. As a wearable exoskeleton, the motion range for each DOF of HUALEX is designed according to human kinematics with some slight differences due to flexible connections between the exoskeleton and wearer. In the sagittal plane, the designed ranges of motion at the hip, knee and ankle joints are -45° to +45°, 0° to -135° and -30° to +30°, respectively. The exoskeleton links are made with ideal design (minimum weight and inertia). The Link lengths are adjustable respect to various pilots.

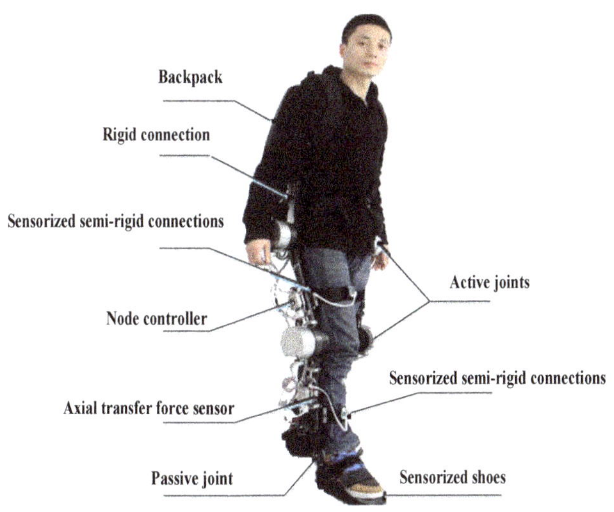

Figure 1: The prototype of HUALEX.

A lot of researches are conducted as a series and continuous modifications for HUALEX control and performance developments. Fuzzy-based impedance regulation for control of the coupled human-exoskeleton systems [22]. The learning approach of the relationship between physical human-exoskeleton interaction and dynamic factors [23]. A modification for BLEEX SAC successfully achieved and applied for HUALEX control [24], Radial Basis Function Neural Network (RBFNN) designed to compensate for the dynamic uncertainty error and minimize the physical human-robot interaction force [25]. Force sensing technology is also an important feature in human-exoskeleton systems for monitor the interaction between pilot and exoskeleton and controls these systems. Because this sensor is additional control device for the human-robot system, it is important for the sensor to be small, lightweight, and non-invasive. Climbing stairs requires active knee extension i.e. additional torque must be applied [3]. The degree of knee extension depends on the height of stairs which is stochastic value. A higher stair will require you to bend your knees more deeply and the greater the amount of flat terrain walking knee flexion. The stairs for this work are designed with step height, and fixed tread length. The proper interaction force sensors of HUALEX make successfully investigation of admittance controller, therefore perfect angle correction. The practical measurements of joint angles, encoders on HUALEX and inclinometers on human limbs beside interaction forces between them are used to investigate the joint flexions for different stair height. Two-dimensional Interaction Force Sensors (TIFS) are developed to measure quasi-interaction force resulting from human on the exoskeleton. The information from this sensor aimed to obtain the desired change in the input walking trajectory when human intend to change gait type. It is especially useful when controlling the desired admittance between the human and HUALEX for transit from flat terrain walking to stair ascent. Since our proposed control strategy for gait transition is a model-based control strategy, the dynamic model of HUALEX project must be given.

$$M(\theta)\ddot{\theta}+C(\theta,\dot{\theta})\dot{\theta}+G(\theta)=\tau_{Exo}+\tau_h \tag{1}$$

in which θ is the vector of each joint angle, τ_{Exo} and τ_h represent the input torques from HUALEX and human wearer, respectively. $M(\theta)$ is the inertia matrix and a function of θ, $C(\theta,\dot{\theta})$ is the Coriolis matrix and a function of θ and $\dot{\theta}$, and $G(\theta)$ is a vector of gravitational torques. During human-exoskeleton system navigation τ_h is changing according to human intentions

B. Admittance Control

The admittance of the human leg shank can be given as:

$$Y_h(S)=\frac{1}{Z_h(S)} \tag{2}$$

Which characterized by inertia moment J_h, damping B_h, and stiffness K_h. The desired set of admittance parameters is required to be achieved online so that the dynamics of human-exoskeleton system interaction behaviour controlled and transferred to appropriate joint flexion correction. The interaction force on shank is measured as described in details for HUALEX with specified wearer [22]. Through inverse dynamics the end-effector measured interaction force f_i

transferred to joint space interaction torque τ_i. The admittance of the coupled human-exoskeleton can be described as follows:

$$\Delta\theta(s)=\frac{\tau_i(s)}{Js^2+Bs+K} \tag{3}$$

where J, B and K are systems inertia moment, damping and stiffness respectively. Admittance function determines the joint angular deviation through inverse kinematics. Hence the purpose of the admittance controller is to keep $\Delta\theta$ the difference between θ_h and θ_{Exo} as small as possible during system navigation. The resulting new reference position $\theta_d^*=\theta_h+\Delta\theta$ is then fed into PD controller. Experimental simulations are conducted to obtain admittance parameters. Torque or moment of force is the tendency of a force to rotate the link about the joint. Using this truth we easily can translate end effector interaction force f_i to joint space interaction torque τ_i consider the distance from the interaction force point to the joint l=(0.366). The regulation of parameters of admittance controller in Eqn. (3) depends on human intention. The human intention resulted in interaction force which must be minimized during admittance controller by perfect shadow for human intended movements. During human-exoskeleton system navigation the variation on field terrains will lead to different intentions, in other words to variations in interaction force. As we can measure the interaction force between HUALEX and wearer we chose admittance control scheme to develop gait transitions methodology. The schematic diagram of 1-DOF platform considered in this paper is shown in Figure 2 and the admittance model is presented in Figure 3.

During human-exoskeleton system navigation the interaction force in shank strap online transformed to equivalent joint torques at knee joint through trajectory modification. Therefore, for human performance augmentation purpose, the admittance controller drives HUALEX to shadow human intended motion in the sagittal plane. Human walking process is three dimensional but the largest motions are in the sagittal plane. For calculation simplification in power augmentation human-robot systems performance analysis will take in consideration the human biomechanics in the sagittal plane. Especially for stair ascent gait the trajectories and joints variables can be obtained easily on the sagittal plane. The main goal of coupled human-exoskeleton systems control is the absolute harmony and smooth flow of human intentions acquired through different sensors and sensors system in general to the joint motors. The optimal reference for this process is human motion, how great creator made the perfect sensors and delay free transmissions to transmit mind intended motions to specified joint. Referred to Figure 3, the inertias of the wearer leg and the exoskeleton are coupled by a damper B_c and spring K_c representing the coupled human and exoskeleton shank brace. The dynamic model of 1-DOF coupled human-exoskeleton system (shank with knee joint) can be represented as in Eqn. (4):

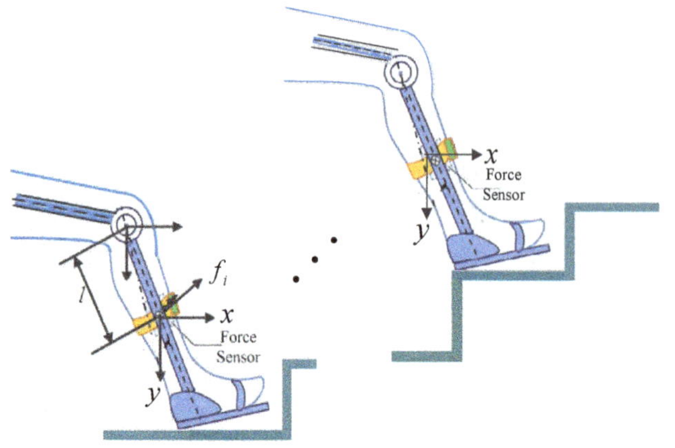

Figure 2: The schematic diagram of simple 1-DOF exoskeleton platform and resulting interaction force with the wearer when going upstairs.

$$J\ddot{\theta}(t)+B\dot{\theta}(t)+mgl\sin\theta(t)+C_F\text{sign}\dot{\theta}(t)=\tau(t)+J^T f_i(t) \tag{4}$$

where J, B, m, l represent inertial moment, viscous friction coefficient, shank mass and length of the one DOF exoskeleton, respectively, $(\theta,\dot{\theta},\ddot{\theta})$ represent the angle, angular velocity, and angular acceleration of the knee joint, C_F represents Coulomb friction coefficient around knee joint, (t) is actuation torque, J^T is the jacobian transpose of the platform and $f_i(t)$ is the interaction force between human wearer and exoskeleton. We use stiffness K, mainly as a linearization of the gravitational torque acting on the shank, such that $\theta\approx K\theta$ [26]. Consider the model of the coupled human-exoskeleton admittance control in Figures 3 and 4 we can rewrite Eqn. (4) as follows:

$$J_c\ddot{\theta}(t)+B_c\dot{\theta}(t)+K_c\theta(t)+C_F\text{sign}\dot{\theta}(t)=\tau(t)+J^T fi(t) \tag{5}$$

The estimation of the dynamic model parameters in Eqn. (4) was defined for HUALEX walking speed control by T. H. Toan et al. [22]. In this paper we define it for gait transitions control by ordinary admittance control method (OAC) we keep C_F = 1.724 as estimated in [22]. To show the efficiency of our proposed variable admittance controller we demonstrate ordinary admittance controller first to evaluate ordinary performance. On the system navigation process the interaction force on shank contact is proportional to the wearer intended motion as follows:

$$\Delta\theta(t)=J^T fi(t) \tag{6}$$

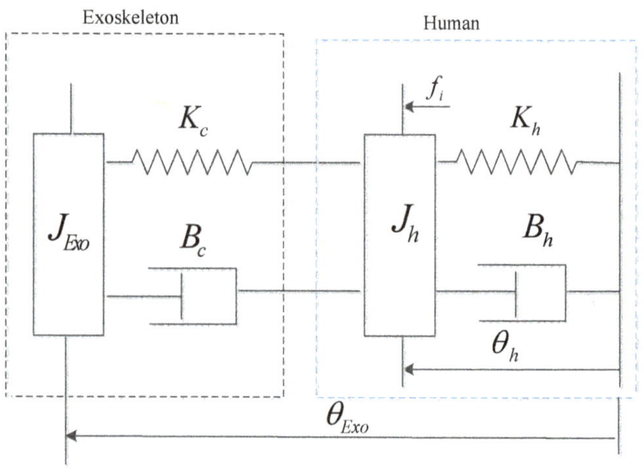

Figure 3: The ordinary admittance control model of the coupled human-exoskeleton system.

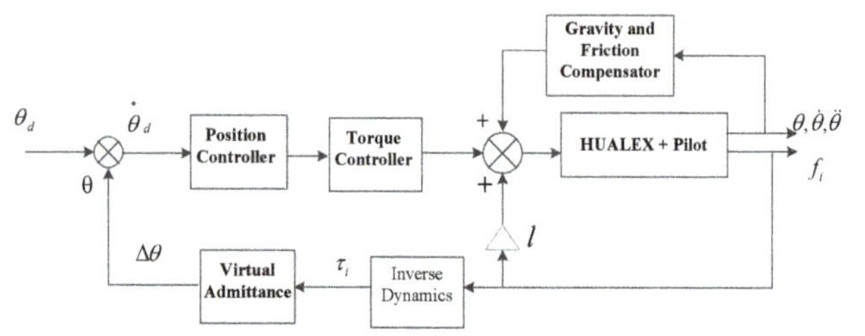

Figure 4: The main block diagram of admittance controller for coupled human-exoskeleton system interaction control.

Therefore, if desired admittance parameters of the controller are invariant; the dynamic relationship between the exoskeleton and the human wearer through the admittance according to interaction force modeled as shown in Eqn. (6). According to the experimental simulation, the resulted tracking control of OAC also indicates performance illness in the gait transition situations. For defining *J*, *BandK* in Eqn.(6) define the vectors:

$$\Phi(t)=\left[\ddot{\theta}(t)\ \dot{\theta}(t)\ \theta(t)\right]^T \qquad (7)$$
$$\Psi=\left[J\ \ B(t)\ \ K\right]^T$$

Assume relaxed wearer muscles operation i.e. $J^T f_i = 0$, with the definitions in Eqn. (12) we can present Eqn. (4) in matrix form:

$$\tau(t)=\Phi^T(t)\Psi \qquad (8)$$

Where, Φ^T is called the regression vector. The objective is to estimate the unknown parameter vector Ψ from observations of $\tau(t)$ and the regression vector $\Phi(t)$. The best estimate of Ψ can be obtained by minimizing the $L(t)$ least-square criterion which defined as

$$L(t)=1/2\left[\tau(t)-\Phi^T(t)\hat{\Psi}\right]^T\left[\tau(t)-\Phi^T(t)\hat{\Psi}\right] \qquad (9)$$

The least-square method is the basic technique for parameters estimation [27]. The minimum of Eqn. (9) is our goal for parameter vector estimation:

$$\frac{\partial \mathrm{L(t)}}{\partial \Psi}=0, \qquad \Psi \to \hat{\Psi} \qquad (10)$$

We can say $\hat{\Psi}(t)$ is optimal if and only if following condition satisfied:

$$\hat{\Psi}(t)\Phi T(t)\Phi(t)=\Phi^T(t)\tau(t) \qquad (11)$$

The condition in Eqn.(13) is called the normal equation. The online best estimation of parameter vector Ψ based on recursive least-square (RLS) estimation can be achieved when assumed that the matrix $\Phi(t)$ has full rank [27]. Then the least-square estimate of $\hat{\Psi}(t)$ satisfies the recursive equations:

$$\hat{\Psi}(t)=\hat{\Psi}(t-1)+K(t)\left[\tau(t)-\Phi^T(t)\hat{\Psi}(t-1)\right]$$
$$K(t)=P(t-1)\Phi^T(t)\left[1+\Phi^T(t)P(t-1)\Phi(t)\right] \qquad (12)$$
$$P(t)=1-\left[K(t)\Phi^T(t)\right]P(t-1)$$

From the experimental trails for flat terrain walking, stair ascent 170 mm height and 200 mm height the mean values for 10 trails are drawn in Table 1. The performance of OAC demonstrated separately in different gait is pretty good on tracking the human wearer intentions. But the tracking error is a little bit unacceptable especially during transition time. In other words, OAC faces the problem of rapid change in physical interaction force when human intend to change gait which lead to the performance illness as shown in the Figure 5. The interaction force and OAC trajectory error for 2 cases of stairs height are shown.

Parameter	Flat terrain walking	Stair ascent (170 mm)	Stair ascent (200 mm)
Inertia *J* (*Kgm*^{2})	0.133	0.122	0.123
Viscous friction coefficient *B* (*Nms/rad*)	6.282	7.371	7.446
Stiffness parameter *K* (*Nm/rad*)	33.921	31.120	29.414

Table 1: The estimated parameters of lower shank and knee joint.

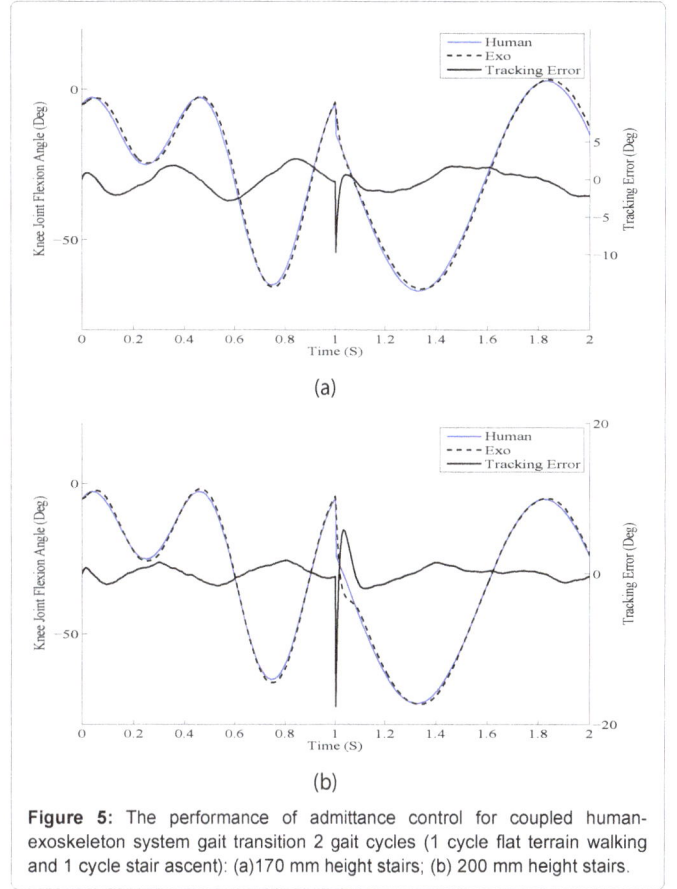

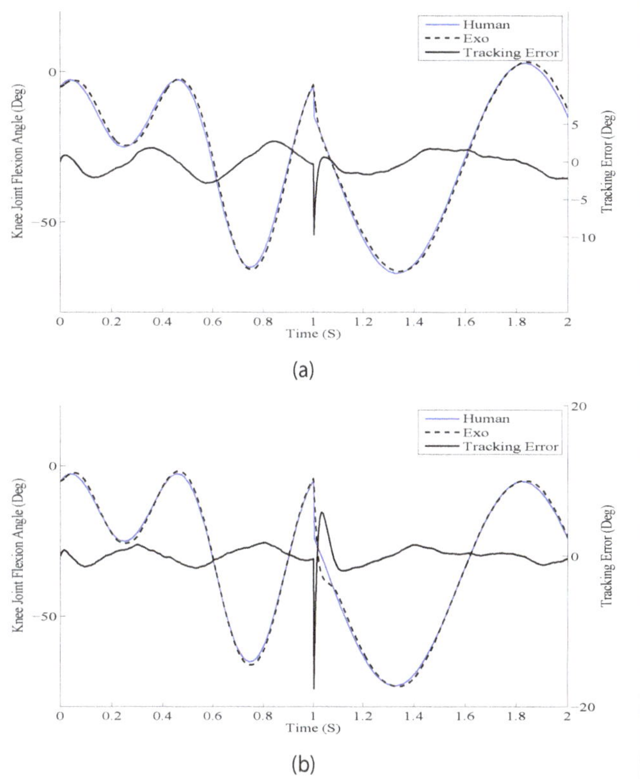

Figure 5: The performance of admittance control for coupled human-exoskeleton system gait transition 2 gait cycles (1 cycle flat terrain walking and 1 cycle stair ascent): (a)170 mm height stairs; (b) 200 mm height stairs.

Variable Admittance Control

A. Experimental simulations

The modifications of OAC is needed due to ill effects on control process on the gait transition case, the overshoot and undershoots resulted on the measured interaction force must have special treatments for smooth transition. Our aim is to regulate admittance parameters to find an appropriate input command (trajectory reference change). It is evident that this input depends on wearer desired flexion angle and desired admittance parameters. To improve such OAC to pretty handle the uncertainties in dynamics model when human intend to change gait we introduce dynamic parameters estimation technique called variable admittance controller (VAC). The investigated VAC algorithm attempts to minimize the interaction force during whole navigation process of the coupled human-exoskeleton system, i.e. better tracking performance.

Force impulse and instantaneous end-effector oscillations are perfect system inputs to estimate VAC parameters. The Cartesian space intended position depends on the stair height in this case can be transformed to the joint space as the mean value of the difference between the desired and the actual trajectories. This difference $\Delta\theta\ (t)$ will be used to modify the input trajectory to get to the optimized trajectory:

$$\theta_h^*(t)=\theta_h(t)+\Delta\theta(t) \qquad (13)$$

The block diagram of proposed variable admittance controller is shown in Figure 6 and the variable viscoelastic model of proposed VAC is drawn in Figure 7.

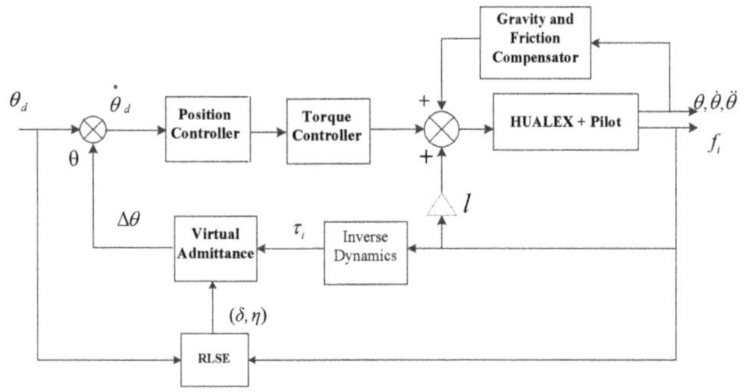

Figure 6: The proposed block diagram for variable admittance control scheme.

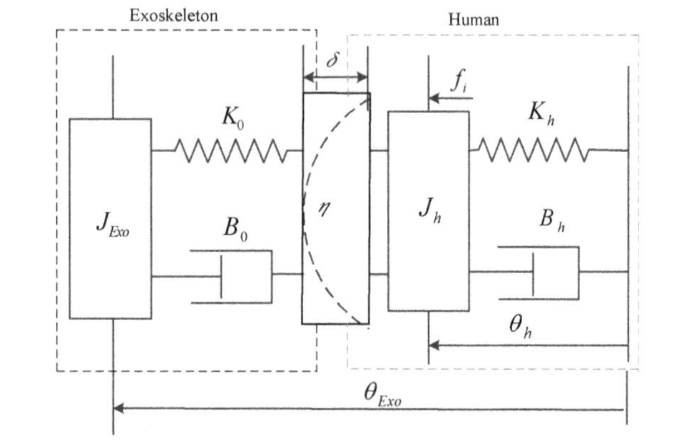

Figure 7: The variable visco-elastic model model of the variable admittance controller.

B. Estimation of viscoelastic model parameters

The dynamic process of input trajectory correction needs dynamic viscoelastic admittance parameters while keeping inertia constant:

$$K_{dyn} = \delta K_0$$
$$B_{dyn} = \eta B_0 \tag{14}$$

Where K_0 and B_0 represent the initial stiffness and viscous friction coefficient of OAC, depends on the current gait type as estimated in section 5. The main goal of proposed VAC scheme is to achieve smooth trajectory tracking during and after gait transitions in other words is to minimize the tracking error when human intend to change gait type. We investigate a relation between VAC parameters and interaction force for accurate and perfect human trajectory tracking even when sudden change in current trajectory happens (gait transition). The dynamic stiffness and viscous are obtained on-line for smooth gait transitions. The linearity of the relation Eqn. (15) is experimentally proved for acceptable range of interaction force, this range is variable depends on human-exoskeleton system function.

$$J_C\left(\ddot{\theta}_h(t)-\ddot{\theta}_{Exo}(t)\right)+B_C\left(\dot{\theta}_h(t)-\dot{\theta}_{Exo}(t)\right)+K_C\left(\theta_h(t)-\theta_{Exo}(t)\right)=J^T(t)f_i(t) \tag{15}$$

Experimentally we estimate the dynamic visco-elastic model parameters from collected data applying recursive least square estimator (RLSE). The considered model for the estimation of dynamic visco-elastic parameters is:

$$\eta B_0 \Delta\dot{\theta}(t) + \delta K_0 \Delta\theta(t) = \tau_i(t) \tag{16}$$

The initial value of unknown parameters vector $\left[\delta K_0 \;\; \eta B_0\right]^T$ are $\left[\delta \;\; \eta\right]^T = \left[1 \; 1\right]^T$, While $\left[K_0 \;\; B_0\right]^T$ are variable through navigation process depend on the current gait. The collected samples are for 2 different gait cycle contains the transition. The inertia coefficient assumed constant. The changing rate of the interaction force between human wearer and exoskeleton are indicating the future changing of viscoelastic model parameters. The relations for efficient parameter estimation are investigated in this paper. The experimental results limits the values of δ and η in the application on VAC for gait transition control, consider the variation of the stair height as follows:

$$0.0259 \le \delta_1 \le 2.3775 \quad 0.0248 \le \delta_2 \le 2.3272$$
$$0.1560 \le \eta_1 \le 2.5197 \quad 0.1549 \le \eta_2 \le 2.5595 \tag{17}$$

where δ_1 and η_1 are scaling coefficients of visco-elastic model for transition to 170 mm stairs, δ_2 and η_2 for transition to 200 mm. The method for adjusting admittance viscoelastic model parameters of the coupled human-exoskeleton system according to different gait types is promising towards fully autonomous human-exoskeleton system navigation. The on-line modification of input trajectory achieved successfully based VAC with dynamic viscoelastic parameters.

Vac Performance Analysis

The main advantage of VAC is the simple structure and implementation process, in other words has an availability for real applications and developing researches (low cost). VAC synchronizes the gait transition for human-exoskeleton system based on human intentions and trajectory construction, the performance is promising for further developments and applications. As shown in the simulation results VAC performs well and with more developing will be promising in Autonomous wearable Exoskeleton field. We take as inputs for control algorithm the pilot and exoskeleton joint angular positions, accelerations and interaction force between them. The stiffness and viscous friction parameters adjusted corresponding to the current gait type in order to keep the tracking error and interaction force as minimum as possible as in Eqn. (15). The proposed VAC scheme takes care of the transitions and allowing time for identification and adaptation to optimize performance. The performance of proposed control strategy for gait transitions is shown in Figure 8. The transition values of viscoelastic model parameters during transition for the different stairs height is shown in Table 2. The performance of proposed VAC algorithm is to determines the input needed to produce the desired plant performance (minimum tracking error). The comparison between OAC and VAC performance for gait transitions interaction force is shown in Figure 9, we compares the interaction force minimization and tracking error in 200 mm stairs case. The error correction appears big in 200 mm height stairs, as a result for the interaction force when human wearer intends to transit from flat terrain walking to stairs ascent.

The small disadvantages appear during proposed scheme simulations is the overshoots of interaction force, even though it was within the acceptable thresholds must have some treatments in future work. The transition process was achieved with high performance accuracy as depicted in Figure 8, with minimum tracking error and minimum interaction force.

Parameter	170 mm Stairs height	200 mm Stairs height
Ktr.	10.238	10.535
Btr.	16.353	14.647

Table 2: The transition values of visco-elastic model parameters.

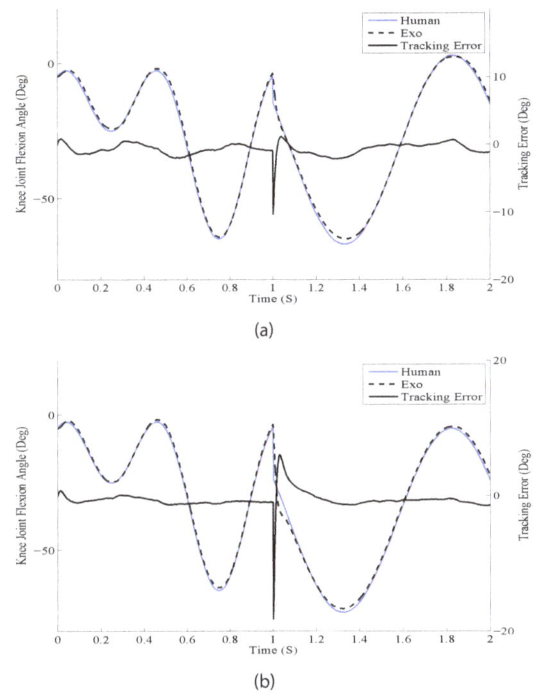

Figure 8: The performance of VAC for coupled human- exoskeleton system gait transition 2 gait cycles (1 cycle flat terrain walking and 1 cycle stair ascent): (a) 170 mm height stairs; (b) 200 mm height stairs.

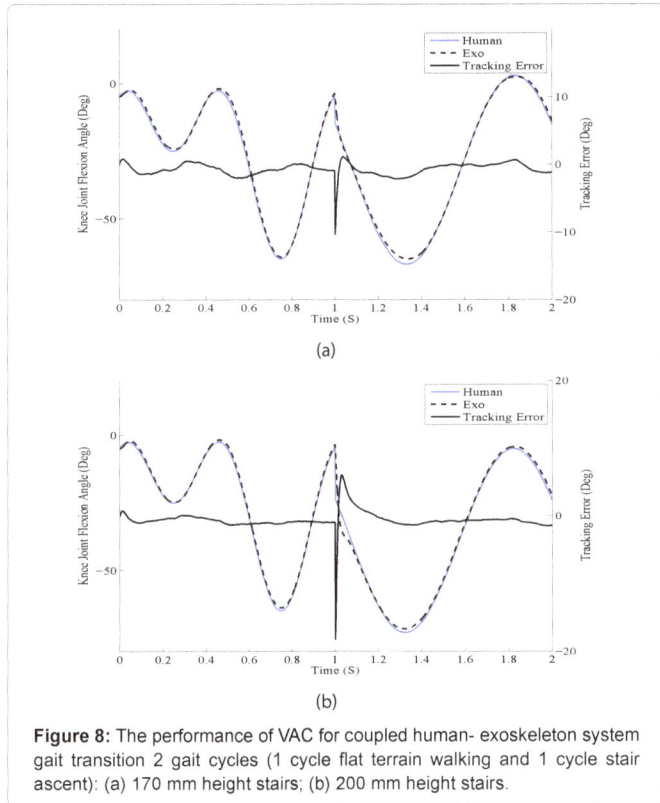

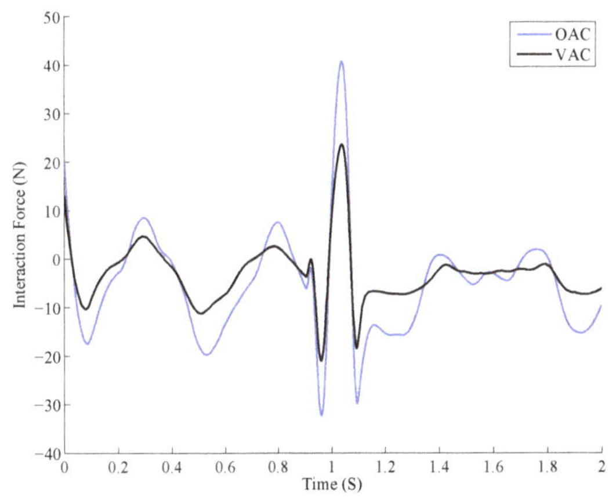

Figure 9: The proposed VAC strategy interaction force compared to OAC.

Conclusions and Future Works

Our control method synchronizes the gait transition for human-exoskeleton system based on human intentions and trajectory construction, the performance is promising for further developments and applications. The proposed VAC scheme for online gait transition control system with limited gait modes (flat terrain walking and stairs ascent) will find a more attentions in future developments forwards fully autonomous HUALEX robotic system. As shown in results our technique performs well and more developing will be promising work in Autonomous wearable Exoskeleton field. In future work will focus on the system response optimization to minimize the transition error, then develop our algorithm to work in obstacle avoidance in general and adapt to difference stairs height and tread length this will overcome the proposed technique limitations, then will go forward for practical implementation in our lab Exoskeleton. The interaction force convex (overshoots or undershoots) will find some attention in our future work to be minimized.

Acknowledgments

This work was support by NSFC (No. 61503060, 6157021026), Fundamental Research Funds for the Central Universities (ZYGX2014Z009) and SRF for ROCS, SEM.

References

1. Kazerooni H, Racine JL, Huang L, Steger R (2005) On the control of the berkeley lower extremity exoskeleton (BLEEX). International Conference of Robotics and Automation pp: 4353-4360.
2. Riener R, Rabuffetti M, Frigo C (2002) Stair ascent and descent at different inclinations. Gait and Posture 15: 32-44.
3. Hoover CD, Fulk GD, Fite KB (2013) Stair ascent with a powered trans femoral prosthesis under direct myoelectric control. IEEE/ASME Transaction on Mechatronics 18: 1191-1200.
4. Amirudina AN, Parasuramanb S, AhmedKhanc AK, Elamvazuthid I (2014) Biomechanics of hip, knee and ankle joint loading during ascent and descent walking. International Conference on Medical and Rehabilitation Robotics and Instrumentation pp: 336-344.
5. Andriacchi TP, Galante JO, Fermier RW (1982) The influence of total knee-replacement design on walking and stair climbing. The Journal of Bone and Joint Surgery 64: 1328-1335.
6. Bergmann G, Graichen F, Rohlmann A (1995) Is staircase walking a risk for the fixation of hip implants?. Journal of Biomechanics 28: 535-553.
7. Amor HB, Neumann G, Kamthe S, Kroemer O, Peters J (2014) Interaction primitives for human-robot cooperation tasks. IEEE International Conference on Robotics and Automation pp: 2831-2837.
8. Ikemoto S, Amor HB, Minato T, Jung B, Ishiguro H (2012) Physical human-robot interaction: mutual learning and adaptation. IEEE Robotics and Automation Magazine 19: 24-35.
9. De Rossi SMM, Lenzi T, Vitiello N, Donati M, Persichetti A, et al. (2011) Development of an in-shoe pressure-sensitive device for gait analysis. Engineering in Medicine and Biology Society, EMBC, Annual International Conference of the IEEE pp: 5637-5640.
10. Liu T, Inoue Y, Shibata K (2010) A wearable ground reaction force sensor system and its application to the measurement of extrinsic gait variability. Sensors 10: 10240-10255.
11. Hassan M, Kadone H, Suzuki K, Sankai Y (2014) Wearable Gait Measurement System with an Instrumented Cane for Exoskeleton Control. Sensors 11: 1705-1722.
12. Crea S, Donati M, De Rossi SMM, Oddo CM, Vitiello N (2014) A wireless flexible sensorized insole for gait analysis. Sensors 14: 1073-1093.
13. Tonietti G, Schiavi R, Bicchi A (2005) Design and Control of a Variable Stiffness Actuator for Safe and Fast Physical Human-Robot Interaction. International Conference on Robotics and Automation pp: 526-531.
14. Hogan N (1984) Impedance Control: An Approach to Manipulation. American Control Conference, pp: 304-313.
15. Miller LM, Rosen J (2010) Comparison of multi-sensor admittance control in joint space and task space for a seven degree of freedom upper limb exoskeleton. Proceedings of the 3rd IEEE RAS and EMBS International Conference on Biomedical Robotics and Biomechatronics, pp: 70-75.
16. Okunev V, Nierhoff T, Hirche S (2012) Human-preference-based control design: adaptive robot admittance control for physical human-robot interaction. The 21st IEEE International Symxposium on Robot and Human Interactive Communication pp: 443-448.
17. Carmichael MG, Liu D (2013) Admittance control scheme for implementing model-based assistance-as-needed on a robot. 35th Annual International Conference of the IEEE EMBS pp: 870-873.
18. Lee BK, Lee HD, Lee JY, Shin K, Han JS, et al. (2012) Development of dynamic model-based controller for upper limb exoskeleton robot. 2012 IEEE International Conference on Robotics and Automation pp: 3173-3178.

19. Yu W, Rosen J, Li X (2011) PID Admittance control for an upper limb exoskeleton. American Control Conference pp: 1124-1129.

20. Augugliaro F, Andrea RDS (2013) Admittance control for physical human-quadrocopter interaction. European Control Conference pp: 1805-1810.

21. Oda M, Zhu C, Suzuki M, Luo X, Watanabe H, et al. (2010) Admittance based control of wheelchair typed omnidirectional robot for walking support and power assistance. 19th IEEE International Symposium on Robot and Human Interactive Communication pp: 159-164.

22. Tran HT, Cheng H, Duong MK, Zheng H (2014) Fuuzy-based Impedance Regulation for Control of the Coupled Human-Exoskeleton System. IEEE International Conference on Robotics and Biomimetics pp: 986-992.

23. Tran HT, Cheng H, Lin X, Huang R (2014) The relationship between physical human-exoskeleton interaction and dynamic factors: using a learning approach for control applications. Science China Information Science 57: 1-13.

24. huang R, Cheng H, Chen Q, Tran Ht, Lin X (2015) Interactive Learning For Sensitivity Factors Of A Human-Powered Augmentation Lower Exoskeleton. IEEE/RSJ International Conference on Intelligent Robots and Systems, pp: 6409-6415.

25. Ka M, Cheng H, Toan TH, Jing Q (2015) Minimizing human-exoskeleton interaction force using compensation for dynamic un-certainty error with adaptive rbf network. Journal of Intelligent and Robotic Systems, pp: 1-21.

26. Colgate JE, Ollinger GA, Peshkin MA, Goswami A (2007) A 1-DOF assistive exoskeleton with virtual negative damping: effects on the kinematic response of the lower limbs. IEEE/RSJ International Conference on Intelligent Robots and Systems, pp: 1938-1944.

27. Astrom KJ, Wittenmark B (1995) Adaptive control, (2ndedn) Addison Wesley, Reading.

Survey of On-line Control Strategies of Human-Powered Augmentation Exoskeleton Systems

Ahmed AIA[1]*, Cheng H[1], Lin X[1], Omer M[2] and Atieno JM[3]

[1]*Center for Robotics, School of Automation, University of Electronic Science and Technology of China, 611731 Chengdu, China*
[2]*School of Automation, University of Electronic Science and Technology of China*
[3]*School of Electronic Engineering, University of Electronic Science and Technology of China, 611731 Chengdu, China*

Abstract

On-line gait control in human-powered exoskeleton systems is still rich research field and represents a step towards fully autonomous, safe and intelligent indoor and outdoor navigation. It is still a big challenge to develop a control strategy which makes the exoskeleton supply an efficient tracking for pilot intended trajectories on-line. Considering the number of degrees of freedom the lower limb exoskeletons are simpler to design, compared to upper limb. The comparison between lower limb and upper limb is useless when consider the control issues, because of the differences in missions and applications. Based on the literature, we aim to give an overview about control strategies of some famous lower limb human power exoskeleton systems. In the state of the art, different control strategies and approaches for different types of lower limb exoskeletons will be compared consider the efficiency and economic issues. Exact estimation of needed joints torques to execute human intended motions on-line with efficient performance, low cost and reliable way is the main goal of studied system's control strategies. We have study different control strategies used for wide known human power augmentation exoskeletons and compare between them in graphs and tables.

Keywords: Coupled human-exoskeleton system; Power augmentation; Autonomous navigation; Interaction force; Gait transition; Control cost

Introduction

On-line motion control of human power augmentation exoskeleton systems is still a big challenge specially for the applications in complicated and dynamic terrains. By this we mean indoors mission conditions with frequent changing between flat terrain walking and stairs ascent or vice versa. Human-exoskeleton systems designed for constraining human movements to allow people to operate more easily or more efficiently in a variety of situations, required consideration of efficient control and economical issues. Here we mean these systems to be available for the public must have an efficient and available control strategy so can have further application developments. The efficiency of motion control strategy will be measured in according to the some performance features, such as interaction force and tracking error. In human-powered exoskeleton systems the real-time adaptive motion control is yet complex process. This complexity is proportional to the system's maneuverability.

The fully autonomous human power augmentation exoskeleton system must be controlled to behave and interact as human being, with the sensory system, communication media, muscles, joints, and main controller. These parts are in loop for standing alone human as well as for human power exoskeleton system (Human-in-the-Loop), with interactive behaviour for different situations and transitions [1]. The main feature needs to pay attention of such system with discrete time events (sensory feedback signals) according to human intentions and continuous time during system's navigation is response time. The human navigation process is unique without time delay and can't be achieved 100% on human power augmentation exoskeleton system, still researches are conducted to achieve high performance efficiency [2]. Looking at the human body from an engineering point of view, we notice that it has many types of inputs and outputs. The human Central Nervous System (CNS) is the part of the nervous system consisting of the brain and spinal cord and represented by the main controller in our system. Because the human body is a very complex structure and and just consider the lower limbs, there are endogenous or internal signals that can be perceived by the senses to control lower limbs motion. The main outputs of the lower limbs from the feedback signals is the movements and maneuvers. Information primary travels from the sensors to the main controller to identify and decide the next coming maneuver, these signals are the reflection of current human intentions.

The aim of this paper is to provide an overview of the most effective motion control strategies for the lower limb human power augmentation exoskeleton systems. Lower limbs Exoskeletons, which have a wide range of possible applications include physical support and facilitating labor-intensiveness by decreasing the load action on the operator. The physical support field of above application form three main groups of powered lower-limb devices: rehabilitative, assistive, and empowering devices. The paper is organized as follows: Section 4 shows the description and integration of the famous human-exoskeleton system, and the our lab exoskeleton HUALEX. We describe motion control strategies in Section 5. Section 6 shows the performance efficiency of the studied strategies on a single DoF exoskeleton platform. Finally, conclusions and some perspective on future uses and further development of these strategies drawn in section 7.

***Corresponding author:** Abusabah IA Ahmed, Center for Robotics, School of Automation, University of Electronic Science and Technology of China, Chengdu, China. E-mail: abusabah22@hotmail.com

Typical Lower-Limb Exoskeleton Systems

Legged locomotion has many advantages comparing to wheels, it's efficient performance on rough and unpredictable terrain [3]. Wheels are useful for flat surfaces applications and simple to control but they have worse performance on rough and unpredictable terrain navigation. In the opposite Legs can adapt to a wide variety of environments, such as rough terrains and staircases, which are impassable by wheeled vehicles [4,5]. In other words legs are maneuverable and effective for wide range of applications. Parallel-limb exoskeleton designed for load transfer to the ground in parallel with the human lower limb. It features the DoFs of the human to be compatible with lower leg dynamics. The exoskeleton robots can keep the balance on its own, while shadow the human wearer intended motion during navigation [6].

Many different exoskeleton robots have been developed from the early 1960s [7]. They can be categorized in several ways: by power source, by actuators, by structure, by function, and by application. Lower limb exoskeletons have been built for augmenting human performance, assisting with disabilities, studying human physiology and reactivate motor deficiencies [8-10]. For purpose of discussion, exoskeletons are divided into two categories here. The first type of exoskeleton is one used to help gait disorder persons or aged people to walk. The second type of exoskeleton is used to help people walking who have to travel long distances by feet with heavy loads. For power augmentation exoskeletons, the power of the robot joints must be generated with active joints such as electric motors or hydraulic cylinders. Power Augmentation exoskeletons refers to exoskeletons that can give neurologically intact, healthy human capabilities above and beyond the normal ones. This has usually been focused on increasing strength and enhancing endurance for difficult conditions applications. Direct interaction of power augmentation exoskeleton with the human neuromotor system during locomotion requires adequate design of the components, both the bio-mechanical and functional aspects for safety issues. The on-line motion of such systems control is challenge especially in outdoors applications, with un-known terrain's condition. The length of the thigh and the shank links of such exoskeletons are adjustable by a mechanism of two telescopic bars that are fixed in different positions by screws to fit different wearer's height in limited range.

A. Berkeley lower extremity exoskeleton (BLEEX)

Berkeley Lower Extremity Exoskeleton (BLEEX) designed around primarily supporting a large load in the form of a backpack, this limited it's direct interaction with the human body [11-13]. BLEEX was funded by the Defence Advanced Research Project Agency (DARPA) at the University of California, Berkley, and beginning in the year 2000 [8]. It has been designed for the specific task of allowing the human wearer to bear a large load on their back. The device is composed of three parts, the two powered robotic legs, the power and computing unit, and a backpack frame as shown in Figure 1. BLEEX is the first load carrying autonomous exoskeleton [14]. With an anthropomorphic design, BLEEX has a left and right three-segment leg, being analogous to the human thigh, shank and foot. Each leg has seven DoFs: hip flexion/extension and abduction/adduction, knee flexion/extension and ankle dorsi/plantar flexion are active. BLEEX provides the operator with load-carrying capability and endurance through versatile legged locomotion. Possible applications include helping soldiers, disaster relief workers, wildfire fighters and other emergency personnel to carry major loads without the strain typically associated with demanding labor [3].

(a) BLEEX [3] (b) HULC [16] (c) HAL [17] (d) HUALEX [26]

Figure 1: The most famous lower limbs human power augmentation exoskeletons.

B. Human universal load carrier (HULC)

From the public information, we know that Exohiker is very good at variety of locomotion and robustness, and can almost follow any action of the human body, or even intentionally sudden step, squat, creeping and other movements. The construction of ExoHiker was completed in February 2005. After finishing the Exohiker study, the team packaged all technology to the Lockheed Martin, who carried out technical transformation of military engineering. In 2009, the system was renamed Human Universal Load Carrier (HULC) and released. On the other hand, it had some upper limb function and was expected to form the Army soldiers' equipment in a few years. Since then, the progress of the research went into secrecy. Exohiker (or HULC) has been the most outstanding one of all lower extremity exoskeleton systems, although there is still a certain gap to the ideal in the weight and working time etc. But the appearance of Exohiker (or HULC) makes the exoskeleton really be out of the laboratory and people see the possibility of its practical application.

Dismounted fighters often carry heavy combat loads that increase the stress on the body leading to potential injuries. With a HULC exoskeleton, these heavy loads are transferred to the ground through powered titanium legs without loss of mobility. The HULC is a completely un-tethered, hydraulic powered anthropomorphic exoskeleton that provides users with the ability to carry loads of up to 200 pounds for extended periods of time and over all terrains. Its flexible design allows for deep squats, crawls and upper-body lifting. An onboard micro-computer ensures the exoskeleton moves in concert with the individual. The HULC's modularity allows for major components to be swapped out in the field. Additionally, its unique power-saving design allows the user to operate on battery power for extended missions. When battery power is low, the HULC system continues to support the loads and does not restrict mobility. HULC can also support a maximum load, with or without power. Lockheed Martin is also exploring exoskeleton designs for industrial use and a wider variety of military mission specific applications [15,16].

HULC specifications are as follows:

- Total weight without batteries is 53 lb.

- The power is Lithium polymer batteries.

- Electronics are Flexible, expandable electronics architecture for future applications, custom single-board microelectronics housed in a sealed enclosure and heat sinks on actuators.

- Hydraulic system is efficient low-flow, high pressure hydraulic system uses standard hydraulic fluid.

The HULC exoskeleton operates on lithium polymer batteries.

The power-saving feature enables the system to support maximum load even when the battery power is low. When equipped with an extended mission power supply with recharge capability, would enable dismounted Soldiers on these missions to carry fewer batteries.

C. Hybrid assistive limb (HAL)

Sankai et al. began developing the exoskeleton Hybrid Assistive Limb (HAL) in the mid of 1990s [17]. Their first prototype had active joints with single DoF at the hips and the knees, as well as a passive joint at the ankles. This model was followed by other versions of HAL-3 as shown in Figure 1. It is one of the most closely integrated with the human body. However, there are many improvements that need to be addressed in order to help user (for rehabilitation or force augmentation) learn to walk [18,19]. These include specific size, physical connections, powered control, software, and communication needs. The links that make up the lower body of the exoskeleton will need to be redesigned to allow for the inclusion of additional motors, and the electric motors may need to be replaced with hydraulics all together to be able to perform with new and additional software requirements. HAL-3 system is composed of three main parts skeleton and actuator, controller, and sensor [20]. Exoskeletal frame consists of a four-link, three-joint mechanism with the links corresponding to the hip the thigh, the lower thigh and the foot, and the joints corresponding to the hip, the knee and the ankle joints of the human body. The actuators of HAL-3 provide assist torque for knee and hip joints. Each actuator has a DC-motor with harmonic drive to generate the assist torques at each joint. The total weight of the skeleton system with the actuators is about 15 Kg [21]. HAL-3 controlled by Cybernic Control System which consists of Cybernic Voluntary Control (CVC) system and Cybernic Autonomous Control (CAC) System [22]. When a person attempts to move their body, nerve signals are sent from the brain to the muscles through the motor neurons, moving the musculoskeletal system. When this happens, small bio signals can be detected on the surface of the skin. The HAL suit registers these signals through a sensor attached to the skin of the wearer. Based on the signals obtained, the power unit moves the joint to support and amplify the wearer's motion. The HAL suit possesses a cybernic control system consisting of both a user-activated voluntary control system known as CVC and a "robotic autonomous control system" known as CAC for automatic motion support.

HAL can be used even if no bio-electrical signals are detected, due to problems, say, in the central nervous system or the muscles. The Japanese-built, wearable, HAL exoskeleton, for example, offers improved endurance and strength for users, but is not mind-controlled. Instead of brain waves, HAL picks up EMG signals, the electrical pulses generated by the muscles and uses these to send commands to the exoskeleton. It also partially relies on the users existing mobility and balance to operate.

Meanwhile, Berkeley Bionics in the US has developed an EMG-based "medical exoskeleton" that enables people with reduced mobility and strength to walk upright. However, it requires a supporting device to keep users balanced.

The latest version of HAL has remained brain-controlled but evolved to a full body robot suit that protects against heavy radiation without feeling the weight of the suit. Eventually it could be used by workers dismantling the crippled Fukushima nuclear plant. The new type of HAL is on display today at the Japan Robot Week exhibition in Tokyo. It will be used by workers at nuclear disaster sites and will be field tested at Fukushima, where a tsunami in March 2011 smashed into the power plant, sparking meltdowns that forced the evacuation of a huge area of north-eastern Japan [22].

HAL uses electrical signals sent to the muscles from the brain to anticipate the wearer's movement. HAL's ability to anticipate movement allows it to move fractions of a second before the wearer, providing seamless interaction between human and robot [23].

D. Human power augmentation lower exoskeleton (HUALEX)

The Human power Augmentation Lower Exoskeleton (HUALEX) was demonstrated at the Centre for Robotic, School of Automation at University of Electronic Science and Technology of China. HUALEX has an an ergonomic design, robust and lightweight equipment for load carrier [24]. As shown in Figure 1, HUALEX has total four active joints to provide torques, which is activated by DC servo motors. The ankle joints of HUALEX is designed as an energy-storage mechanism which can store energy in stance phase and deliver in swing phase during walking. Besides the joints and rigid links, many compliant connections at waist, thighs, shanks and feet are provided for semi-rigid connecting HUALEX to the wearer. HUALEX system designed as lower limbs human power augmentation exoskeleton, in the hip structure has two DoFs performing functions of flexion/extension actuated by Maxon DC motors, two non-actuated DoFs abduction/adduction. At the knee joint, there is one DoF performing flexion/extension actuated by Maxon DC motors. At the ankle joint, there is one DoF performing dorsiflexion/planter flexion and another one at the metatarsophalangeal joint for flexion/extension, both are non-actuated. Thus, there are only four actuated DoFs in total in our system using Maxon DC motors attached with a harmonic drive gear because the flexion/extension DoFs of hip and knee play an important role during normal walking and its energy [24].

HUALEX control schemes are based on compliance method that relied on the measurement of interaction force resulted from the wearer intention to change gait mode or gait speed [25]. The learning approach of the relationship between physical human-exoskeleton interaction and dynamic factors [26,27] allows efficient application of Admittance Control (AC). Ordinary AC [28,29] can't be applied for maneuverable human-powered augmentation exoskeleton systems (Figure 1).

Motion Control Strategy

The human provides an intelligent control system for the exoskeleton, while the exoskeleton actuators provide most of the strength necessary for walking [30]. The control algorithms ensure that the exoskeleton moves in concert with the wearer depending on interaction force between them [31-33]. In this paper we address a distinctive control strategies for above mentioned exoskeleton systems and the implementation with a given system to show their performance efficiency. A model-based Control system [34] is one of the exoskeleton control system categories. According to the model used, the control strategy for the exoskeleton can be divided into two types: the dynamic model and the muscle model based control. The dynamic model can be obtained through the mathematical model, the system identification and the artificial intelligent method. Beside model-based control system other control system suitable for human power augmentation exoskeleton control are Hierarchy based Control System and Physical parameters based control system. The utilized and control system must meet the needed develop on the next step such as the assist as needed, the user's intention detection, the safety and the stability to give better performance.

E. Sensitivity amplification control (SAC)

Sensitivity Amplification Control (SAC) algorithm was first

proposed in the augmentation applications of Berkeley Lower Extremity Exoskeleton (BLEEX). The SAC algorithm is widely used in human augmentation applications since it just need the information from the exoskeleton robot, so that the complexity of exoskeleton system can be reduced greatly. BLEEX control methodology SAC, efficient control way to shadow human motion but so expensive and resource consumer (hardware and software) [13].

Racine proposed a method named virtual joint torque control, this method also be called SAC [8], and apply it for BLEEX control. SAC needs no direct measurements from the pilot or from the human-machine interface (e.g. no force sensors between the them); instead, the controller estimates, based on measurements from the exoskeleton suits only, how to move so the pilot feels very little force. This control scheme, which has never before been applied to any robotic system, is an effective method of generating locomotion when the contact location between the pilot and the exoskeleton is unknown and unpredictable.

SAC is a control method seriously relies on the dynamic model of the system. The variety of the parameter will have an important influence on the system. Simulation research can help us to solve this problem. But the most difficult is exoskeleton suit is a human-machine system and the human-machine interface model is hard to describe which results in little simulation results can be found in present. In this paper, we take the human-machine interface model as a PID controller, and combined the controller with exoskeleton model and SAC controller to simulate the interactive cooperation between the human and the exoskeleton suit. The swing phase of normal gait is the situation in which the foot is not in contact with the ground as shown in Figure 2. Consider the swing phase of a single DoF exoskeleton platform (Knee joint), SAC algorithm can applied to the swing phase, Position Control successfully can drive the stance phase with it's small flexion angles [11]. The ring-based networked control architecture (ExoNET) that together enables BLEEX to support payload while safely moving in concert with the human pilot. The main controller with external Graphical User Interface (GUI) is complex and expensive systems see Figures 2 and 3.

The inverse dynamics of BLEEX in sagittal plane is modeled differently depending on three gait phases: a full 7-DOF serial link mechanism for the single-support phase; two 3-DOF serial link mechanisms with one connection DOF along their uppermost link for the double-support phase; a 3-DOF serial link mechanism for the support leg and a 4-DOF mechanism for the redundant leg during late stance phase. The gait phases are distinguished by foot insole

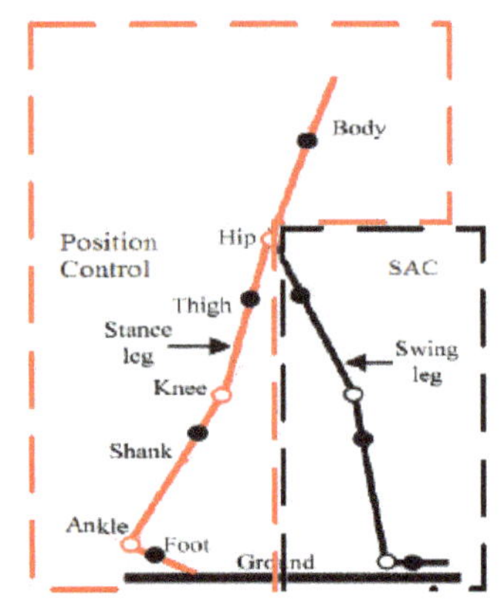

Figure 2: The gait phases considered for BLEEX control.

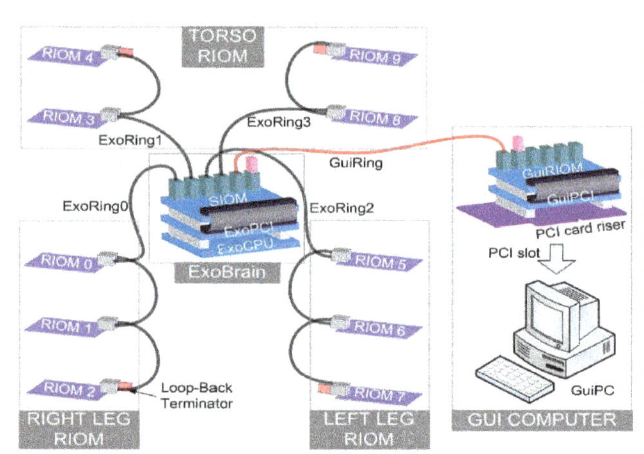

Figure 3: The overall view of ExoNET networked control system and external GU I debug terminal demonstrated from [12].

sensors. Let's consider the dynamic equation for single DoF platform exoskeleton with human wearer represented as:

$$J\ddot{\theta}(t)+B\dot{\theta}(t)+mgl\sin\theta(t)=\tau_a(t)+\tau_h(t) \quad (1)$$

where *J*, *B*, *m*, *l* represent inertial moment, viscous friction coefficient, shank mass and length of the one DoF exoskeleton, respectively, $(\theta,\dot{\theta},\ddot{\theta})$ represent the angle, angular velocity, and angular acceleration of the knee joint, *g* is the gravity τ_a and τ_h represent the input torque from the actuator and the pilot, respectively. However, for different transition, the dimension and value of these parameters are quiet different. To apply SAC to Eqn. (1) for system control we can write:

$$\tau_a = mgl\sin\theta+\left(1-\alpha^{-1}\right)\left(J\ddot{\theta}+B\dot{\theta}\right) \quad (2)$$

where τ_a represents actuator torque, αrepresents the sensitivity amplifier factor (greater than unity). Then the interaction torque exerted by human on the exoskeleton τ_h can be expressed as in the following equation:

$$\tau_h = k\left(\theta_h-\theta_{Exo}\right)+b\left(\dot{\theta}_h-\dot{\theta}_{Exo}\right) \quad (3)$$

where k and b are positive quantities, θ_h and $\dot{\theta}_h$ are predefined angle and angular velocity of the pilot trajectory, θ_{Exo} and $\dot{e}_{Exo}$ are the actual angle and angular velocity of the system. During locomotion of the pilot, the interaction torque τ_h is changed over the time based on the wearer intentions for gait transitions [15]. If we take $\alpha = 10$, the torque exerted by human will changed to be minimum. When human intend to change gait SAC will increase the closed loop system sensitivity to this intention (forces and torques) [13]. SAC strategy, the overall view of ExoNET networked control system and external GUI debug terminal.

F. Cybernic autonomous control

The Cybernics Autonomous Control (CAC) autonomously provides a desired functional motion generated according to the wearer's body constitution, conditions and purposes of motion support. HAL-3 with the Cybernic Autonomous Control successfully enhances a healthy person's walking, stair-climbing, standing up from a sitting posture and cycling, synchronizing with human body condition. Floor Reaction Forces (FRFs) and joint angles are used as motion information to detect a wearer's conditions. Posture control, as well as sensing and recognition for an environment including a wearer are essential technologies for an entire autonomous physical support, but they remain to be solved as shown in Figure 4. Human Intention Estimator (HIE) is used for wearer's intentions detection to use for robot motion control. Instead of the bioelectrical signals used for the control of the conventional HAL, the FRF is used for an intention estimation of the wearer who can control his weight balance using two canes with his hands. HAL estimates which leg supports a wearer's weight, when a wearer begins to swing a right or left leg and when wearer wants to stop walking. This control consists of the PD control using reference walking patterns based on healthy person's walk, in the swing phase and the constant-value control in the landing and support phase. Figure 4 shows a block diagram for this tracking control and phase synchronization. The HIE located in the upper-left part in the Figure 4 has the FRF as inputs for the estimation algorithms described below. Three blocks under the HIE are a library of the reference patterns in the swing phase and the reference values in the landing and support phase. The HIE allocates these references to two legs during walking. There are six ordinary PD control blocks on the right side of the HIE and the library. The upper three blocks are controllers for the right leg and the lower ones are for the left leg. The reference walking patterns, should be adjusted according to the wearer's intentions, for example a walking cycle and amplitude of each joint trajectory in swinging leg, while the stance phase is position controlled. The command torque for a single DoF τ is calculated by:

$$\tau = K\left(C\theta_{ref} - \theta\right) + \hat{K}\left(C\dot{\theta}_{ref} - \dot{\theta}\right) \tag{4}$$

where θ is the actual wearer's joint angle, $\dot{\theta}$ is angular velocity, respectively. In addition, θ_{ref} and $\dot{e}_{ref}$ are the reference joint angle and the reference angular velocity, respectively. On the other hand, K is the feedback gain of the joint angle error, and $\hat{K}$ is the feedback gain of the joint angular velocity error. The different feedback gains are used in the swing, landing or support phase independently by adopting this control architecture. In addition, C is the gain to the reference joint angle and angular velocity.

G. Variable admittance control (VAC)

The modified Admittance Control AC performs well on the gait transitions and walking speed changing cases, the overshoot and undershoots resulted on the measured interaction force have been treated for smooth transition [35-37]. The regulated admittance parameters help to find an appropriate input command (trajectory reference change). It is evident that this input depends on wearer desired flexion angle and desired admittance parameters [38-40]. To improve such AC to pretty handle the uncertainties in dynamics model when human intend to change gait dynamic parameters estimation technique called variable admittance controller (VAC) was introduced. The investigated VAC algorithm attempts to minimize the interaction force during whole navigation process of the coupled human-exoskeleton system, i.e. better tracking performance [41,42].

Force impulse and instantaneous end-effector oscillations are perfect system inputs to estimate VAC parameters [43]. The Cartesian space intended position depends on the stair height in this case can be transformed to the joint space as the mean value of the difference between the desired and the actual trajectories. This difference $\Delta\theta(t)$ will be used to modify the input trajectory to get to the optimized trajectory:

$$\theta_h^*(t) = \theta_h(t) + \Delta\theta(t) \tag{5}$$

The block diagram of VAC is shown in Figure 5 and the variable viscoelastic model of proposed VAC is drawn in Figure 6.

The dynamic process of input trajectory correction needs dynamic viscoelastic admittance parameters while keeping inertia constant:

$$\begin{aligned} K_{dyn} &= \delta K_0 \\ B_{dyn} &= \eta B_0 \end{aligned} \tag{6}$$

where K_0 and B_0 represent the initial stiffness and viscous friction coefficient of OAC, depends on the current gait type as estimated in section III. The main goal of proposed VAC scheme is to achieve smooth trajectory tracking during and after gait transitions in other words is to minimize the tracking error when human intend to change gait type. We investigate a relation between VAC parameters and interaction force for accurate and perfect human trajectory tracking even when sudden change in current trajectory happens. The dynamic stiffness and viscous are obtained on-line for smooth gait transitions. The linearity of the relation Eqn. (7) is experimentally proved for acceptable range of interaction force, this range is variable depends on human-exoskeleton system function.

$$J_C\left(\ddot{\theta}_h(t) - \ddot{\theta}_{Exo}(t)\right) + B_C\left(\dot{\theta}_h(t) - \dot{\theta}_{Exo}(t)\right) + K_C\left(\theta_h(t) - \theta_{Exo}(t)\right) = J^T(t) f_i(t) \tag{7}$$

Experimentally we estimate the dynamic viscoelastic model parameters from collected data applying recursive least square estimator (RLSE). The considered model for the estimation of dynamic viscoelastic parameters is:

$$\eta B_0 \Delta\dot{\theta}(t) + \delta K_0 \Delta\theta(t) = \tau_i(t) \tag{8}$$

The initial value of unknown parameters vector $\left[\delta K_0 \quad \eta B_0\right]^T$ are $\left[\delta \quad \eta\right]^T = \left[1 \quad 1\right]^T$, While $\left[K_0 \quad B_0\right]^T$ are variable through navigation process depend on the current gait (Figures 5and 6).

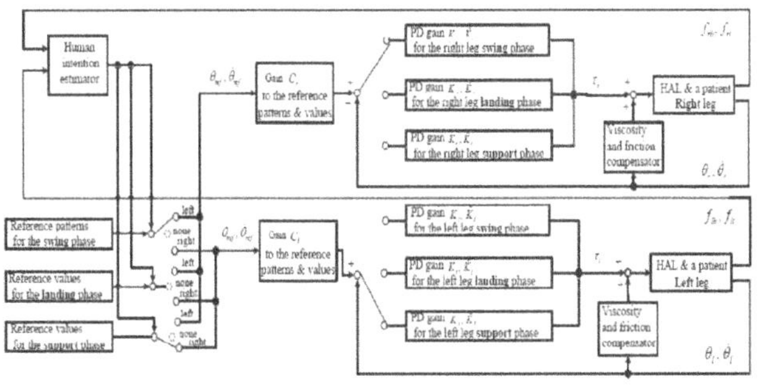

Figure 4: The main block diagram for HAL-3 tracking control demonstrated from [17].

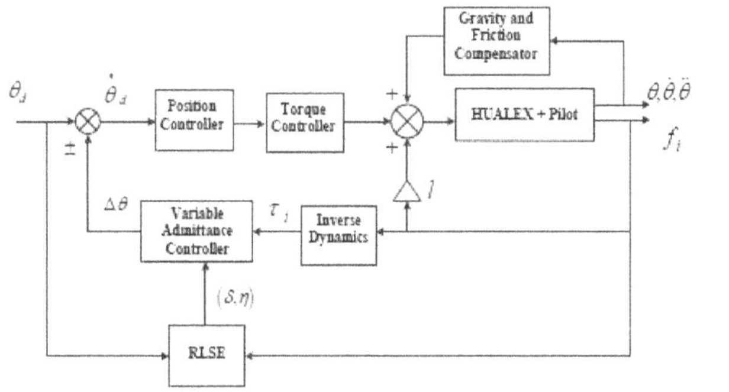

Figure 5: The proposed block diagram for variable admittance control scheme.

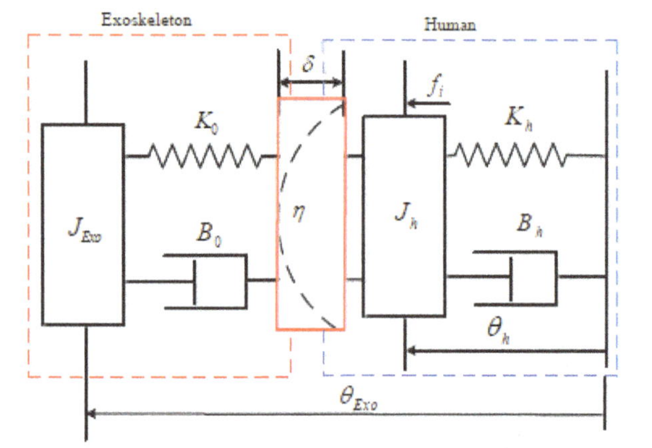

Figure 6: The variable viscoelastic model of the variable admittance controller.

Simulation and Discussion

A. Single DoF Exoskeleton

The all mentioned above Control strategies are simulated on same coupled human-exoskeleton system dynamic model. In the simulation of gait transition and walking speed changing, we consider normal stair height 200mm and for walking speed changing 1.5m/s is changed to 2m/s. The dynamic model of single DoF exoskeleton system (shank with knee joint) can be represented as in Eqn. (9) [44]:

$$J\ddot{\theta}(t)+B\dot{\theta}(t)mgl\sin\theta(t)+C_F\text{sign}\dot{\theta}(t)=\tau(t)+J^T f_i(t) \quad (9)$$

where J, B, m, l represent inertial moment, viscous friction coefficient, shank mass and length of the one DOF exoskeleton, respectively, $(\theta,\dot{\theta},\ddot{\theta})$ represent the angle, angular velocity, and angular acceleration of the knee joint, CF represents Coulomb friction coefficient around knee joint, $\tau(t)$ is actuation torque, J^T is the jacobian transpose of the platform and $f_i(t)$ is the interaction force between human wearer and exoskeleton.

The sudden changing in motion trajectories for dynamic obstacle avoidance during human-powered exoskeleton systems navigation needs special care additional efforts. As identified by computer simulations a large overshoots and undershoots are happen when transit to stairs ascent from flat terrain walking. The smooth behaviour of feedback signal illustrates the graduated correction of current trajectory which grantee smooth transit from flat terrain walking to stair ascent. The resulted overshoots and undershoots are proportional to stairs height. The above mentioned control strategies already have some measures to to keep the tracking error as minimum as possible. flexion. The stairs for this work are designed with step height, and fixed tread length. For performance evaluation we demonstrate simulation trails on the single DoF exoskeleton for the stairs ascent maneuver. The resulted tracking performance for stair climbing and walking speed changing is shown in Figure 7 and 8.

The proper interaction force sensors of HUALEX make successfully investigation of admittance controller, therefore perfect angle correction. The practical measurements of joint angles, encoders on HUALEX and inclinometers on human limbs beside interaction forces between them are used to investigate the joint flexions for different stair height. Beside this reduced interaction force overshoot technique is applied. With high sensitivity of SAC which mean expensive control (complicated calculations and additional blocks) still the performance suffers from feedback overshoots according to pilot's intention. The separate treatment of the different gait phases (swing and stance) leads to modification in control efficiency and minimization in control cost in SAC and CAC.

For the CAC control algorithm K taken from 100.0 to 200.0 and K^ was from 0.10 to 0.20 at less than 1.0 Hz on the knee joint.

HUALEX control method synchronizes the walking speed for human-exoskeleton system based on human intentions and CPGs frequency control as new kind of hybrid control. The CPGs are used to be applied to control the locomotion of humanoid robots, here adaptive CPGs applied to control trajectory frequency according to pilot intentions. This leads to good performance on tracking error minimization beside reduced interaction forces between pilot and exoskeleton. The performance efficiency of gait transition control algorithm can be measured from accuracy of tracking trajectory during transition and walking speed changing. The control algorithm minimizes the tracking error, as a natural result that transforming the interaction force to appropriate joint correction angle flexion.

VAC developed a technique aimed to keep feedback signal controlled within required range for smooth feedback signal. Consider the problem of tracking error minimization, subject to the bound feedback signal (interaction force) being placed on the overshoot. That is the maximum permitted overshoot value is limited to some thresholds. Take as inputs for control algorithm the experimentally calculated interaction force convex threshold and the current interaction force value to generate reliable and smooth feedback signals. Locally Weighted Scattersite Smoothing (LOWESS and LOESS) are strongly related non-parametric regression method that combine multiple regression models in a k-nearest-neighbour-based meta-model. LOESS is a later generalization of LOWESS [45,46]. LOESS and LOWESS thus build on classical methods, such as linear and nonlinear least squares regression. LOESS combines much of the simplicity of linear least squares regression with the flexibility of nonlinear regression. It does this by fitting simple models to localized subsets of the data to build up a function that describes the deterministic part of the variation in the data, point by point.

B. Performance analysis

The results of the commented computer simulations on a single DoF exoskeleton platform shown in Figures 7 and 8 can be used to highlight the differences between mentioned above control strategies. Since the exoskeleton is in contact with a pilot, a stable behaviour of the human-robot interaction has to be guaranteed during navigation and transitions. We check the ability of each strategy to adapt fast changing dynamics from the pilot (sudden maneuvers), the tracking error during transitions. The all mentioned above control strategies applied hybrid assistive control aims to control the exoskeleton by applying different assistive strategies adopts a force controller in swing phase and a position controller in stance phase. The small disadvantages appear during proposed scheme simulations is the overshoots of interaction force, even though it was within the acceptable thresholds must have some treatments in future work. The reviewed control strategies beside model-based property they are intention estimation-based too, partially for specified mission (CAC and VAC) or for all kind of gait (SAC). The treating of:

1) SAC: SAC is comfortable for pilot, no measurements needed from the pilot, the controller estimates, based on measurements from the exoskeleton. The cost of control process in this strategy is extremely high, in both hardware and software, the complicated systems for

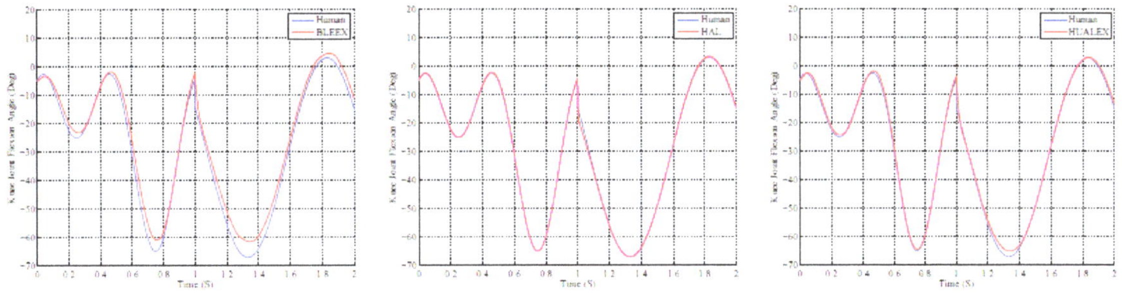

(a) (b) (c)

Figure 7: The performance of studied control strategies for gait changing from flat terrain to stair ascent: (a) SAC; (b) CAC; (c) VAC.

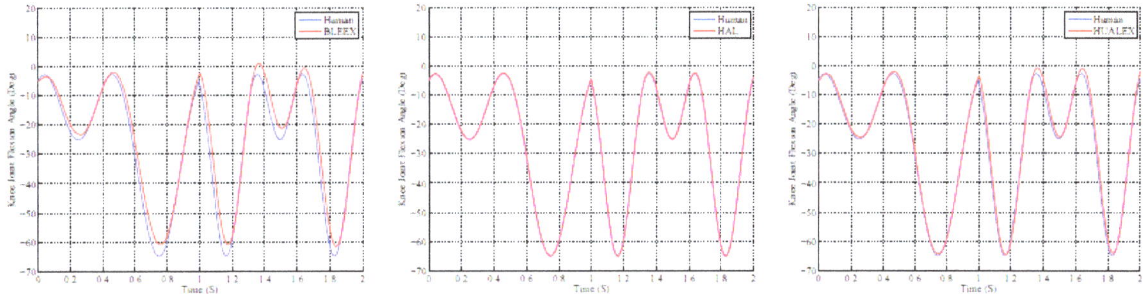

(a) (b) (c)

Figure 8: The performance of studied control strategies for walking speed acceleration: (a) SAC; (b) CAC; (c) VAC.

signals conditioning and useful feedback extraction make heavy calculations. The high level of sensitivity in response to the forces and torques imposed by the pilot is counted as negative time and resource consuming process. The accuracy of dynamic model must be provided in the SAC algorithm so, a complex model identification process is necessary (heavy calculations). The position control method which employed for the stance phase is not efficient since the dynamic model in this phase is complicated.

2) CAC: HAL intention estimation efficiency of a swing leg depends on the predefined thresholds, The performance efficiency depends on the efficiency of thresholds definition. The HAL utilizes a big number of sensors for control: skin-surface Electro-Myo-Graphic (EMG) electrodes placed below the hip and above the knee on both the front and the back sides of the pilot's body, potentiometers for joint angle measurement, ground reaction force sensors, and a gyroscope and accelerometer mounted on the backpack for torso posture estimation. These sensing modalities are used in two control systems that together determine user intent and operate the suit: an EMG-based system and a walking-pattern-based system. It takes a long time to calibrate the HAL for a specific user while BLEEX and HUALEX exoskeletons and sensing systems are adjustable. The sensors variety specific leads to proper tracking performance as validated and depicted in Figures 7 and 8. The HAL sensing method fine wire or surface EMG can be counted as an advantage, as the pilot intention can be predicted before muscle movement. There is a fraction of second between muscle movement and joint movement, so ordinary force sensors in SAC and VAC getting the human intention later than CAC.

3) VAC: The main advantage of VAC is the simple structure and implementation process, in other words has an availability for real applications and developing researches (low cost). VAC synchronizes the gait transition for human-exoskeleton system based on human intentions and trajectory construction, the performance is promising for further developments and applications. As shown in the simulation results VAC performs well and with more developing will be promising in Autonomous wearable Exoskeleton field.

Conclusions and Future Works

The autonomous navigation of human power augmentation exoskeleton systems demand high adaptability especially for complicated terrains applications. The accurate estimation of wearer intentions and quick response of such systems is another great challenge. Maintaining reference trajectory tracking of on-line navigation is complex and challenge, to be provided in stable manner need reliable feedback systems with sensitive sensors. Optimized response for feedback signals (timely) is important issue to minimize the transition tracking error and interaction force between the wearer and exoskeleton. The important future research point for such systems is effective obstacle avoidance while changing gait or during stairs ascent. We mean that, the high maneuverable exoskeleton systems must maintains difference maneuvers beside gait transitions and speed changing. The methodology to detect motion intention varies according to the different targeted motion tasks, different control strategy and typology of wearers. As shown in the results control strategies performance is varied for different intended motion, with more developing the navigation of human power augmentation exoskeletons will be from mind to joint control.

Acknowledgments

This work was support by NSFC (No. 61503060, 6157021026), Fundamental Research Funds for the Central Universities (ZYGX2014Z009) and SRF for ROCS, SEM.

References

1. Schirner G, Erdogmous D, Chowdhurt K, Padir T (2013) The future of human-in-the-loop cyber-physical systems. Computer 46: 36-45.
2. Winter DA (2009) Biomechanics and motor control of human movement, (4thedn) New Jersey: John Wiley and Sensor Inc.
3. Zoss A, Kazerooni H (2006) Design of an electrically actuated lower extremity exoskeleton. Advanced Robotics 20: 967-988.
4. Andriacchi TP, Galante JO, Fermier RW (1982) The Influence of Total Knee-replacement Design on Walking and Stair Climbing. The Journal of Bone and Joint Surgery 64: 1328-1335.
5. Amirudin AN, Parasuraman S, AhmedKhan MKA, Elamvazuthi I (2014) Biomechanics of Hip, Knee and Ankle joint loading during ascent and descent walking. International Conference on Medical and Rehabilitation Robotics and Instrumentation 42: 336-344.
6. Tran HT, Cheng H (2014) Learning the Relation of Physical Interaction to Dynamic Factors during Human-Exoskeleton Collaboration. IEEE International Conference on Multiple Fusion and Information Integration, Beijing.
7. Mosher RS (1967) Handyman to Hardiman. Society of Automation Engineering (SAE) International.pp: 1-10.
8. Racine JLC (2003) Control of a Lower Extremity Exoskeleton for Human Performance Amplification.
9. Kawamoto H, Lee S, Kanbe S, Sankai Y (2003) Power assist method for HAL-3 using EMG-based feedback controller. IEEE International Conference on Systems, Man and Cybernetics, pp: 1648-1653.
10. Huang R, Cheng H, Chen Q, Tran HT, Lin X (2015) Interactive Learning for Sensitivity Factors of a Human-powered Augmentation Lower Exoskeleton. IEEE/RSJ International Conference on Intelligent Robots and Systems, pp: 6409-6415.
11. Kazerooni H, Steger R, Huang L (2006) Hybrid Control of the Berkeley Lower Extremity Exoskeleton (BLEEX). International Journal of Robotics Research 25: 561-573.
12. Kazerooni H, Chu A, Steger R (2007) That which does not stabilize, will only make us stronger. The International Journal of Robotics Research 26: 75-89.
13. Kazerooni H, Racine JL, Huang L, Steger R (2005) On the Control of the Berkeley Lower Extremity Exoskeleton (BLEEX). International Conference of Robotics and Automation, pp: 4353-4360.
14. Zoss A, Kazerooni H, Chu A (2005) On the Mechanical Design of the Berkeley Lower Extremity Exoskeleton (BLEEX). International Conference on Intelligent Robots and Systems, pp: 3132-3139.
15. Ghan J, Steger R, Kazerooni H (2006) Control and system identification for the berkeley lower extremity exoskeleton. Advanced Robotics 20: 989-1014.
16. LockHeedMartin, http://www.lockheedmartin.com/content/dam/lockheed/data/mfc/pc/hulc/mfc-hulc.
17. Kawamoto H, Sankai Y (2005) Power assist method based on phase sequence and muscle force condition for HAL. Advanced Robotics 19: 717-734.
18. Lee S, Sankai Y (2002) Power assist control for walking aid with HAL-based on EMG and impedance adjustment around knee joint. IEEE/RSJ International Conference on Intelligent Robots and Systems (IROS 2002), pp: 1499-1504.
19. Okamura J, Tanaka H, Sankai Y (1999) EMG-based prototype powered assistive system for walking aid. Asian Symposium on Industrial Automation and Robotics (ASIAR'99), pp: 229-234.
20. Lee S, Sankai Y (2005) Virtual Impedance Adjustment in Unconstrained Motion for Exoskeletal Robot Assisting Lower Limb. Advanced Robotics 19: 773-795.
21. Suzuki K, Kawamura Y, Hayashi T, Sakurai T, Hasegawa Y et al. (2005) Intention-Based Walking Support for Paraplegia Patients with Robot Suit HAL. IEEE International Conference on Systems, Man and Cybernetics, pp: 2707-2027.
22. Japan Robot Week (2016) International Robot Exhibition (iREX)
23. Kawamoto H, Sankai Y (2005) Power assist method based on phase sequence and muscle force condition for HAL. Advanced Robotics 19: 717-734.
24. Ka M, Cheng H, Toan TH, Jing Q (2015) Minimizing Human- Exoskeleton Interaction Force Using Compensation for Dynamic Un- certainty Error with Adaptive RBF Network. Journal of Intelligent and Robotic Systems, pp: 1-21.
25. Tran HT, Cheng H, Duong MK, Zheng H (2014) Fuuzy-based Impedance Regulation for Control of the Coupled Human-Exoskeleton System. IEEE International Conference on Robotics and Biomimetics, pp: 986-992.
26. Tran HT, Cheng H, Lin X, Huang R (2014) The Relationship between Physical Human-Exoskeleton Interaction and Dynamic Factors: Using a Learning Approach for Control Applications. Science China Information Science 57: 12.
27. Ikemoto S, Amor HB, Minato T, Jung B, Ishiguro H (2012) Physical Human-Robot Interaction Mutual Learning and Adaptation. IEEE Robotics and Automation Magazine 19: 24-35.
28. Augugliaro F, DeAndrea R (2013) Admittance Control for Physical Human-Quadrocopter Interaction. European Control Conference, pp: 1805-1810.
29. Oda M, Zhu C, Suzuki M, Luo X, Watanabe H, et al. (2010) Admittance Based Control of Wheelchair Typed Omnidirectional Robot for Walking Support and Power Assistance. 19th IEEE International Symposium on Robot and Human Interactive Communication, pp: 159-164.
30. Astrom KJ, Wittenmark B (1995) Adaptive control (2ndedn) Addison Wesley, Reading.
31. Hassan M, Kadone H, Suzuki K, Sankai Y (2014) Wearable Gait Measurement System with an Instrumented Cane for Exoskeleton Control. Sensors 11: 1705-1722.
32. De Rossi SMM, Lenzi T, Vitiello N, Donati M, Persichetti A, et al. (2011) Development of an in-shoe pressure-sensitive device for gait analysis. Engineering in Medicine and Biology Society, EMBC, Annual International Conference of the IEEE, pp: 5637-5640.
33. Liu T, Inoue Y, Shibata K (2010) A Wearable Ground Reaction Force Sensor System and Its Application to the Measurement of Extrinsic Gait Variability. Sensors 10: 10240-10255.
34. Pons JL (2008) Wearable robots: bio mechatronic exoskeletons. Wiley Online Library 70.
35. Crea S, Donati M, De Rossi SMM, Oddo CM, Vitiello N (2014) Wireless Flexible Sensorized Insole for Gait Analysis. Sensors 14: 1073-1093.
36. Hoover CD, Fulk GD, Fite KB (2013) Stair Ascent With a Powered Trans femoral Prosthesis Under Direct Myoelectric Control. IEEE/ASME Transaction on Mechatronics 18: 1191-1200.
37. Riener R, Rabuffetti M, Frigo C (2002) Stair Ascent and Descent at Different Inclinations. Gait and Posture 15: 32-44.
38. Hogan N (1984) Impedance Control: An Approach to Manipulation. American Control Conference, pp: 304-313.
39. Tonietti G, Schiavi R, Bicchi A (2005) Design and Control of a Variable Stiffness Actuator for Safe and Fast Physical Human-Robot Interaction. International Conference on Robotics and Automation, pp: 526-531.
40. Miller LM, Rosen J (2010) Comparison of Multi-Sensor Admittance Control in Joint Space and Task Space for a Seven Degree of Freedom Upper Limb Exoskeleton. Proceedings of the 3rd IEEE RAS and EMBS International Conference on Biomedical Robotics and Biomechatronics, pp: 70-75.
41. Okunev V, Nierhoff T, Hirche S (2012) Human-preference-based Control Design: Adaptive Robot Admittance Control for Physical Human- Robot Interaction. The 21st IEEE International Symposium on Robot and Human Interactive Communication, pp: 443-448.
42. Carmichael MG, Liu D (2013) Admittance Control Scheme for Implementing Model-based Assistance-As-Needed on a Robot. 35th Annual International Conference of the IEEE EMBS, pp: 870- 873.
43. Colgate JE, Ollinger GA, Peshkin MA, Goswami A (2007) A 1-DOF Assistive Exoskeleton with Virtual Negative Damping: Effects on the Kinematic Response of the Lower Limbs. IEEE/RSJ International Conference on Intelligent Robots and Systems, pp: 1938-1944.
44. Lee BK, Lee HD, Lee JY, Shin K, Han JS (2012) Development of Dynamic Model-based Controller for Upper Limb Exoskeleton Robot. 2012 IEEE International Conference on Robotics and Automation, pp: 3173-3178.
45. Laboratory NB (1988) Anthropometry and mass distribution for human analogues. Naval Biodynanlics Laboratory.
46. Fox J (2002) Nonparametric Simple Regression: Smoothing Scatterplots. Smoothing Scatterplots. Thousand Oaks, CA.

Hand-assisted Robotic Approach for Ovarian Cancer Management

Cindy Chan[1], Li-Hsuan Chiu[1,2], Ching-Hui Chen[1,2] and Wei-Min Liu[1,2]*

[1]*Department of Obstetrics and Gynecology, Taipei Medical University Hospital, Taipei, Taiwan*
[2]*Department of Obstetrics and Gynecology, School of Medicine, College of Medicine, Taipei Medical University, Taipei, Taiwan*

Abstract

Robotic assisted staging surgery has been increasingly employed for a variety of gynecological malignancies such as ovarian, endometrial, and cervical cancers. Here we demonstrate a hand-assisted robotic approach for managing ovarian cancer with large tumor mass and predominantly solid components, where mini-laparotomy is performed followed by robotic surgical staging procedures. In this retrospective descriptive analysis of 29 ovarian cancer patients, admitted from December 2011 to May 2014, who had a large tumor mass (≥ 7 cm) and received laparoscopic surgical staging, traditional robotic surgical staging or hand-assisted robotic procedures, we reviewed for patient demographics, surgical procedures, and perioperative parameters. The results were comparable and we conclude the hand-assisted robotic approach offers a safe and feasible way to perform ovarian cancer surgical staging for patients with large tumor masses.

Keywords: Robotic surgery; Hand-assisted surgery; Ovarian cancer; Staging surgery; Laparoscopic surgery; Mini-laparotomy

Introduction

Since its introduction in 2005, robotic assisted surgical procedures have been widely adopted by gynecologists to perform surgical procedures such as cystectomies, sacral colpopexies, myomectomies, radical hysterectomies and cancer staging surgeries [1-6]. Recent studies show that, as compared to a conventional laparoscopic procedure or laparotomy, robotic surgery is associated with decreased blood loss; lower conversion rate; less intraoperative complications and shorter hospital stay [4,7-10]. Robotic assisted staging surgery has also been increasingly employed for a variety of gynecological malignancies such as ovarian, endometrial, and cervical cancers.

Ovarian cancer is a peritoneal disease, with a higher risk of peritoneal spread, compared to other gynecological cancers. Ovarian cancer is one of the ten leading cancer types and accounts for 3% of all new cancer cases among women in the United States in 2013 [11]. The overall 5-year survival rate of ovarian cancer is 43% [11], and when the disease is diagnosed in the early stages, the survival rate can be as high as 94% [12]. However, only 15% of women are diagnosed during the early stages, at which robotic surgery is considered an effective treatment [13]. For management at advanced stages, exploratory or extensive dissection of lesion sites in the abdomen and pelvis is required. With such cases, comprehensive surgical staging using robotic or laparoscopic procedures are thought to be difficult [13]. Robotic assisted surgery for these advanced cases faces obstacles such as difficulty removing large tumor masses without rupture into the peritoneal cavity and limited access to upper abdominal quadrants when disease is diffuse.

Here, we introduce a hand-assisted robotic cancer staging surgery technique, where a small midline laparotomy incision is first made caudal to the umbilicus. Large tumor masses may be aspirated and removed within an endo-bag under direct vision and upper abdominal procedures are done with more convenient access. Palpation of abdominal surfaces can also be done. The incision is then closed and the remaining robotic-assisted surgical staging procedures are performed as usual, with better field of vision. This technique for the management of large ovarian tumor masses and upper abdominal disease may save time and be an option where robotic staging surgery for ovarian cancer cases is considered.

Materials and Methods

From December 2011 to May 2014, a total of 32 ovarian cancer patients were treated by laparoscopy, traditional robotic-assisted surgery or hand-assisted robotic surgery for surgical staging procedures at our hospital. Among them, 29 patients had a large tumor mass, measuring greater than 7 cm, and these 29 patients were retrospectively analyzed for review of patient demographics, surgical procedures and peri-operative parameters.

Before surgery, all cases were evaluated by an expert meeting consisting of gynecologists, pathologists, radiologists, oncologists and robotic surgical team members. Among the enrolled cases, 12 patients received laparoscopic surgical staging, 6 patients received traditional robotic-assisted surgical staging, and 11 underwent hand-assisted robotic surgical staging. Robotic-assisted surgery was done sing a da Vinci Surgical Si System by a single surgeon. All cases reviewed were considered to have received optimal de-bulking and surgical staging. Intraoperative and postoperative parameters that were reviewed included operation time, blood loss, post-operative pain scores, and the time for hospital stay.

Technique: All patients underwent general anesthesia and were set in lithotomy position. Each patient was then draped in a sterile manner and a uterine manipulator was placed. All surgeries were performed by a single surgeon.

For patients who received laparoscopic or traditional robotic surgical staging surgery, pneumoperitoneum was first obtained and trocar placement was done. Laparoscopic surgical staging was done

***Corresponding author:** Dr. Wei-Min Liu, Department of Obstetrics and Gynecology, Taipei Medical University Hospital and Taipei Medical University, No.252, Wu-Hsing St., inyi District, Taipei 11031, Taiwan
E-mail: weimin@tmu.edu.tw

with camera port set at the umbilicus and three trocars placed at the following sites: 8-10 cm caudal-lateral to the scope at the left side of the patient; 8-10 cm caudal to the scope on the midline; and assistant port at 8-10 cm caudal-lateral to the scope at the right side of the patient, as needed. Laparoscopic surgery was performed mainly using mono-polar curved scissors and bipolar forceps. Grasper and suction irrigation was used via the assistant port to assist the surgical procedures.

For traditional robotic surgical staging using the da Vinci Surgical Si System, three trocars were docked at the following sites: 6 cm along the midline above the umbilicus for the scope and 8 to 10 cm bilateral to the scope for the two side arms. In addition, one 10 mm trocar was placed 6 to 8 cm caudo-lateral to the left arm for the accessory port. The side cart was set between the patient's legs and the robotic arms were docked. The robotic surgery was performed with monopolar curved scissors and fenestrated bipolar forceps. Grasper and suction irrigation was used via the accessory port to assist the surgical procedures.

Surgical staging procedures included total hysterectomy, bilateral salpingo-oophrectomy, bilateral pelvic lymph node dissection (including obturator, internal iliac, external iliac and common iliac lymph nodes), para-aortic lymph node dissection, omentectomy and appendectomy. Large ovarian tumor masses were placed in an endo-bag and then aspirated to avoid peritoneal spillage. Suspected malignant masses were removed and subjected for frozen section analysis for confirmation before the remaining procedures were done. To minimize bleeding, uterine arteries were also cauterized before the hysterectomy procedure. After all procedures were completed, the trocars were removed and the intra-abdominal gas was released. The trocar sites were closed with sutures and the patient was subjected for further evaluation.

For the hand-assisted robotic surgical procedures, a 4 to 5 cm in length midline abdominal incision beginning at the umbilicus was first performed. Unilateral salpingo-oophorectomy was carried out for frozen section analysis, and ascites was collected for cytology analysis. For removal of the large tumor masses, suction was applied to drain the cyst fluid within an endo-bag to avoid spillage. After deflating the cyst, it was removed and subjected for frozen section analysis. The wound was then sutured closed, and, if malignancy was confirmed, docking of the da Vinci Surgical Si System followed. Then, the remaining surgical procedures, as described before, including contralateral salpingo-oophrectomy, omentectomy, peritoneal washings, and appendectomy were performed.

Patient demographics and operative parameters: The medical charts of the enrolled cases were reviewed for patient demographics, surgical approach and procedures, peri-operative parameters, pathological staging, and surgical-related complications. The reviewed baseline characteristics of the enrolled patients included age, body mass index (BMI), percentage of cases with positive lymph nodes, and pathological staging in accordance with the International Federation of Gynecology and Obstetrics. The assessed intraoperative parameters included operation time, estimated blood loss, and lymph node yield. The post-operative parameters included 24-hour pain scores, amount of time before the patient was able to resume a full diet after surgery, and the length of hospital stay.

The volume of blood loss was defined as the total volume of fluids collected by suction during surgery. The operation time was measured from the time of skin incision to skin closure. All patients received pain control with patient-controlled analgesia (PCA) and non-steroidal anti-inflammatory drugs (NSAIDs) during postoperative care. The 24-hour pain score was measured 24 hours after the operation. The pain scores were self-reported and routinely evaluated for each patient during postoperative care using an adult pain score numerical rating scale (NRS-11). For reference of the pain score, scoring 0 indicates no pain, and scoring 10 indicates the worst pain imaginable. The amount of time before full diet resumption was defined as the number of postoperative days until the patients could tolerate regular intake of solid food. The length of hospital stay was defined as the number of postoperative days until the patient was discharged.

Statistical Analysis

For statistical analysis, the mean, standard deviation (SD), median, and range of each parameter were reported. To examine the differences between the surgical groups, statistical analysis was performed with one-way ANOVA with Turkey HSD post-hoc analysis or Chi-Square analysis; a p-value of less than .05 was considered statistically significant between the groups. All data were analyzed using SPSS statistics (version 21.0, IBM). The research protocols were approved by the Taipei Medical University Joint Institutional Review Board (TMUJIRB-201301047).

Results

Twenty-nine cases were evaluated in this study. All patients were reviewed for patient demographics, surgical procedures, and peri-operative parameters (Table 1). The mean ages were 42.2 ± 11.0 (traditional robotic surgical staging), 39.6 ± 12.3 (hand-assisted robotic surgical staging), and 45.5 ± 8.5 (laparoscopic surgical staging) for each surgical group. The mean BMIs were 22.8 ± 3.3 (traditional robotic surgical staging), 20.9 ± 3.1 (hand-assisted robotic surgical staging), and 22.9 ± 4.7 (laparoscopic surgical staging) kg/m2. The percentages of cases with positive lymph node findings from each group were 33.3% (traditional robotic surgical staging), 27.3% (hand-assisted robotic surgical staging), and 16.7% (laparoscopic surgical staging). The percentage and case number of each disease stage and histological type were also reported in the Table 1. The differences in age, BMI, disease

	Robotic surgical staging (n=6)	Hand-assisted robotic surgical staging (n=11)	Laparoscopic Surgical staging (n=12)	p-value
Age (years)	42.2 (11.0)	39.6 (12.3)	45.5 (8.5)	0.899
BMI (kg/m^2)	22.8 (3.3)	20.9 (3.1)	22.9 (4.7)	0.824
Tumor size (cm)	9.2 (2.5)	12.8 (4.7)	10.8 (3.4)	0.167
History of prior pelvic surgeries, % (n)	33.3% (2/6)	36.4% (4/11)	30.8 (4/13)	0.959
Cases with positive lymph nodes, % (n)	33.3% (2/6)	27.3% (3/11)	16.7% (2/12)	0.744
Pathological Stage				
Stage I	66.6% (4/6)	54.5% (6/11)	75.0% (9/12)	0.698
Stage II	0% (0/6)	9.1% (1/11)	8.3% (1/12)	0.754
Stage III	33.3% (2/6)	36.4% (4/11)	16.7% (2/12)	0.562

Table 1: Baseline characteristics of the enrolled patients.

stage, cases with positive lymph nodes, and pathological staging were found to be insignificant between the groups, indicating that the study population of each group was comparable.

The intra-operative and post-operative parameters for all patients are shown in Table 2. The operation time was significantly reduced in the traditional robotic and hand-assisted robotic surgical staging group (mean 151.8 ± 28.8 min and 185.8 ± 45.3 min, respectively) compared with the laparoscopic surgical staging group (266.7 ± 96.9 min). The volume of blood loss during the operation was also significantly decreased in the traditional robotic and hand-assisted robotic surgical staging group (mean 66.7 ± 40.8 mL and 145.5 ± 119.3 mL, respectively) compared with the laparoscopic surgical staging group (412.5 ± 371.2 mL). The average number of dissected lymph nodes appeared comparable as seen in Table 2. Intra-operative complications were not observed in any cases. No patient in the traditional robotic surgical staging or laparoscopic surgical staging group underwent conversion to laparotomy during the operation.

Post-operative outcomes including pain scores, amount of time before resumption of full diet and days of hospital stay for all surgical groups were evaluated as well (Table 2). The average 24-hour postoperative pain score and amount of time before resumption of full diet for the three groups showed no significant difference. Furthermore, compared with the laparoscopic surgical staging group (6.3 ± 2.7 days), the traditional robotic surgical staging group (3.03 ± 0.6 days) and hand-assisted robotic surgical staging group (3.4 ± 1.5 days) showed significantly decreased duration of hospital stay. Postoperative complications were not observed in any of the cases.

Discussion

The hand assisted robotic approach offers a safe and feasible way to perform ovarian cancer staging surgery for patients with large tumor masses and solid components. Our results show that many peri-operative parameters of the hand-assisted robotic staging surgery group are comparable to the traditional robotic-assisted staging surgery group. Considering the hand-assisted robotic approach includes a laparotomy wound, it was expected that operation time, pain scores or length of hospital stay would differ from the traditional robotic approach, but our data showed that both groups had similar results in these categories. This may be due to small case number, surgeon experience or use of patient controlled analgesia postoperatively. Operation time may also have been comparable due to direct and clear view and access after docking provided by removal of the tumor mass first, so other surgical procedures could be done smoothly. Without a tumor mass hindering the surgical field, an experienced surgeon could easily and swiftly perform the necessary procedures. In the end, the hand-assisted robotic approach still showed to have improved results compared to the laparoscopic approach, such as decreased operative time and blood loss. Therefore, the hand-assisted robotic surgical staging is an option when deciding a surgical approach for large ovarian tumor masses with predominantly solid components.

Though a laparotomy incision may be contradictory to minimally invasive surgery, it allows the surgeon to overcome previous limitations seen with robotic assisted surgical staging surgery. Tumor rupture and spillage into the peritoneal cavity can be avoided and access to upper abdominal quadrants is achieved. In our experience, tumor masses 7 cm or larger are more difficult to manipulate and remove through the trocar wound within an endo-bag when the traditional robotic surgical staging method is employed. The hand-assisted method's small laparotomy wound would provide easy access and controlled removal of these tumor masses. All the while, the advantages of robotic surgery, including better visualization and more precise surgical manipulation, can be reserved for the deep retroperitoneal spaces of the pelvic cavity where it is most needed, such as for lymph node dissection.

Large tumor masses are difficult to remove without rupture while

	Robotic surgical staging (n=6)	Hand-assisted robotic surgical staging (n=11)	Laparoscopic surgical staging (n=12)	Post-hoc analysis	p-value
Operation time (min)					
Mean	151.8 (28.8)	185.8 (45.3)	266.7 (96.9)	R=H<L	0.013
Median	153	195	285		
Range	(116-185)	(95-240)	(100-390)		
Estimated blood loss (mL)					
Mean	66.7 (40.8)	145.5 (119.3)	412.5 (371.2)	R=H<L	0.015
Median	50	50	300		
Range	(50-150)	(50-450)	(50-1200)		
Lymph node yield a					
Mean	30.3 (27.6)	32.2 (10.4)	20.2 (8.2)	R=H=L	0.27
Median	19	27	18.5		
Range	(15-71)	(21-47)	(7-33)		
24-hour pain score					
Mean	1.7 (1.2)	1.7 (1.0)	4.3 (2.4)	R=H=L	0.053
Median	1	1	3		
Range	(1-3)	(1-3)	(1-8)		
Receiving full diet (days)					
Mean	1.8 (1.0)	1.7 (0.8)	1.9 (1.1)	R=H=L	0.968
Median	1.5	2	1		
Range	(1-3)	(1-3)	(1-3)		
Hospital stay (days)					
Mean	3.0 (0.6)	3.4 (1.5)	6.3 (2.7)	R=H<L	0.001
Median	3	3	6.5		
Range	(2-4)	(1-7)	(3-10)		

Table 2: Intra-operative and post-operative parameters of the enrolled patients.

performing robotic surgery. Use of laparoscopic bags is common but may still pose a risk if integrity of the bag is not maintained. Previous data show conflicting conclusions about the effect of intraoperative capsular rupture on the prognosis of ovarian cancer patients, but should still be avoided, as rupture or spillage will upstage the disease and may cause spread of tumor cells in the peritoneal cavity. Hand-assisted laparoscopic surgery for ovarian cancer patients was previously described by Krivak et al. where traditional laparoscopy was combined with placement of a hand intraperitoneally through a 6-7 cm midline vertical incision, providing the surgeon with tactile sensation during the procedure and ability to palpate peritoneal surfaces and retroperitoneal structures [14]. But here we have introduced a hand-assisted robotic staging surgery technique for ovarian cancer with a smaller vertical midline incision that can provide an opening to better wholly remove large ovarian tumor masses. In addition, upper abdominal procedures of staging surgery can be performed through this incision as well.

Robotic surgery was previously not suggested for patients with advanced ovarian cancer due to limited upper abdominal access with the standard trocar set-up for pelvic surgery. However, through the years, robotic surgery techniques have improved and new methods have been developed. In 2013, Nehzat et al. described a hybrid technique of combined conventional and robotic-assisted laparoscopy for staging and debulking of ovarian cancer, allowing better access to all four abdominal quadrants [15]. Here, we offer another option to resolve limited upper abdominal access through a midline incision, which most surgeons are familiar with and may decrease operating time. Through this incision, the surgeon is also able to palpate the abdominal cavity walls for peritoneal seeding and other tumor masses [16].

In our study, we have retrospectively reviewed ovarian cancer cases that have undergone hand-assisted robotic staging surgery. The hand-assisted robotic approach offers a safe and feasible way to perform ovarian cancer surgical staging for patients with large tumor mass and predominantly solid components. Though an additional midline incision seems contradictory to the principles of minimally invasive surgery, this technique provides a method to possibly decrease operating time and overcome current limitations of robotic surgery. The limitations of our study include small case number, the retrospective nature, and only short-term follow-up done. Further studies to compare outcomes, both short term and long term, between this technique and traditional robotic assisted staging surgery should be done to fully understand the benefits and disadvantages.

References

1. Knox ML, El-Galley R, Busby JE (2013) Robotic versus open radical cystectomy: identification of patients who benefit from the robotic approach. Journal of Endourology 27: 40-44.
2. Antosh DD, Grotzke SA, McDonald MA, Shveiky D, Park AJ, et al. (2012) Short-term outcomes of robotic versus conventional laparoscopic sacral colpopexy. Female Pelvic Med Reconstr Surg 18: 158-161.
3. Bedient CE, Magrina JF, Noble BN, Kho RM (2009) Comparison of robotic and laparoscopic myomectomy. American Journal of Obstetrics and Gynecology 201: 566 e561-565.
4. Tinelli R, Malzoni M, Cosentino F, Perone C, Fusco A, et al. (2011) Robotics versus laparoscopic radical hysterectomy with lymphadenectomy in patients with early cervical cancer: a multicenter study. Ann Surg Oncol 18: 2622-2628.
5. Magrina JF, Cetta RL, Chang YH, Guevara G, Magtibay PM (2013) Analysis of secondary cytoreduction for recurrent ovarian cancer by robotics, laparoscopy and laparotomy. Gynecol Oncol 129: 336-340.
6. Magrina JF, Zanagnolo V, Giles D, Noble BN, Kho RM, et al. (2011) Robotic surgery for endometrial cancer: comparison of perioperative outcomes and recurrence with laparoscopy, vaginal/laparoscopy and laparotomy. European Journal of Gynaecological Oncology 32: 476-480.
7. Cardenas-Goicoechea J, Soto E, Chuang L, Gretz H, Randall TC (2013) Integration of robotics into two established programs of minimally invasive surgery for endometrial cancer appears to decrease surgical complications. J Gynecol Oncol 24: 21-28.
8. Coronado PJ, Herraiz MA, Magrina JF, Fasero M, Vidart JA (2012) Comparison of perioperative outcomes and cost of robotic-assisted laparoscopy, laparoscopy and laparotomy for endometrial cancer. Eur J Obstet Gynecol Reprod Biol 165: 289-294.
9. ElSahwi KS, Hooper C, De Leon MC, Gallo TN, Ratner E, et al. (2012) Comparison between 155 cases of robotic vs. 150 cases of open surgical staging for endometrial cancer. Gynecol Oncol 124: 260-264.
10. Wright JD, Herzog TJ, Neugut AI, Burke WM, Lu YS, et al. (2012) Comparative effectiveness of minimally invasive and abdominal radical hysterectomy for cervical cancer. Gynecol Oncol 127: 11-17.
11. Siegel R, Naishadham D, Jemal A (2013) Cancer statistics. CA Cancer J Clin 63: 11-30.
12. Jemal A, Siegel R, Xu J, Ward E (2010) Cancer statistics. CA Cancer J Clin 60: 277-300.
13. Tusheva OA, Gargiulo AR, Einarsson JI (2013) Application of robotics in adnexal surgery. Rev Obstet Gynecol 6: e28-34.
14. Krivak TC, Elkas JC, Rose GS, Sundborg M, Winter WE, et al. (2005) The utility of hand-assisted laparoscopy in ovarian cancer. Gynecol Oncol 96: 72-6.
15. Nezhat FR, Pejovic T, Finger TN, Khalil SS (2013) Role of minimally invasive surgery in ovarian cancer. J Minim Invasive Gynecol 20: 754-65.
16. Nezhat FR, Finger TN, Khalil SS (2013) Staging and cytoreductive surgery for early, advanced and recurrent ovarian cancer via a combination of conventional and robotic-assisted laparoscopy: a hybrid technique. J Minim Invasive Gynecol 20.

22

Modeling of an Autonomous Underwater Robot with Rotating Thrusters

Jebelli A*, Yagoub MCE and Dhillon BS

Faculty of Engineering, University of Ottawa, Canada

Abstract

Mathematical modeling, simulation and control of an underwater robot are a very complex task due to its non-linear dynamic structure. In this paper, the authors present kinematic and dynamic modeling of an underwater robot with two rotating thrusters. Through a virtual environment implemented in MATLAB and LabVIEW, the performance of the proposed robot under real operating conditions was demonstrated.

Keywords: AUV; Hydrodynamics coefficients; Added mass; Graphical user interface (GUI); Virtual reality (VR)

Introduction

These days, underwater vehicles have wide range of applications in different marine industries. Controllability and maneuverability of these vehicles which are strongly affected by various forces including hydrodynamic forces applied on them. At the design stage, knowing and calculating these forces are of very high importance which a mistake in calculations and analysis of forces can have a lot of damage to be followed including time, the cost of re-design and total change in the body and following to that some changes in internal and external, mechanical and electronic components of that. Therefore, simulating a robot in a real system and implementation and real monitoring of a robot performance can significantly help the designer in observing the robot performance in different condition. But it needs a good understanding of the robot environment modeling body coordinate and accurate calculation of the external forces applied on the body. In this project, we implement a virtual simulation of the robot in real conditions after the accurate analysis of the applied forces on the designed robot body and coding achieved equations in MATLAB and the environment of LabVIEW which is able to give us the whole performance of the robot for the moment which this display control in the actual operation of the robot in the water is a very accurate indicator of the robot performance in tests that enables us to manually apply necessary orders to robot through a wireless transceiver of the robot by changing the parameters of this display.

Design the Body

To successfully design the body, one should establish a primary plan of the body structure with the location of each sub-part, knowing that the device will be made of one piece with a camera on the top and engines attached on the sides. In this work (Figure 1); in this case, all thrusters should be put "on" at the same time to create a simultaneous vertical-horizontal movement; it will then lead to high power consumption. Our first contribution was to use a pair of mobile thrusters instead of fixed vertical/horizontal ones, thus saving energy. In practice, the two thrusters shown in Figure 2 could be oriented within a specific angle based on the vertical and horizontal forces needed to move the device in a predefined direction. Any change of angle should be made possible by the instant movement of a servo motor which consumes much less energy than the constant movement of a thruster.

Mass Shifter

As shown in Figure 3, a mass shifter was included, first because of the thruster's movement and second, to make the maneuver possible in the direction of the pitch. The whole set can have two movement modes. In the first mode, the body will have, by the help of the mass shifter, a constant horizontal movement and its movement towards vertical and horizontal directions should be performed through a change in the

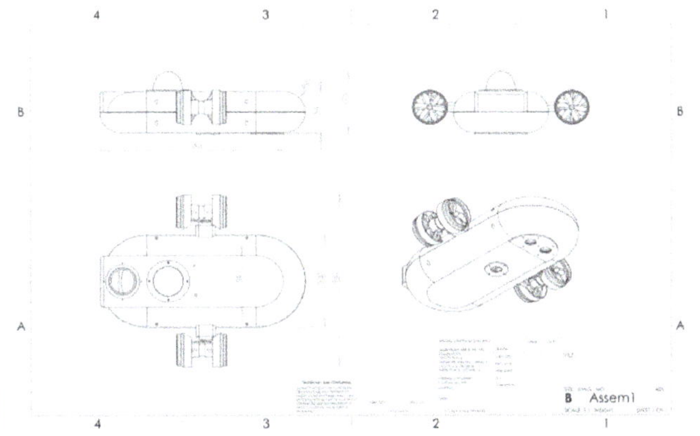

Figure 1: Isometric scheme of the designed robot.

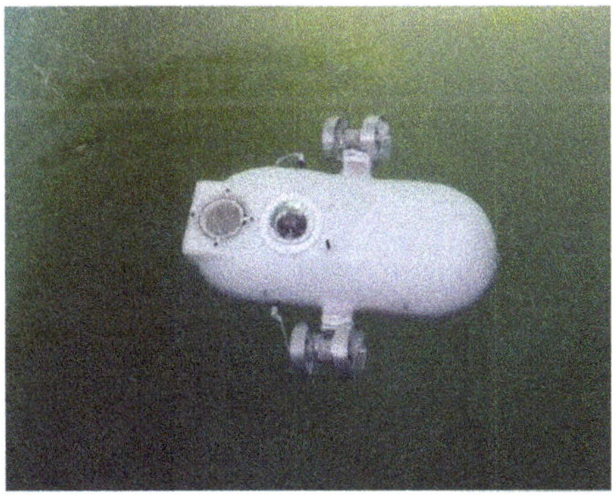

Figure 2: The designed robot in the sea.

***Corresponding author:** Jebelli A, Faculty of Engineering, University of Ottawa, Canada, E-mail: ajebelli@uottawa.ca thruster's angle. In the second mode, the thrusters should be kept fixed in a horizontal direction and the vertical movement should be made possible through a change in the body angle in the pitch direction. The first mode is used to maneuver or pass obstacles while the second mode is used to preserve and store energy as well as to change direction in relatively large depths.

Physical Device

According to Figure 4, the center of the physical device is matched with the reference point of the vehicle. Its axis comes out from the front end of the vehicle and its y axis extends to the right; the z axis completes the right hand rule. Let this system be called B and its axes b_1, b_2 and b_3.

Gravity Model

Assuming the Earth as a perfect sphere, the gravity in terms of height is determined by the following equation:

$$g = \frac{\mu}{R^2} \tag{1}$$

where μ is the earth constant (3.986005×10^{14} m^3/s^2). R, the distance between the two masses, depends on the earth's radius R_e [1,2].

$$R = R_e + h \tag{2}$$

Modeling the Forces

In this work, it has been assumed that the fluid in which the vehicle is moving has a steadily and fixed density $\rho = 1000$ kg/m^3. The external forces applied on the vehicle can be then evaluated [3].

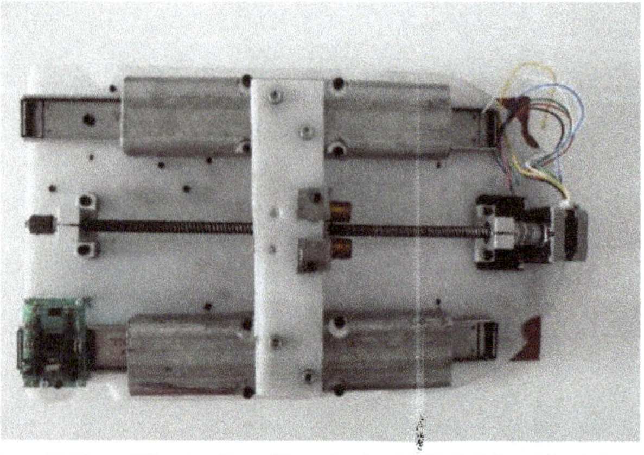

Figure 3: Mass shifter: location of the cube dumbbells, ball Screw, and stepper motor.

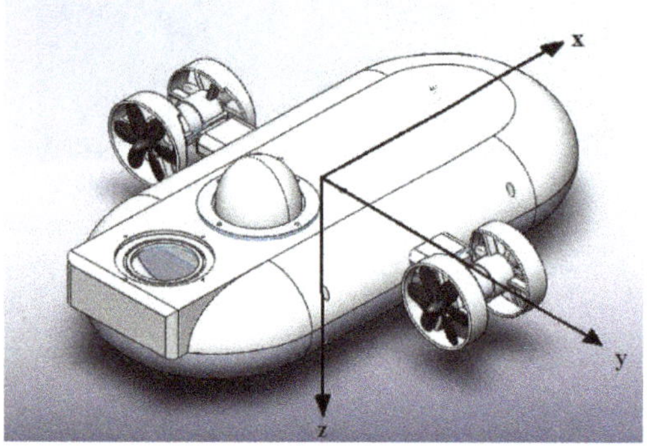

Figure 4: Physical coordinates system.

Buoyancy force

According to Archimedes' principle, the buoyancy force, applied vertically to the body and in the upward direction, equals the weight of the displaced fluid by the body:

$$F_{Buoyancy} = \rho V g \tag{3}$$

where ρ is the density of the fluid and V it's volume.

Weight force

The weight force is calculated as

$$F_W = m_B g \tag{4}$$

where m_B is the vehicle mass.

Thrust force

As shown in Figure 5, two engines on either side of the vehicle are used to provide the thrust force. The forces produced by the left and right engines are noted F_r and F_L, respectively.

Added mass force

As the device moves within the fluid, a certain amount of liquid will move with it [4]. As inertial and Coriolis matrices relate the accelerations and angular/linear velocities of the body to the forces applied to the device, the added mass matrices and added Coriolis relate the accelerations and angular/linear velocities to the hydrodynamic force arising from the liquid displacement and applied to the device.

When a vehicle moves inside the fluid, a dynamic pressure distribution is created around it. Bernoulli's law states that the pressure applied on a differential surface dS depends on the fluid particle velocity on the differential surface and also to the water column height above it and this pressure applies a differential force dF and differential moment dM on the differential surface dS. The force and differential moment are called the force and added mass moment.

When a force is applied to the fluid particles, its adjacent particles are accelerated due to a viscosity. That is why when a device moves in the fluid, the fluid particles which are located exactly on the device move at the same rate of the vehicle and the particles that are far from the vehicle move with different velocities. In fact, there is a velocity distribution on the differential surface dS and the farthest particles

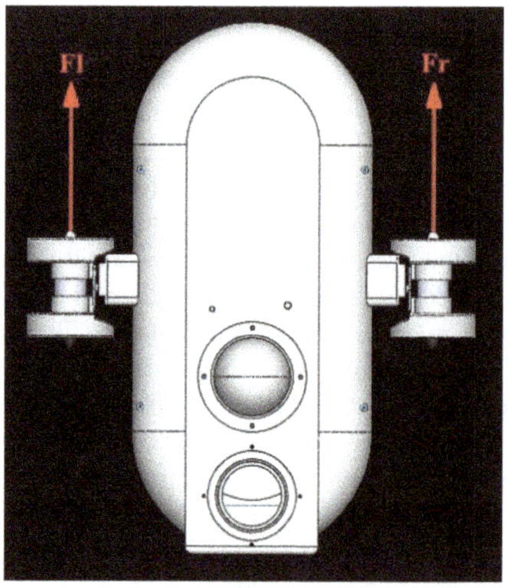

Figure 5: Propulsion.

have a zero velocity [5]. According to Regan et al., [1] the force due to the added mass is

$$F_A = M_A\dot{v} + C_A(v)v \tag{5}$$

where F_A is the added mass, M_A the added mass matrix, v the velocity and C_A the added Coriolis matrix.

Hydrodynamic force

As stated refs. the main hydrodynamic forces that are applied on a subsurface vessel include Quadratic drag forces and lineal skin friction forces. Although the equations of a vehicle with six degrees of freedom are highly non-linear, a series of simplifications are usually used.

The most common simplification is to assume that the linear and quadratic forces in the direction i depend on the velocity vector components in that direction. Thus, the hydrodynamic forces are calculated as follows :

$$F_{hyd} = -\begin{bmatrix} X_u u \\ Y_v v \\ Z_w w \\ K_p p \\ M_q q \\ N_r r \end{bmatrix} - \begin{bmatrix} X_{u|u|}|u|u \\ Y_{v|v|}|v|v \\ Z_{w|w|}|w|w \\ K_{p|p|}|p|p \\ M_{q|q|}|q|q \\ N_{r|r|}|r|r \end{bmatrix} \tag{6}$$

where

- u, v and w are the components of the velocity vector in the respective directions x, y and z.
- p, q and r are physical components of the angular velocity vector on the respective directions x, y and z.).
  - X_u and $X_{u|u|}$ are the linear viscosity and quadratic coefficients along the x direction.
  - Y_v and $Y_{v|v|}$ are linear viscosity and quadratic coefficients along the y direction.
  - Z_w and $Z_{w|w|}$ are linear viscosity and quadratic coefficients along the z direction.
  - K_p and $K_{p|p|}$ are linear viscosity and quadratic coefficients along the p direction.
  - $M_{q|q|}$ and $M_{q|q|}$ are linear viscosity and quadratic coefficients along the q direction.
  - N_r and $N_{r|r|}$ are linear viscosity and quadratic coefficients along the r direction.

Kinematics

To derive a vector V vs. time in any coordinates system (Table 1), the following operator D=d/dt is used:

$$D^A V = D^B V + \omega^{BA} \times V \tag{7}$$

Then, the transfer matrix form a point B to a given vehicle W can be stated as in ref:

$${}^W_B R = \begin{bmatrix} cos\theta cos\psi & sin\phi sin\theta cos\psi - cos\phi sin\psi & cos\phi sin\theta cos\psi + sin\phi sin\psi \\ cos\theta sin\psi & sin\phi sin\theta sin\psi + cos\phi cos\psi & cos\phi sin\theta sin\psi - sin\phi cos\psi \\ -sin\theta & sin\phi cos\theta & cos\phi cos\theta \end{bmatrix} \tag{8}$$

To calculate the angular speeds, the following equation are used [2]:

$$\dot{\phi} = p + qsin\phi tan\theta + rcos\phi tan\theta \tag{9}$$

$$\dot{\theta} = qcos\phi - rsin\phi \tag{10}$$

$$\dot{\psi} = \frac{qsin\phi + rcos\phi}{cos\theta} \tag{11}$$

Physical speeds p, q and r are calculated from the following equation [6]:

$$p = \dot{\phi} - \dot{\psi} sin\theta \tag{12}$$

$$q = \dot{\theta} cos\phi + \dot{\psi} sin\phi cos\theta \tag{13}$$

$$r = -\dot{\theta} sin\phi + \dot{\psi} cos\phi cos\theta \tag{14}$$

To model the rotation in three dimensions, the concept in classical mechanics is to use rotational kinematics. One of the ways to describe rotation is the quaternion method. The four elements of the quaternion, q_0, q_1, q_2 and q_3, are related to the physical speed components p , q and r through the following relation:

$$\begin{Bmatrix} \dot{q}_0 \\ \dot{q}_1 \\ \dot{q}_2 \\ \dot{q}_3 \end{Bmatrix} = \frac{1}{2} \begin{bmatrix} 0 & -p & -q & -r \\ p & 0 & r & -q \\ q & -r & 0 & p \\ r & q & -p & 0 \end{bmatrix} \begin{Bmatrix} q_0 \\ q_1 \\ q_2 \\ q_3 \end{Bmatrix} \tag{15}$$

Dynamics Analyze

The Linear momentum of a m_B mass relative to an arbitrary vehicle, denoted I, is given by

$$p_B^I = m_B v_G^I, \tag{16}$$

The system under discussion contains two objects namely, the floating vehicle, considered as m_1 and the weights, considered as m_2 that move relatively to m_1. According to the Newton's second law, the resultant forces acting on a system m_B are equal with the linear momentum changes over time in the inertia system (Figure 6).

Assuming that both objects have constant mass, we get:

$$f_B = D^I\left(m_1 v_{G_1}^I + m_2 v_{G_2}^I\right) \tag{17}$$

with $v_{G_1}^I$ the velocity of the center of the 1st object and $v_{G_2}^I$ the velocity of the center of the 2nd object mass. Thus,

$$D^I v_{G_1}^I = D^I D^I s_{G_1 I}$$

Degrees of freedom	Movement direction	Position and Euler angle	Linear and angular speed	Force and moment
1	Movement along X	X	u	X
2	Movement along Y	Y	v	Y
3	Movement along Z	Z	w	Z
4	Rotation along X	φ	p	K
5	Rotation along Y	θ	q	M
6	Rotation along Z	ϕ	r	N

Table 1: Variables for subsurface floating vessels with: (1) Surge, (2) Sway, (3) Heave, (4) Roll, (5) Pitch, and (6) Yaw.

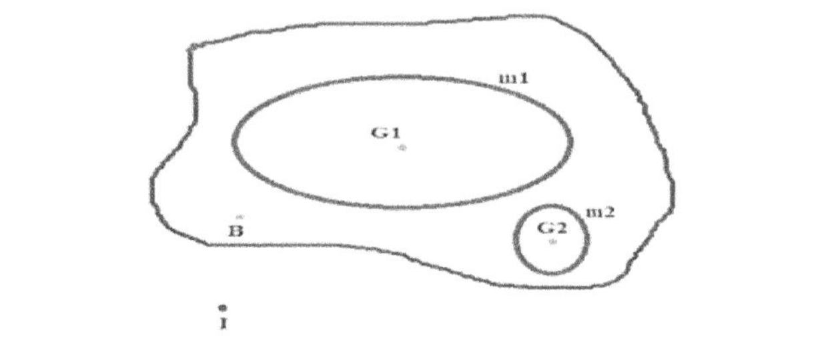

Figure 6: An example of the System B.

$$= D^I\left(D^B s_{G_1B} + \omega^{BI} \times s_{G_1B}\right) + D^I D^I s_{BI}$$

$$= D^B D^B s_{G_1B} + \left(D^B\omega^{BI}\right) \times s_{G_1B} + 2\omega^{BI} \times \left(D^B s_{G_1B}\right) + \omega^{BI} \times \left(\omega^{BI} \times s_{G_1B}\right) + a_B^I \tag{18}$$

$$D^I v_{G_2}^I = D^I D^I s_{G_2I}$$

$$D^I v_{G_2}^I = D^B D^B s_{G_2B} + \left(D^B\omega^{BI}\right) \times s_{G_2B} + 2\omega^{BI} \times \left(D^B s_{G_2B}\right) + \omega^{BI} \times \left(\omega^{BI} \times s_{G_2B}\right) + a_B^I \tag{19}$$

Since m_2 is moving compared to the reference system B connected to the mass m_1, some of the terms removed in equation (18) cannot not be removed in this last equation, thus leading to:

$$f_B = m_1\left(D^B\omega^{BE}\right) \times s_{G_1B} + m_1\omega^{BI} \times \left(\omega^{BE} \times s_{G_1B}\right) + m_1\dot{a}_B^I + m_2 a_{G_2}^B + m_2\left(D^B \quad ^{BI}\right) \times \quad _{G_2B} + 2m_2\omega^{BI} \times \left(v_{G_2}^B\right) + m_2\omega^{BI} \times \left(\omega^{BI} \times s_{G_2B}\right) + m_2 a_B^I \tag{20}$$

Using the relation $m_B = m_1 + m_2$, and reorganizing the equations, gives

$$f_B = \left(D^B\omega^{BE}\right) \times \left(m_1 s_{G_1B} + m_2 s_{G_2B}\right) + \omega^{BE} \times \left(\omega^{BI} \times \left(m_1 s_{G_1B} + m_2 s_{G_2B}\right)\right) + m_2 a_{G_2}^B + 2m_2\omega^{BI} \times v_{G_2}^B + m_B a_B^I \tag{21}$$

$$a_B^I = D^E\left(v_B^E + \omega^{EI} \times s_{BE}\right) + \omega^{EI} \times \left(v_B^E + \omega^{EI} \times s_{BE}\right) = D^E v_B^E + \left(D^E\omega^{EI}\right) \times s_{BE} + 2\omega^{EI} \times v_B^E + \omega^{EI} \times \left(\omega^{EI} \times s_{BE}\right) \tag{22}$$

Then,

$$a_B^I = D^B v_B^E + \omega^{BE} \times v_B^E + 2\omega^{EI} \times v_B^E + \omega^{EI} \times \left(\omega^{EI} \times s_{BE}\right) \tag{23}$$

After substitution,

$$f_B = \left(D^B\omega^{BE}\right) \times \left(m_1 s_{G_1B} + m_2 s_{G_2B}\right) \omega^{BI} \times \left(\omega^{BI} \times \left(m_1 s_{G_1B} + m_2 s_{G_2B}\right)\right) + m_2 a_{G_2}^B + 2m_2\omega^{BI} \times v_{G_2}^B + m_B D^B v_B^E + m_B\omega^{BE} \times v_B^E + 2m_B\omega^{EI} \times v_B^E + m_B\omega^{EI} \times \left(\omega^{EI} \times s_{BE}\right) \tag{24}$$

According to the definition of the center of mass:

$$s_{GB} = \frac{m_1 s_{G_1B} + m_2 s_{G_2B}}{m_B} \rightarrow m_B s_{GB} = m_1 s_{G_1B} + m_2 s_{G_2B} \tag{25}$$

with

$$\omega^{BI} = \omega^{BE} + \omega^{EI} \tag{26}$$

and

$$f_B = \left(D^B\omega^{BE}\right) \times m_B s_{GB} + \left(D^B\omega^{EI}\right) \times m_B s_{GB} + \omega^{BE} \times \left(\omega^{BE} \times m_B s_{GB}\right) + \omega^{BE} \times \left(\omega^{EI} \times m_B s_{GB}\right) + \omega^{EI} \times \left(\omega^{BE} \times m_B s_{GB}\right) + \omega^{EI} \times \left(\omega^{EI} \times m_B s_{GB}\right) + m_2 a_{G_2}^B + 2m_2\omega^{BE} \times v_{G_2}^B + 2m_2\omega^{EI} \times v_{G_2}^B + m_B D^B v_B^E + m_B\omega^{BE} \times v_B^E + 2m_B\omega^{EI} \times v_B^E + m_B\omega^{EI} \times \left(\omega^{EI} \times s_{BE}\right) \tag{27}$$

So f_B can be expressed as:

$$f_B = m_B D^B v_B^E + m_B\left(D^B\omega^{BE}\right) \times s_{GB} + m_B\omega^{BE} \times v_B^E + 2m_B\omega^{EI} \times v_B^E + m_B\omega^{BE} \times \left(\omega^{BE} \times s_{GB}\right) + m_B\omega^{BE} \times \left(\omega^{EI} \times s_{GB}\right) + m_B\omega^{EI} \times \left(\omega^{BE} \times s_{GB}\right) + 2m_2\omega^{BE} \times v_{G_2}^B + m_B\left(D^B\omega^{EI}\right) \times s_{GB} + m_B\omega^{EI} \times \left(\omega^{EI} \times s_{GB}\right) + m_2 a_{G_2}^B + 2m_2\omega^{EI} \times v_{G_2}^B + m_B\omega^{EI} \times \left(\omega^{EI} \times s_{BE}\right) \tag{28}$$

Knowing that

$$D^B\omega^{EI} = \omega^{EI} \times \omega^{BE} \tag{29}$$

$$f_B = m_B D^B v_B^E + m_B\left(D^B\omega^{BE}\right) \times s_{GB} + 2m_B\omega^{EI} \times v_B^E + m_B\omega^{BE} \times \left(\omega^{BE} \times s_{GB}\right) + m_B\omega^{BE} \times \left(\omega^{EI} \times s_{GB}\right) + m_B\omega^{EI} \times \left(\omega^{BE} \times s_{GB}\right) + 2m_2\omega^{BE} \times v_{G_2}^B + m_B\left(\omega^{EI} \times \omega^{BE}\right) \times s_{GB} + m_B\omega^{EI} \times \left(\omega^{EI} \times s_{GB}\right) + 2m_2\omega^{EI} \times v_{G_2}^B + m_B\omega^{EI} \times \left(\omega^{EI} \times s_{BE}\right) \tag{30}$$

which leads to

$$f_B = m_B D^B v_B^E + m_B\left(D^B\omega^{BE}\right) \times s_{GB} + m_B\omega^{BE} \times v_B^E + 2m_B\omega^{EI} \times v_B^E + m_B\omega^{BE} \times \left(\omega^{BE} \times s_{GB}\right) - 2m_B\omega^{EI} \times \left(\omega^{BE} \times s_{GB}\right) + 2m_2\omega^{EI} \times v_{G_2}^B + m_B\omega^{EI} \times \left(\omega^{EI} \times s_{GB}\right) - m_2 a_{G_2}^B + 2m_2\omega^{EI} \times v_{G_2}^B + m_B\omega^{EI} \times \left(\omega^{EI} \times s_{BE}\right) \tag{31}$$

Rotational Dynamics Equations

Angular momentum

The (absolute) angular momentum of m_B can be defined as follows:

$$l_P^{BI} = I_G^B\omega^{GI} + s_{GP} \times m v_G^I \tag{32}$$

On the other hand, the relative angular momentum of m_B is defined according to:

$$l_{P(rel)}^{BI} = I_P^B\omega^{PI} + s_{GP} \times m v_G^P \tag{33}$$

According to White [3], the angular momentum of m_B around its center of mass is given by,

$$l_G^{BI} = I_G^B\omega^{GI} \tag{34}$$

and around the point P, we have

$$l_P^{BI} = I_G^B \omega^{GI} + s_{GP} \times m v_G^I \tag{35}$$

leading to the following equality:

$$l_{P(rel)}^{BI} = I_G^B \omega^{GI} + s_{GP} \times m D^I s_{GP} \tag{36}$$

$$l_B^{BI} = \left(I^B\right)\omega^{BI} + s_{G_1B} \times m_1 v_B^I + s_{G_2B} \times m_2 v_{G_2}^I \tag{37}$$

A. Euler law:

The Euler law says that the resultant momentum exerted on m_B is equal to the time derivative of angular momentum in the inertia system [7]

$$M_B^P = D^I l_P^{BI} \tag{38}$$

$$M_B^B = D^I\left(\left(I^B\right)\omega^{BI} + s_{G_1B} \times m_1 v_B^I + s_{G_2B} \times m_2 v_{G_2}^I\right) \tag{39}$$

which leads to:

$$D^I\left(I^B\omega^{BI}\right) = I^B\left(D^B\omega^{BE}\right) + I^B\left(\omega^{EI} \times \omega^{BE}\right) + \omega^{BE} \times \left(I^B\omega^{BE}\right) + \omega^{EI} \times \left(I^B\omega^{BE}\right) + \omega^{BE} \times \left(I^B\omega^{EI}\right) + \omega^{EI} \times \left(I^B\omega^{EI}\right) \tag{40}$$

Each term can be calculated separately,

$$D^I\left(I^B\omega^{BI}\right) = I^B\left(D^B\omega^{BE}\right) + I^B\left(\omega^{EI} \times \omega^{BE}\right) + \omega^{BE} \times \left(I^B\omega^{BE}\right) + \omega^{EI} \times \left(I^B\omega^{BE}\right) + \omega^{BE} \times \left(I^B\omega^{EI}\right) + \omega^{EI} \times \left(I^B\omega^{EI}\right) \tag{41}$$

$$\left(D^I s_{G_1B}\right) \times m_1 v_B^I = \left(\omega^{BI} \times s_{G_1B}\right) \times m_1\left(D^E s_{BE} + \omega^{EI} \times s_{BE}\right)$$

$$\begin{aligned}
\left(D^I s_{G_1B}\right) \times m_1 v_B^I &= \left(\omega^{BI} \times s_{G_1B}\right) \times m_1\left(D^E s_{BE} + \omega^{EI} \times s_{BE}\right) \\
&= \left(\left(\omega^{BE} + \omega^{EI}\right) \times s_{G_1B}\right) \times m_1\left(v_B^E + \omega^{EI} \times s_{BE}\right) \\
&= m_1\left(\omega^{BE} \times s_{G_1B}\right) \times v_B^E + m_1\left(\omega^{EI} \times s_{G_1B}\right) \times v_B^E + \\
&\quad m_1\left(\omega^{BE} \times s_{G_1B}\right) \times \left(\omega^{EI} \times s_{BE}\right) + m_1\left(\omega^{EI} \times s_{G_1B}\right) \times \left(\omega^{EI} \times s_{BE}\right)
\end{aligned} \tag{42}$$

$$\begin{aligned}
s_{G_1B} \times m_1\left(D^I v_B^I\right) &= \\
&= m_1 s_{G_1B} \times D^B v_B^E + m_1 s_{G_1B} \times \left(\omega^{BE} \times v_B^E\right) + \\
&\quad 2m_1 s_{G_1B} \times \left(\omega^{EI} \times v_B^E\right) + m_1 s_{G_1B} \times \left(\omega^{EI} \times \left(\omega^{EI} \times s_{BE}\right)\right)
\end{aligned} \tag{43}$$

$$\begin{aligned}
\left(D^I s_{G_2B}\right) \times m_2 v_{G_2}^I &= \\
&m_2\left(D^B s_{G_2B} + \omega^{BI} \times s_{G_2B}\right) \times \left(D^I s_{BE}\right) \\
&= m_2 v_{G_2}^B \times v_B^E + m_2\left(\omega^{BE} \times s_{G_2B}\right) \times v_B^E + \\
&\quad m_2\left(\omega^{EI} \times s_{G_2B}\right) \times v_B^E + m_2 v_{G_2}^B \times \left(\omega^{EI} \times s_{BE}\right) + \\
&\quad m_2\left(\omega^{BE} \times s_{G_2B}\right) \times \left(\omega^{EI} \times s_{BE}\right) + m_2\left(\omega^{EI} \times s_{G_2B}\right) \times \left(\omega^{EI} \times s_{BE}\right)
\end{aligned} \tag{44}$$

$$v_{G_2}^B + \left(\omega^{BE} + \omega^{EI}\right) \times \left(\left(\omega^{BE} + \omega^{EI}\right) \times s_{G_2B}\right) + D^B v_B^E + \omega^{BE} \times$$

$$\begin{aligned}
D^I v_{G_2}^I &= a_{G_2}^B + \left(D^B\omega^{BE} + \omega^{EI} \times \omega^{BE}\right) \times s_{G_2B} + 2\left(\omega^{BE} + \omega^{EI}\right) \times \\
&\quad v_{G_2}^B + \left(\omega^{BE} + \omega^{EI}\right) \times \left(\left(\omega^{BE} + \omega^{EI}\right) \times s_{G_2B}\right) + D^B v_B^E + \omega^{BE} \times \\
&\quad v_B^E + 2\omega^{EI} \times v_B^E + \omega^{EI} \times \left(\omega^{EI} \times s_{BE}\right)
\end{aligned}$$

$$\begin{aligned}
&= a_{G_2}^B + D^B\omega^{BE} \times s_{G_2B} + \left(\omega^{EI} \times \omega^{BE}\right) \times s_{G_2B} + 2\omega^{BE} \times v_{G_2}^B + \\
&\quad 2\omega^{EI} \times v_{G_2}^B + \omega^{BE} \times \left(\omega^{BE} \times s_{G_2B}\right) + \omega^{BE} \times \left(\omega^{EI} \times s_{G_2B}\right) + \\
&\quad \omega^{EI} \times \left(\omega^{BE} \times s_{G_2B}\right) + \omega^{EI} \times \left(\omega^{EI} \times s_{G_2B}\right) + D^B v_B^E + \omega^{BE} \times \\
&\quad v_B^E + 2\omega^{EI} \times v_B^E + \omega^{EI} \times \left(\omega^{EI} \times s_{BE}\right)
\end{aligned} \tag{45}$$

$$\begin{aligned}
s_{G_2B} \times m_2\left(D^I v_{G_2}^I\right) &= m_2 s_{G_2B} \times a_{G_2}^B + m_2 s_{G_2B} \times \left(D^B\omega^{BE} \times s_{G_2B}\right) + \\
&\quad m_2 s_{G_2B} \times \left(\left(\omega^{EI} \times \omega^{BE}\right) \times s_{G_2B}\right) + 2m_2 s_{G_2B} \times \left(\omega^{BE} \times v_{G_2}^B\right) \\
&\quad + 2m_2 s_{G_2B} \times \left(\omega^{EI} \times v_{G_2}^B\right) + m_2 s_{G_2B} \times \left(\omega^{BE} \times \left(\omega^{BE} \times s_{G_2B}\right)\right) + \\
&\quad m_2 s_{G_2B} \times \left(\omega^{BE} \times \left(\omega^{EI} \times s_{G_2B}\right)\right) + m_2 s_{G_2B} \times \\
&\quad \left(\omega^{EI} \times \left(\omega^{BE} \times s_{G_2B}\right)\right) + m_2 s_{G_2B} \times \left(\omega^{EI} \times \left(\omega^{EI} \times s_{G_2B}\right)\right) \\
&\quad + m_2 s_{G_2B} \times D^B v_B^E + m_2 s_{G_2B} \times \left(\omega^{BE} \times v_B^E\right) \\
&\quad + 2m_2 s_{G_2B} \times \left(\omega^{EI} \times v_B^E\right) + m_2 s_{G_2B} \times \left(\omega^{EI} \times \left(\omega^{EI} \times s_{BE}\right)\right)
\end{aligned} \tag{46}$$

$$\begin{aligned}
D^I\left(m_B s_{GB} \times v_B^I\right) &= \\
&= m_B\left(v_G^B + \omega^{EI} \times s_{GB}\right) \times \left(v_B^E + \omega^{BI} \times s_{BE}\right) + m_B s_{GB} \times a_B^I \\
&= m_B v_G^B \times v_B^E + m_B v_G^B \times \left(\omega^{EI} \times s_{BE}\right) + m_B\left(\omega^{BE} \times s_{GB}\right) \times \\
&\quad v_B^E + m_B\left(\omega^{EI} \times s_{GB}\right) \times \left(\omega^{EI} \times s_{BE}\right) + m_B\left(\omega^{EI} \times s_{GB}\right) \times v_B^E \\
&\quad + m_B\left(\omega^{EI} \times s_{GB}\right) \times \left(\omega^{EI} \times s_{BE}\right) + m_B s_{GB} \times D^B v_B^E + m_B s_{GB} \\
&\quad \times \left(\omega^{BE} \times v_B^E\right) + 2m_B s_{GB} \times \left(\omega^{EI} \times v_B^E\right) \\
&\quad + m_B s_{GB} \times \left(\omega^{EI} \times \left(\omega^{EI} \times s_{BE}\right)\right)
\end{aligned} \tag{47}$$

By combining the above terms, we obtain:

$$\begin{aligned}
M_B^B &= I^B\left(D^B\omega^{BE}\right) + I^B\left(\omega^{EI} \times \omega^{BE}\right) + \omega^{BE} \times \left(I^B\omega^{BE}\right) + \\
&\quad \omega^{EI} \times \left(I^B\omega^{BE}\right) + \omega^{BE} \times \left(I^B\omega^{EI}\right) + \omega^{EI} \times \left(I^B\omega^{EI}\right) + \\
&\quad m_B v_G^B \times v_B^E + m_B v_G^B \times \left(\omega^{EI} \times s_{BE}\right) + m_B\left(\omega^{BE} \times s_{GB}\right) \times \\
&\quad v_B^E + m_B\left(\omega^{BE} \times s_{GB}\right) \times \left(\omega^{EI} \times s_{BE}\right) + m_B\omega^{EI} \times \\
&\quad \left(s_{GB} \times v_B^E\right) + m_B\left(\omega^{EI} \times s_{GB}\right) \times \left(\omega^{EI} \times s_{BE}\right) + \\
&\quad m_B s_{GB} \times D^B v_B^E + m_B s_{GB} \times \left(\omega^{BE} \times v_B^E\right) + m_B s_{GB} \times \\
&\quad \left(\omega^{EI} \times v_B^E\right) + m_B s_{GB} \times \left(\omega^{EI} \times \left(\omega^{EI} \times s_{BE}\right)\right) + \\
&\quad m_2 s_{G_2B} \times a_{G_2}^B + m_2 s_{G_2B} \times \left(\left(D^B\omega^{BE}\right) \times s_{G_2B}\right) + \\
&\quad 2m_2 s_{G_2B} \times \left(\omega^{BE} \times v_{G_2}^B\right) + 2m_2 s_{G_2B} \times \left(\omega^{EI} \times v_{G_2}^B\right) + \\
&\quad m_2 s_{G_2B} \times \left(\omega^{BE} \times \left(\omega^{BE} \times s_{G_2B}\right)\right) + 2m_2 s_{G_2B} \times \\
&\quad \left(\omega^{EI} \times \left(\omega^{BE} \times s_{G_2B}\right)\right) + m_2 s_{G_2B} \times \left(\omega^{EI} \times \left(\omega^{EI} \times s_{G_2B}\right)\right)
\end{aligned} \tag{48}$$

Thus,

$$\begin{aligned}
m_B v_G^B &= m_B D^B s_{GB} = D^B\left(m_1 s_{G_1B} + m_2 s_{G_2B}\right) = \\
&m_1 D^B s_{G_1B} + m_2 D^B s_{G_2B} = m_2 v_{G_2}^B
\end{aligned} \tag{49}$$

Making the main equation as:

$$
\begin{aligned}
&\omega^{EI} \times \left(I^B \omega^{EI}\right) + m_2 v_{G_2}^B \times v_B^E + m_2 v_{G_2}^B \times \\
&\left(\omega^{EI} \times s_{BE}\right) + m_B\left(\omega^{BE} \times s_{GB}\right) \times v_B^E + \\
&m_B\left(\omega^{BE} \times s_{GB}\right) \times \left(\omega^{EI} \times s_{BE}\right) - m_B \omega^{EI} \times \\
&\left(s_{GB} \times v_B^E\right) + m_B\left(\omega^{BE} \times s_{GB}\right) \times \left(\omega^{BE} \times s_{BE}\right) + \\
&m_B s_{GB} \times D^B v_B^E + m_B s_{GB} \times \left(\omega^{BE} \times v_B^E\right) - m_B s_{GB} \times \\
&\left(\omega^{EI} \times v_B^E\right) + m_B s_{GB} \times \left(\omega^{EI} \times \left(\omega^{BE} \times s_{BE}\right)\right) + \\
&m_2 s_{G_2B} \times a_{G_2}^B + m_2 s_{G_2B} \times \left(\left(D^B \omega^{BE}\right) \times s_{G_2B}\right) + \\
&2m_2 s_{G_2B} \times \left(\omega^{EI} \times v_{G_2}^B\right) - 2m_2 s_{G_2B} \times \left(\omega^{EI} \times v_{G_2}^B\right) - \\
&m_2 s_{G_2B} \times \left(\omega^{BE} \times \left(\omega^{EI} \times s_{G_2B}\right)\right) + 2m_2 s_{G_2B} \times \\
&\left(\omega^{EI} \times \left(\omega^{BE} \times s_{G_2B}\right)\right) + m_2 s_{G_2B} \times \left(\omega^{EI} \times \left(\omega^{EI} \times s_{G_2B}\right)\right)
\end{aligned}
\tag{50}
$$

which can be expressed as a set of dynamic equations

$$
\begin{aligned}
&[f_B]^B = \mathrm{m_B}\left[\dot{v}_B^E\right]^B - m_B[s_{GB}]_\times^B\left[\dot{\omega}^{BE}\right]^B - \\
&\mathrm{m_B}\left[v_B^E\right]_\times^B\left[\omega^{BE}\right]^B - m_B\left\{\left[\omega^{BE}\right]_\times^B[s_{GB}]^B\right\}_\times\left[\omega^{BE}\right]^B \\
&+2\mathrm{m_B}\left[\omega^{EI}\right]_\times^B\left[v_B^E\right]^B - 2m_B\left[\omega^{EI}\right]_\times^B[s_{GB}]_\times^B\left[\omega^{BE}\right]^B \\
&-2\mathrm{m_2}\left[v_{G_2}^B\right]_\times^B\left[\omega^{BE}\right]^B + m_B\left[\omega^{EI}\right]^B \times \left(\left[\omega^{EI}\right]^B \times [s_{GB}]^B\right) \\
&+\mathrm{m_2}\left[a_{G_2}^B\right]^B - 2\mathrm{m_2}\left[\omega^{EI}\right]^B \times \left[v_{G_2}^B\right]^B + \mathrm{m_B}\left[\omega^{EI}\right]^B \times \\
&\left(\left[\omega^{EI}\right]^B \times [s_{BE}]^B\right)
\end{aligned}
\tag{51}
$$

and,

$$
\begin{aligned}
&\left[M_B^B\right]^B = m_B[s_{GB}]_\times^B\left[\dot{v}_B^E\right]^B + \left[I^B\right]^B\left[\dot{\omega}^{BE}\right]^B + \\
&m_B[s_{GB}]_\times^B\left[\omega^{BE}\right]_\times^B\left[v_B^E\right]^B - \left\{\left[I^B\right]^B\left[\omega^{BE}\right]^B\right\}_\times\left[\omega^{BE}\right]^B + \\
&\left[I^B\right]^B\left[\omega^{EI}\right]_\times^B\left[\omega^{BE}\right]^B + \left[\omega^{EI}\right]_\times^B\left[I^B\right]^B\left[\omega^{BE}\right]^B - \\
&\left\{\left[I^B\right]^B\left[\omega^{EI}\right]^B\right\}_\times\left[\omega^{BE}\right]^B - m_2\left[v_{G_2}^B\right]_\times^B\left[v_B^E\right]^B + \\
&m_B\left\{\left[\omega^{BE}\right]_\times^B[s_{GB}]^B\right\}_\times\left[v_B^E\right]^B + m_B\left\{\left[\omega^{EI}\right]_\times^B[s_{BE}]^B\right\}_\times \\
&[s_{GB}]_\times^B\left[\omega^{BE}\right]^B - m_B\left[\omega^{EI}\right]_\times^B[s_{GB}]_\times^B\left[v_B^E\right]^B + \\
&m_B[s_{GB}]_\times^B\left[\omega^{EI}\right]_\times^B\left[v_B^E\right]^B + m_2\left[s_{G_2B}\right]_\times^B\left[s_{G_2B}\right]_\times^B\left[\dot{\omega}^{BE}\right]^B \\
&-2m_2\left[s_{G_2B}\right]_\times^B\left[v_{G_2}^B\right]_\times^B\left[\omega^{BE}\right]^B + m_2\left[s_{G_2B}\right]_\times^B\left[\omega^{BE}\right]_\times^B\left[s_{G_2B}\right]_\times^B\left[\omega^{EI}\right]^B \\
&-2m_2\left[s_{G_2B}\right]_\times^B\left[\omega^{BI}\right]_\times^B\left[s_{G_2B}\right]_\times^B\left[\omega^{BE}\right]^B \\
&+\left[\omega^{EI}\right]^B \times \left(\left[I^B\right]^B\left[\omega^{EI}\right]^B\right) + m_2\left[v_{G_2}^B\right]^B \times \\
&\left(\left[\omega^{EI}\right]^B \times [s_{BE}]^B\right) - m_B\left(\left[\omega^{EI}\right]^B \times [s_{GB}]^B\right) \times \\
&\left(\left[\omega^{EI}\right]^B \times [s_{BE}]^B\right) + m_B[s_{GB}]^B \times \left(\left[\omega^{EI}\right]^B \times \left(\left[\omega^{EI}\right]^B \times [s_{BE}]^B\right)\right) \\
&+m_2\left[s_{G_2B}\right]^B \times \left[a_{G_2}^B\right]^B + 2m_2\left[s_{G_2B}\right]^B \times \left(\left[\omega^{EI}\right]^B \times \left[v_{G_2}^B\right]^B\right) \\
&+m_2\left[s_{G_2B}\right]^B \times \left(\left[\omega^{EI}\right]^B \times \left(\left[\omega^{EI}\right]^B \times \left[s_{G_2B}\right]^B\right)\right)
\end{aligned}
$$

Simulation Platform

The equations obtained in the previous section and describing the movement of the vehicle were coded in MATLAB. The Runge-Kutta method was used to solve them with a time step size of 20 ms. This value was tuned to simultaneously assure relatively fast convergence and acceptable accuracy. After completing the coding in MATLAB, the Lab View software was used to design a Graphical User Interface (GUI) as well as a virtual reality (VR) to virtually observe the maneuvers of the vehicle.

Initial conditions

As shown in Figure 7, this part should be set before running the simulation. Here the user can set the initial conditions of the vehicle including the following cases:

- Initial latitude and longitude and height
- Initial velocity
- Initial physical angle
- Initial angular velocity

Simulation monitoring

As displayed in Figure 8, it is possible to perform some of the settings related to the platforms including setting the frequency of the loops or choosing the processor.

Setting the simulation parameters

As seen in Figure 9, this part includes four main components:

Setting the environmental parameters (Figure 10). In this part, the simulation parameters include:

- Determining the Greenwich longitude and the radius of the earth.
- Determining the rotational velocity of the earth.
- Determining the earth chamfer: If this value is zero, the Earth is assumed a perfect sphere.
- Determining the fluid density in which the vehicle moves.
- Setting the volume mass parameters of the vehicle: this part, as seen in Figure 11, consists in determining the total volume of the vehicle.

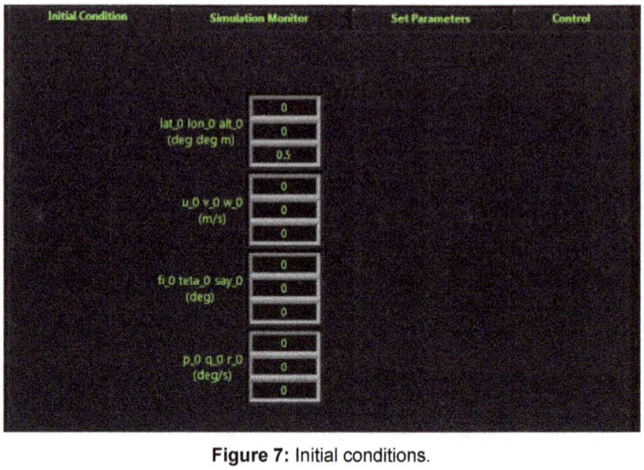

Figure 7: Initial conditions.

Figure 8: Platform settings.

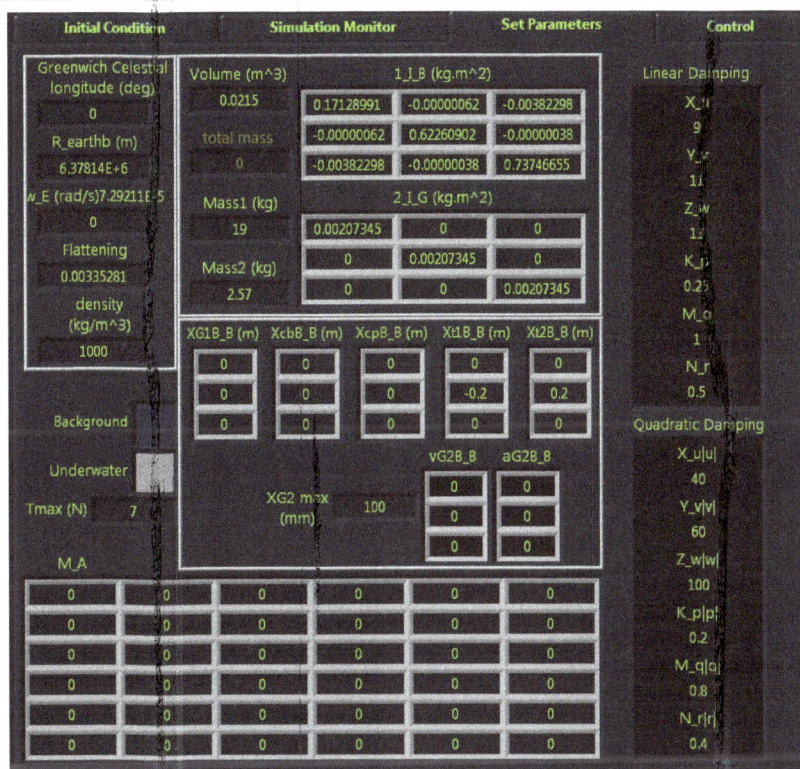

Figure 9: Settings parameters.

- Determining the total volume and the moment of inertia of the entire vehicle regardless of the moving mass.

- Determining the mass and the moment of inertia of the moving mass.

Setting the hydrodynamic parameters

In this part, the hydrodynamic parameters of the device are set by the user. This part includes three main components:

- Determining linear drag coefficients (Figure 12a)

- Determining quadratic drag coefficients (Figure 12b)

- Setting the added mass elements (Figure 13).

Locating the critical points versus the reference point

In this part the user should determine the location vector, which contains the critical points of the vehicle versus the reference point B. These important points include (Figure 14):

- The location of the center of the moving mass.

- The buoyancy location,

- The pressure location.

- The left and right engine force location.

Control and displays

This part is composed of three sets that include (Figure 15):

- The physical angular velocity.

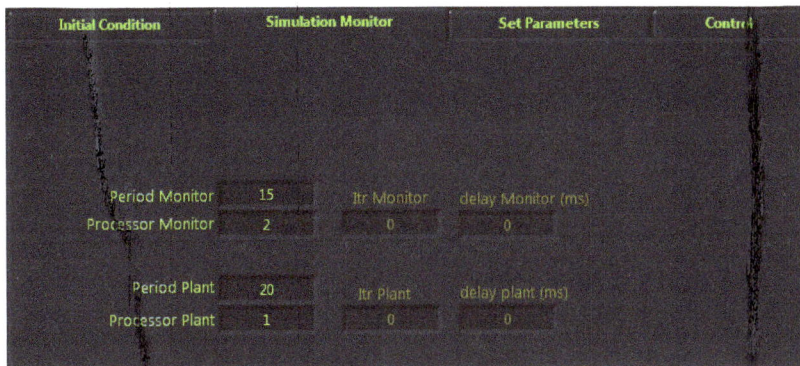

Figure 10: Environmental parameters.

Figure 11: Setting device parameters.

Figure 12: Linear drag and Quadratic drag.

Figure 13: Added mass.

Figure 14: The critical points' location vector.

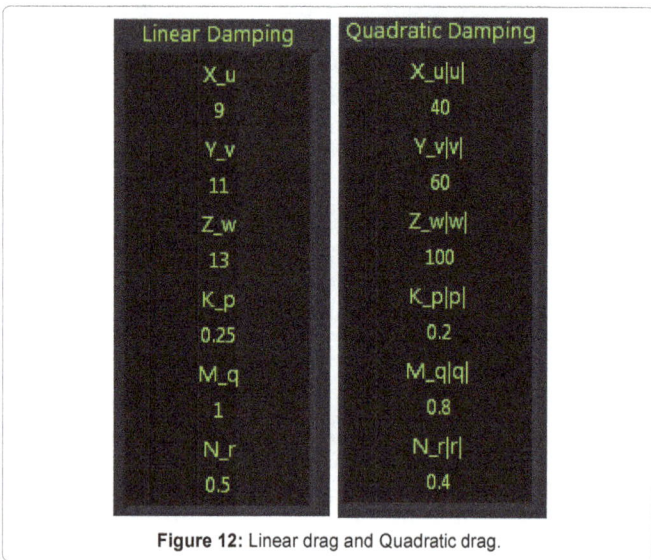

Figure 15: Control.

- The latitude and longitude.

- The physical angle versus the North East Down (NED) coordinate system.

- The vertical velocity that resents the rate of reduced or increased height.

- The height.

VR control

Two controllers have been implemented to control the VR environment, including the perspective and setting the distance between the two cameras located on the back of the device (eye) versus the body.

Vehicle control tools

Using these tools, the user can control the device in the GUI environment, i.e., determining:

- The right engine speed.

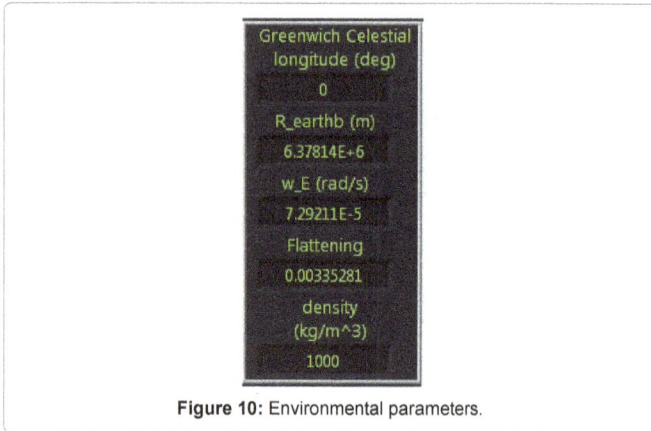

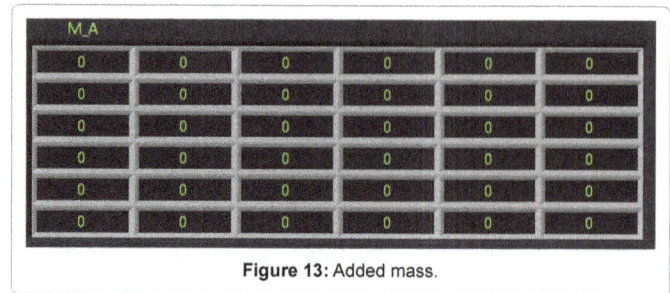

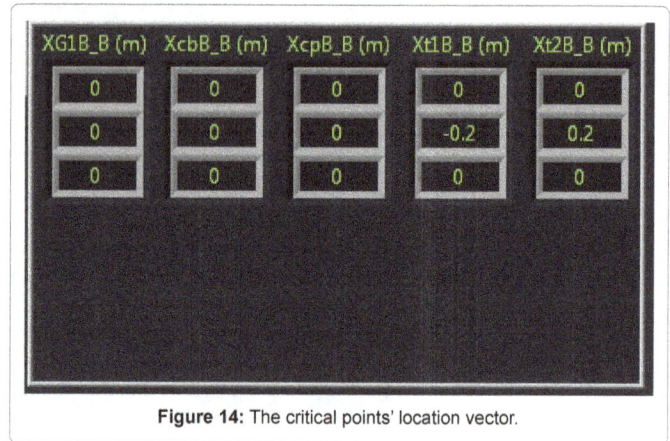

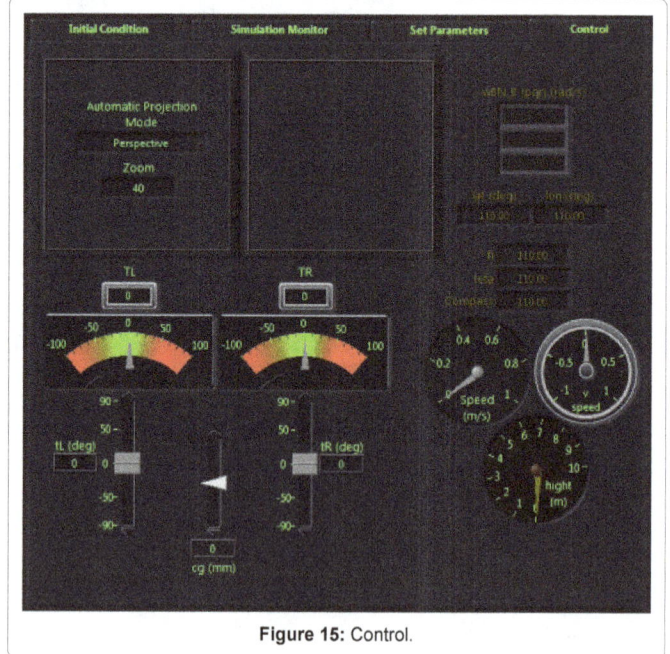

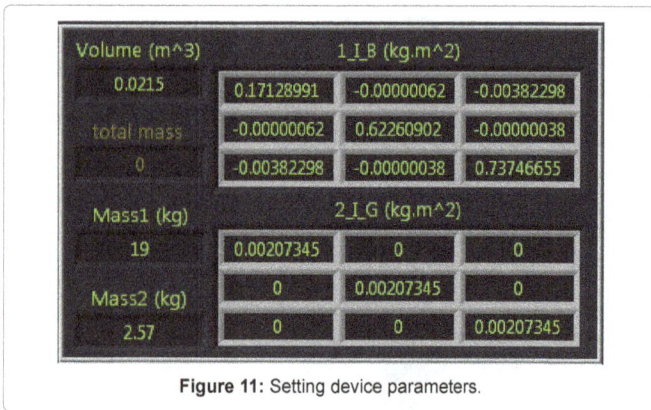

- The right engine speed relative to the body.
- The left engine speed.
- The left engine speed relative to the body.
- The moving mass location.

Data

In this part, the diagrams of the movement data are visible to the operator. These data include:

- The velocity.
- The rotational velocity.
- The body angle to the magnetic north and horizon.
- The acceleration.

Figure 16: VR Environment.

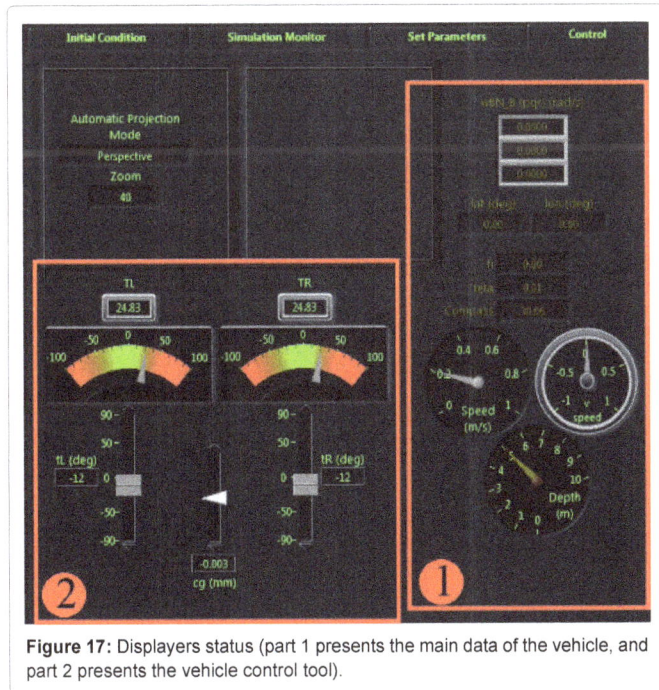

Figure 17: Displayers status (part 1 presents the main data of the vehicle, and part 2 presents the vehicle control tool).

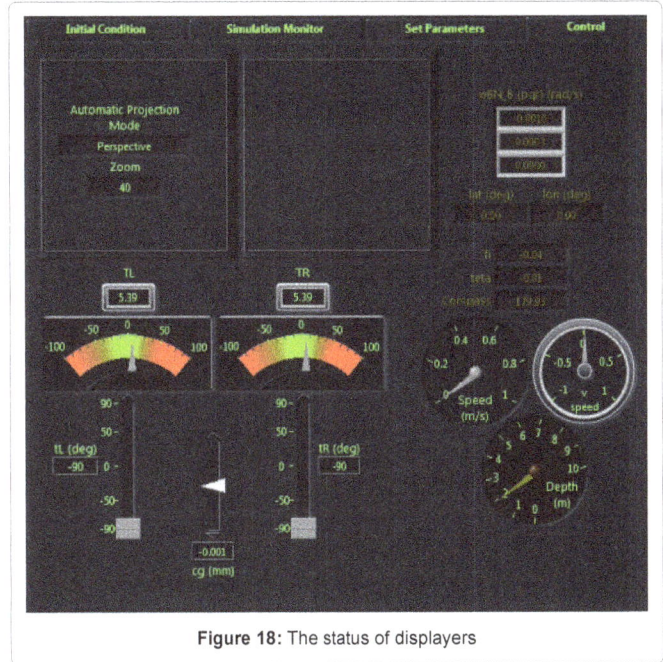

Figure 18: The status of displayers

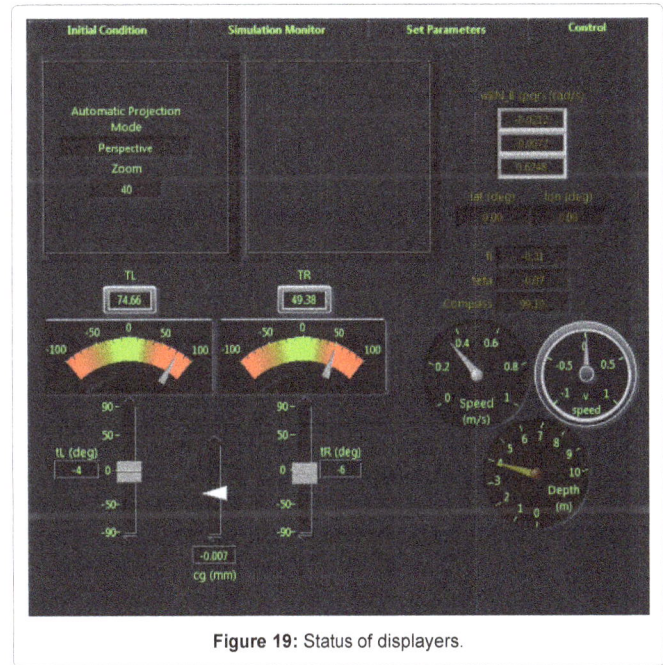

Figure 19: Status of displayers.

VR environment

In this part, a three-dimensional model of the body is created in 1-scale. This graphic model is directly connected to the equation solution subsystem and displays the movement changes of the vehicle including the translational and rotational movements (Figure 16).

Results

Figure 17 shows the status of displayers in a position where the robot depth's is 5 m, with a constant speed of 0.2 m/s and with a direction angle of 30° towards the north and with an horizontal movement.

Figure 18 shows the displayers in a status where the robot is at rest, at a depth of 2 m, and with a 180° direction.

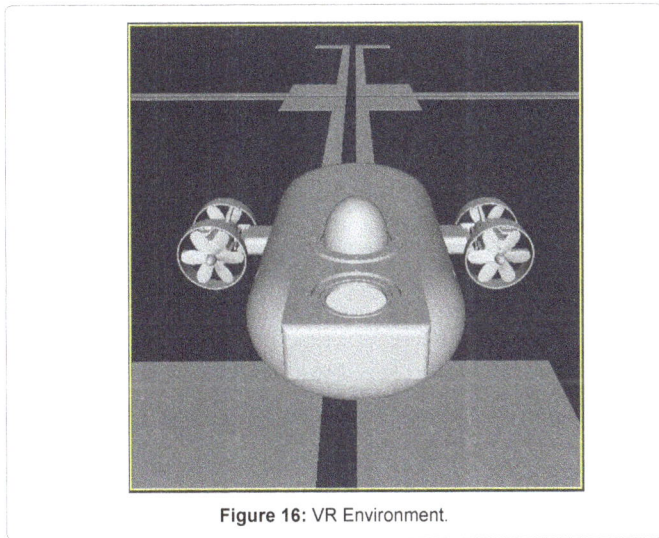

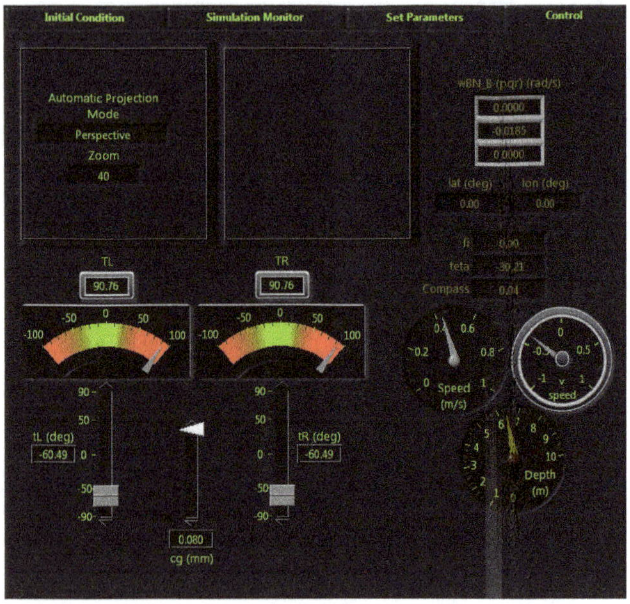

Figure 20: Status of displayers.

Figure 19 shows the displayers in a position where the device is moving at a depth of 4 m, a speed of 0.35 m/s, and at the same time spinning to the right side to adjust the direction angle to 160°. The image shows the moment where the current direction angle of the device is 99.1°.

Figure 20 shows the status of the displayers in a position where the robot is changing its depth from 5 m to 7 m. The image shows the moment when it has reached the depth of 6.4 m.

Conclusion

This work deals with the modeling of the operation of an underwater robot with two rotating thrusters. To efficiently control the device movement at different depths with the help of a mass shifter, an accurate monitoring system was implemented. This virtual environment gives the user the ability to observe and adjust, in real time, the required parameters for controlling the robot in different situations. It also plays an effective role in enhancing the performance of the underwater by using the obtained results to plan/prepare the robot for future tasks.

References

1. Regan FJ, Anandakrishnan SM, Regan NSF, Anandakrishnan S (1993) Dynamics of Atmospheric. American Institute of Aeronautics and Astronomy.
2. Goodman R, Zavorotniy (2009) Newton's Law of Universal Gravitation, New Jersey Center for Teaching and Learning, NJ, USA.
3. White FM (2015) Fluid Mechanics, (8th edn). McGraw-Hill Education, USA.
4. Lee SK, Joung TH (2014) Evaluation of the added mass for a spheroid-type unmanned underwater vehicle by vertical planar motion mechanism test. Int J Naval Architect Ocean Eng 3: 174-180.
5. Galdi GP (2011) An Introduction to the Mathematical Theory of the Navier-Stokes Equations. Springer Monographs in Mathematics.
6. Ridao P, Batlle J, Carreras M (2001) Dynamics Model of an Underwater Robotic Vehicle. University of Girona, Spain.
7. Chuan ST (1999) Modeling and simulation of the autonomous underwater vehicleAutolycus. Master's Thesis, Massachusetts Institute of Technology.

The Current Scope of Robotic Surgery in Colorectal Cancer

Al Bandar MH, Al Sabilah J, Kim NK*

Division of Colorectal Surgery, Yonsei University College of Medicine, Seoul, Republic of Korea

Abstract

Robotic surgical systems have dramatically overcome laparoscopic surgery limitation, which show great touch on the scope of minimum invasive surgery. Robotic surgery has great influence on the surgeon performance and comfort during surgery, in which can handle the procedure with lesser extent of fatigability. Implantation of three-dimension magnified stable camera, articulated instruments, and ability to omit physiologic tremors help to extent scope of dexterity and ergonomics. Therefore, robotic platforms could potentially assist to improve overall patient outcome with highly sophisticated technique. However, the success of Robotic oncological outcome has not addressed well in the literature with on-going controversies. In order to weight and balance the advantages and the cost of robotic surgery, further resources are required to validate the true value of Robotic surgery in colorectal field. The aim of this review is to summarize the current evidence of robotic surgery in clinical and oncological outcomes in colorectal cancer.

Keywords: Review; Robotic surgery; Colorectal surgery; Current state

Introduction

Laparoscopic colorectal surgery for cancer has gained popularity and widely used as the most widespread approach for colorectal surgery with improved short outcome and comparable long-term oncologic outcomes to those of open surgery [1]. Laparoscopic surgery has several limitation and barriers including hand tremors, loss of human wrist's motion, and loss of three-dimensional vision, the need to use longer instruments, loss of dexterity, long steep learning curve and surgeon exhaustion [2,3]. Robotic surgery has emerged into the territory of gastrointestinal surgery to highlight its additional features that could mitigate the obstacles of laparoscopic surgery in colorectal cancer. This new advent of robotic colorectal surgery had started first in 2001, which had been remarkable with lots of promises in the colorectal field [4,5]. Currently, the only commercially available robotic platform, the da Vinci system (Intuitive Surgical, Inc., Sunnyvale, CA, USA), has many advantages such as three-dimensional vision, 7° of wrist-like motion, tremor filtering, motion scaling, better ergonomics, and less fatigue help to overcome laparoscopic limitations However, robotic colorectal surgery (RCS) has several drawbacks such as the lack of haptic sense, bulky robotic cart, higher cost, potential risk of external collisions, the limited range of movement of the robotic arms and increased operative time [6].

Furthermore, oncological outcome are almost likewise to laparoscopic surgery, beside the possibility of faster urinary and functional outcome in robotic surgery. More controversial, however, is to prove the superiority of robotic surgery in colorectal cancer compared to laparoscopic technique in terms of oncologic outcome. Thus this review is to summarize the comprehensive evidences of the current state of robotic surgery and to assess safety, feasibility, and outcomes of this newly emerging technology of robotic surgery.

Techniques of Robotic Surgery in Rectal Cancer

There are different techniques described in the literature with various robotic sets up. We experienced rectal surgery on the most recent versions of robotic machine Da Vinci Xi and Si system as described in the next paragraph.

Da Vinci si system

Setting up Da Vinci si system: There are several techniques in setting robotic system up, which are single, dual docking, hybrid technique and single port robotic surgery. Recently, Bae et al. [7] described the two stage robotic dual docking technique in 61 patients with left sided colon cancer, succeeded to mobilize splenic flexure fully in all patients without the need for conversion with an efficient oncological outcome. However, this technique might end with longer operating time [8], that compensated by upgrade learning curve, knowledge and robotic penetration in the medical field.

To shorten our journey in robotic surgery, we follow a single docking technique in our institute. This technique aims to bypass the need of frequent docking of the robotic machine with faster preparation, especially if robotic system had installed in experience hands [9] which suggested first by Hellan et al. [10]. The drawback of singles docking technique is the possibility of collision, which could be avoided by following certain pathway and measure that we experience in our institute. Ports placement and patient position discussed in details in our previous report [11], as we experience this technique without troublesome external collisions as illustrated in Figure 2. However, the disturbance of workflow by external collision during splenic flexure mobilization is a common obstacle in the beginner's hands. Well skillful surgeons and selecting the proper port site are the primary concern to avoid such obstacles. For better illustration, procedure videos available in the following attached link; (http://www.davincisurgerycommunity.com/playvideo?type=AM&fileEntryId=2357671).

Hybrid technique: Hybrid technique is a technique that required

***Corresponding author:** Kim NK, Division of Colorectal Surgery, Yonsei University College of Medicine, Seoul, Republic of Korea
E-mail: mhb3001@hotmail.com

two minimums invasive systems in a single patient, thus surgeon have to be adapted and skillful in both laparoscopic and robotic systems. The first part of procedure starts with laparoscopic system to facilitate splenic flexure mobilization as well as mobilization of left colon and IMA ligation branches. Then robotic system comes afterword to pelvic side, as the main advantages maximize during rectal dissection by robotic system utilization [11]. Despite higher cost of the procedure due to using laparoscopic instruments on top of robotic system cost, might help beginners to fasten up the procedure, particularly splenic flexure part. Moreover, in order to compensate the cost of the procedure, improve skills in laparoscopic surgery help to carry out the first stage of the procedure faster, in which you would be able to cut down operating time as much as possible.

We recommend fully understanding each technique to handle each case by case accordingly. For example; if you need to take down splenic flexure in fatty mesentery will be easier in Hybrid technique. In addition, it takes good place for training improvement in initial surgeon series in robotic surgery.

Single port robotic surgery: Efforts are challenging to further concentrate on the cosmetic outcome of robotic surgery as well as reduce port-related morbidities. Single incision laparoscopic surgery (SILS) was first described for laparoscopic appendectomy [12] then successfully implanted in colon procedure [13]. First record of SILS right colectomy was in 2008 [14,15] with several limitations such as instrument angulations, working in a different direction side with narrow field vision and encountering dissection difficulty due to axis orientation. Robotic surgery has emerged to overcome all of these complexities in SILS techniques, beside cosmetic outcome ensured, optimizing visualization and handling tissue in the right track which help to gain adequate oncological specimen quality, less postoperative pain and shorter operation time. Single robotic port surgery has described recently in details by Bae et al. [16] and Spinoglio et al. [17] in left and right colon procedures respectively. we experience single port robotic surgery with great success and feasibility. It maintained adequate surgical outcome without a record of conversion to open surgery in our practice. Interestingly, Bae et al. [16], studied outcome of 11 patients with left colon cancer, operated by single port robotic surgery, shown le ss operative time compared to other robotic techniques. Single port robotic surgeries facilitate adequate dissection in an excellent cosmetic outcome without encountering struggles reported in SILS technique.

Da Vinci xi System: Several limitations in robotic Si version in colorectal surgery, for instance: inability to perform multi-quadrant operation, fixed heavy arms, need of re-docking and risk of collisions which disrupt working channel and might extent operative time further. Recently, a new innovation of Da Vinci Xi has admitted in the market, which contributed to overcome obstacles and limitations of the previous platform. Rectal cancer surgery is a good example to look at how Da Vinci Xi platform works in multi-quadrant areas smoothly, however potential risk of collision is possible, since totally robotic pelvic procedure hasn't standardized in Da Vinci Xi yet.

Moreover, Da Vinci Xi docking is simple, designed slim and flexible with movable top roof, without draping. New platform of Da Vinci Xi implanted with a light camera scope, has autofocus, camera lens at the tip of the scope and lastly, camera scope can be placed in any of robotic arms freely. Interestingly, Universal Port Placement Guidelines Manual in which a surgeon can follow the recommended trocar position depending on the type of procedure. Nevertheless, this guideline has not provided with multi-quadrant targets approach, which is required in rectal surgery to approach splenic flexure and pelvic at the same operation in a single docking technique.

Luca Morelli et al. [18], follow Left Lower Abdominal Procedures Universal Port Placement Guidelines from intuitive surgery, he stated the ability of single docking totally robotic surgery with dual target approach. In our experience, we follow keywords to avoid troublesome during the operation. First, linear configuration of part site insertion with 2-3 cm distances from umbilicus as demonstrated in Figure 3. Secondly, targeting the new platform of Da Vinci Xi at the sigmoid colon, in which we able to mobilize splenic flexure completely as well as dissection down to the pelvic floor easily without changing patient position or altering platform target. Further experience is strongly recommended to standardize the technique and to appreciate and clarify the role of Da Vinci Xi in rectal surgery and its real advantages over the Da Vinci Si system.

Application of Robotic Surgery in Rectal Cancer

Total mesorectal excision [TME]

TME procedure is the gold standard for rectal cancer surgery, in order to preserve pelvic plexus and to avoid presacral bleeding, we should optimize visual accuracy to stay in avascular plane along the fascia propria of the rectum without causing injury to adjacent structures [19,20]. New mission of robotic machine has come to approve its safety and feasibility as it is illustrated by kim et al. [9]. Since 2007, we performed TME using robotic system with comparative oncological outcome to laparoscopic surgery. Few keywords to maintain integrity and quality of TME in several steps:

1. Caution dissection at inferior mesenteric artery (IMA) root, where superior hypogastric plexus network lied there. If injured, might end with retrograde ejaculation

2. Mobilization of the rectosigmoid colon from the gonadal vessels and ureters, the hypogastric nerves are at risk at this level. Therefore, the correct surgical plane should be between the rectal proper fascia and prehypogastric nerve fascia

3. Caution at inferior mesenteric vain (IMV) ligation, as collateral vessel crossing IMV root, if injured, could contribute in blood supply cut down then increase risk of anastomotic leakage

4. Avoid blunt dissection in the posterior pelvic side, particularly at recto-sacral fascia to avoid fascia avulsions and presacral bleeding

5. Anterior liner incision at the peritoneum reflection with intensive caution to 3 important structure, which are seminal vesicles in men or vaginal wall in women, watch neurovascular bundles from the pelvic plexus run along the tip of the seminal vesicle (2 o'clock and 10 o'clock directions), [11] and lastly, as deeper you proceed with anterior dissection, as better recognition of Denonviliers fascia will be, where posterior dissection is recommended to avoid troublesome bleeding and nerves damage, unless if the tumor located anteriorly or threating up front structure, then consider taken down Denonviliers fascia with the specimen

6. Final step is to keep circumferential dissection all around the rectum to avoid coning of the mesorectum at the pelvic floor

Cho et al. [21] compared an overall outcome between Robotic TME (R-TME) and laparoscopic TME (L-TME), illustrated similar pathological and oncological outcome, beside faster voiding function in R-TME group. In addition, Petriti et al. [22] recorded 0% conversion rate in R-TME compared to 19% in L-TME. Furthermore, Saklani et al. [23] conducted a comparative retrospective study in 138 patients operated by R-TME and L-TME, found less conversion rate in robotic arm rated at 1.4% vs. 6.3% than laparoscopic arm but didn't reach statistical significant (P=0.183), while long and short term outcome were similar in both groups (Table 1).

The inter-sphincteric resection (ISR)

It is an extended procedure to TME steps with further dissection on the pelvic floor. This procedure required knowledge of the pelvic floor anatomy and fusion lines between the muscles and rectum. Adequate skills required to identify ISR plane starting from abdominal phase between the pubococcygeus or puborectalis and internal anal sphincter (IAS) muscle [24]. Secondly, transanal phase which is started by tumor localization to decide how extent you would be in surgery. As ISR is classified to partial, subtotal, and total ISR, according to the level of incision placement at the white line of Hilton, as above the dentate line, between the dentate line and the inter-sphincteric groove and total excision of the IAS respectively [6]. Excision of the deep external anal sphincter (EAS) muscles could be performed whenever tumor infiltration suspected. Lastly, and before coloanal anastomosis, we ensured four important parameters to avoid complications in our practice, which are obtain healthy bowel, maintain free tension anastomosis, reassert vascularity status and to maintain adequate tension in order to prevent mucosal prolapse later on as demonstrated in Figure 1.

Apparently robotic surgery has potential advantages in better identification of pelvic floor structures through three dimensions camera, proper magnification, and camera controlled by surgeon and robotic function to eliminate physiological tremors. A recent muticentric study conducted by Park et al. [25], compared robotic-ISR to L-ISR in 334 patients, demonstrated less conversion rate, reduced need of long stay stoma, less hospital stay, less complications than L-ISR beside higher cost and longer operation time in R-ISR that required further evidence to justify high cost in robotic surgery.

Abdominoperineal resection (APR)

It is procedure that follows TME techniques with perineum excision as described by Bae et al. [26] using robotic surgery. In our institute, we consider levator eni muscle excision if invaded or threaten by the tumor in order to minimize risk of positive circumferential resection margin (CRM + ve). Recently Kim et al. [27] compared 48 patients underwent APR either by Robotic or laparoscopic technique, which showed larger number of lymph nodes retrieved in robotic arm than laparoscopic APR (P=0.035), in addition, four CRM+ recorded in open APR compared to robotic one. Interestingly they reported the mean depth of CRM was more than three times greater in the robotic than in the open arm (P=0.017) and higher incidence of non-cylindrical resection in open arm. Robotic system can visualize deeper structures in the pelvis without troublesome obstacles in laparoscopic surgery. In turn, robotic surgery could maintain higher quality of specimen and oncological outcome anticipated in near future.

Hemi-elevator excision

Our experience in robotic procedure explained in details by SF AlAsari et al. [28]. Certainly, we consider hemi- elevator excision procedure if we suspect tumor invasion at the level of levator eni.

Authors	Country	Study Type	Type of Surgery	Study Sample	Conversion Rate	Leak Rate	Operation Time (min)	LOS
Cho et al. [21]	South Korea	Retrospective PSM	L-TME vs. R-TME	556 patients	S	10.8%vs10.4%, (P=1.000)	Longer in R-TME**	S
D'Annibale et al. [74]	Italy	Retrospective	R-TME vs. L-TME In rectal caner	100 patients	Lower in R-TME (P= 0.011).	S	S	--
Baik et al. [40]	South Korea	RCT	R-TSME vs. L-TSME	36 patient	S	--	13 min longer in R-TME (217 vs. 204.3)	Significantly shorter in RCS
Baik et al. [44]	South Korea	RCT	R-LAR vs. L-LAR	113 patients	Less in R-LAR (0 vs. 10.5%) (P = 0.013)	--	R-LAR favor	Shorter in RCS (5.7 ± 1.1 vs. 7.6 ± 3.0 d, P = 0.001)

1a: Short-term outcome in low rectal cancer surgery.

Author	Country	Study Type	Type of Surgery	Study Sample	Conversion Rate	Leak Rate	Operation Time
Mak et al. [56]	China				(0% to 8.0%) in RS Vs. (1.8% to 22%) in LS (P>0.05)	6.4% in RS vs. 7.4% in LS	
Saklani et al. [23]	South Korea	Retrospective	Proctectomy*	138 Patients	Favor Robotic 6.3% vs.1.4% (P=0.183)	S	Longer in Robotic (p=0.033)
Kang et al. [60]	South Korea	Retrospective PSM	Proctectomy*	495 Patients	--	(P=0.126)	Longer in Robotic (P=0.012)
Memon et al. [97]	Australia	Meta-analysis	Proctectomy*	73 articles	Risk reduction of 7% favoring RCS	S	43 min more in RCS
Patriti A et al. [22]	Italy	RCT	Proctectomy*	66 Patients	Favor Robotic (0 vs. 19%) P<0.05	Favor Robotic (2.7% vs. 6.8%) P>0.05	Longer in Robotic P<0.05

1b: Short-term outcome in proctectomy surgery.

Abbreviations: OT: operation time, LOS: length of stay, EBL: estimated blood loss, S: similar, LCS: laparoscopic colorectal surgery, RCS: robotic colorectal surgery, N: patients number, Lap: laparoscopic surgery, d:day, TME: total mesorectal excision, LAR: low anterior resection, CA: coloanal anastomosis, APR: abdominopreneal resection. Proctectomy (Rectal cancer operations) (TME, APR, LAR,CA). PSM: propensity score match study. Longer operation time (361.6_91.9 vs 272.4_83.8 min) P<0.001).

Table 1: Short-term outcome in robotic vs. laparoscopic surgery in rectal cancer.

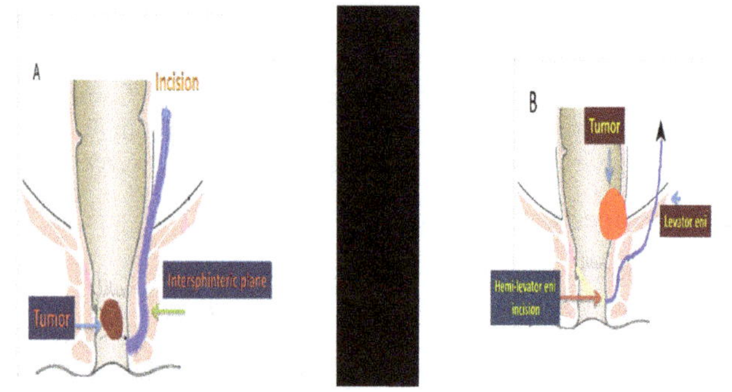

Figure 1: Illustration of Extensive Perianal Procedure with Sphincter Preserving Surgery,

A: Inter-sphincter Resection of Very Low Rectal Tumor, B: Hemi-elevator Excision in Low Rectal Tumor Invading Levator eni Muscle.

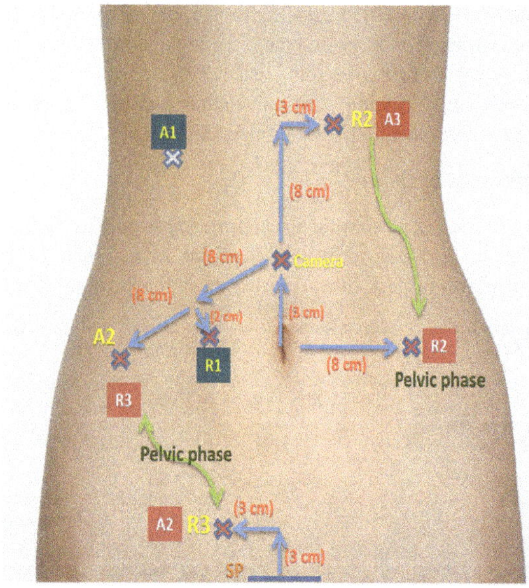

Robotic single docking ports placement in the abdomen and pelvic stages for rectal cancer procedure. **In the abdomen phase (yellow mark);** A: Assistance port located in the mid-clavicle line 8 – 10cm away from other port, at least 2 cm away from bone, Camera port located at 3 cm superior tothe umbilicus, R1: robotic arm No.1 placed at 8 cm from camera port, R2: robotic arm No.2 incised at 8 cm from camera port then 3 cm laterally away from subcostal bone, R3: robotic arm No.3 placed at 3cm superior and laterally to symphysis pubisbone (SP) as illustrated. **In the pelvic phase (red mark);** 2 arms move only as shown in green arrow, otherwise resemble to abdominal phase; R3: moved to 16 cm from camera port toward anterior superior iliac spine in exchange with assistance port No2, R2: moved to 8 cm laterally from umbilicus as shown in the picture, A3: 3rd assistance port for rectum retraction during pelvic phase.

Figure 2: Illustration of Ports Site Insertion in Robotic Da Vinci Si System, Single Docking Technique in Rectal Surgery.

Robotic surgery has great help to visualize tumor location and relation to adjacent muscles on the pelvic floor. Since advent of robotic system in our institute, we successfully divide levator eni muscle through abdominal phase, which help to avoid blunt or blind dissection in perineum phase. Nevertheless, lack of comparative study or RCT trial to approve the effectiveness and superiority of robotic hemi-elevator excision on laparoscopic surgery, has made robotic surgery less popular, along with higher cost and longer operating time with similar outcome reported in similar procedures.

Robotic-assisted lateral pelvic lymph node dissection [LPLND]

We have successfully performed robotic LPLND in our practice with tremendous outcome [29] shortly; Patients placed in the Trendelenberg position at 30° and tilted right side down at an angle of 10°-15°. LPND performed after TME had completed, thus port placement would be as it's in rectal surgery without additional port requirement. The first step in LPND was dissection and isolation of the ureters with a silastic loop. Lymph nodes and fatty tissue were dissected from the bifurcation of the aorta extending down to internal iliac vessel to identify obturator canal, lymphatic tissue cleared at a safe distance from the lateral side of the pelvic plexus, obturator nerve and vessels were identified medial to the external iliac vein and lateral to the superior vesical artery. The obturator lymph nodes were resected leaving the obturator nerve and vessel in the obturator fossa preserved.

Whether robotic surgery has succeeded to approve its theory over laparoscopic surgery or not? Yet lack of supportive evidence to answer this question in colorectal field. However few studies published with optimistic vision in minimum invasive surgery. As in Bae et al. [29], compared 21 patients underwent LPLN dissection by minimum invasive technique [robotic and laparoscopic] compared to open way, revealed higher success in minimum invasive approach, whereas no trials in robotic vs. laparoscopic approach in LPLN dissection.

Robotic single docking ports placement in the abdomen and pelvic stages for rectal cancer procedure. In the abdomen phase [yellow mark]; A: Assistance port located in the mid-clavicle line 8-10 cm away from other port, at least 2 cm away from bone, Camera port located at 3 cm superior to the umbilicus, R1: robotic arm No.1 placed at 8 cm from camera port, R2: robotic arm No.2 incised at 8 cm from camera

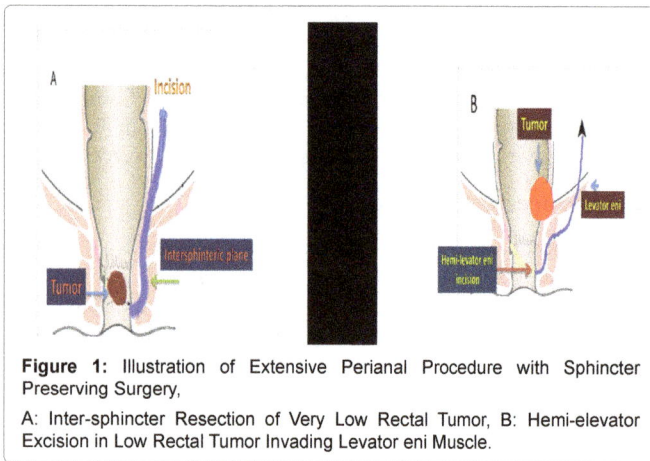

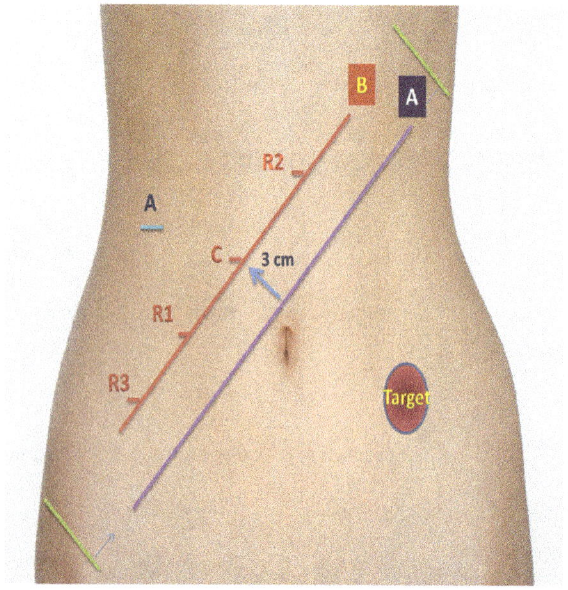

Procedure: started by drawing 2 imaginary lines, first line between femoral head and 8th rib, marked as (A) line. Second imaginary line, marked as (B), is parallel to the (A) line with 3 cm apart from (A) line. Camera (c) port is 3 cm away and perpendicular from (A) line. R1 (robotic arm 1), R2 (robotic arm 2), R3 (robotic arm 3) are located at the (B) line with 8 cm apart from each other. A (assistance port), located at the middle abdominal quadrant with at least 6-8 cm from robotic arms. Target point is where to point out and target Da Vinci Xi system toward it, in which the system will recognize the target to configure the shape of the robotic arm according to the target point.

Figure 3: Port Site Insertion in Rectal Surgery for DaVinci Xi System.

port then 3 cm laterally away from subcostal bone, R3: robotic arm No.3 placed at 3 cm superior and laterally to symphysis pubis bone [SP] as illustrated. In the pelvic phase [red mark]; 2 arms move only as shown in green arrow, otherwise resemble to abdominal phase; R3: moved to 16 cm from camera port toward anterior superior iliac spine in exchange with assistance port No2, R2: moved to 8 cm laterally from umbilicus as shown in the picture, A3: 3rd assistance port for rectum retraction during pelvic phase.

Procedure: started by drawing 2 imaginary lines, first line between femoral head and 8th rib, marked as (A) line. Second imaginary line, marked as (B), is parallel to the (A) line with 3 cm apart from (A) line. Camera (c) port is 3 cm away and perpendicular from (A) line. R1 (robotic arm 1), R2 (robotic arm 2), R3 (robotic arm 3) are located at the (B) line with 8 cm apart from each other. A (assistance port), located at the middle abdominal quadrant with at least 6-8 cm from robotic arms. Target point is where to point out and target Da Vinci Xi system toward it, in which the system will recognize it, then to configure out the shape of the robotic arm according to the target point.

Critical Landmark to Prevent Autonomic Nerves Plexus Damages

Sympathatic nerves plexus in pelvic cavity

Originated tenth thoracic (T10) to the second lumbar (L2) spinal segments, T12 -L2, or L1-L3 [30-32]. The course of these nerves branches down into 3 divisions in the pelvic cavity, they are bilateral hypogastric nerves, sacral sympathetic chain, and superior rectal plexus, branched from the inferior mesenteric plexus. The superior rectal plexus accompanies the superior rectal artery is sacrificed during IMA dissection. The first and foremost important nerve to save is superior hypogastric plexus, which has direct effect on urinary and sexual function [33]. Kinugasa et al. emphasized the presence of hypogastric fascia that cover hypogastric nerve (HGN) as a sandwich layers, by two fascial structures; the ventral fascia seemed to correspond to the mesorectal fascia, whereas the dorsal fascia corresponded to the presacral fascia. These fasciae or the HGN sheaths extended laterally along the ventral aspects of the great vessels and associated lymph follicles. The ventral fascia is, to some extent, fused with the mesocolon on the left side of the body. In addition, he notified the lateral continuation of these two fascia's to sandwich the left ureter, but not the right ureter, due to modifications by the left-sided fusion fascia. He made an effort to discover fascia embryology and morphology in order to preserve HGN. The paired hypogastric nerves run 1-2 cm medial to the ureters and enter the pelvis by crossing the common iliac arteries at the level of the first sacrum and then, run along the posterolateral wall of the pelvis [34]. These nerves located between prehypogastric nerve fascia and parietal presacral fascia [35], where you keep your surgical dissection between prehypogastric fascia and the rectal proper fascia to prevent damage of these nerves. Injury to the unilateral hypogastric nerve causes retrograde ejaculation, and bilateral damage may result in urinary incontinence, retrograde ejaculation in men, and decreased orgasm in women [36].

Parasympathetic nerves in the pelvic cavity

Raised from 2nd to 4th sacral spinal nerves, referred as the pelvic splanchnic nerves or nervi erigentes. Nervi erigentes meet hypogatric nerve to form pelvic plexus at the anterolateral side of the pelvic. The unique landmark to identify these plexus is the tip of seminal vesicles bilaterally [37]. Injury to the pelvic plexus may cause voiding disorder, erection, ejaculation, or lubrication dysfunction. The branching and confluence pattern of the inferior hypogastric nerve, pelvic plexus, and neurovascular bundles form a 'Y' or 'T' shape [38].

Neuro-vascular bundles key point

Originated from the pelvic plexus and descend to the urogenital organ at the lateral corner of the seminal vesicle in the 2 o'clock and 10 o'clock directions. Injury to the neurovascular bundles may cause erection, ejaculation, or lubrication dysfunction.

Robotic Surgery Outcome

Short-term outcome

Operating time: Majority of recent robotic studies demonstrated longer operative time in robotic colorectal surgery (RCS) groups compared to laparoscopic colorectal surgery (LCS) [39]. Hybrid technique targeted to reduce robotic operating time and to compensate the early training phase in robotic surgery, however, cost might be raised without proper justification till now. Moreover, hybrid technique can jeopardize the benefit of robotic system in visualizing and preserving autonomic nerves at the IMA root, which can misinterpret outcome of robotic hybrid surgery in rectal cancer. In the other hand, several reports illustrated similar operative time between RCS and LCS regardless the technique or procedure [22,40]. However, lack of strong comparative evidence between different type of robotic technique and docking including operative time, short and long term to add further maturity for each robotic technique in different field of surgery.

Randomized clinical trial comparing robotic to laparoscopic TME surgery showed minimum longer operating time in robotic side [40], Likewise in Trinch et al. [41] showed only 38.4 min longer operating time in RCS compared to LCS. A recent meta-analysis of 4 randomized controlled trial [42], compared short outcome of RCS to LCS in colorectal cancer, showed RCS has a tendency to take longer operating time than LCS, but this difference wasn't statistical significant (P=0.06). As the surgeon's robotic experience increases, the techniques improved and the operation times will be reduced consequently. In addition, we have to consider the unequal comparison between robotic and laparoscopic operating time, since the former has gained popularity among the medical stuff and get used to its set up compared to early experience of robotic machine that contributed in longer operating hours [43] as demonstrated in Table 1. The operating time still represents an obstacle of robotic surgery in early stage of robotic training; however, this might be overcome with increased experience and knowledge of the robotic installations.

Estimated blood loss: EBL ranges between 90 ml and 320 ml for LCS and between 20 ml and 486 ml for RCS according to a recently published review [44]. Liao et al. [42] estimated EBL was significantly lower in RCS compared to LCS group that may significantly reduce the probability of transfusion and might prevent the recurrence of cancer group. Patriti et al. [22] and Several other studies showed resemble or favor at bleeding control in RCS as demonstrated in Table 1. Surprisingly, patients who receive more perioperative transfused blood are at greater risk for cancer recurrence [45] which emphasize to closely monitor potential area of bleeding and utilize the proper device. Dexterity and ergonomic of Da Vinci system might help to reduce bleeding rate, particularly in those whom bleeding tendency is the highest where robot can visualize minute bleeding points and assist to control it then a potential to reduce local recurrence afterwards in the future.

Intraoperative conversion to open: Conversion rates ranged from

1% to 7.3% in robotic rectal procedures [46], which is way less when compared to LCS in the CLASSIC trial which was rated at 29% [2] and 17% in COLORII [47]. Liao et al. [42] in a recent meta-analysis, revealed that the conversion rate was significantly lower in the RCS group than in the LCS group [P=0.04]. Moreover, Tam et al. [48] demonstrated conversion rate in favor of RCS (7.8 vs. 21.2%), (p<0.001) (Table 1).

Conversion rate is paramount valuable factor in surgical quality. Lower conversion rates associated with fewer postoperative complications [2], less hospital stay reduce total hospital charges, and decrease morbidity and mortality [49]. A recent systemic review [8] demonstrated 2.8% conversion rate in robotic surgery and illustrated reasons for conversion included obesity with heavy mesentery, inability to identify important vascular structures, vascular injury, adhesions, and narrow pelvis, technical difficulties that included stapler misfiring, inappropriate robotic arm placement, as well as robotic malfunction. Thus, dramatic reduction in conversion rate is one of key benefits of robotic system. Therefore, robotic surgery may be indicated in patients with previous abdominal surgery, lower rectal cancers and previous chemo-radiotherapy [46].

Duration of hospitalization: Reduction of hospital stay will be directly impact on the patient's fast recovery, return to normal activity and possible justification of cost effectiveness of robotic surgery. Indeed, length of stay recorded in meta-analysis and several trial showed shorter length of stay in RCS than LCS group, except in patriti A et al. [22] reported a longer hospital stay in robotic group compared to laparoscopic surgery as shown in Table 1.

Bowel function recovery: Defined as first flatus after surgery or a number of days to start peristalsis, which had defined by Park [50], and Baik et al. [40] respectively. Baik et al. [44], found quicker return of bowel function in RCS (4.7 ± 1.1 vs. 5.5 ± 1.5 days in LCS, (P=0.008). In addition Liao et al. [42] revealed that RCS group exhibited shorter times to bowel recovery than LCS group (P=0.008). Patel et al. [51] commented that RCS technique might resulted in reduced trauma and subsequent less postoperative pain, leading to earlier bowel return and discharge home earlier than LCS. These all can be used as evidences for the feasibility and safety of RCS in colorectal field, in addition to shorter LOS and faster recovery which could interpreted as a source of cost effectiveness of RCS in the field of colorectal surgery.

Pathological finding: The integrity of the mesorectum envelope, clear circumferential resection margin (CRM) and adequate distal resection margin [DRM] are important oncological and surgical end points. CRM<1 mm is predictive of an increased risk of distant metastases and shorter survival, whereas CRM<2 mm is a risk for increased local recurrence [52,53]. Recent studies suggested DRM of at least 2 cm is a therapeutic goal [54]. Baik [38] and Park et al., [48] reported proximal and distal resection margin indices were similar in both RCS and LCS (P>0.05). A meta-analysis [42] showed equivalent pathological outcome in both arms. Saklani et al. [23], found higher incidence of CRM+ in robotic group compared to laparoscopic, however it wasn't significant (3.4% vs. 1.6%; P=0.384). Throughout several studies, number of harvested lymph nodes ranged from (10.3-20) in robotic group compared to (11.2-21) lymph nodes in the laparoscopic group with no significant difference in both groups [46]. Take in consideration, the finding of discrepancies between RCS and LCS in the tumor level and depth, as lower tumor and advance cases had seen in robotic surgery, which might justify robotic surgery safety and feasibility without compromising oncological outcome despite the worse features of the tumor in RCS patients.

Furthermore, quality of the TME dissection is paramount, as a break in TME envelope would increase local and distant recurrence. Two comparative studies found robotic dissection is superior to LCS and may offer additional advantage in the future [47,55]. Baik et al. [40], prospective randomized study with 14.3 months follow up, found a significant different of mesorectal grade between RCS and LCS, rated at complete TME 52 vs. 43 patients respectively with (P=0.033), however no statistical significant difference shown in CRM, DRM or proximal resection margin [PRM] as shown in Table 2.

Robot-assisted surgery allowed us to achieve a complete and oncological adequate resection of the cancer with superior TME quality preferred in most of the studies, which in turn robotic TME could reduce, local recurrence and enhance overall.

Postoperative complications: In a recent systemic review found overall complication rates were similar between robotic and laparoscopic group in colorectal cancer [42,56]. Liao et al. [39] illustrated the complication rates were similar across studies, and there was no significant heterogeneity. Cho et al. [21] demonstrated comparative results of early and late complications of R-TME and L-TME group at 25.9% vs. 23.7% and 23.7% vs. 20.1% respectively

Articles	HLN	CRM	DRM	TME quality
Cho et al. [21]	**S*(P=**0.069)	S (4.7% vs. 5.0%) (P=1.000)	S	--
Mak et al. [56]	S**	S	S	Superior in RCS
Kang et al. [60]	S (Favor RS)	S (P=0.77)	S	RCS favor
Memon et al. [97]	S (P=0.94).		S (P=0.84) Except Patriti et al. who Reported a high standard deviation for LCS (7.2 cm)	
Baik et al. [44]	S	S	S	RCS>LCS
RCT	P=0.825	P=0.749	P=0.497	P=0.033
Baik et al. [40]	Favor RCS	S	Favor RCS	Favor RS (17 vs.13 cTME)
D'Annibale et al. [74]	Favor RCS P=0.053	Favor RCS P=0.022	S P=0.908	--

Abbreviations: HLN: Harvested lymph nodes, CRM: Circumferential resection margin, DRM: Distal resection margin, TME: Total mesorectal excision, RCS: Robotic colorectal surgery, LCS: laparoscopic colorectal surgery, S: similar. cTME: complete TME.* HLN in L-TME vs R-TME: 16.2_8.1 nodes vs 15.0_8.1 nodes. ** HLN (10.3 to 20) in RCS vs. (11.2 to 21) in LCS.

Table 2: Pathological outcome of robotic vs. laparoscopic surgery in rectal cancer

P > 0.05. Interestingly, Saklani et al. [23] included 138 patients in a comparative study between robotic and laparoscopic rectal cancer surgery after long course chemoradiotherapy, revealed higher complication rate in laparoscopic procedures anastomotic leaks and pelvic abscess but didn't reach statistical significant. Moreover, most of the studies reported favorable or similar complications rate in robotic than laparoscopic surgery as shown in Table 1.

Hence then, advantages of robotic system might be associated with lower postoperative complication rates that justify robot cost effectiveness in the future.

Anastomotic leakage: One of the most dreaded complications following rectal cancer surgery is anastomotic leak. Overall, the median anastomotic leakage reported at 7.6% (range, 1.8-13.5%) for RCS compared with a median anastomotic leakage was 7.3% (range, 2.4-11.2%) for LCS [46]. Cho et al. [21] reported similar anastomotic leakage as illustrated in Table 1. Surprisingly, a recent systemic review [24,35] reported a lower leakage rate in robotic ISR arm compared to laparoscopic surgery. In contrary Baek et al. [57] reported a leakage rate of 8.6% for the robotic procedures versus a rate of 2.9% for laparoscopic surgery with no statistical difference (p=0.62). Throughout review several articles, found robotic anastomotic leakage are either similar or lesser than laparoscopic surgery, which support feasibility and threshold toward lesser complication in robotic as shown in Table 1.

Long term outcome of robotic surgery: Recent emerge of robotic surgery in the field of colorectal surgery; long-term oncology outcome has not addressed well. Few studies reported their robotic surgery experience in colorectal field. Baek et al. [57] demonstrated a long-term oncologic outcomes of robotic TME for rectal cancer at 3-year overall and disease-free survival rates of 96.2% and 73.7%, respectively. Cho et al. [58] illustrated likewise results with comparable long term outcome between both groups, rates at 5-year overall survival, disease free survival, and local recurrence rates (93.1% vs. 92.2%, P.0.422; 79.6% vs. 81.8%, P.0.538; 3.9% vs. 5.9%, P.0.313, respectively).

Additionally Kwak et al. [59], showed no significant differences between robotic and laparoscopic-assisted group in terms of loco-regional recurrence, distant. Furthermore Kang et al. [60] found no difference in 2-year survival between robotic assisted group (83.5%), laparoscopy group (81.9%) and open surgery (79.7%) (P=0.855). Moreover, a comparative study by Lim et al. [61] between RCS of sigmoid resection and LCS in term of oncologic outcomes, showed a 3-year overall and disease-free survival rate at 92.1% versus 93.5% (P=0.735) and 89.2% versus 90.0%, respectively (P=0.873). Lastly baik et al. [40] found no different between RCS and LCS in term of local or systemic recurrence. Innovation of robotic surgical system technology is safe and effective to maintain and achieve a complete TME in a convenient way without compromising oncological outcome.

Rule of Robotic Surgery in Specific Field

Robotic inter-sphinectric resection [R-ISR] outcome

Since ISR introduced in colorectal field, APR has remarkably reduced, which has facilitated by robotic system through adequate sharp dissection and proper visualization of pelvic muscles and anatomical planes. Leong et al. [62], conducted a prospective study in robotic ISR outcome, stated complete resection (R0) achieved for (90%) of the study sample, acceptable hospital stay, adequate CRM achievement with no major consequences, apart from 10% anastomotic leak which had treated conservatively. Moreover, R-ISR morbidities were comparable to robotic or laparoscopic TME [57,63]. Park et al. [64] commented on the feasibility of R-ISR to achieve an adequate short and long-term outcome compared to laparascopic ISR, however operative intra-abdominal time was longer but perineal phase was significantly shorter in the R-ISR group than L-ISR.

Recently, retrospective study by Yoo et al. [65], compared R-ISR with L-ISR, demonstrated similar operative, oncological, and functional outcomes beside unfavorable tumor features in robotic arm. Lastly, there were no significant differences in the 3-year OS (88.5 vs. 95.2%; p=0.174), 3-year RFS (75.0 vs. 76.7%; p=0.946) [65]. Likewise in park et al. [25] reported a comparable oncological outcome to L-ISR apart from higher cost recorded in R- ISR group. Whereas baek et al. [66] showed similar surgical outcome in both groups but favor R-ISR in term of shorter hospital stay, lower conversion rate and higher level of comfort during surgery. In a recent prospective study by kim et al. [24] compared open ISR to R-ISR, revealed a Moderate to severe sexual dysfunction and greater fecal incontinent in open surgery, (p=0.023) and (p<0.05) respectively. Despite infancy stage of R-ISR, we could record few advantages of R-ISR over conventional methods, however further studies and longer follow up required evaluating the true value of Robotic surgery.

Is Robotic right colectomy outcome superior to laparoscopic surgery?

New innovation of robotic system in the field of colon cancer has gained popularity in the surgical field due to it is safety and feasibility in colorectal cancer, which was reported initially by weber et al. [5] in benign disease, then several reports published afterward [67]. Despite higher cost of robotic surgery, there are ongoing clinical trials to answer the actual oncological benefit of robotic surgery in colon cancer. Yet few studies have published to compare between robotic and laparoscopic right colectomy. A retrospective study by deSouza et al. [68] showed similar outcome in both approach, however higher cost and longer operation time recorded in robotic arm. Park et al. [50] showed similar results in both arms, but higher cost in robotic surgery, reached US $12 235 versus $10 320; (P=0.013) as shown in Table 3.

Interestingly Trastulli et al. [69] showed faster return of bowel function and shorter hospital stay in robotic surgery. In addition Lujan et al. [70] studied outcome of 47 patients underwent robotic and laparoscopic right colectomy retrospectively, found significant difference in blood loss, favoring robotic arm at range of 10-200 ml vs. 10-300 ml in laparoscopic arm, P=0.037), otherwise other parameters were equal. In 2015, a recent meta-analysis by Rondelli et al. [71], reviewed 8 studies comparing R-RC and L-RC, stated a significant lower incidence of intra-operative blood loss and faster bowel function in robotic arm, however longer operating time and higher cost found in robotic group, that explained by docking and reset robotic machine as well as considering early learning in intra-corporeal suturing that could affect the overall operative time. Morpurgo et al. [72] studied 48 patients R-RC and compared them to 48 L-RC, demonstrated several advantage of robotic over laparoscopic surgery which were faster bowel function (3.0-1.0 days vs. 4.0-1.2 days; P<0.05), shorter hospital stay (7.5-2.0 days vs. 9.0-3.2 days; P<0.05) respectively in additional to four anastomotic complications and four incisional hernias reported in L-RC and none in R-RC (P<0.05). These trials could potentially answer the inquired questions about robotic surgery, however further trials are required to weight and balanced the true advantages of robotic right colectomy in future.

Author	Country	Study Type	Type of Surgery	Study Sample	Conversion Rate	Leak Rate	Operation Time	Length of Stay
Rondelli et al. [71]	Italy	Meta-analysis and systemic review	R-RC vs. L-RC	8 Studies	S	S	--	S
Trastulli et al. [69]	Italy	Retrospective Multicentric study	R-RC vs. L-RC	236 Patients	S Favor R-RC	S P = 0.845	Longer in R-RC p<0.001	Shorter in R-RC P<0.001
Park et al. [50]	South Korea			71 Patients	S	1 case in R-RHC	Longer in R-RC	S
		RCT	R-RC vs. L-RC				P<0.001	
deSouza et al. [68]	USA	Retrospective	R-RC vs. L-RC	175 Patients	S	No leak in both arms	Longer in R-RC Operative time (P=.001)	S

Abbreviation: R-RC: robotic right colectomy, L-RC: laparoscopic right colectomy, OT: operative time, S: similar

Table 3: Short-term outcomes of robotic right colectomy vs. laparoscopic surgery.

Author (Year)	County	Study type	Sample N.	Operation Time (Min).	Conversion	Complication	LOS (SD)	EBL (ml)
Spinoglio et al. [43]	Italy	Prospective In LC cancer	R – 10 L - 73	Longer in RLH P<0.00	S	S	S	S
Shin et al. [98]	South Korea	Retrospective in LC cancer	R – 7 L – 12	337 vs. 265	No conversion	--	1.7 vs. 2.1	106 vs. 167
				Long in RLH	1 case in LLC	Favor RLH	Favor RLH	S
Lim et al. [61]	South Korea	Retrospective in LC cancer	R - 34 L-146	P = 0.016	No conversion in RLC	(5.9% vs. 10.3%) p=0.281		P=0.546
Casillas et al. [73]	USA	Prospective in colorectal procedure	R – 68 L - 82	188 vs. 109	Favor RLH (5.8% vs. 10.9%)	Favor RLH 11.7% vs. 20.7%)	Favor RLH 3.6d vs.6.5d	Favor RLH 89 vs.110

Abbreviation: R: robotic, L: laparoscopic, OR: operation room, SD: standard deviation, ml: milliliter, RLH: robotic left colectomy. LC: left colon, S: similar

Table 4: Short-term outcome of robotic left colectomy vs. laparoscopic surgery.

Is Robotic left colectomy outcome superior to laparoscopic surgery?

Multiple reports have reasserted the feasibility and safety of robotic left colectomy [57]. Few articles published in robotic left colectomy (R-LC) and compared to laparoscopic left colectomy (L-LC) in term of short and long-term outcome. Most of these articles revealed similar rate of surgical outcome except longer operating hours in R-LC which could be managed by encourage training system and education in robotic system as demonstrated in Table 4. A recent retrospective study by Lim et al. [61], compared robotic to laparoscopic left colectomy, revealed no significant difference between R-LC and L-LC in estimated blood loss, pathological and oncological outcome with favorable shorter hospital stay but longer operating time compared to laparoscopic surgery, rated at 252.5 ± 94.9 min in RLH and 217.6 ± 70.7 min in LLH (P=0.016).

In 2014, retrospective study by Casillas et al. [73], compared postoperative outcome between robotic and laparoscopic technique in colorectal procedures, included 68 patients underwent robotic and 81 patients laparoscopic left colectomies, found R-LC associated with longer operative time (188 min vs. 109 min, P<0.01), but significant shorter length of hospital stay (3.6 days vs. 6.5 days, P=0.01), lower conversion rate, less complication rate and bleeding rate than L-LC. Moreover Spingoli et al. [43] a prospective study, stated initial experience in 50 robotic cases in colorectal cancer, reported that robotic surgery is convenient, safe and feasible technology in the field of colorectal procedures without badly influenced on the oncological outcome with known time obstacle in robotic arm, which would be shorten by enhancing level of experiences and skills in robotic installations and procedures. Robotic colectomy is safe and promising technology in colorectal field with promising future.

Urogenital Function after Robotic TME for Rectal cancer

Identify pelvic autonomic plexus and neurovascular bundles during deep pelvic dissection are critical in order to preserve sexual and voiding function after TME in rectal cancer especially in young men [30,38]. Although up front chemo-radiation therapy (CRT) or adjuvant chemotherapy (AC) may deteriorate postoperative function, still intraoperative nerve crushed is the primary reason for sexual and urinary dysfunction [30,31]. Up scaling technical part and understanding the anatomy are a must in order to gain complete TME envelop with preserve pelvic plexus. However, TME principles are very challenging in a narrow or deep pelvis, therefore innovation of robotic system installed to assist surgeons with 3-dimensional surgical view, surgeon-operating camera system, filtering of tremor, and ergonomic instrumentation that facilitate fine dissection and stable traction to watch out these critical structures as well as to maintain integrity of TME envelop.

Sexual dysfunction

Overall sexual dysfunction after TME for rectal cancer rated at 11%-55% [74-76]. The main causes of genitourinary dysfunction are superior hypogastric plexus or sacral splanchnic nerves damages during surgery, resulted in urinary incontinence, retrograde ejaculation in men, and decreased orgasmic intensity in women [38,77,78]. In order to prevent sexual and urinary complications avoid common and potential sites of pelvic nerve damage, first, superior hypogastric plexuses that located close to IMA root, ejaculation dysfunction on male patients and impaired lubrication in females if injury occurred [2], second is pelvic splanchnic nerves or the pelvic plexuses located at posterolateral region of mesorectum, if injured will end with erectile dysfunction in men. Our experience in robotic rectal in term of earlier erectile recovery, sexual desire and urinary function compared to the laparoscopic group, nevertheless there was no significant difference in long-term follow-up.

Erectile dysfunction

Patriti et al. [22] reported erectile dysfunction rate of 5.5% and 16.6% in the robotic and laparoscopic group respectively with no statistical difference (p>0.05) along with worse dysfunction found in bulky tumors. In the Park et al. [50], patients asked to fill up a questionnaire before surgery, 3 and 6 months postoperatively, stated worse erectile dysfunction in laparoscopic group than robotic one, whereas similar urinary function. D'Annibale et al. [79] is prospective trial, reported 1-year follow-up assessment of erectile dysfunction, found marked erectile dysfunction in laparoscopic (13 out of 23; 56.5%) compared to robotic group (1 out of 17; 5.6%) (p=0.045), however loss of follow up in (LCS=23.3% vs. RCS=40.0%) should be considered carefully.

Interestingly Kim et al. [50] compared erectile dysfunction of robotic with laparoscopic TME [80], revealed faster recovery of sexual function in robotic than laparoscopic TME [6 months vs. 12 months] (p=0.036). A recent meta-analysis [81] compared LCS and RCS in sexual active patients postoperatively, showed better erectile function in RCS at 3 and 6 months follow up with p=0.002 vs. p=0.001 respectively. These characteristics of RCS can facilitate certain steps in rectal cancer such as: autonomic nerve preservation, ureter and gonadal vessel identification, dissection in the narrow pelvis, and dynamic suturing [82]. Quah et al. [76] suggested that autonomic nerve preservation is challenging in laparoscopic surgery, due to inadequate traction, whereas, a magnified view of R-TME could permit accurate observation of Denonvilliers fascia without injury of the neurovascular bundle.

Urinary retention

In general, 0%-27% is urinary dysfunction reported after TME for rectal cancer [75]. Throughout web sites, most of comparative studies have not showed significant differences yet in urinary or voiding dysfunction [10,50,59,83,84]. However kim et al. [60], found recovery of the urinary dysfunction after robotic TME faster (3 months) than laparoscopic TME (6 months), which could explain the rule and function of robotic system in proper visualization of hypogastric and pelvic plexus during critical points.

Fecal incontinent

Patriti et al. [22] reported 2.7% vs. 6.8% of fecal incontinence rate in laparoscopic and robotic groups respectively, without significant differences. We believe that enhance surgical view with 3-dimensional magnification (surgeon control) and ergonomic robotic instruments can facilitate preservation of the pelvic autonomic nerve which help to achieve favorable sexual, fecal and voiding functioning after rectal cancer surgery.

Limitation of Robotic Technique in Colorectal Surgery

Robotic setting

Docking and patient positioning, collisions are well known reason for unnecessary longer operation time and disrupting workflow [10]. Also, repeated docking and undocking of the robot is often needed when using the robot to perform surgical procedure in different compartments in the abdominal cavity, result in prolonged operating time and delayed conversion in case of massive bleeding [85].

Cost effectiveness of robotic surgery

Installation of robotic machine is expensive compared to laparoscopic surgery. In South Korea, national insurance covers most of the patient hospitality and surgery except robotic surgery because of lack of supportive evidence in robotic utilization. Therefore, penetration of robotic system in South Korea would be steady unless has reimbursed by national insurance. Park et al. [50] reported that overall hospital costs were higher in the RCS group (US $12235 vs. $10319.7) compared to LCS. Halabi et al. [86] illustrated significant higher total hospital fees in RCS, reached 12,965$US (P<0.001). kim et al. [87] studied cost effectiveness in R-TME compared to L-TME in 468 patients, reported higher cost and longer hospital stay in R-TME than L-TME, rated at ($9756.10 vs. $1724.80). Furthermore, the cost of robotic rectal surgery recorded as three times more than laparoscopic surgery [59,62]. Indeed, Robotic surgery is unable clarify the cost-effectiveness at this time, which has impact of robotic system penetration [88].

Despite early admission and lack of robotic justification and cost effectiveness, robot tracks the same channel where laparoscopic surgery was on. At the time of LCS admission in colorectal field, was costly without supportive resources, however, currently overall hospital cost of laparoscopic surgery has shown comparable to that of OCS due to reduction in the cost of post-operative care, hospital stay and faster return to activity. Therefore, the initial trial of robotic surgery would cost higher than LCS as it is new advent in the colorectal field without sufficient support. However, faster training curve, faster bowel recovery, lesser conversion rate and better function outcome would probably help to reduce the overall cost in the future. So, the cost is still an on-going obstacle in robotic surgery, cross this obstacle in the robotic road will enhance robotic sound in the field of colorectal surgery.

Lack of both tactile sensation and tensile feedback

This obstacle might result unexpected complication that can occur easily by excessive traction or accidental use of different robotic paddle which could cauterize ureter or vessels unintentionally [10]. Therefore, surgeon has to improve visual skills and accuracy to estimate the adequate amount of tension needed in several procedure steps. Likewise, caution should be taken during robotic suturing as suture could cut down with excessive tension [88]. Therefore, great care must be taken to avoid traumatic injuries when handling tissue.

Surgeon's experience and learning curve

Laparoscopic approach in colorectal surgery is challenging with relatively long learning curve [89]. Maggiori et al. [90] suggested to start laparoscopic training on stepwise manner, such as to start with benign tumor, then female T1, T2 rectal cancer till you gain adequate skills in L-TME afterwards. Moreover, 30 to 100 cases suggested overcoming difficult laparoscopic TME patients for instance; male, obese, narrow pelvic or radiated field. On the other hand, three-dimension view, dynamic movement, fines instruments and ergonomic shorten the journey in the learning curve of robotic surgery. The learning curve could be divided into three level as illustrated by Bokhari et al. [91] in a large retrospective study [CUSUM], includes 15 cases initially then additional 10 cases and putting hand on more complex condition afterword. Hence then, surgeon would achieve higher level of maturity and confidence in 15 to 25 operation [91].

Interestingly, surgeon adaptation for robotic surgery is very fast even with lack of laparoscopic skills; showed operative time may reduce in the first 20 cases [92]. Park et al. [93] found the learning curve after 17 cases. D'Annibale et al. [79], found mean operative time decreased from 312.5 min in the first 25 procedures to 238.2 min in the last 10 procedures (P=0.002). These results suggest robotic rectal surgery has a shorter learning curve than laparoscopic once, however park et al. [94] found similar learning curve for both laparoscopic and robotic surgery.

Future Aspect of Robotic Surgery

New release of robotic Da Vinci Xi stapler

These staplers designed to provide surgeons with natural dexterity, flexibility with 360 rotation and articulation. Da Vinci Xi stapler approved from the FDA in October 2012 for the Si version, and in 2014 for the Xi version. Currently, robotic stapler Si version approve in South Korea during the 1st quarter. Endo Wrist Stapler designed to ensure the function at its optimum, in term of resection, transection and anastomoses that provided with 3 staplers' lines. These staplers have several important functions in our practice, stapler estimate tissue thickness in which could help to select the proper stapler size and depth. Stapler has ability to study bowel viability and vascularity. Robotic stapler has a safety mark where you place tissue in between these marks.

Release of robotic stapler in the medical market is an evidence of smartness of robotic system and controlled completely by surgeon. Although, DaVinci Xi staplers are smart and effective, they brought to market in 45 mm size only which probably several staplers might use in a single operation which might increase the cost even further. DaVinci Xi stapler advent in the market should be carefully controlled and weight the risk and benefit of using such device in the future.

Indo-cyanine green [icg] dye

Intraoperative near infrared fluorescence (INIF) imaging uses laser technology to show an intravenously delivered agent. ICG is rapidly bound to plasma proteins, which allows ICG to remain longer in the blood vessels to facilitate its appearance clearly in vascular structures. Administration of INIF imaging system (Firefly) installed on the previous platform of Da Vinci Si has shown great success in our practice in several parameters. Firefly techniques assist to visualize and identify hidden vessels (arc of Riolan), evaluate vascularity status of bowel segments, hidden lymph nodes and determine tumor location. Robot is able to change normal visual system to the fluorescent mode that could identify ICG dye in the patient tissue within 50 seconds. ICG has a half-life of 2-5 minutes and is excreted mainly though the biliary system, making it impossible to visualize the ureters. The maximum dosage of ICG is 2 mg/kg. Utility of INIF imaging in performing robotic-assisted colorectal procedures is safe and effective to delineate vascular structure in simple pattern and mode switch [95].

Ongoing major clinical trails

Due to the limited evidence from RCT to support the use of robotic-assisted surgery for rectal cancer, the RO-botic versus LAparoscopic Resection for Rectal cancer (ROLARR) trial has been designed to address this issue [96]. Trial to Assess Robot-assisted Surgery and Laparoscopy-assisted Surgery in Patients with Mid or Low Rectal Cancer (COLRAR) is another ongoing trial [97,98]. This is an international, multicentric, prospective, randomized, controlled, unblinded, parallel-group trial of robotic-assisted versus laparoscopic surgery for the curative treatment of rectal cancer. The study will perform a detailed analysis of robotic-assisted rectal cancer surgery against conventional laparoscopic rectal cancer resection by means of a randomized, controlled trial.

Conclusion

Robotic colorectal surgery has just begun its primitive stage with great ability to approve its safety and feasibility in colorectal surgery. Robotic system is clearly an exciting technology with ability to overcome laparoscopic limitation in the field of colorectal surgery and may ensure improvements in postoperative outcome, enhancing the number of harvested lymph nodes, shorter hospital stay and faster urinary and sexual function. Nevertheless there is an increase of the procedure cost and longer operative time compared to laparoscopic surgery but there is a future prospective vision to approve cost effectiveness with upscale training level, upgrade skills and knowledge curve and popularity of robotic installations among medical stuffs. Adaptation to robotic system setting would help to compensate longer procedure time and facilitate better outcome in the field of colorectal surgery. Randomized clinical trials are needed to assert the true impact of robotic surgery in oncological outcome in colorectal surgery.

References

1. Nelson H, Sargent D, Wieand HS (2004) A comparison of laparoscopically assisted and open colectomy for colon cancer. N Engl J Med 350: 2050-2059.
2. Guillou PJ, Quirke P, Thorpe H, Walker J, Jayne DG, et al. (2005) Short-term endpoints of conventional versus laparoscopic-assisted surgery in patients with colorectal cancer (MRC CLASICC trial): multicentre, randomised controlled trial. Lancet 365: 1718-26.
3. Sammour T, Kahokehr A, Srinivasa S, Bissett IP, Hill AG (2011) Laparoscopic colorectal surgery is associated with a higher intraoperative complication rate than open surgery. Ann Surg 253: 35-43.
4. Makin GB, Breen DJ, Monson JR (2001) The impact of new technology on surgery for colorectal cancer. World J Gastroenterol 7: 612-21.
5. Weber PA, Merola S, Wasielewski A, Ballantyne GH (2002) Telerobotic-assisted laparoscopic right and sigmoid colectomies for benign disease. Dis Colon Rectum 45: 1689-94.
6. Ito M, Saito N, Sugito M, Kobayashi A, Nishizawa Y (2009) Analysis of clinical factors associated with anal function after intersphincteric resection for very low rectal cancer. Dis Colon Rectum 52: 64-70.
7. Bae SU, Baek SJ, Hur H, Baik SH, Kim NK (2015) Robotic left colon cancer resection: a dual docking technique that maximizes splenic flexure mobilization. Surg Endosc 29: 1303-9.
8. Alasari S, Min BS (2012) Robotic colorectal surgery: a systematic review. Surg 2012: 293-894.
9. Kwak J, Kim S (2011) The technique of single-stage totally robotic low anterior resection. J Robotic Surg 5: 25-8.
10. Hellan M, Anderson C, Ellenhorn JD, Paz B, Pigazzi A (2007) Short-term outcomes after robotic-assisted total mesorectal excision for rectal cancer. Ann Surg Oncol 14: 3168-73.
11. Kim NK, Kim YW, Cho MS (2015) Total mesorectal excision for rectal cancer with emphasis on pelvic autonomic nerve preservation: Expert technical tips for robotic surgery. Surg Oncol 24: 172-80.
12. Esposito C (1998) One-trocar appendectomy in pediatric surgery. Surg Endosc12: 177-8.
13. Valadez DIR, Patel CB, Ragupathi M, Pickron TB, Haas EM (2010) Single-incision laparoscopic right hemicolectomy: safety and feasibility in a series of consecutive cases. Surg Endosc 24: 2613-6.
14. Remzi FH, Kirat HT, Kaouk JH, Geisler DP (2008) Single-port laparoscopy in colorectal surgery. Colorectal Dis 10: 823-6.
15. Bucher P, Pugin F, Morel P (2008) Single port access laparoscopic right hemicolectomy. Int J Colorectal Dis 23: 1013-6.
16. Bae SU, Jeong WK, Bae OS, Baek SK (2015) Reduced-port robotic anterior resection for left-sided colon cancer using the Da Vinci single-site platform. The international journal of medical robotics computer assisted surgery.
17. Spinoglio G, Lenti LM, Ravazzoni F, Formisano G, Pagliardi F, et al. (2015) Evaluation of technical feasibility and safety of Single-Site robotic right colectomy: three case reports. The international journal of medical robotics computer assisted surgery 11: 135-40.
18. Morelli L, Guadagni S, Di Franco G, Palmeri M, Caprili G, et al. (2015) Use of the new Da Vinci Xi® during robotic rectal resection for cancer: technical considerations and early experience. Int J Colorectal Dis 30: 1281-3.

19. Heald RJ (1979) A new approach to rectal cancer. British journal of hospital medicine 22: 277-81.

20. Enker WE, Thaler HT, Cranor ML, Polyak T (1995) Total mesorectal excision in the operative treatment of carcinoma of the rectum. J Am Coll Surg 181: 335-46.

21. Cho MS, Baek SJ, Hur H, Min BS, Baik SH, et al. (2015) Short and long-term outcomes of robotic versus laparoscopic total mesorectal excision for rectal cancer: a case-matched retrospective study. Medicine 94: e522.

22. Patriti A, Ceccarelli G, Bartoli A, Spaziani A, Biancafarina A, et al. (2009) Short- and medium-term outcome of robot-assisted and traditional laparoscopic rectal resection. JSLS 13: 176-83. 22

23. Kim JC, Lim SB, Yoon YS, Park IJ, Kim CW (2014) Completely abdominal intersphincteric resection for lower rectal cancer: feasibility and comparison of robot-assisted and open surgery. Surg Endosc 28: 2734-44.

24. Park JS, Kim NK, Kim SH, Lee KY, Lee KY, et al. (2015) Multicentre study of robotic intersphincteric resection for low rectal cancer. Br J Surg 102: 1563-1573.

25. Bae SU, Saklani AP, Hur H, Min BS, Baik SH (2014) Robotic interface for transabdominal division of the levators and pelvic floor reconstruction in abdominoperineal resection: a case report and technical description. The international journal of medical robotics + computer assisted surgery.

26. Kim JC, Kwak JY, Yoon YS, Park IJ, Kim CW (2014) A comparison of the technical and oncologic validity between robot-assisted and conventional open abdominoperineal resection. Int J Colorectal Dis 29: 961-9.

27. AlAsari SF, Lim D, Kim NK (2013) Robotic hemi-levator excision for low rectal cancer: A novel technique for sphincter preservation. OA Robotic Surgery 16: 1-3.

28. Bae SU, Saklani AP, Hur H, Min BS, Baik SH, et al. (2014) Robotic and laparoscopic pelvic lymph node dissection for rectal cancer: short-term outcomes of 21 consecutive series. Ann Surg Treat Res 86: 76-82.

29. Lange MM, Velde CJ (2011) Urinary and sexual dysfunction after rectal cancer treatment. Nat Rev Urol 8: 51-7.

30. Havenga K, Enker WE (2002) Autonomic nerve preserving total mesorectal excision. Surg Clin North Am 82: 1009-18.

31. Bleier JI, Maykel JA (2013) Outcomes following proctectomy. Surg Clin North Am 93: 89-106.

32. Kinugasa Y, Niikura H, Murakami G, Suzuki D, Saito S, et al. (2008) Development of the human hypogastric nerve sheath with special reference to the topohistology between the nerve sheath and other prevertebral fascial structures. Clin Anat 21: 558-67.

33. Nagpal K, Bennett N (2013) Colorectal surgery and its impact on male sexual function. Curr Urol Rep 14: 279-84.

34. Kinugasa Y, Murakami G, Suzuki D, Sugihara K (2007) Histological identification of fascial structures posterolateral to the rectum. Br J Surg 94: 620-6.

35. Büchler M (2005) Rectal cancer treatment.

36. Walsh PC, Schlegel PN (1988) Radical pelvic surgery with preservation of sexual function. Ann Surg 208: 391-400.

37. Kim NK (2005) Anatomic basis of sharp pelvic dissection for curative resection of rectal cancer. Yonsei Med J 46: 737-49.

38. Jimenez RM, Diaz JM, Portilla F, Prendes E, Hisnard JM (2011) Prospective randomised study: robotic-assisted versus conventional laparoscopic surgery in colorectal cancer resection. Cir Esp 89: 432-8.

39. Baik SH, Ko YT, Kang CM, Lee WJ, Kim NK, et al. (2008) Robotic tumor-specific mesorectal excision of rectal cancer: short-term outcome of a pilot randomized trial. Surg Endosc 22: 1601-8.

40. Trinh BB, Jackson NR, Hauch AT, Hu T, Kandil E (2014) Robotic versus laparoscopic colorectal surgery. JSLS 18: e187.

41. Liao G, Zhao Z, Lin S, Li R, Yuan Y, et al. (2014) Robotic-assisted versus laparoscopic colorectal surgery: a meta-analysis of four randomized controlled trials. World J Surg Oncol 12: 122-124.

42. Spinoglio G, Summa M, Priora F, Quarati R, Testa S (2008) Robotic colorectal surgery: first 50 cases experience. Dis Colon Rectum 51: 1627-32.

43. Baik SH, Kwon HY, Kim JS, Hur H, Sohn SK, et al. (2009) Robotic versus laparoscopic low anterior resection of rectal cancer: short-term outcome of a prospective comparative study. Ann Surg Oncol 16: 1480-7.

44. Amato A, Pescatori M (2006) Perioperative blood transfusions for the recurrence of colorectal cancer. Cochrane Database Syst Rev.

45. Scarpinata R, Aly EH (2013) Does robotic rectal cancer surgery offer improved early postoperative outcomes? Dis Colon Rectum 56: 253-62.

46. Vander MH, Haglind E, Cuesta MA, Furst A, Lacy AM, et al. (2013) Laparoscopic versus open surgery for rectal cancer (COLOR II): short-term outcomes of a randomised. Lancet Oncol14: 210-8.

47. Tam MS, Kaoutzanis C, Mullard AJ, Regenbogen SE, Franz MG, et al. (2015) A population-based study comparing laparoscopic and robotic outcomes in colorectal surgery. Surg Endosc.

48. White I, Greenberg R, Itah R, Inbar R, Schneebaum S (2011) Impact of conversion on short and long-term outcome in laparoscopic resection of curable colorectal cancer. Jsls 15: 182-7.

49. Park JS, Choi GS, Park SY, Kim HJ, Ryuk JP (2012) Randomized clinical trial of robot-assisted versus standard laparoscopic right colectomy. Br J Surg 99: 1219-26.

50. Patel CB, Ragupathi M, Ramos-Valadez DI, Haas EM (2011) A three-arm (laparoscopic, hand-assisted, and robotic) matched-case analysis of intraoperative and postoperative outcomes in minimally invasive colorectal surgery. Dis Colon Rectum 54: 144-50.

51. Nagtegaal ID, van Krieken JH (2002) The role of pathologists in the quality control of diagnosis and treatment of rectal cancer-an overview. European journal of cancer 38: 964-72.

52. Nagtegaal ID, Marijnen CA, Kranenbarg EK, van de Velde CJ, van Krieken JH (2002) Circumferential margin involvement is still an important predictor of local recurrence in rectal carcinoma: not one millimeter but two millimeters is the limit. The American journal of surgical pathology 26: 350-7.

53. Park IJ, Kim JC (2010) Adequate length of the distal resection margin in rectal cancer: from the oncological point of view. Journal of gastrointestinal surgery 14: 1331-7.

54. Jayne DG, Guillou PJ, Thorpe H, Quirke P, Copeland J, et al. (2007) Randomized trial of laparoscopic-assisted resection of colorectal carcinoma: 3-year results of the UK MRC CLASICC Trial Group. J Clin Oncol 25: 3061-8.

55. Mak TW, Lee JF, Futaba K, Hon SS, Ngo DK, et al. (2014) Robotic surgery for rectal cancer: A systematic review of current practice. World J Gastrointest Oncol 6: 184-93.

56. Baek JH, McKenzie S, Garcia-Aguilar J, Pigazzi A (2010) Oncologic outcomes of robotic-assisted total mesorectal excision for the treatment of rectal cancer. Ann Surg 251: 882-6.

57. Choi PW, Kim HC, Kim AY, Jung SH, Yu CS, et al. (2010) Extensive lymphadenectomy in colorectal cancer with isolated para-aortic lymph node metastasis below the level of renal vessels. J Surg Oncol 101: 66-71.

58. Kwak JM, Kim SH, Kim J, Son DN, Baek SJ, et al. (2011) Robotic vs laparoscopic resection of rectal cancer: short-term outcomes of a case-control study. Dis Colon Rectum 54: 151-6.

59. Kang J, Yoon KJ, Min BS, Hur H, Baik SH, et al. (2013) The impact of robotic surgery for mid and low rectal cancer: a case-matched analysis of a 3-arm comparison--open, laparoscopic, and robotic surgery. Ann Surg 257: 95-101.

60. Lim DR, Min BS, Kim MS, Alasari S, Kim G, et al. (2013) Robotic versus laparoscopic anterior resection of sigmoid colon cancer: comparative study of long-term oncologic outcomes. Surg Endosc 27: 1379-85.

61. Leong QM, Son DN, Cho JS, Baek SJ, Kwak JM, et al. (2011) Robot-assisted intersphincteric resection for low rectal cancer: technique and short-term outcome for 29 consecutive patients. Surg Endosc 25: 2987-92.

62. Kim SH, Park IJ, Joh YG, Hahn KY (2006) Laparoscopic resection for rectal cancer: a prospective analysis of thirty-month follow-up outcomes in 312 patients. Surg Endosc 20: 1197-202.

63. Lim SB, Yu CS, Kim CW, Yoon YS, Park SH, et al. (2013) Clinical implication of additional selective lateral lymph node excision in patients with locally advanced rectal cancer who underwent preoperative chemoradiotherapy. Int J Colorectal Dis 28: 1667-74.

64. Yoo BE, Cho JS, Shin JW, Lee DW, Kwak JM, et al. (2015) Robotic versus

laparoscopic intersphincteric resection for low rectal cancer: comparison of the operative, oncological, and functional outcomes. Ann Surg Oncol 22: 1219-25.

65. Baek SJ, Al-Asari S, Jeong DH, Hur H, Min BS, et al. (2013) Robotic versus laparoscopic coloanal anastomosis with or without intersphincteric resection for rectal cancer. Surg Endosc 27: 4157-63.

66. Ruurda JP, Draaisma WA, van Hillegersberg R, Borel Rinkes IH, Gooszen HG, et al. (2005) Robot-assisted endoscopic surgery: a four-year single-center experience. Digestive surgery 22: 313-20.

67. DeSouza AL, Prasad LM, Park JJ, Marecik SJ, Blumetti J, et al. (2010) Robotic assistance in right hemicolectomy: is there a role? Dis Colon Rectum 53: 1000-6.

68. Trastulli S, Desiderio J, Farinacci F, Ricci F, Listorti C, et al. (2013) Robotic right colectomy for cancer with intracorporeal anastomosis: short-term outcomes from a single institution. Int J Colorectal Dis 28: 807-14.

69. Lujan H, Maciel V, Romero R, Plasencia G (2013) Laparoscopic versus robotic right colectomy: a single surgeon's experience. J Robotic Surg 7: 95-102.

70. Rondelli F, Balzarotti R, Villa F, Guerra A, Avenia N, et al. (2015) Is robot-assisted laparoscopic right colectomy more effective than the conventional laparoscopic procedure? A meta-analysis of short-term outcomes. International journal of surgery 18: 75-82.

71. Morpurgo E, Contardo T, Molaro R, Zerbinati A, Orsini C (2013) Robotic-assisted intracorporeal anastomosis versus extracorporeal anastomosis in laparoscopic right hemicolectomy for cancer: a case control study. Journal of laparoendoscopic and advanced surgical techniques 23: 414-7.

72. Casillas MA, Leichtle SW, Wahl WL, Lampman RM, Welch KB, et al. (2014) Improved perioperative and short-term outcomes of robotic versus conventional laparoscopic colorectal operations. American journal of surgery 208: 33-40.

73. D'Annibale A, Pernazza G, Monsellato I, Pende V, Lucandri G, et al. (2013) Total mesorectal excision: a comparison of oncological and functional outcomes between robotic and laparoscopic surgery for rectal cancer. Surg Endosc 27: 1887-95.

74. Kim NK, Aahn TW, Park JK, Lee KY, Lee WH, et al. (2002) Assessment of sexual and voiding function after total mesorectal excision with pelvic autonomic nerve preservation in males with rectal cancer. Dis Colon Rectum 45: 1178-85.

75. Quah HM, Jayne DG, Eu KW, Seow-Choen F (2002) Bladder and sexual dysfunction following laparoscopically assisted and conventional open mesorectal resection for cancer. Br J Surg 89: 1551-6.

76. Heald RJ (1988) The 'Holy Plane' of rectal surgery. Journal of the Royal Society of Medicine 81: 503-8.

77. Lee JF, Maurer VM, Block GE (1973) Anatomic relations of pelvic autonomic nerves to pelvic operations. Arch Surg 107: 324-8.

78. D'Annibale A, Morpurgo E, Fiscon V, Trevisan P, Sovernigo G, et al. (2004) Robotic and laparoscopic surgery for treatment of colorectal diseases. Dis Colon Rectum 47: 2162-8.

79. Kim JY, Kim NK, Lee KY, Hur H, Min BS (2012) A comparative study of voiding and sexual function after total mesorectal excision with autonomic nerve preservation for rectal cancer: laparoscopic versus robotic surgery. Ann Surg Oncol 19: 2485-93.

80. Broholm M, Pommergaard HC, Gogenur I (2015) Possible benefits of robot-assisted rectal cancer surgery regarding urological and sexual dysfunction: a systematic review and meta-analysis. Colorectal Dis 17: 375-81.

81. Mirnezami AH, Mirnezami R, Venkatasubramaniam AK, Chandrakumaran K, Cecil T (2010) Robotic colorectal surgery: hype or new hope? Colorectal Dis 12: 1084-93.

82. Ng KH, Lim YK, Ho KS, Ooi BS, Eu KW (2009) Robotic-assisted surgery for low rectal dissection: from better views to better outcome. Singapore Med J 50: 763-7.

83. Park JS, Choi GS, Lim KH, Jang YS, Jun SH (2010) Robotic-assisted versus laparoscopic surgery for low rectal cancer: case-matched analysis of short-term outcomes. Ann Surg Oncol 17: 3195-202.

84. Kim NK, Kang J (2010) Optimal Total Mesorectal Excision for Rectal Cancer: the Role of Robotic Surgery from an Expert's View. J Korean Soc Coloproctol 26: 377-87.

85. Halabi WJ, Kang CY, Jafari MD, Nguyen VQ, Carmichael JC, et al. (2013) Robotic-assisted colorectal surgery in the United States: a nationwide analysis of trends and outcomes. World J Surg 37: 2782-90.

86. Kim CW, Baik SH, Roh YH, Kang J, Hur H, et al. (2015) Cost-effectiveness of robotic surgery for rectal cancer focusing on short-term outcomes: a propensity score-matching analysis. Medicine 94: e823.

87. Baik SH (2008) Robotic colorectal surgery. Yonsei Med J 49: 891-6.

88. Aly EH (2013) Have we improved in laparoscopic resection of rectal cancer: critical reflection on the early outcomes of COLOR II study. Translational Gastrointestinal Cancer 2: 175-8.

89. Maggiori L, Panis Y (2013) Is it time for a paradigm shift: laparoscopy is now the best approach for rectal cancer? Translational Gastrointestinal Cancer 3: 1-3.

90. Bokhari MB, Patel CB, Ramos-Valadez DI, Ragupathi M, Haas EM (2011) Learning curve for robotic-assisted laparoscopic colorectal surgery. Surg Endosc 25: 855-60.

91. Pigazzi A, Luca F, Patriti A, Valvo M, Ceccarelli G, et al. (2010) Multicentric study on robotic tumor-specific mesorectal excision for the treatment of rectal cancer. Ann Surg Oncol 17: 1614-20.

92. Park SY, Choi GS, Park JS, Kim HJ, Ryuk JP (2013) Short-term clinical outcome of robot-assisted intersphincteric resection for low rectal cancer: a retrospective comparison with conventional laparoscopy. Surg Endosc 27: 48-55.

93. Park EJ, Kim CW, Cho MS, Kim DW, Min BS, et al. (2014) Is the learning curve of robotic low anterior resection shorter than laparoscopic low anterior resection for rectal cancer? a comparative analysis of clinicopathologic outcomes between robotic and laparoscopic surgeries. Medicine 93: e109.

94. Bae SU, Baek SJ, Hur H, Baik SH, Kim NK (2013) Intraoperative near infrared fluorescence imaging in robotic low anterior resection: three case reports. Yonsei Med J 54: 1066-9.

95. Collinson FJ, Jayne DG, Pigazzi A, Tsang C, Barrie JM, et al. (2012) An international, multicentre, prospective, randomised, controlled, unblinded, parallel-group trial of robotic-assisted versus standard laparoscopic surgery for the curative treatment of rectal cancer. Int J Colorectal Dis 27: 233-41.

96. Choi GSPJ, Kim SH, Kim NK (2011) A trial to assess robotic assisted surgery and laparoscopy-assisted surgery in patients with mid or low rectal cancer. A prospective randomized trial.

97. Memon S, Heriot AG, Murphy DG, Bressel M, Lynch AC (2012) Robotic versus laparoscopic proctectomy for rectal cancer: a meta-analysis. Ann Surg Oncol19: 2095-101.

98. Shin JY (2012) Comparison of Short-term Surgical Outcomes between a Robotic Colectomy and a Laparoscopic Colectomy during Early Experience. J Korean Soc Coloproctol 28: 19-26.

Robotic Pancreatectomy: What is the Current Evidence?

Lee SY[1,2] and Goh BK[1,2*]

[1]Department of Hepatopancreatobiliary and Transplant Surgery, Singapore General Hospital, Singapore

[2]Duke -National University of Singapore (NUS) Medical School, Singapore

[*]**Corresponding author:** Goh BK, Department of Hepatopancreatobiliary and Transplant Surgery, General Hospital, Singapore
E-mail: bsgkp@hotmail.com

Abstract

Laparoscopic pancreatectomy has evolved from resection of benign lesions to the treatment of malignant lesions without compromising patient safety and oncologic principles. Driven by the technical shortcomings of laparoscopic surgery, robotic pancreatectomy is the latest development in this evolution. Presently, there are limited but increasing amount of data comparing the outcomes of the various approaches for pancreatectomy: robotic versus laparoscopic and open pancreatectomy. Most studies to date are single large institutional retrospective case series or case-control studies reporting on the safety and feasibility of robotic pancreatectomies but most fail to address key issues like cost-benefit ratio and selection biases. Hence, presently, there is only low level evidence from retrospective studies supporting the use of robotic pancreatectomy. These studies have demonstrated that robotic pancreatectomy is safe and feasible with outcomes at least comparable to conventional laparoscopy and open surgery. There is some evidence suggesting that robotic surgery may decrease the learning curve and conversion rate in minimally invasive pancreatic surgery. Further research is needed to evaluate and compare the effectiveness of robotic pancreatectomy with conventional laparoscopy and open surgery.

Keywords: Robotic pancreatectomy; Minimally invasive surgery; DaVinci; Robot; Laparoscopic pancreatectomy

Introduction

Laparoscopic pancreatic resection remains one of the newer developments in pancreatic surgery although it was first reported 2 decades ago in 1994 by Cushieri [1]. Following the introduction of robotic abdominal surgery, the first series of 13 robotic pancreatic resections was published by Giulianotti et al. [2]. Laparoscopic pancreatic resections were initially performed for benign lesions requiring left-sided pancreatosplenectomies. However as experience, surgical techniques and equipment improved; its indications have been expanded to malignant neoplasms and more technically demanding procedures such pancreaticoduodenectomies and spleen-preserving pancreatectomies. Despite the advantages associated with laparoscopic surgery such as decreased pain, diminished blood loss, decreased hospital stay and improved cosmesis [3]; the jury is still out there with regards to the definite role and cost-benefit ratio of laparoscopy for complex surgical procedures such as pancreatectomies. Presently, the role of robotic pancreatic surgery remains even more poorly defined as there is limited evidence available in the current literature demonstrating its clinical utility.

As custodians of our patients' health, questions will always arise when new technology or innovation enters our clinical practice. One key question would be: Does it really benefit our patients and improve outcomes? The technology from which the robotic surgical platform arose from military research and research for space exploration. Currently, the DaVinci® surgical system by Intuitive Surgical® is the dominant system. It consists of a three or four-armed bedside robot which is operated by a surgeon who sits at a console [4,5]. In theory, robotic surgery should retain the benefits of laparoscopic techniques with regard to smaller incisions, shorter hospitalization periods, less physiological stress induced by surgery and quicker patient recovery. Additionally, there are several potential technical advantages of robotic surgery over conventional laparoscopy [6]. This includes a high definition 3-dimensional (3D) view, tremor filtration, motion scaling, improved surgeon ergonomics and significant increased range of motion due to an internal articulated EndoWrist [7,8]. These features are obviously attractive to pancreatic surgeons due to the intricate nature and the complexity of pancreatic resections. However, this new platform has also raised concerns among surgeons on the lack of tactile feedback, higher costs and longer surgical time as compared to conventional laparoscopy or open surgery [9].

To objectively examine the evidence for robotic pancreatectomy, one should ideally review the different forms of resections separately, namely distal pancreatectomy (DP) with and without splenic preservation, pancreatoduodenectomy (PD) and others procedures such as central pancreatectomy (CP) and total pancreatectomy (TP). To date, there have been several relatively large surgical series which may shed some light on this issue [10-13].

The adoption of the laparoscopic approach for DP especially when performed with a splenectomy has become more widely accepted compared to PD or CP. This is because it is a relatively technically less demanding procedure as it does not require any reconstruction or anastomosis. Ever since the first reported robotic DP; several institutions have published their experience and outcomes with robotic DP [9,14]. Currently, there is limited evidence comparing open DP and/or laparoscopic DP vs. robotic DP. To date, there are only 5 retrospective and 1 prospective comparative studies published in the literature [9,15-20] (Table 1). Some potential benefits of the robotic approach over conventional laparoscopic DP which have been reported include a higher rate of splenic preservation (vessel preserving) and a lower conversion rate [10,15,21]. The increased rate of splenic vessel preservation after DP has been attributed to the improved dexterity and precision of the robotic arms when working in tight spaces [10,15,21]. However, despite its many theoretical technical advantages, robotic DP remains a complex procedure associated with a significant learning curve. In a recent study of 100 consecutive robotic DP, the learning curve of robotic DP with regards to operation time was reported to be optimized at about 40 cases [22]. This relatively long learning curve for a relatively rare surgical procedure poses a major obstacle to the widespread adoption of robotic DP world-wide. However, it is highly likely that the learning curve could potentially be shortened in future adopters with increasing familiarity with the platform and standardization of surgical techniques [22]. The Memorial Sloan Kettering Cancer Center recently published their 14 year experience of over 800 DP comparing open, laparoscopic and robotic DP and found that robotic and laparoscopic DP were comparable with respect to most perioperative outcomes, with no clear advantage of one approach over the other [9]. Today, DP remains a surgical procedure associated with a high morbidity rate especially from postoperative pancreatic fistula regardless of the surgical approach [23].

Author (Year)	Country	Approach	N	Mean operative time (min)	EBL (ml)	Conversion rate, (%)	Spleen preservation rate (%)	R0 resection, %	Morbidity (%)	Mortality (%)	Pancreatic fistula (%)	Length of stay (days)	Cost (USD)
Waters et al. [16]	USA	Open	32	245*	279	N.A	14*	100	18	0	18	8*	$16059
		Laparoscopic	28	222*	667	11	28*	82	33	0	11	6*	$12986
		Robotic	17	298*	681	12	65*	100	18	0	0	4*	$10588
Kang et al. [17]	Korea	Laparoscopic	25	258*	-	NR	64*	NR	16	0	NR	7.1	$3861
		Robotic	20	348*	-	NR	95*	NR	10	0	NR	7.3	$8304
Daouadi et al. [18]	USA	Laparoscopic	94	372*	150	16*	18	64	64	1.1	41	7.1	NR
		Robotic	30	293*	150	0*	7	100	66	0	46	6.1	NR
Lai et al. [19]	Hong Kong, China	Laparoscopic	17	172*	282	NR	41.2	NR	41.2	0	35.3	14.2	NR
		Robotic	9	242*	104	NR	66.7	NR	55.6	0	44.4	8	NR
Lee et al. [9]	USA	Open	637	185*	596*	N.A	14	88	40	0.6	12	7	NR
		Laparoscopic	131	193*	262*	41(31)	22	100	32	0	8	5	NR
		Robotic	37	213*	193*	14(38)	8	100	32	0	8	5	NR
Butturini et al. [20]	Italy	Laparoscopic	21	195	BT: 0%	4.7	19	NR	71.4	0	57.1	7	1500*
		Robotic	22	265	BT: 13%	4.5	27.3	NR	68.2	0	50	7	3000*

Abbreviations: EBL: Estimated blood loss; $^*p < 0.05$; NR: Not reported; BT: Blood Transfusion (Intraoperative); #: Euros; NA: Not Applicable

Table 1: Summary of studies comparing Robotic distal pancreatectomies vs. Open and/or Laparoscopic distal pancreatectomies

In contrast to laparoscopic DP, the adoption of laparoscopic PD has been limited. From 2005-2010, it was estimated that more than a quarter of DP were performed laparoscopically in North America [24]. In contrast, unlike the ubiquitous adoption of laparoscopic DP, laparoscopic PD had a much slower adoption rate most likely due to its steeper learning curve. Laparoscopic PD was first reported by Garner and Pomp in 1994 [9]. In 2012, it was reported that only 7 centers had an experience of more than 30 patients who had undergone laparoscopic PD, with very few surgeons in the world acquiring a significant experience with the procedure [25,26]. Interestingly, the adoption of robotic PD in comparison with laparoscopic PD has increased at a relatively more rapid rate compared to the frequency of adoption of robotic DP versus laparoscopic DP. In 2001, Giulianotti et al. first reported robotic PD in an initial series of 8 patients with a morbidity rate of 37.5%. More recently, investigators from the University of Pittsburgh published the largest series of robotic pancreatic resections to-date, which included 132 PD [2,14]. Although it is impossible to determine the exact reasons behind this observation; it is not implausible to postulate that a major reason could be due to the enabling effect of the robotic system over conventional laparoscopy for surgeons resulting in a gentler learning curve for complex pancreatic operations. Most surgeons would agree that the robot is superior with regards to performing laparoscopic maneuvers such as precise suturing and fine dissection which is essential when performing a PD. The morbidity of PD is largely associated with the incidence of pancreatic fistula and the assistance of the robot may potentially improve the precision and dexterity when performing the pancreato-enteric reconstruction, although this have yet to be proven [15]. A recent meta- analysis comparing open and robotic pancreatectomy favored the latter approach with a risk difference of 12% in both re-operation and morbidity rate. However, a major limitation of this meta-analysis was that the study population was heterogeneous and included PD, DP as well as CP cohorts [11]. Another 2 systematic reviews on robotic pancreatectomies were published in 2013. Nigri et al. analyzed 8 studies that compared minimally invasive surgery (MIS) (n = 204) vs open PD (n = 419), they concluded that there were no significant differences in complications such as pancreatic fistula, delayed gastric emptying, wound infection, reoperation and mortality rates and found that the MIS approach was associated with a greater lymph node harvest with better margins but with longer operating times. However, the authors combined both laparoscopic and robotic approaches as a single MIS group and did not compare between the 2 groups [12]. The other systematic review by Cirocchi et al. which focussed solely on robotic PD included 13 studies with 207 patients, they reported an average R1 resection rate of 9% and concluded that for highly selected patients, robotic PD is feasible with similar morbidity and mortality compared to open or purely laparoscopic approaches [13] (Table 2).

Author (Year)	Country	Approach	N	Age (Years)	Mean operative time (min)	EBL (ml)	Conversion rate, n (%)	R0 resection	Harvested lymph nodes, mean (n)	Morbidity (%)	Mortality (%)	Pancreatic fistula (%)	Length of stay (days)
Buchs [33]	USA	Open	39	56*	559*	827*	N.A	81.5	11*	48.7	2.6	20.5	14.6
		Robotic	44										
Zhou [34]	China	Open	8	57	420*	210*	N.A	83.3	NR	75*	0	37.5	24.3*
		Robotic	8	65	718*	153*	N.A	100	NR	25*	12.5	50	16.4*
Chalikonda [35 ¶	USA	Open	30	61	366.4*	775	N.A	87*	11.8	43	0	7(Grade B)	13.26*
		Robotic	30	62	476.2*	485.8	3(12)	100*	13.2	30	4	7(Grade B/C)	9.79*
Lai et al. [19]	Hong Kong	Open	67	62.1	264.9*	774.8*	N.A	64.1	10	49.3	3	17.9	25.8*
		Robotic	20	66.4	491.5*	247*	5	73.3	10	50	0	35	13.7*
Hammil et al. [36 §	China	Open	69	55	398	450	N.A	NR	NR	23	1.4	NR	7
		Robotic	8	62.5	648	256	1 (11)	NR	NR	25	0	NR	14
Walsh [37 ¶	USA	Open	25	62	364*	840	N.A	73	NR	44	0	NR	14
		Robotic	25	63	488*	537	3 (12)	100	NR	32	4	NR	10

Abbreviations: EBL: Estimated blood loss; *statistically significant, $p < 0.05$; NA: Not Applicable; § Published ab NR stracts. ¶: Possible overlap of patient cohorts; NR: Not reported

Table 2: Summary of studies comparing Open vs. Robotic pancreaticoduodenectomies

The data available for other pancreatic resections via the robotic platform remains limited to small case series and reports. Cheng et al. published their initial experience in 7 patients matched with 36 patients who undergone robotic and open CP respectively and concluded that patients in the robotic group experienced faster gastrointestinal tract recovery [27]. Similarly, Kang et al. presented a small series of 5 robot-assisted CP and compared it to their open CP experience. The robotic surgery group was associated with a significantly longer operating time but decreased blood loss [28]. In their experience of 250 robotic pancreatic resections, surgeons from the University of Pittsburgh reported a variety of pancreatic procedures such as 13 CPs, 5 TPs, 10 enucleations, 4 Appleby resections and 3 Frey procedures [14]. Based on their experience which is the largest robot pancreatectomy experience to-date, they concluded that robotic pancreatectomy was safe and feasible. They further added that there were no unanticipated risks inherent to this new technology [14]. More recently, there have even been anecdotal reports of robot-assisted pancreatic transplants with 3 cases reported in a recent study [29].

The evidence for robotic pancreatic surgery is culminating but still lacks robust data. Most case series and case control comparison studies unfortunately have failed to address several main issues related to the robotic technology. One of the most important issues is the considerable start-up and high recurring costs associated with this new technology. Presently, the cost-benefit ratio of robotic pancreatic surgery has not been well-studied. Kang et al. reported that the robotic surgery cost was almost 2.5 times that of conventional laparoscopic surgery in Korea [17]. Presently, there is some evidence from retrospective studies that laparoscopic DP is more cost effective than open DP [30]. However, this does not apply to robotic pancreatectomy at present whereby the absolute cost is almost certainly higher than conventional laparoscopy or open surgery. Nonetheless, it is important to note that it is almost inevitable that the cost of robotic surgery would decrease significantly in the near future as it becomes more widely and readily available. Furthermore, the emergence of competing companies and robotic systems would almost certainly result in the lowering of costs of robotic surgery which at present is monopolized by a single commercial firm.

A common limitation in all the studies to date reporting on robotic pancreatectomies was that the analyses were all conducted in retrospective patient cohorts. Hence, the reported findings were likely to be limited by selection bias in these non-prospective, non-randomized studies. The advantages of robotic surgery over laparoscopy and open surgery could be due to selection bias or other confounding factors. For example, the shorter learning curve and decreased conversion rates observed with robotic surgery could be confounded by the fact that more frequently than not, surgeons who begin performing robotic pancreatectomies had already acquired a significant amount of experience in open and conventional laparoscopy before embarking on robotic surgery [31].

Finally, it is important to note that the introduction and adoption of new surgical technology is a complex event and often poorly studied. Unlike other research questions; randomized controlled trials in this aspect are rarely feasible nor realistic due to the inability to truly blind the subjects and investigators and due to the inherent lack of true clinical equipoise [32]. Thus far, the majority of such innovations today are established on the findings of large retrospective experiences. The crux of the matter is that until the cost, accessibility and experience of robotic surgery becomes equitable to laparoscopic surgery, defining the exact role of robotic pancreatectomy today will continue be an ongoing challenge for pancreatic surgeons.

Conclusion

Presently, there is only low level evidence from retrospective cases series and case-control studies supporting the use of robotic pancreatectomy. These studies have demonstrated that robotic pancreatectomy is safe and feasible with outcomes at least comparable to conventional laparoscopy and open surgery. There is some evidence suggesting that robotic surgery may decrease the learning curve and conversion rate in MIS pancreatic surgery. Further research is needed to evaluate and compare the effectiveness of robotic pancreatectomy with conventional laparoscopy and open surgery.

References

1. Cuschieri A (1994) Laparoscopic surgery of the pancreas. J R Coll Surg Edinb 39: 178-184.
2. Giulianotti PC, Coratti A, Angelini M, Sbrana F, Cecconi S et al. (2003) Robotics in general surgery: personal experience in a large community hospital. Arch Surg 138: 777-784.
3. Ricci C, Casadei R, Taffurelli G, Toscano F, Pacilio CA, et al. (2015) Laparoscopic Versus Open Distal Pancreatectomy for Ductal Adenocarcinoma: A Systematic Review and Meta-Analysis. J Gastrointest Surg 19: 770-781.
4. Mack MJ (2001) Minimally invasive and robotic surgery. Jama 285: 568-572.
5. Wilson TG (2014) Advancement of Technology and Its Impact on Urologists: Release of the da Vinci Xi. A New Surgical Robot. Eur Urol 66: 793-4.
6. Lendvay TS, Hannaford B, Satava RM (2013) Future of robotic surgery. Cancer J 19: 109-119.
7. Ballantyne GH (2007) Telerobotic gastrointestinal surgery: Phase 2 safety and efficacy. Surg Endosc 21: 1054-1062.
8. Talamini MA, Chapman S, Horgan S, Melvin WS (2003) A prospective analysis of 211 robotic-assisted surgical procedures. Surg Endosc 17: 1521-1524.
9. Lee SY, Allen PJ, Sadot E, D'Angelica MI, DeMatteo RP (2014) Distal pancreatectomy: A single institution's experience in open, laparoscopic and robotic approaches. Journal of the American College of Surgeons 22: 18-27.
10. Joyce D, Morris-Stiff G, Falk GA, El-Hayek K, Chalikonda S (2014) Robotic surgery of the pancreas. World J Gastroenterol 20: 14726-14732.
11. Zhang J, Wu WM, You L, Zhao YP (2013) Robotic versus open pancreatectomy: a systematic review and meta-analysis. Ann Surg Oncol 20: 1774-1780.
12. Nigri G, Petrucciani N, Torre LM, Magistri P, Valabrega S (2014) Duodenopancreatectomy: Open or minimally invasive approach? Surgeon 12: 227-234.
13. Cirocchi R, Partelli S, Trastulli S, Coratti A, Parisi A (2013) A systematic review on robotic pancreaticoduodenectomy. Surg Oncol 22: 238-246.
14. Zureikat AH, Moser AJ, Boone BA, Bartlett DL, Zenati M et al. (2013) 250 Robotic pancreatic resections: Safety and feasibility. Ann Surg 258: 554-559.
15. Zureikat AH, Hogg ME, Zeh HJ (2014) The utility of the robot in pancreatic resections. Adv Surg 48: 77-95.
16. Waters JA, Canal DF, Wiebke EA, Dumas RP, Beane JD (2010) Robotic distal pancreatectomy: Cost effective? Surgery 148: 814-823.
17. Kang CM, Kim DH, Lee WJ, Chi HS (2011) Conventional laparoscopic and robot-assisted spleen-preserving pancreatectomy: Does da Vinci have clinical advantages? Surg Endosc 25: 2004-2009.
18. Daouadi M, Zureikat AH, Zenati MS, Choudry H, Tsung A (2013) Robot-assisted minimally invasive distal pancreatectomy is superior to the laparoscopic technique. Ann Surg 257: 128-132.
19. Lai EC, Tang CN (2013) Current status of robot-assisted laparoscopic pancreaticoduodenectomy and distal pancreatectomy: A comprehensive review. Asian J Endosc Surg 6: 158-164.
20. Butturini G, Damoli I, Crepaz L, Malleo G, Marchegiani G (2015) A prospective non-randomised single-center study comparing laparoscopic and robotic distal pancreatectomy. Surg Endosc 29: 3163-3170.
21. Goh BK, Wong JS, Chan CY, Cheow PC, Ooi LL (2016) First experience with robotic spleen-saving vessel preserving distal pancreatectomy in Singapore: report of 3 consecutive cases. SMJ.
22. Shakir M, Boone BA, Polanco PM (2015) The learning curve for robotic distal pancreatectomy: An analysis of outcomes of the first 100 consecutive cases at a high-volume pancreatic centre. HPB 17: 580-6.
23. Goh BK, Tan YM, Chung YF (2008) Critical appraisal of 232 consecutive distal pancreatectomies with emphasis on risk factors, outcome, and management of the postoperative pancreatic fistula: A 21-year experience at a single institution. Arch Surg 143: 956-65.
24. Velderrain AR, Bowers SP, Goldberg RF, Clarke TM, Buchanan MA (2012) National trends in resection of the distal pancreas. World J Gastroenterol 18: 4342-4349.
25. Gagner M, Pomp A (1994) Laparoscopic pylorus-preserving pancreatoduodenectomy. Surg Endosc 8: 408-410.
26. Kendrick ML (2012) Laparoscopic and robotic resection for pancreatic cancer. Cancer J 18: 571-576
27. Cheng K, Shen B, Peng C, Deng X, Hu S (2013) Initial experiences in robot-assisted middle pancreatectomy. HPB 15: 315-21.
28. Kang CM, Kim DH, Lee WJ, Chi HS (2011) Initial experiences using robot-assisted central pancreatectomy with pancreaticogastrostomy: a potential way to advanced laparoscopic pancreatectomy. Surg Endosc 25: 1101-6.
29. Tzvetanov I, D'Amico G, Bejarano-Pineda L, Benedetti E (2014) Robotic-assisted pancreas transplantation: where are we today? Curr Opin Organ Transplant 19: 80-2.
30. Mehrabi A, Hafezi A, Arvin J (2015) A systematic review and meta-analysis of laparoscopic versus open distal pancreatectomy for benign and

malignant lesions of the pancreas: Its time to randomize. Surgery 157: 45-55.

31. Goh BK (2014) Robot-assisted minimally invasive distal pancreatectomy is superior to the laparoscopic technique. Ann Surg.
32. Wilson CB (2006) Adoption of new surgical technology. Bmj 332: 112-114.
33. Buchs NC, Addeo P, Bianco FM, Ayloo S, Benedetti E et al. (2011) Robotic versus open pancreaticoduodenectomy: a comparative study at a single institution. World J Surg 35: 2739-46.
34. Zhou NX, Chen JZ, Liu Q (2011) Outcomes of pancreatoduodenectomy with robotic surgery versus open surgery. Int J Med Robot 7: 131-7.
35. Chalikond (2012) Laparoscopic robotic-assisted pancreaticoduodenectomy: a case-matched comparison with open resection. Surg Endosc 26: 2397-402.
36. Hammill C, Cassera M, Swanstrom L, Hansen P (2010) Robotic assistance may provide the technical capability to perform a safe, minimally invasive pancreaticoduodenectomy. HPB 12: 198-200.
37. Walsh M, Chalikonda S, Saavedra JRA, Lentz G, Fung J (2011) Laparoscopic robotic assisted Whipple: early results of a novel technique and comparison with the standard open procedure. Surg Endosc 25: S221.

Development of an Autonomous Mobile Health Monitoring System for Medical Workers with High Volume Cases

Ajiroghene O[1], Obiei-uyoyou O[1], Chuks M[1], Ogaga A[2], Chukwumenogor O[2], Elvis R[2], Udoka ED[3], Bright AE[3], Ofualagba G[1] and Ejofodomi OA[1*]

[1]*Department of Electrical and Electronics Engineering, Federal University of Petroleum Resources (FUPRE), Nigeria*
[2]*Department of Mechanical Engineering, Federal University of Petroleum Resources (FUPRE), Nigeria*
[3]*Department of Marine Engineering, Federal University of Petroleum Resources (FUPRE), Nigeria*

Abstract

Mobile health care monitoring system have the capacity to provide critical assistance in areas of medical care, especially in situations where there are limited number of health workers needed to attend to patients. This project describes the development of a mobile health care monitoring system capable of monitoring the heart rate of patients and recording this information electronically. The system will provide much needed relief for medical workers with high volume cases. The system was constructed using a mobile robot capable of traversing rugged terrain, a motor shield, a Global Positioning System (GPS) shield for autonomous navigation, an ultrasound sensor for obstacle avoidance, and a heart rate sensor for measuring patient heart rate. The mobile system was able to successfully avoid obstacles and navigate to the predetermined locations of three hospital beds and obtain the heart rate of three patients. Future improvements to the system include the addition of wireless capability so patient data can be transmitted wirelessly to physician's PC, incorporation of speech capability to enhance interaction between robot and patient and addition of more sensors to measure other vital signs such as blood glucose.

Keywords: Global positioning system (Gps); Mobile health; Medical workers

Introduction

Mobile health care monitoring system have the capacity to provide critical assistance in areas of medical care, especially in situations where there are limited number of health workers needed to attend to patients, for instance in a united nation refugee camp where there is shortage of medical staff and resources are scarce. It is important to always monitor the vital signs of patients as this gives a very good idea of their medical condition.

Currently mobile health monitoring systems typically consists of wearable biomedical sensors that are attached to a person's physical body. Relevant biological signals are acquired by these sensors and the data is transferred wirelessly to a PC. Shahriyar et al. [1] developed an Intelligent Mobile Health Monitoring System (IMHMS) that uses the Wearable Wireless Body/Personal Area Network for collecting data from patients, mining the data, intelligently predicts patient's health status and provides feedback to patients through their mobile devices. The acquired biological data is stored in the patient's mobile device and is then sent to an intelligent automated medical server for analysis. The automated server will then provide feedback to the patient based on the data it receives. This system encourages patients to participate in the health care process using their mobile devices and patients are able to access their health information from anywhere and at any time. IMHMS is intended to be used by patients who have been discharged from the hospital but still need to constantly monitor vital biological signals. It is not intended to be used in the hospital setting.

Ziyu Lv et al. [2] have developed a mobile health monitoring system called iCare, which is to be used by older people anytime anywhere. The iCare uses body sensors and devices to collect physiological signals from the elderly and transmit them to a smart phone that will process physiological data locally and send an alarm automatically to the emergency centre and pre-assigned people when the data exceeds the threshold of the fixed device. The emergency centre will call an ambulance to the current location of the elderly. The location information can be gained from the alarm message sent to the emergency centre. Sensors and devices can be tailored depending on the old people's health condition. The physiological data will be not only analyzed locally, but also sent to the server in bulk to construct the personal health information system. With viewing the current and history condition of the old people in the personal health information system, doctors remotely set the thresholds and give advices which can guide the old people to adjust themselves to health mode [2]. The iCare also offers auxiliary functions such as the life assistant, which provides services like reminders and location tracking. The iCare system is to be used to provide remote medical monitoring services to older patients in their homes, and not for patients in the hospital setting.

Many other mobile health care monitoring systems have been developed [3-11]. However, the major design concept remains the same: patient is fitted with wearable biomedical sensors outside the hospital setting and vital signs are measured and transferred wirelessly either to a mobile device or to a server. Acquired data is either reviewed by medical professionals for remote medical monitoring, or by an intelligent medical server as is the case for IMHMS. All these health care monitoring systems have been developed to be used outside the hospital setting.

***Corresponding author:** Ejofodomi OA, Department of Electrical and Electronics Engineering, Federal University of Petroleum Resources (FUPRE), Nigeria
E-mail: tegae@yahoo.com

The mobile health care monitoring system presented in this paper takes a different approach. It has been designed specifically to be used in the hospital setting to provide assistance to medical workers dealing with high volume cases. The biomedical sensors used in this system are not are mounted on a mobile robot, which autonomously moves from patient to patient. This system affords the opportunity to monitor multiple patients and will provide much needed relief for medical workers, especially in developing countries.

Blood pressure is a very vital health parameter that is directly related to the soundness of the human cardiovascular system. High blood pressure is often an important risk factor for many diseases [12]. The mobile health care monitoring system presented in this paper monitors the heart rate of patients and records this information electronically. It uses a technique of measuring heart rate through the fingertip by means of an Arduino development board in conjunction with a mobile robot. When the heart beats, it pumps blood throughout the body, causing the blood volume inside the arteries of the fingers to change too. This fluctuation of blood can be detected through a pressure sensing mechanism on which the fingertip is placed. The signal can be amplified further for a microcontroller to count the rate of fluctuation, which is actually the heart beat rate. The arduino heart rate meter computes the heart beat rate by processing the analog pulse signal output from the sensor. Heart rate values are stored electronically in a memory card that can be accessed by the physician once the mobile system has completed its rounds. The design and construction of the automated mobile health monitoring system is presented in the materials and methods section. Analysis of the system's design, its advantages and limitations are discussed in the discussion and conclusion section. Future improvements to the system are also mentioned briefly.

Materials and Methods

Materials

Robot chassis: The robot chassis used in this research project was a four wheel drive chassis from Dagu Electronics (Figure 1). It is designed to drive on rough terrain, making it a great platform for robots that need to perform tasks in outdoor environments. The size of the robot is 11 by 12 by 5 inches. The movement of the robot was provided by four DC motors with brass brushes and 75:1 steel gearboxes that drive 120 mm diameter spiked tires. These motors are designed to run between 2V and 7.2V. The chassis can reach a top speed of approximately 3 kilometers per hour (km/hr) when powered at 7.2V. The stall torque of each motor is roughly around 11 kg-cm. The chassis has a spring suspension system which keeps each wheel in contact with the ground for maximum traction. The chassis is made from a 2 mm thick corrosion-resistant anodized aluminum plate, and the nuts and bolts are made from stainless steel [13].

Arduino Uno R3: The arduino board (Figure 2) serves as the brain for this mobile health monitoring system. It processes all input and output system of the robotic system as well as all the integrated circuits connected for specific operation of the robotic system such as GPS, heart rate sensor and ultrasonic sensor. The USB cable was used to burn the code developed in a computer into the microcontroller on the arduino board.

Figure 1: Chassis for Autonomous Mobile Health Monitoring System.

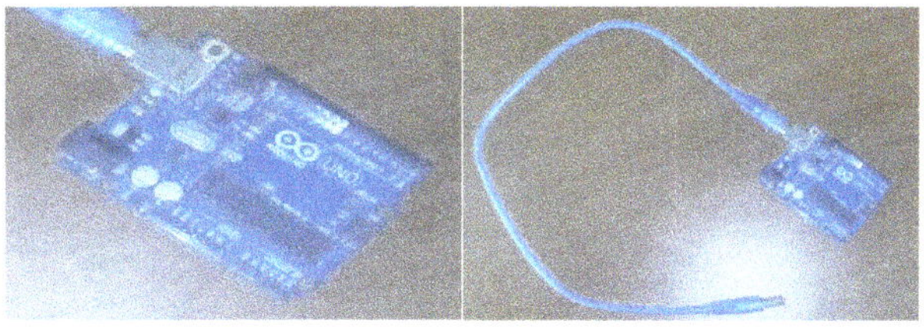

Figure 2: Arduino board with USB cable attached.

(a) MegaMoto Shield

(b) Moto shield mounted on robot chassis.

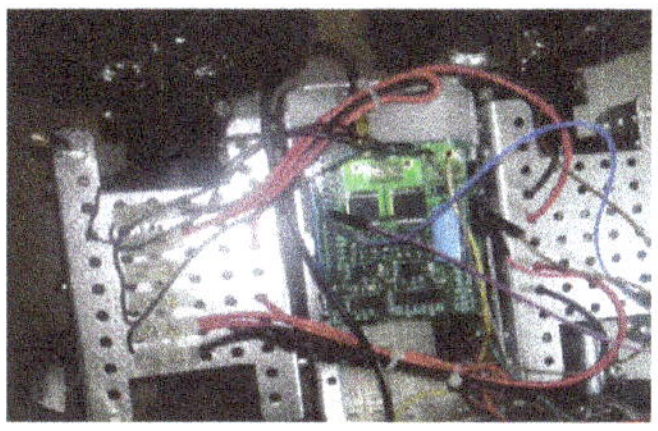

Figure 3: Robot Power Mega Moto Shield.

Robot power mega moto shield: The motor shield was used to control the geared motor system, which in turn controls the wheels of the robot (Figure 3). The motor shield was powered by an external battery because the Arduino microcontroller supplies a maximum current of 40 mA which is insufficient to drive the motor of the mobile robot. The MegaMoto circuit is two independent half-bridge circuits mounted on a single PCB. The two motor outputs may work together to form a full H-bridge for bi-directional drive of a single motor or other load or each half-bridge may be used independently. When connected in half-bridge mode the MegaMoto may be used for driving uni-directional loads such as lamps, heaters, solenoids or DC motors in one direction. In this research project, a single mega moto shield was connected in half-bridge mode to drive to two pairs of DC motor. In this configuration, the motor is not enabled to do a reverse movement. To enable a reverse movement, two mega moto shields will be needed [14].

GPS shield: The Arduino GPS shield is a GPS module breakout board designed for Global Positioning System receiver with SD interface (Figure 4). It can be used to read position data of the fire detection system and can also store the data in the SD card. The 5V/3.3V compatible operation voltage level makes it compatible with Arduino boards. The GPS shield was used for determining the GPS

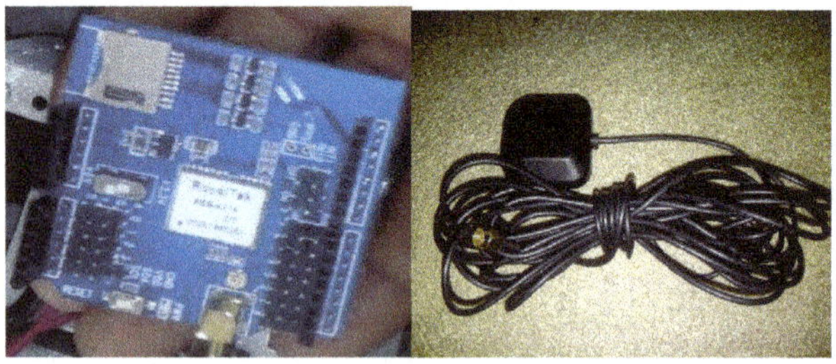

Figure 4: Global Positioning System (GPS) shield for Arduino.

Figure 5: Ultasonic sensor mounted in front of robot for autonomous navigation.

location of the position where any operation will be carried out by the robot. A GPS antenna magnetic mount SMA was also used for sending and transmitting information containing fire location. The GPS shield is suitable for automotive navigation, personal positioning, fleet management and marine navigation [15].

Ultrasonic sensor: The ultrasonic sensor was used to detect the obstacles at a certain distance from the mobile robot (Figure 5). An ultrasonic pulse is sent out. The time taken to receive an echo signal by the sensor is used to estimate the distance between the robot and an obstacle. If the obstacle is close to the robot, the program instructs the robot to take an alternate path.

Hear rate sensor: The heart rate sensor is an integrated circuit device that detects heat rate in a human.

Buzzer: A piezo buzzer was used to give out a sound corresponding to a particular frequency. This served as the alarm signal if the measured heart rate was above a certain threshold.

Methods

All hardware device and circuitry used were connected to the Arduino Uno R3 microcontroller board. The different component parts connected were made as shown in the block diagram below (Figure 6).

The megamoto shield was stacked on the arduino uno and the gps shield was mounted on the megamoto shield (Figure 7). The arduino uno was powered by a 9 volts battery supply system and served as the power supply for the heart rate monitoring circuit, buzzer circuit and GPS shield. The megamoto shield was powered by a rechargeable 12 volts lead acid battery and was used to control for the geared motors which were responsible for movement of the mobile robot. Ultrasonic sensor was connected to the arduino and mounted in front of the robot (Figure 5) to aid obstacle avoidance. The heart rate sensor and the piezo buzzer were also connected to the arduino and mounted at the front of the robot. The schematic of how the mobile health monitoring system functions is shown in Figure 8. The robot is programmed to navigate a pre-defined route. For testing purposes it was assumed that the system would be recording the heart rate of only three patients in a hospital setting. The GPS locations of the three hospital beds are stored within the system's program. The mobile system begins its rounds from the physician's office. The ultrasonic sensor turns ON as the robot moves. The ultrasonic sensor detects any object at a distance and sends signals to the microcontroller which sends signals to the motor shield to turn the robot in a different direction.

The robot navigates until it arrives at the GPS location of the first hospital bed. It then gives a buzzing sound to prompt the patient to place his or her finger tip on the sensor. After sensor measures the heart rate of the patient, the value is stored in the SD card in the GPS shield. If the patient fails to place his or her finger tip on the sensor, a specific value is recorded, indicating the absence of a reading. The entire process is repeated for the second and third hospital bed. Once the heart rates of the three patients have been recorded, the robot navigates back to the physician's office. The physician can then extract the SD card from the system to view the heart rate log file.

Results and Discussion

The mobile health monitoring system was assembled and tested at the Student's Center in the Federal University of Petroleum Resources.

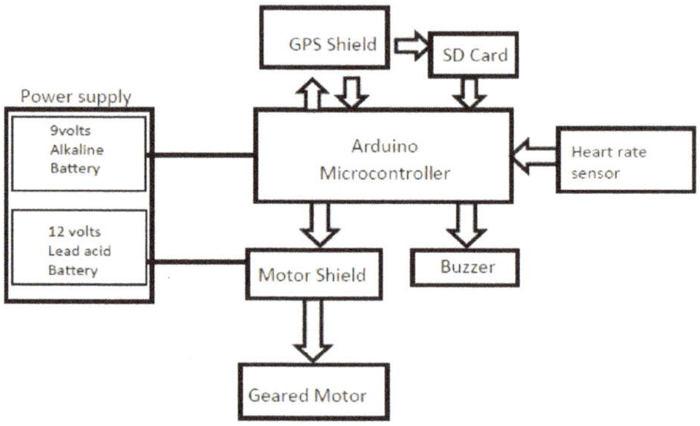

Figure 6: Block Diagram of Mobile Health Monitoring System.

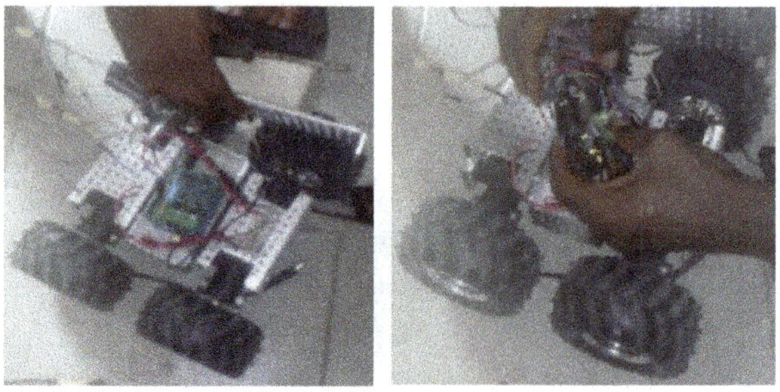

Figure 7: Assembling the autonomous mobile health monitoring system.

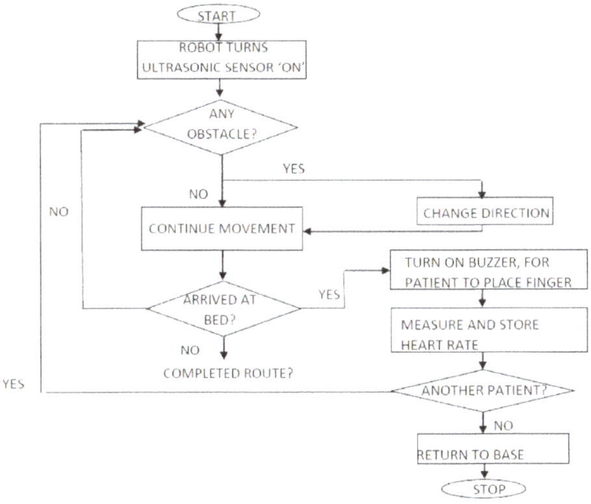

Figure 8: Flow Chart of Mobile Health Monitoring System.

(a) Mobile system avoiding obstacles as it navigate to the GPS location of patient

(b) Mobile robot measuring the heart rate of a patient.

(c) Viewing Patient data stored in SD card of the system

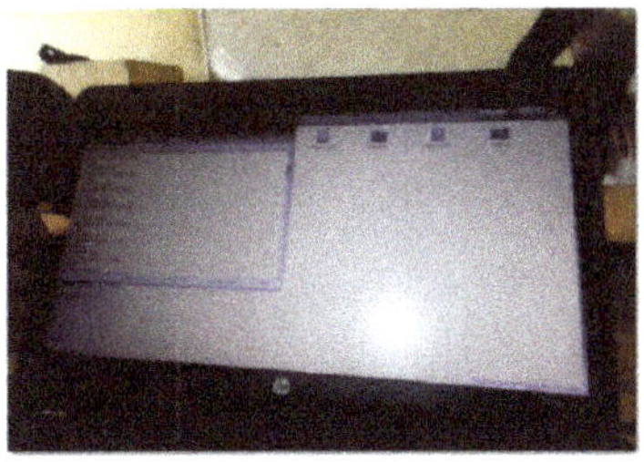

Figure 9: Autonomous mobile health monitoring system.

The ability to avoid obstacles was successfully demonstrated (Figure 9a) by placing a set of brick blocks in the path of the robot. The robot turned left away from its straight path, moved along that direction for a second before turning right towards its normal path, continuing its movement before avoiding the second obstacle in a similar pattern. It then moved to the place where the three patients were standing and gave a buzz sound for the first patient to place his fingertip, measured and recorded the heart rate, moved to the second patient, measured and recorded the heart rate and then did the same for the third patient (Figure 9b). After this the robot returned to its starting point, The memory card was removed and inserted into a computer where the results of the heart rate were indicated as high, normal and low with corresponding values of their heart rate in volts (Figure 9c).

To our knowledge, this is the first mobile health monitoring system that is being developed to be used within a hospital setting. As mentioned in the previous section, all mobile health monitoring systems currently being developed are being used outside of the hospital setting for remote health care monitoring by the patient. This system, on the other hand, has been designed to be used by the medical professionals in hospitals. To our knowledge it is also the first health monitoring system that has locomotive capabilities. The results presented in this paper are from the initial design and testing of this new health monitoring system. It is expected that subsequent versions of this system will demonstrate a more robust design and higher complexity.

The mobile robot used as the base of the health monitoring system is only about 5 inches high. This is not ideal for obtaining blood pressure measurements from patients lying in hospital beds. This limitation can be addressed in one of two ways. The first is to use a larger mobile robot with about the same height as the hospital bed. The second alternative is to install long rods on the existing robot on which the biomedical sensors can be attached. The height of the rods can be adjusted to the position that is most comfortable for the patient.

The primary focus for the initial testing of the system was to ensure that the robot could autonomously go to the patient's bedside, obtain the patient's blood pressure, save that data in its onboard memory, and return back to the physician with the acquired data. This was successfully demonstrated by the system. However, there are several improvements that will be made to the mobile health monitoring system. Wireless capability will be incorporated into the system so that the patient data are transmitted wirelessly to the physician's PC without the physician needing to remove the SD card to acquire the data. This could improve the response time of the physician in cases where urgent attention is needed. The system will also be given speech capability to enhance interaction between robot and patient so as to ensure accurate readings are obtained. The ultimate goal is to have a humanoid robot capable of assisting medical workers in the hospital setting and so robot-patient interaction should be identical to doctor-patient interaction.

There will be inclusion of more sensors and other devices to obtained more detailed medical information about the patient. Presently, the mobile health monitoring system is only capable of measuring and detecting patient heart rate, but it could be upgraded to measure other vital signs such as blood glucose content for diabetic patients. A video camera will also be integrated into the system to provide visual inspection of the patient. Future tests on the automated health monitoring system will be conducted inside of a hospital, as it is intended to be used in hospital settings.

Conclusion

Mobile health care monitoring systems developed to date are meant to be used outside the hospital settings. Patients are fitted with wearable biomedical sensors, and vital signs are measured and transferred wirelessly either to a mobile device or to a server for remote medical monitoring. The mobile health care monitoring system presented in this paper has been designed specifically to be used in the hospital setting to provide assistance to physicians dealing with high volume cases. The biomedical sensors used in this system are not are mounted on a mobile robot, which autonomously moves from patient to patient. This system affords the opportunity to monitor multiple patients and will provide much needed relief for medical workers, especially in developing countries. To our knowledge, this is the first health monitoring system being developed to be used in the hospital setting. It is also the first automated health monitoring system with locomotive capabilities. The system was able to successfully measure and store heart rates from three patients in three different locations. The ultimate goal is to develop a humanoid robot capable of assisting physicians with high volume cases in developing countries.

References

1. Shahriyar R, Bari F, Kundu G, Ahamed SI, Akbar M (2009) Intelligent Mobile Health Monitoring System (IMHMS). International Journal of Control and Automation 2: 13-28.
2. Lv Z, Xia F, Wu G, Yao L, Chen Z (2010) I Care: A Mobile Health Monitoring System for the Elderly. Physical and Social Computing (CPSCom) 1: 699-705.
3. Bourouis A, Feham M, Bouchachia A (2011) Ubiquitous Mobile health Monitoring System for Elderly (UMHSE). International Journal of Computer Science and Information Technology (IJCSIT) 3: 1-3.
4. Jovanov E, Milenkovic A, Otto C, De Groen PC (2005) A wireless body area network of intelligent motion sensors for computer assisted physical rehabilitation. Journal of NeuroEngineering and Rehabilitation 2: 1-6.
5. Konstantas D, Halteren AV, Bults R, Wac K, Jones V, et al. (2004) Mobile health: Ambulant Patient Monitoring Over Public Wireless Networks. Mediterranean Conference on Medical and Biological Engineering.
6. Choi JM, Choi BH, Seo JW, Sohn RH, Ryu MS, et al. (2004) System for Ubiquitous Health Monitoring in the Bedroom via a Bluetooth Network and Wireless LAN. Engineering in Medicine and Biology Society 2: 3362-3365.
7. Farella E, Pieracci A , Brunelli D, Benini L, Ricco B, et al. (2005) Design and implementation of WiMoCA node for a body area wireless sensor network. Proceedings of the 2005 Systems Communications 2: 342-347.
8. Morón MJ, Luque JR, Botella AA, Cuberos EJ, Casilari E (2007) A Smart Phone-based Personal Area Network for Remote Monitoring of Biosignals. 4th International Workshop on Wearable and Implantable Body Sensor Networks 13: 116-121.
9. Dai S, Zhang Y (2006) Wireless Physiological Multi-parameter Monitoring System Based on Mobile Communication Networks. 19th IEEE Symposium on Computer-Based Medical Systems Based on Mobile Communication Networks, Washington.
10. Lee JW, Jung JY (2007) ZigBee Device Design and Implementation for Context-Aware UHealthcare System. The IEEE 2nd International Conference on Systems and Networks Communications, Cap Esterel, French.
11. Yang G (2006) Body Sensor Networks. (1stedn), Springer.
12. Lloyd-Jones D, Adams R, Carnethon M (2009) Heart disease and stroke statistics-update: a report from the American heart association statistics committee and stroke statistic subcommittee. Circulation 119: 40-41.
13. www.dagurobot.com
14. www.robotpower.com
15. www.iteadstudio.com

26

Asymmetrical Performance and Abnormal Synergies of the Post-Stroke Patient Wearing SCRIPT Passive Orthosis in Calibration, Exercise and Energy Evaluation

Yun Qin*, Naila Rahman and Farshid Amirabdollahian

School of Computer Science, University of Hertfordshire, Hatfield, Hertfordshire, UK

Abstract

In the context of therapeutic human-robot interaction, it is important to detect human contribution in interaction with robots, thus to auto-tune a robot to compensate or resist based on such input. A passive orthosis is used to evaluate interaction based on kinematic data and energy flow model to identify human-contributions during interaction experiments with healthy subject and stroke patient. The results identified presence of abnormal synergies between wrist and fingers, showed a skewness apparent in stroke patient performance which seemed to decrease over-time after the rehabilitation practice and indicated lack of fine control. We hypothesise that the presented methods can be used as potential performance benchmarks allowing to identify subject's contribution during an interaction session but also to observe extent of fine motor control over time.

Keywords: Energy flow; Passive orthosis; Stroke rehabilitation; Abnormal synergies

Introduction

Stroke can cause damages to motor regions in the brain resulting in loss of control to upper and lower limbs. The impairment after a stroke event including weakness of muscles, abnormal muscle tone, damage of motion range, abnormal movement synergies and loss of sensation, has an influence on quality of life and activities of daily living [1-3].

Using robot-mediated rehabilitation has witnessed twenty-five years of development [4-7]. The functions of the robotic rehabilitation systems have varied among delivering repetitive trainings to relearning lost motor skills. Studies have mostly focused on repetitive training of reaching tasks for upper limb, while wrist and hand training has been more limited due to the inherent complexity of designing grasping tools. Hesse et al. developed a robot-assisted arm trainer for the passive and active practice of forearm and wrist movements in hemiparetic subjects and this research showed promising results while making intensive training of elbow and wrist possible [8]. In a different study, a wrist extension has been designed to complement the MIT-MANUS robotic device for the proximal arm (InMotion ARMTM) [9]. This study showed promising results highlighting added benefits of wrist exercise after a period of 6 weeks training. A further study from this group investigated if administering six-weeks of hand and wrist therapy before six-weeks of shoulder and elbow had a different outcome compared to a reverse order (shoulder and elbow first, then hand and wrist). While the clinical scores between the two approaches were similar, the study showed that in terms of generalizability and transfer of skills, training of the distal limbs led to twice as much carryover effects [10]. After stroke, wrist, hand and fingers extensions are often especially affected, and these are often the last symptoms to show some improvement. Therefore, a training environment for distal control for grasping and manipulation of objects has a large potential in contributing to such improvements.

Different methods including unactuation, passive actuation and active actuation have been widely used in the robotic post-stroke rehabilitation systems. Carmeli et al. proposed an unactuated device HandTutor which also provided visual and auditory feedback [11]. Passive actuators have also been used in rehabilitation systems. Springs were involved in the orthotic aided training SaeboFlex device developed by Farrell et al. [12]. Functional electrical stimulator which triggers muscles in the hand and wrist has been used in neuro-rehabilitation [13]. Loureiro and Harwin presented a 9-DOF reach and grasp device for neuro-rehabilitation [14]. A hand-wrist robotic manipulator with electric motors was developed by Takahashi et al. [15]. These studies focused on multiple approaches to hand and wrist rehabilitation, while with the exception of SaeboFlex, none were developed with the particular aim of supporting home-based rehabilitation.

The SCRIPT (Supervised Care & Rehabilitation Involving Personal Tele-robotics) project, partially funded by the European Community focuses on improving recovery gains of hand and wrist for chronic stroke survivors through larger repetitions and frequent exercises [16]. A home-based device SCRIPT passive orthosis for the hand and wrist rehabilitation [17] is delivered together with several interactive therapeutic exercise games operated by different movements such as flexion and extension of the wrist and fingers [18].

In the SCRIPT project, we utilize detailed knowledge on measures of energy and energy flow from the participant to the orthosis or vice versa which is essential to identify the extent of 4 human contribution during an interaction session. We rely that awareness of human contribution makes the interaction more meaningful and personalized. We hypothesise that such knowledge provides a good indicator for changes in ability to isolate movements of wrist and fingers during hand and wrist articulation and can thus be used as an indicator in assessing recovery. Mak, Gomes and Johnson proposed a model to

***Corresponding author:** Yun Qin, Adaptive Systems Research Group, School of Computer Science, University of Hertfordshire, Hatfield, Herts AL10 9AB, UK
E-mail: qinyun19850803@hotmail.com

study the interaction between a patient and a rehabilitation robot and their study showed the potential for interpreting changes during rehabilitation [19]. Tzemanaki et al. developed a robotic system for stroke and post hand-surgery patient rehabilitation which allows patients to gradually regain flexibility in their finger-joints by passive extension and flexion of their fingers and dynamic models of a human hand have been derived by their research [20].

In order to auto-tune an orthosis to compensate or resist for the therapeutic human-orthosis interaction, identifying the human contribution in interaction with orthosis is crucial. In this paper, the evaluation of the SCRIPT passive orthosis interaction based on raw kinematic measures, temporal aspects of movement and an energy flow model is used as a proof of concept highlighting applicability of these approaches in identifying human-contributions. An interaction experiment is conducted and discussed in support of this proof of concept study. The structure of the paper is as follows. In the Introduction, the SCRIPT system and its training environment is briefly introduced. The experiment plan is introduced in the Material and Methods. The motion comparison results for the healthy subject and stroke patient during calibration and an exercise game are discussed and the energy flow evaluation results for both subjects are also presented in the Results and Analysis. Then discussion is given in Discussions.

Script system

The SCRIPT tele-robotic upper-limb rehabilitation system as shown in Figure 1 consists of a passive hand, wrist and forearm exoskeleton customised to patient's hand-size [1] mounted on a Saebo Mobile Arm Support [21] and attached to a Windows based personal computer (PC) which allows patients to interact with a range of therapeutic motivational games using their affected upper-limb. The patient's therapy is planned and monitored offline remotely by a healthcare professional using a dedicated interface.

SCRIPT device

Leaf springs and elastic cords are used to apply passive torques to assist extension of the five fingers. Use of elastic bands allows the passive torques to be varied depending on the degree of weakness exhibited by the patient. A potentiometer is used to measure the angle of wrist extension and flexion. Flex sensors are used to measure the flexion of the fingers. An Arduino Nano microprocessor board is used to sample and convert data from the potentiometer and flexion sensors from analog to digital values. An inertial measurement unit (IMU) is used to measure pronation, supination and three-dimensional position

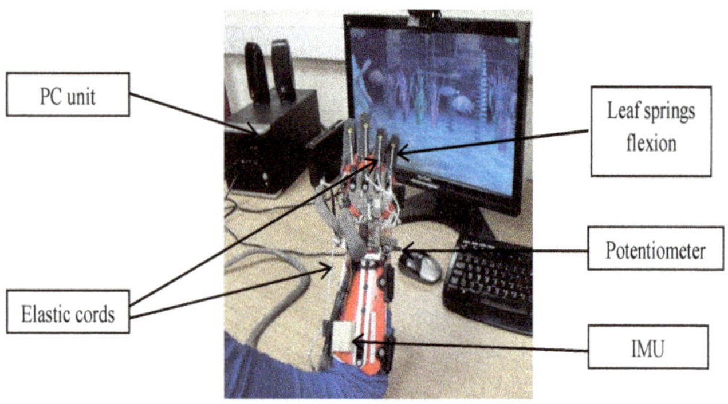

Figure 1: SCRIPT system

of the hand.

In order to allow standardisation, each device is calibrated to ensure that raw digital data is converted into normalised values that provide joint angles. The range for normalised angles for the wrist and fingers is from zero to ninety degrees.

SCRIPT patient PC training environment

The patient training environment allows users to complete a training plan prescribed by their therapist, to monitor their own progress and to communicate with the therapist. The training plan consists of interactive therapeutic motivational games [22] which are controlled by the patient while wearing the device.

A server process on the PC acquires data from the SCRIPT device at 30 Hz and processes it to recognise hand, wrist and arm movements or gestures. Due to the passive nature of the device, the sampling rate of 30 Hz was seen as adequate for this interaction. The time-stamped device data and recognised gestures are stored in a normalised MySQL database. The gesture data is sent to a game server process and is used to control games.

Game calibration: The range of motion exhibited by people after stroke varies significantly from healthy individuals and is mainly dependent on the degree of loss of motor function. However, the range of motion for a patient is not static, as there may be an increase due to recovery or a decrease due to fatigue or lack of motivation.

Before each session of the game, the participant is required to perform a few repetitions of a movement such as fingers and wrist flexion and extension under a calibration procedure. The purpose is to record a baseline measure prior to each exercise session, while providing a chance to make the game adaptive to the level of subject's available range of motion on the day. The procedure analyses the duration and amplitude of the ranges recorded. During these calibration steps, the user is instructed to perform, in isolation, gestures used to control a game. The device data readings are then used to determine minimum and maximum values of joint angles and to recognize valid gestures during game play. During the exercise game, the fingers flexion and extension postures can be detected according to the subject's personalized 7 range of motion measured during calibration. On the other hand, the calibration also makes sure that the participant exercises in a range close to 90% of his/her active range of motion at an achievable speed which prevents the game from being too easy or too challenging. The gestures calibrated are the hand/wrist flexion and extension, forearm lateral movements (left and right movement of the forearm), forearm pronation and supination and hand drop and lift.

Motivational therapeutic games: The overall objective of the SCRIPT project is to encourage patients to perform multiple repetitions of functional movements with their affected arm and hand for at least 180 minutes each week. In order to keep the patient motivated and engaged, the system provides therapeutic games for training. The current phase of the project provides three games: Sea Shell; Super Crocco and Labyrinth. There are three variants of the Super Crocco and Labyrinth games, designed to encourage the patient to perform increasingly difficult functional movements as they recover motor function. This follows the Gentile's taxonomy [23] starting with movement with simple functions, such as fingers flexion and progressing towards gestures performed in motion i.e. lateral movement of the hand while grasping a key. The architecture also allows new games to be added to the system. Figure 2 presents the SCRIPT calibration screen and the Sea Shell game.

Before playing each game, the user is required to perform a calibration step. The game is then played in five minute blocks and the user is advised to rest between blocks.

Material and Methods

As a proof of concept study, one healthy participant and one patient recovering from stroke were considered for this initial comparative study. The healthy participant was recruited from the University of Hertfordshire, United Kingdom under the ethical approval number COM/ST/UH/00006. The stroke patient was selected from the pool of participants currently undertaking a 6 weeks summative evaluation ethically approved by the Medical Ethics Committee under approval number NL42483.044.12. Patients included in the summative study were chronic stroke (>6 months after stroke) with unilateral ischemic or haemorrhagic stroke between 18-80 years old, clinically diagnosed with central paresis of the hand. They had a minimum of 15 degrees of elbow flexion, with about a quarter range of finger flexion with a fair cognitive level to allow them to read, understand and follow instruction. Both participants provided a signed informed consent.

The main objective of this experiment was to compare the calibration and game play data recorded during one interaction session, between the two participants. The healthy participant, coded as S1 in this study only completed one interaction session with the SCRIPT device as required. Participating patient, coded as S2, was involved in the summative evaluation and played the Sea Shell game in different sessions among the six weeks. Session 1 of the game is the first session among the six-week period when S2 started to play the game for the first time. The reason for picking up this session is to make sure both participants had the same extent of familiarity with the system within a session, which correspondingly lasted for five minutes. Another set of S2 game playing data was collected during session 11 after the subject had played ten sessions of Sea Shell game. This was to monitor if any changes in interaction parameters could be observed over time and with increased familiarity. The 9 choice of first and last session is similar to a large number of clinical studies in robotics for rehabilitation where pre and post intervention indicators are compared. The spatial features, temporal features and flow of energy are compared by the numerical data deriving from bending sensor and the numerical result of energy flow calculation. Table 1 highlights study schedule and sessions used to extract data in support of this experiment.

Results and Analysis

Recorded interaction data for this experiment was extracted from the MySQL database used to store data. These were used to conduct the next three analyses.

Comparison during calibration phase

The calibration procedure provided the user's baseline range of motion. The comparable experiment was done between the two subjects when they did the calibration for the first time. The calibration for S1 lasted for 8.5 seconds, while the calibration for S2 in session 1 lasted for 25 seconds until the SCRIPT system was able to recognize the range of motion. Due to the high similarity among five fingers, the result for one finger is illustrated in this paper. Figure 3 provides an observation of the index finger and wrist flexion angle when the participant was asked to perform only finger flexion and extension.

From Figures 3a and 3b, we could see that S1 is able to achieve 90 degrees range of motion and on the contrary to that, S2 had a limited range of motion up to 60 degrees. Regarding the temporal aspects, if we only focus on the time between start to the 8.5 seconds, S1 10 accomplished six rounds of finger flexion and extensions and S2 only finished five repetitions. From Figures 3b and 3c, the SCRIPT system took 18.6 seconds to recognise S2's amplitude of the motion range in session 11, compared to 25 seconds in session 1. These two figures highlight that with increased familiarity of the game, S2 has improved in performing the calibration procedure.

Figure 3 also shows the index finger and wrist flexion angles when the participants were directed to do only the finger flexion and extension. From Figure 3a, it is notable that S1 had the ability to near perfectly isolate the movement of the finger and wrist and s/he exactly did the required movement where the index finger flexed and extended while his wrist was kept at a fixed position. Based on Figure 3b, each local maxima and minima of the Index Finger curve was followed by the flexion/extension of the Wrist. The index finger movement of post-stroke patient was accompanied by the wrist movement.

Comparison during game play

The therapeutic objective of the Sea Shell game is to encourage patients to extend and flex their wrist and fingers. The game manoeuvres a sea shell to catch fishes underwater by opening and closing the sea shell. The goal of the game is to catch as many fishes as possible when the fishes arrive near the sea shell. The sea shell opening and closing is operated by flexing and extending fingers. Considering post-stroke patients tendency to flex their wrist due to an inability to relax the flexors or to engage the extensors sufficiently [1], the game is also designed to encourage the patient to extend their wrist before presenting finger flexion/extension sufficient to catch a fish. When the wrist is flexed for a certain threshold duration (50% of the range of motion), the sea shell falls into sleep thus unable to catch any fishes

Figure 2: SCRIPT calibration and Sea Shell game environment

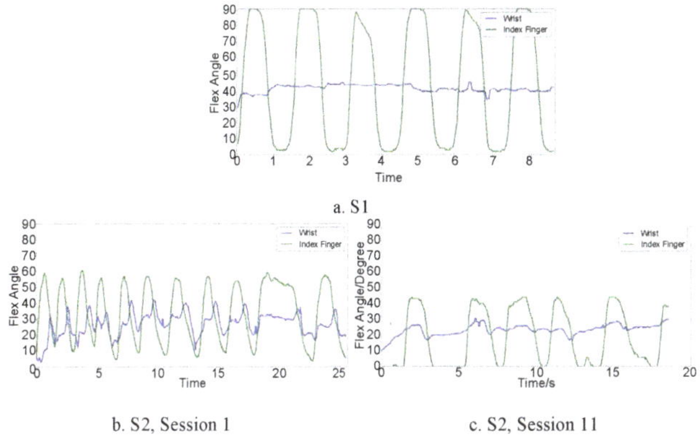

Figure 3: The flexion angle during finger flexion and extension calibration

	Calibration 1	Game 1	...	Calibration 11	Game 11
S1	☑	☑			
S2	☑	☑		☑	☑

Table 1: Participants completed sessions used for this experiment

thus encouraging wrist extension prior to flexion of fingers .

The flexion angles of index finger and wrist during Sea Shell game for the healthy subject (session 1) and post-stroke patient (sessions 1 and 11) are illustrated with respect to time in Figure 4a-4c. When one target (fish) appeared on the screen, subjects might perform several grasping movements until they felt confident for successfully catching the fish. Each grasping movement was corresponding to one visible peak in the graph. The movement of the subjects for a given target hit (fish caught) might include a position profile with multiple convexity.

11 flexion/extension sufficient to catch a fish. When the wrist is flexed for a certain threshold duration (50% of the range of motion), the sea shell falls into sleep thus unable to catch any fishes thus encouraging wrist extension prior to flexion of fingers .

The flexion angles of index finger and wrist during Sea Shell game for the healthy subject (session 1) and post-stroke patient (sessions 1 and 11) are illustrated with respect to time in Figure 4a-4c. When one target (fish) appeared on the screen, subjects might perform several grasping movements until they felt confident for successfully catching the fish. Each grasping movement was corresponding to one visible peak in the graph. The movement of the subjects for a given target hit (fish caught) might include a position profile with multiple convexity.

From Figure 4, it can be found that in a 100-seconds window selected arbitrarily, S1 was able to perform 23 successful grasping and S2 could only perform 11 successful grasping. After the ten sessions training, the total successful grasping achieved by S2 was 18. This presented an increase of 64% in game performance over a period of 10 sessions as reflected by the number of grasping. The abnormal synergies of the index finger flexion movement and wrist flexion movement for S2 can also be observed in Figure 4. Figure 5 presents one of the convexities magnified in a new time-window starting from zero.

From Figure 5a, it is trivial to notice that for S1, his/her index finger flexion angles during flexion and extension were highly symmetrical. Figures 5b and 5c shows that S2's index finger flexion angles were asymmetrical for flexion and extension with respect to its mean. Furthermore, we use skewness to reflect the extent of asymmetry as

By calculating the skewness for the data set of index finger flexion angles of S2 game playing session 1, the Index Finger curve was positively skewed with the skewness of 0.424. While the skewness for the S1 was very close to zero (0.004). After ten training sessions, the skewness for the data set of S2's index finger flexion angles in Sea Shell game playing session 11 was 0.171.

Energy flow evaluation

By using the similar approach as of Mak et al. [19], the energy amount and energy flow for the index finger and wrist for post-stroke patient and healthy subject during Sea Shell game playing is compared in this Results and Analysis. Fingers flexion angles were recorded by the five bending sensors and the wrist flexion angle was measured by the potentiometer in the SCRIPT device and the force applied by the elastic band to the finger tips was calculated using the modulus of elasticity. The known parameters also included the length, velocity, acceleration and mass of the fingers and palm. This allowed the authors to derive complete forward kinematic and dynamic models for the hand-orthosis system. Newton-Euler formulation was used to analyse the forward dynamic behaviour of the hand-orthosis system and balance of linear forces and balance of moments equations of the fingers and wrist were built up, respectively.

In the numerical evaluation, the data were digitally low-pass filtered with a 3rd-order Butterworth filter and passed through the formulation to calculate finger and wrist moments during the Sea Shell game. For the purpose of this comparison, one successful grasping was selected and data from this interaction was used within the calculation to obtain a moment/moment plot for finger and wrist flexion/extension. Pre-post comparison of the energy flow evaluation for the patient was accomplished to investigate his improvements between the first and last sessions. The energy involved during S1 game playing and S2 game playing in sessions 1 and 11 is illustrated in Figure 6.

From Figure 6a, the Flexion curve for S1 was smooth and one crossover happened temporarily when the index finger nearly reached the maximum flexing position and the Extension curve was also relatively smooth for this participant. From Figures 6b and 6c, we could see that there was one more peak in the Flexion curve compared to the

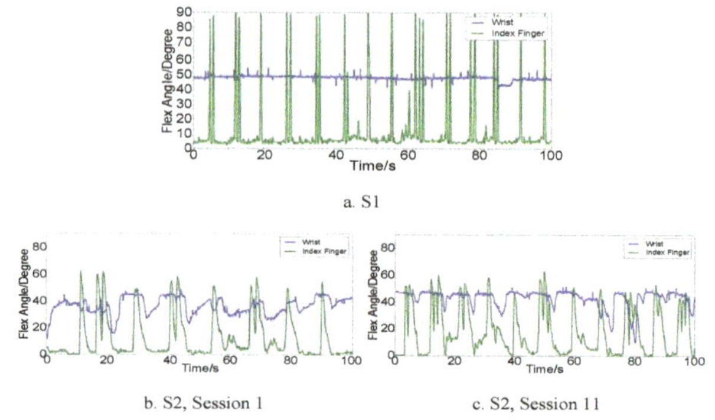

Figure 4: The flexion angles during Sea Shell game

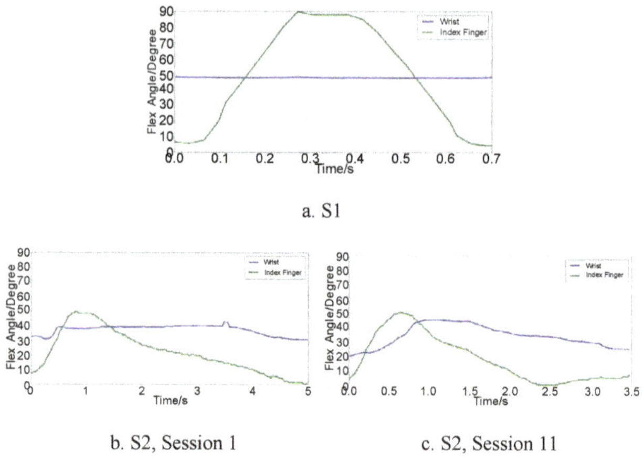

Figure 5: The flexion angles during one grasping in Sea Shell game

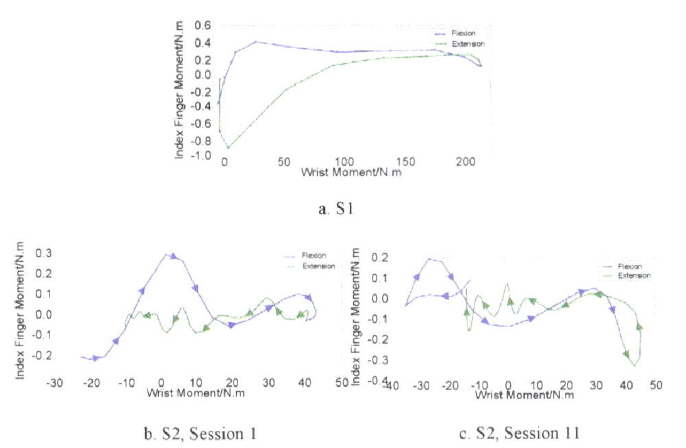

Figure 6: The energy involved in game play

S1 and there were many local maxima and minima in the Extension curve.

Discussions

Comparisons between S1 and S2 during calibration highlighted longer calibration time required by S2, and S2's calibration time became shorter after 10 training sessions compared to the beginning of his exercise, while within a normalised time window, S1 had achieved more repetitions of the flexion and extension task. The calibration data allows us to monitor person's status prior to game play and here highlight patient compliance, either due to becoming more familiar with the SCRIPT1 system and games, or due to changes in their underlying recovery trends.

More interestingly, it was observed that S1 was capable of limiting wrist movements while performing the hand flexion while S2 presented abnormal wrist synergies during the grasping gestures in calibration session 1. Importantly, this seems substantially improved during calibration 11. Session 11 presented a picture closer to that of the healthy subject, with better control over wrist flexion, during flexion of the fingers, highlighting a reduced level of abnormal synergies.

When game play was considered for the comparison, S1 presented better performance as reflected by the number of grasping, while S2 presented an improvement of over 64% in achieving interaction objectives when comparing session 1 to session 11. Again, this could be due to compliance related to familiarity or underlying recovery.

With a closer look into game play based on temporal properties of the movement, it was observed that a positive skewness was apparent in S2 game play, which could potentially effect successful catching of the fishes due to bad timing of the sea shell opening and closing. However, an improvement was also observed based on this parameter when comparing sessions 1 and 11. The skewness parameter indicated a reduction towards that of the healthy subject. This parameter can be applied to each movement thus providing more insights into movement's symmetry. It is thought that movement asymmetry is a good measure of movement coordination [24] thus this parameter can inform us regarding improvement in the extent of movement coordination after successive game play sessions.

When energy flow modelling was used, S1 presents a clearly well-controlled grasping while wrist activities are kept to a minimum, while S2 presents synergistic wrist moments during flexion of the hand. More interestingly, here we can observe a difference between Flexion and Extension, indicative of better controlled flexion of the fingers when compared to the extension. This is in line with observations by Cruz et al. [25] highlighting difficulties in finger extension tasks after stroke, when compared to finger flexion. Surprisingly, this parameter also shows improvement at the level of finger moment vs wrist moment variation at session 11, which is in line with the observations made when comparing spatial and temporal features of the movement.

Another notable observation here is that S1 wrist moments during extension remain on the positive quadrant, while S2 wrist moments vary from positive to negative moment. Here reduction in wrist moment is associated with the extension assistance provided by the wrist extension elastic cord. The assistance provided by this cord is at minimum for a fully extended wrist. Moreover, finger moments present more fluctuations when compared to the wrist. These indicate exchange of energy between the device and the patient during finger extension, clearly indicating device's contribution in assisting a full extension.

Presence of abnormal synergies have been observed and classified by other studies for example work from Dewald et al. [26] presented shoulder and elbow abnormal synergies identified using EMG electrodes while Dipietro et al. hypothesised that changes in the proximal arm abnormal synergies could be potentially used as a measure of recovery [27]. This study focused on the distal arm and identified multiple areas where presence of such synergies is visible. The current study has the limitation of small subject numbers and relatively short follow-up period. In order to identify changes in abnormal wrist and hand synergies over time, the current study is being extended to include a larger sample size of healthy volunteers and stroke patients using the system over time, while also considering session-by-session improvements during a clinical evaluation of 24 stroke patients. The 16 intention is to identify any supporting evidence for reduction of abnormal synergies over time. We intend to consider the characteristics of the moment/moment plots, e.g. number of local minima and maxima and the area under the curve, to further expand on the usefulness of this approach for wrist/hand interactions. We also intend to further explore the use of skeweness as a measure of improvement in movement coordination, as it can reflect on the extent of fine control, leading to successful achievement of game objectives. Furthermore, considering change of moment alongside changes in active range of motion, so moment/angle plots is another dimension for our future investigations.

Conclusions

In this paper, finger motion during calibration and game play as well as the energy involved in the game play were compared between a healthy subject (S1) and a post-stroke patient (S2). By comparing the index finger flexion angle during calibration, we found that the post-stroke patient had a reduced (2/3) range of motion for the index finger compared to the S1 and the speed of the movement for the S2 was slower than S1. After analysing the data during the game play, it was notable that the S2 presented index finger flexion angles that were asymmetrical for flexion and extension motion. Skewness was used to reflect the extent of asymmetry and movement coordination and showed improvements between session 1 and session 11. The index finger movement of the S2 was always accompanied by the wrist movement in both calibration and game playing and this indicated the presence of abnormal synergies.

The energy amount and energy flow of the index finger and

wrist for both subjects during game playing were also evaluated. The resultant moment/moment plot for S1 was generally smooth and only had one temporary crossover when the motion changed from flexion to extension hinting on momentary wrist involvement. Given that many local maxima and minima were present for S2, we concluded that these are linked to the patient's inability to control the finger flexion in isolation with wrist, while highlighting the device effect in extending the fingers using elastic cords. Our results are evidence for the presence of abnormal synergies. Furthermore, we found that S2 had worse control for the extension process than flexion process, which was anticipated.

The asymmetrical performance between flexion and extension and abnormal synergies for higher number of post-stroke patients will be analysed in future work. Furthermore, changes in angle/moments are considered for further explorations towards identifying positive changes in the extent of movement over time. We also aim to explore the lack of ability to control as a benchmark to evaluate the rehabilitation recovery and improvement of using the SCRIPT system.

References

1. Ates S, Lobo-Prat J, Lammertse P, Kooij H, Stienen AHA (2013) SCRIPT passive orthosis: design and technical evaluation of the wrist and hand orthosis for rehabilitation training at home. International Conference on Rehabilitation Robotics, Seattle, Washington, USA.
2. Penfield W, Rasmussen T (1950) The cerebral cortex of man; a clinical study of localization of function. JAMA 144: 1412
3. Elbaum J, Benson D (2007) Acquired brain injury.
4. Kwakkel G, Kollen BJ, van der Grond J, Prevo AJH (2003) Probability of regaining dexterity in the flaccid upper limb - Impact of severity of paresis and time since onset in acute stroke. Stroke 34: 2181-2186.
5. Lum PS, Burgar CG, Kenny DE, Van der Loos HFM (1999) Quantification of force abnormalities during passive and active-assisted upper-limb reaching movements in poststroke hemiparesis. IEEE Trans. Biomedical Engineering 46: 652-662.
6. Coote S, Stokes E (2001) Physiotherapy for upper extremity dysfunction following stroke. Physical therapy reviews 6: 63-69.
7. Reinkensmeyer DJ, Schmit BD, Rymer WZ (1999) Assessment of active and passive restraint during guided reaching after chronic brain injury. Annals of Biomedical Engineering 27: 805-814.
8. Hesse S, Schulte-Tigges G, Konrad M, Bardeleben A, Werner C (2003) Robot-assisted arm trainer for the passive and active practice of bilateral forearm and wrist movements in hemiparetic subjects. Archives of Physical Medicine and Rehabilitation 84: 915-920
9. Krebs HI, Celestino J, Williams D, Ferraro M, Volpe B, et al. (2004) A wrist extension for MIT-MANUS. Advances in Rehabilitation Robotics 306: 377-390.
10. Krebs HI, Volpe BT, Williams D, Celestino J, Charles SK, et al. (2007) Robot-aided neurorehabilitation: A robot for wrist rehabilitation. IEEE Trans. Neural Syst. Rehab. Eng. 15: 327-335.
11. Carmeli E, Peleg S, Bartur G, Elbo E, Vatine JJ (2010) HandTutor (TM) enhanced hand rehabilitation after stroke - a pilot study. Physiother. Res. Int.
12. Farrell JF, Hoffman HB, Snyder JL, Giuliani CA, Bohannon RW (2007) Orthotic aided training of the paretic upper limb in chronic stroke: results of a phase 1 trial. Neuro Rehabilitation 22: 99-103.
13. Sheffler L, Chae J (2007) Neuromuscular electrical stimulation in neurorehabilitation. Muscle Nerve 35: 562-590.
14. Loureiro RCV, Harwin WS (2007) Reach & grasp therapy: design and control of a 9-DOF robotic neuro-rehabilitation system. IEEE 10th International Conference on Rehabilitation Robotics ICORR 2007: 757-763.
15. Takahashi CD, Der-Yeghiaian L, Le V, Motiwala RR, Cramer SC (2008) Robot-based hand motor therapy after stroke. Brain 131: 425-437.
16. Prange GB, Hermens HJ, Schafer J, Nasr N, Mountain G et al. (2012) Script: Tele-robotics at home - functional architecture and clinical application. 6th International Symposium on E-Health Services and Technologies (EHST), Geneva, Switzerland.
17. Schäfer J, Klein P, Prange G, Amirabdollahian F (2013) Script: Hand & wrist tele-reha for stroke patients involving personal tele-robotics. Proceedings of the Technically Assisted Rehabilitation (TAR) 2013 Conference, Berlin.
18. Basteris A, Amirabdollahian F (2013) Adaptive human-robot interaction based on lag-lead modelling for home-based stroke rehabilitation. IEEE Systems, Man. and Cybernetics, Manchester, UK.
19. Mak P, Gomes GT, Johnson GR (2002) A robotic approach to neuro-rehabilitation- interpretation of biomechanical data. 7th International Symposium on the 3D Analysis of Human Movement, Centre for Life, Newcastle upon Tyne, UK.
20. Tzemanaki A, Raabe D, Dogramadzi S (2011) Development of a novel robotic system for hand rehabilitation. 24th International Symposium on Computer-Based Medical Systems (CBMS), Bristol.
21. https://www.saebo.com/products/saebomas/
22. Steffen A, Schäfer J, Amirabdollahian F (2013) Script: usability of hand & wrist tele-rehabilitation for stroke patients involving personal tele-robotics.
23. Gentile AM (1987) Skill acquisition. Foundations for Physical Therapy - Movement Science, Heineman Physiotherapy, London
24. Jeka JJ, Kelso JAS (1995) Manipulating symmetry in the coordination dynamics of human movement. Journal of Experimental Psychology: Human Perception and Performance 21: 360-374.
25. Cruz EG, Waldinger HC, Kamper D G (2005) Kinetic and kinematic workspaces of the index finger following stroke. Brain 128: 1112-1121.
26. Dewald JP, Pope PS, Given JD, Buchanan TS, Rymer, WZ (1995) Abnormal muscle coactivation patterns during isometric torque generation at the elbow and shoulder in hemiparetic subjects. Brain 118: 495-510.
27. Dipietro L, Krebs HI, Fasoli SE, Volpe BT, Stein J, et al. (2007) Changing motor synergies in chronic stroke. Journal of neurophysiology 98: 757-768.

Artificial Neural Network Based Forward Kinematics Solution for Planar Parallel Manipulators Passing through Singular Configuration

Ammar H Elsheikh*, Ezzat A Showaib and Abd Elwahed M Asar

Department of Production Engineering and Mechanical Design, Faculty of Engineering, Tanta University, Tanta, Egypt

Abstract

It is well known that, the main drawback of parallel manipulators is the existence of singularities within its workspace, an Artificial Neural Network (ANN) based solution is proposed in this paper. The proposed approach can certainly learn the input-output data and discover the non-linear relationships which are inherent in the training data. Additionally, the proposed approach can provide solution of the forward kinematic problem with reasonable errors at and in the vicinity of kinematic singularities. The approach is implemented for the 3-RPR, 3-PRR, and 3-RRR planar parallel manipulators.

Keywords: Parallel manipulators; Forward kinematics; Singularities; Artificial Neural Network (ANN)

Introduction

Parallel manipulators, due to its closed-loop structure, posses a number of advantages over traditional serial manipulators such as high rigidity, high load-to-weight ratio, high natural frequencies, high speed and high accuracy [1]. However, they also have a few disadvantages such as a relatively small workspace, relatively complex forward kinematics and the most importantly, existence of singularities inside the workspace [2]. Kinematics analysis of parallel manipulators separate in two types, forward kinematics and inverse kinematics. The inverse kinematics, which maps the task space to joint space, is not difficult to solve. On the other hand, the forward kinematics, which maps the joint space to task space, is so hard to solve. Also, the existence of not only multiple inverse kinematic solutions (or working modes) but also multiple forward kinematic solutions (or assembly modes) is another problem in kinematics analysis [3]. The challenging problem is not to find all possible solutions but to directly determine the unique feasible solutions, the actual physical solution, in among all possible solutions starting from a certain initial configuration [4].

Forward kinematics and singularity analysis of planar parallel manipulators have been investigated by many researchers [5-7]. Efforts to solve the forward kinematics of planar parallel manipulators have concentrated on 3-RPR manipulator due to its inherent simplicity. It is established the forward kinematic solution of general 3 DOF planar parallel manipulators can be lead to a polynomial of degree 8 [8]. However, the forward kinematic problem for the manipulator under study leads to a maximum of 6 real solutions. It is worth taking into considerations, the three manipulators under study are kinematically equivalent to each other and, as a result, we derived the forward kinematics equations for 3-RRR and modified it to the two other manipulators. Additionally, the existence of singularities and uncertainties inside the workspace where the manipulator gains some degrees of freedom and become uncountable. In such configurations, the actuated joints forces of the manipulator will become unacceptably large that often reach their allowable limits. To overcome the problem of kinematics singularities a neural network –based approach is developed which has the ability of generalization and can successfully learn relationships that are not present in the training set in an efficient manner.

There have been increasing research interests of Artificial Neural Networks (ANNs) due to their extreme flexibility and the capability of non-linear function approximation. Many efforts have been made on applications of Neural Networks to various types of parallel manipulators [9-13].

In this paper, a supervised neural network approach is developed to control the motion of the 3-RPR, 3-PRR and 3-RRR planar parallel manipulators. Multiple neural networks are used to overcome the problem of the multiple solution branches of either forward or inverse kinematics. This approach also overcomes the problems of singularities and uncertainties' arising in trajectory planning as it has, like any ANN algorithms, generalization ability. In this approach a network is trained using training data generated from the inverse kinematics. The training is done off-line until reaching acceptable error and a validation test is also done, at each iteration, to avoid model over fitting. It may be noted here that the present work may be considered as an implementation of the artificial neural network approach for serial manipulators passing through singular configuration, as proposed by [14], for planar parallel manipulators.

Kinematics of Parallel Manipulators

Kinematic analysis of parallel manipulators includes solution to forward and inverse kinematic problems. The forward kinematics of a manipulator deals with the computation of the position and orientation of the manipulator end-effector in terms of the active joints variables. Forward kinematic analysis is one of essential parts in control and simulation of parallel manipulators. Contrary to the forward kinematics, the inverse kinematics problem deal with the determination of the joint variables corresponding to any specified position and orientation of the end-effector. The inverse kinematics problem is

***Corresponding author:** Ammar H. Elsheikh, Department of Production Engineering and Mechanical Design, Faculty of Engineering, Tanta University, Tanta, Egypt, E-mail: eng_ammar_sheikh@yahoo.com essential to execute manipulation tasks. Most parallel manipulators can admit not only multiple inverse kinematic solutions, but also multiple forward kinematic solutions. This property produces more complicated kinematic models but allows more flexibility in trajectory planning [15]. In other words, a manipulator configuration can be defined either by actuator coordinates $q=[q_1, .., q_n]^T$ or by Cartesian end-effector coordinates $x= [x_1, .., x_n]^T$ with n the DOF of the manipulator under study. The transformation between actuator coordinates and Cartesian coordinates is an important issue from viewpoint of kinematic control. Computation of the end-effector coordinates from given actuator coordinates (forward kinematics) can be written in the general form

$$x= f_{FKP}(q) \qquad (1)$$

The inverse task which is to establish the actuator coordinates corresponding to a given set of end effector coordinates (inverse kinematics) can be also written in the general form

$$q= f_{IKP}(x) \qquad (2)$$

Then the kinematic constraints imposed by the limbs can be written in the general form

$$f(x,q)=0 \qquad (3)$$

Differentiating Eq.(3) with respect to time, we obtain a relationship between the input joint rates and the end-effector output velocity

$$J_x\dot{x} = J_q\dot{q}$$

Where

$$J_x= \frac{\partial f}{\partial x} \text{ and } J_q= \frac{\partial f}{\partial q}$$

Inverse kinematic singularity occurs when different inverse kinematic solutions coincide that happens usually at the workspace boundary. Hence the manipulator loses one or more degrees of freedom. Mathematically they can detected by det $(J_q)=0$

Forward kinematic singularity occurs when different forward kinematic solutions coincide. Hence the manipulator gains one or more degrees of freedom. That happens inside the workspace so it is a great problem. Mathematically they can detected by det $(J_x)=0$

Manipulators Under Study

The architectures of the planar parallel manipulators under study, 3-RPR, 3-PRR and 3-RRR, are illustrated in Figures 1a-1c, Where R, P, R and P denote revolute, prismatic, actuated revolute and actuated prismatic joints, respectively. For manipulators under study the three fixed pivots A_1, A_2 and A_3 define the geometry of the fixed base, and the three moving pivots C_1, C_2 and C_3 define the geometry of the moving platform, where point O and H are the centroids of the fixed base and moving platform respectively. Three limbs connect the moving platform to the fixed base. Each limb of the 3-RPR is composed of a R, a P, and a R joint in sequence. Each limb of the 3-PRR is composed of a P, a R, and a R joint in sequence. Likewise, each limb of the 3-RRR is composed of three R joints in sequence. The origin of the fixed coordinate frame is located at point A_1.The x-axis points along the direction of A_1A_2 and the y-axis is perpendicular to A_1A_2. We assume that the manipulators under study are symmetrical, manipulators with equilateral base and moving platform [15]. The moving platform pose, i.e., its position and its orientation, is determined by means of the Cartesian coordinates vector $H=[H_x,H_y]^T$ of operation point H and angle φ, namely, the angle between C_1C_2 and the positive direction of x-axis.

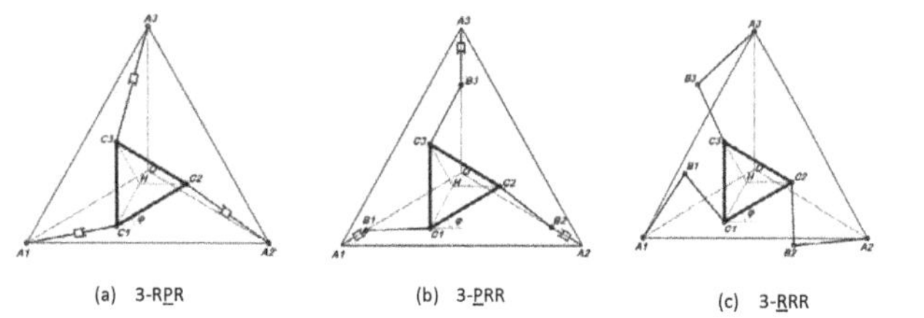

Figure 1: The three planar parallel manipulators under study.

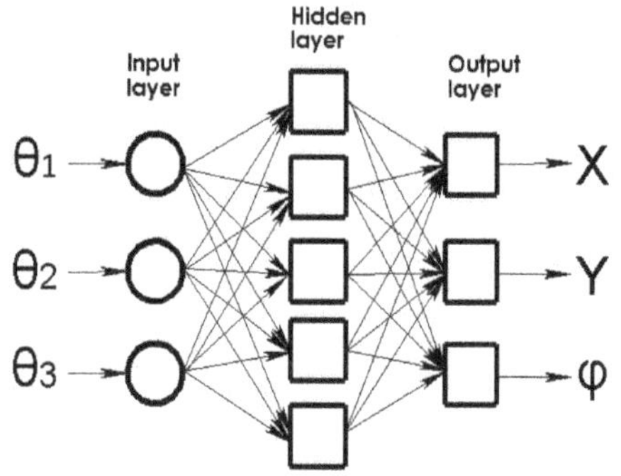

Figure 2: The topology of MLP network with a single hidden layer.

Artificial Neural Networks

Artificial neural network (ANN) is an algorithm that model brain performs a particular task, and is usually implemented using electronic components or simulated in software on digital computers. It has the ability of imitating of the mechanisms of learning and problem solving functions of the human brain which are flexible, powerful, and robust. In artificial neural networks implementation, knowledge is represented as numeric weights, which are used to gather the relationships between data that are difficult to realize analytically, and this iteratively adjusts the network parameters to minimize the sum of the squared approximation errors using a gradient descent method [14]. One category of the artificial neural networks is the multilayer perceptron (MLP) which be considered a supervised back propagation learning algorithm. It consists of an input layer, some hidden layers and an output layer as shown in Figure 2. MLP is trained by back propagation of errors between desired values and outputs of the network using some effective algorithms such as gradient descent algorithm. The network starts training after the weight factors are initialized randomly. Weight adjusting takes place until, we get reasonable errors or no more weight changes occur. There is no available theoretical procedures to choose the appreciate network architecture, i.e. number of hidden layers and number of neurons of each layer. This depends on the problem under investigation and user's experience.

Results of Numerical Simulations

Simulations have been conducted for the 3-RPR, 3-PRR, and 3-RRR planar parallel manipulators to demonstrate the performance of the developed approach. First point H (the centroid of the end-effector) is moved along a given trajectory which passing through singular locus then the correct active prismatic joint or joint angle variables to track this trajectory are calculated using the inverse kinematic model of the simulated manipulator which give a unique solution for a given working mode. Then, those active prismatic joint or joint angle variables are fed

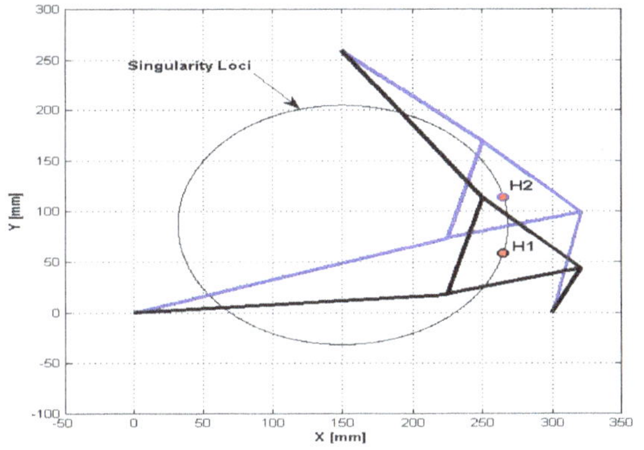

Figure 3: The 3-RPR cross singularity loci at H1(265,58.499) and H2(265,114.706).

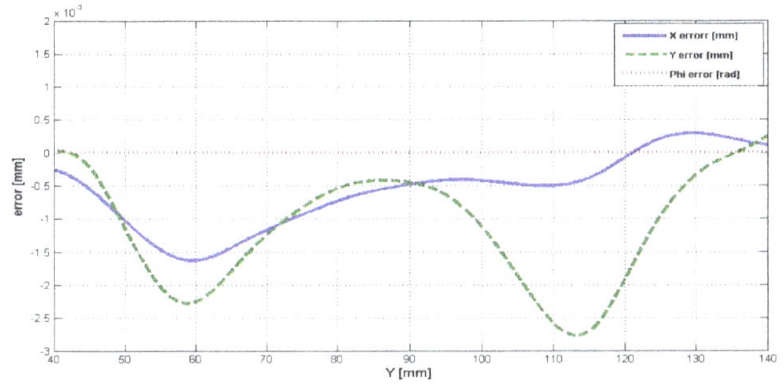

Figure 4: Tracking error for the 3-RPR along a vertical trajectory passing through singularity points H1 (265, 58.499) and H2 (265,114.706).

to the MIP to track the trajectory and the tracking errors are calculated. The simulated manipulators are assumed to be ideal mechanisms with no flexibility and no joint clearance that affect the accuracy of the manipulator. Also, the prismatic joints are assumed to have unlimited length. A two-hidden layer MLP with back propagation learning is considered here. The input layer has as many nodes as the number of inputs to the map, namely three actuator variables. Similarly the output layer will have three nodes which represent the pose of the end-effector. The number of neurons in the hidden layers and its configuration are used as a design parameter. Sigmoid and linear activation functions are used for all hidden and output layer nodes respectively. Supervised learning scheme is used in which the network is taught to learn the map by observing the inputs and outputs. The network is trained by 10,000 training input-output patterns generated, randomly within the workspace of the manipulator, from the inverse kinematic model. Random initialization is used for the weights. For each manipulator, different configurations of the MLP network were tested to get the optimal configuration used for solve the problem. About 36 multi-layer feed forward networks with two hidden layers are trained. All these networks were trained over 1,000 training epochs to ensure the success of the training process and to avoid over fitting the model. Simulation results showed that 40×60 multilayer perceptron neural network with two hidden layers had the best performance when the minimum tracking error is used as performance index. All manipulators under study are symmetric with three identical limbs. Each side of the moving end-effector equilateral triangle is 100 mm, while that of the base is 300 mm. The lengths of the proximal links and the distal links are 120 mm and 80 mm, respectively.

3 RPR planar parallel manipulator

Three end-effector trajectories are specified as straight lines which cross over singularity loci at H1(265, 58.499) mm and H2(265,114.706) mm as shown in Figure 3. The first trajectory is a vertical straight line starting at H_i(265,40) mm with orientation angle φ=15° and ending at point H_f(265,140) mm with the same orientation it is obvious the selected trajectory passes through singular points H1 and H2. The tracking errors in x- and y-directions are depicted in Figure 4. The maximum tracking error along the trajectory points is 0.0027 mm which happens in the vicinity of kinematic singularities. We also note that there is a significant increasing in the tracking error near the singularity points. Anyway, the developed approach can provide solution for the problem with reasonable errors.

The second trajectory is a horizontal straight lines starting at H_i(245,58.499) mm with orientation angle φ=15° and ending at point H_f(275,58.499) mm with the same orientation it is obvious the selected trajectory passes through singular point H1. The tracking errors in x- and y-directions are depicted in Figure 5. The maximum tracking error

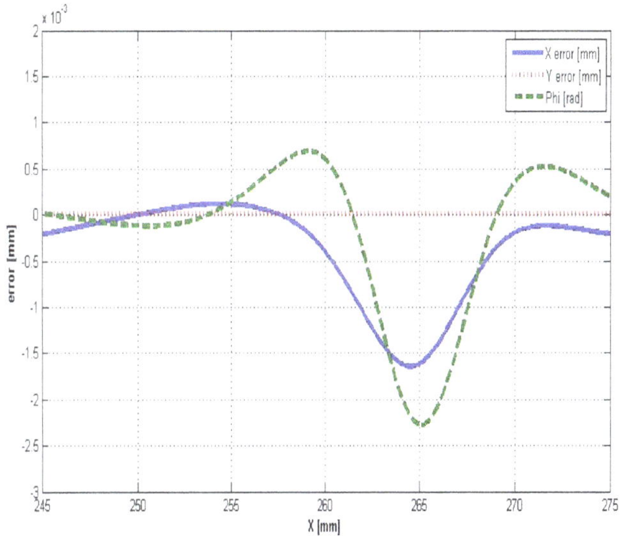

Figure 5: Tracking error for the 3-RPR along a horizontal trajectory passing through singularity point H1 (265, 58.499).

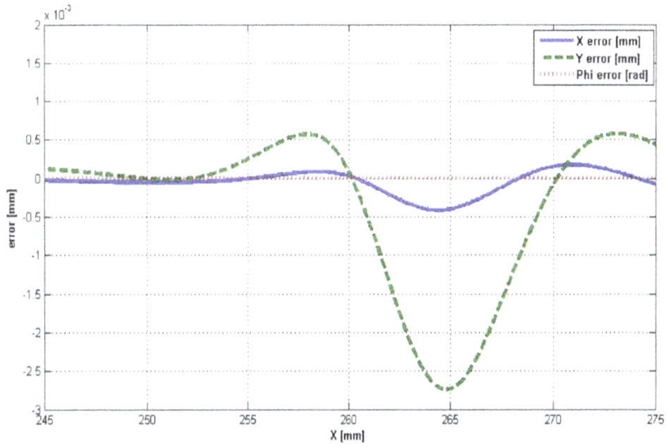

Figure 6: Tracking error for the 3-RPR along a horizontal trajectory passing through singularity point H1 (265, 58.499).

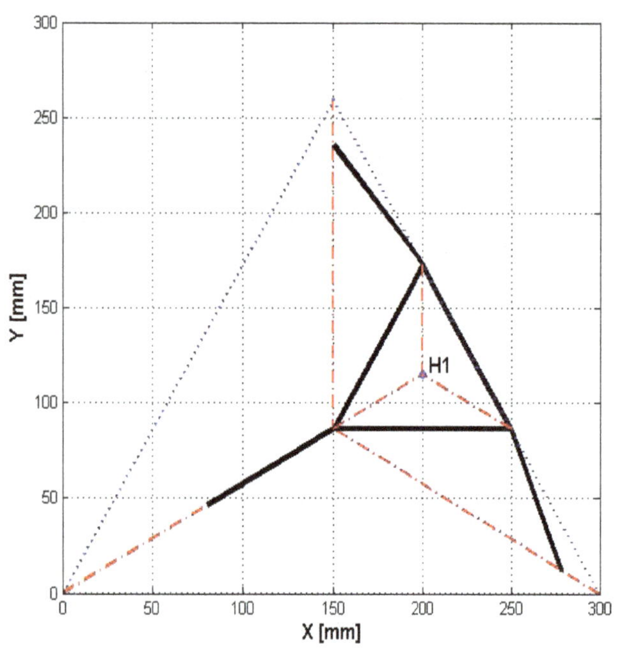

Figure 7: The 3-PRR cross singularity loci at H1(200,115.47).

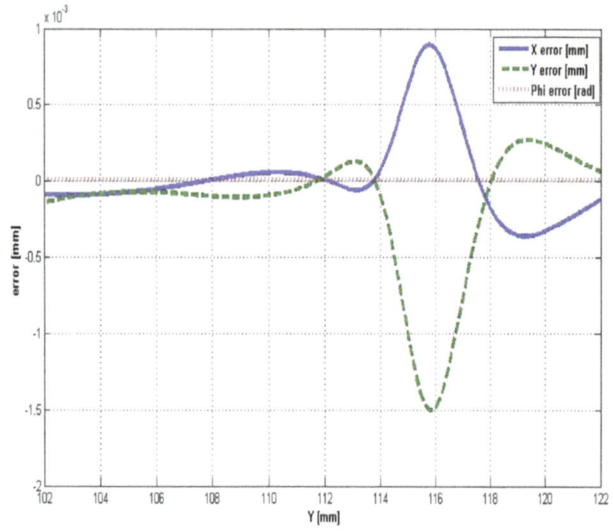

Figure 8: Tracking error for the 3-PRR along a vertical trajectory passing through singularity point H1 (200,115.470).

along the trajectory points is 0.0023 mm which also happens in the vicinity of kinematic singularities.

Finally, the third trajectory is a horizontal straight lines starting at $H_i(245,114.706)$ mm with orientation angle φ=15° and ending at point $H_f(275,114.706)$ mm with the same orientation. The selected trajectory passes through singular point H2. The tracking errors in x- and y-directions are depicted in Figure 6. The maximum tracking error along the trajectory points is 0.0027 mm which also happens in the vicinity of kinematic singularities.

3 PRR planar parallel manipulator

Two end-effector trajectories are specified as straight lines which cross over singularity loci at $H_1(200,115.470)$ mm as shown in Figure 7. The first trajectory is a vertical straight line starting at H_i (200,102) mm with orientation angle φ=0° and ending at point $H_f(200,122)$ mm with the same orientation it is obvious the selected trajectory passes through singular point H1. The tracking errors in x- and y-directions are depicted in Figure 8. The maximum tracking error along the trajectory points is 0.0015 mm which happens in the vicinity of kinematic singularities.

The second trajectory is a horizontal straight lines starting at

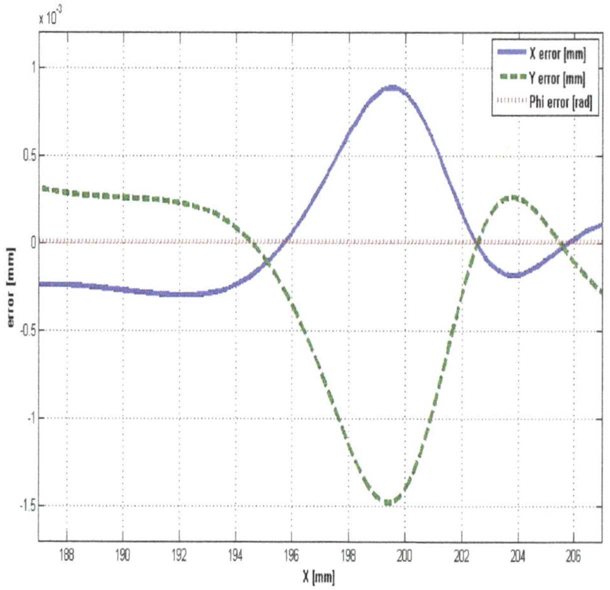

Figure 9: Tracking error for the 3-PRR along a horizontal trajectory passing through singularity point H1 (200,115.470).

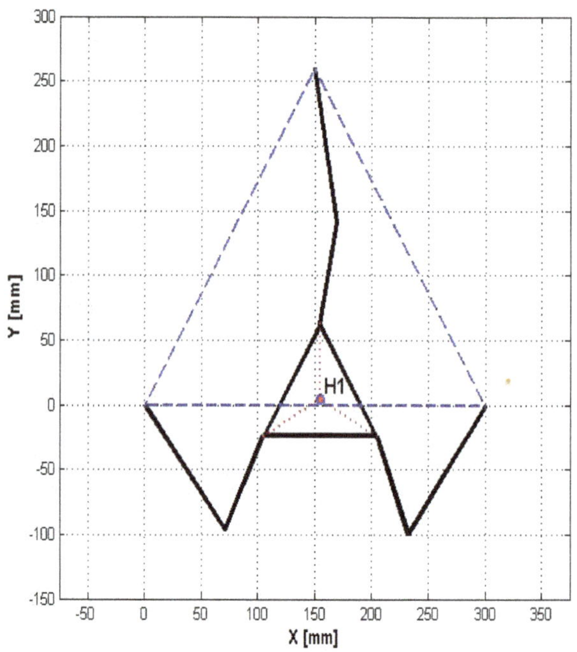

Figure 10: The 3-RRR cross singularity loci at H1(155, 4.845).

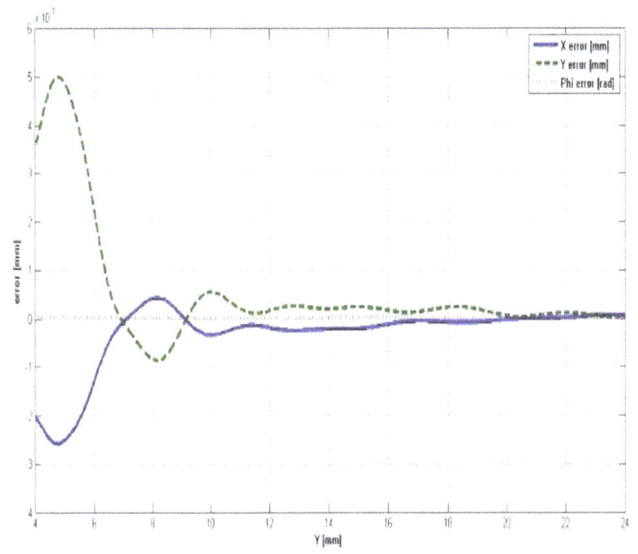

Figure 11: Tracking error for the 3-RRR along a vertical trajectory passing through singularity point H1 (155, 4.845).

Hi(187,115.470) mm with orientation angle φ=0° and ending at point Hf(207,115.470) mm with the same orientation it is obvious the selected trajectory passes through singular point H1. The tracking errors in x- and y-directions are depicted in Figure 9. The maximum tracking error along the trajectory points is 0.0015 mm which also happens in the vicinity of kinematic singularities.

3 RRR planar parallel manipulator

In the same way, two end-effector trajectories are specified as straight lines which cross over singularity loci at H_1(155, 4.845) mm as shown in Figure 10. The first trajectory is a vertical straight line starting at Hi(155,4) mm with orientation angle φ=0° and ending at point H_f(155,24) mm with the same orientation it is obvious the selected trajectory passes through singular point H_1. The tracking errors in x- and y-directions are depicted in Figure 11. The maximum tracking error along the trajectory points is 0.005 mm which happens in the vicinity of kinematic singularities.

The second trajectory is a horizontal straight lines starting at H_i(142,4.845) mm with orientation angle φ=0° and ending at point H_f(162,4.845) mm with the same orientation it is obvious the selected trajectory passes through singular point H_1. The tracking errors in x- and y-directions are depicted in Figure 12. The maximum tracking error along the trajectory points is 0.0053 mm which also happens in the vicinity of kinematic singularities.

Conclusion

In this paper, we proposed to use neural networks for forward kinematic solution of three different architectures of planar parallel manipulators, which can be elaborated to generate the best estimation of forward kinematics of the manipulators under study. Even though the manipulators passing through the kinemaic singularities, the proposed approach can provide solution for the problem with reasonable errors. The results of this paper can be used to find the forward kinematics solutions at critical points (singularity points) which can be then avoided, as long as we specify them, in dynamic control stage.

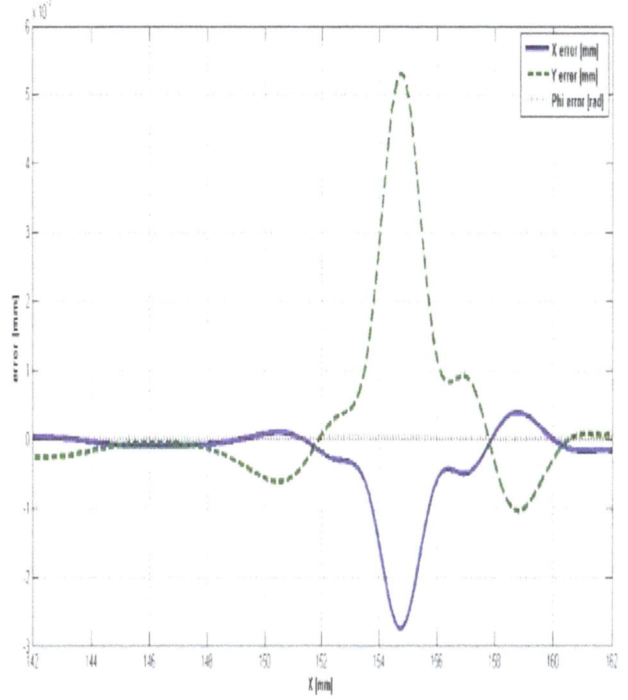

Figure 12: Tracking error for the 3-RRR along a horizontal trajectory passing through singularity point H1 (155, 4.845).

References

1. Cheng H, Liu GF, Yiu YK, Xiong ZH, Li ZX (2001) Advantages and dynamics of parallel manipulators with redundant actuation. IEEE/RSJ International Conference on Intelligent Robots and Systems Proceedings, HI.
2. Hunt KH(1983) Structural Kinematics of In-Parallel-Actuated Robot-Arms. J. Mech. Des 105:705-712.
3. Merlet JP(2006) Parallel Robots. Kluwer Academic, Dordrecht, The Netherlands.
4. Samy FM Assal (2012) Self-organizing approach for learning the forward kinematic multiple solutions of parallel manipulators. Robotica 30: 951-961.
5. Gosselin CM, Sefrioui J (1991) Polynomial solutions for the direct kinematic problem of planar three-degree-of-freedom parallel manipulators. ICAR Fifth International Conference on Robots in Unstructured Environments, Italy.
6. Merlet JP (1996) Direct kinematics of planar parallel manipulators. IEEE International Conference on Robotics and Automation, France.
7. Oetomo D, Hwee C L, Alici G, Shirinzadeh B(2006) Direct Kinematics and Analytical Solution to 3RRR Parallel Planar Mechanisms.ICARCV '06. 9th International Conference on Control, Automation, Robotics and Vision, Singapore.
8. Tsai LW (1999) Robot Analysis: The Mechanics of Serial and Parallell Manipulators.
9. Yee CS, Kah-Bin Lim (1991) Neural network for the forward kinematics problem in parallel manipulator. IEEE International Joint Conference on Neural Networks.
10. Dehghani M, Ahmadi M, Khayatian A, Eghtesad, M (2008) Wavelet Based Neural Network Solution for Forward Kinematics Problem of HEXA Parallel Robot INES International Conference on Intelligent Engineering Systems, Iran.
11. Lee Hyung Sang, Myung-Chul Han (1991) The estimation for forward kinematic solution of Stewart platform using the neural network IROS '99 IEEE/RSJ International Conference on Intelligent Robots and Systems, Kyongju.
12. Parikh PJ, Lam SSY (2005) A hybrid strategy to solve the forward kinematics problem in parallel manipulators IEEE Transactions on Robotics.

13. Dan Zhang, Jianhe Lei (2011) Kinematic analysis of a novel 3-DOF actuation redundant parallel manipulator using artificial intelligence approach. Robotics and Computer-Integrated Manufacturing 27: 157-163.

14. Ali TH, Ismail N Hamouda AMS, Ishak Aris , Marhaban MH et.al (2010) Artificial neural network-based kinematics Jacobian solution for serial manipulator passing through singular configurations. Advances in Engineering Software 41: 359-367.

15. Wenger P, Chablat D (2004) The Kinematic Analysis of a Symmetrical Three-Degree-of-Freedom Planar Parallel Manipulator. CISM-IFTOMM Symposium on Robot Design, Dynamics and Control Montreal.

Design an Arm Robot through Prolog Programming Language

Aram Azad[1] and Tarik Rashid[2*]

[1]Computer Science and Information Technology, UTS: University of Technology, Sydney, Sydney, Australia
[2]Software Engineering Department, College of Engineering, Salahaddin University, Hawler, Kurdistan

Abstract

This paper aims at furnishing basic definitions for robotics in the span of man-made intelligence. Simple robotics and their faculties such as sensors and actuators are included in the definition. Moreover, intelligent and non-intelligent types of robotics are described in this research work. We generally focus on non-intelligent robotics (program-based robotics) and the setbacks designers can encounter .The methods and designs of an arm robot generally encompass three components: namely, the mechanical assembly of the arm robot, the electronic circuit, and software design employing Prolog software programming language. Our design concentrates generally on programming an arm robot to manipulate its movements. In our case study programming an arm robot we delineate probable resolutions to setbacks we encountered without going intensely into theory, and we demonstrate a stable design for resolving these setbacks via a flowchart. The design of the flowchart is next explained and subsequently coded.

Keywords: Robotics; Robot control methods; Intelligent robot; Non intelligent; Arm robot programming

Introduction

General history of robotics

The root of the word robot dates back to Karel Capek's play Rossum Universal Robots (RUR) in 1921. This actually originated from the Czech word for "corvee" [1-11]. Technologically speaking, robotics fall into two categories; these are namely: tele-manipulators, and the ability of numerical control of machines. The first types of Tele-manipulators were made of an arm and a gripper; this machine can be remotely controlled. The human gives instructions via his control device to control the arm and gripper's movements. These types of robotics are used to handle radioactive material.

However, the second type, Numeric control, is used to allow very precise control of machines in relation to a given coordinate system. This was first utilised at MIT in 1952. This type of robotics tipped to the first programming language for machines called Automatic Programmed Tools (APT) [1-11].

The amalgamation of the two types mentioned above resulted in the first programmable telemanipulators; moreover, these principles were first considered and fitted into industrial robots in 1961. Car construction plants are good examples of locations where these types of robotics remain useful. There was a desire to develop automated transportation and autonomous transport systems and production processes. Another type of robots that can move, called mobile robotics, are being built, and there is a particular type which is called insectoid robotics where the robots have many legs. These types are used for more autonomous purposes such working underwater.

In recent years wheel-driven robots have been commercially promoted and utilised for public service and purposes such as in hospitals and other places. In 1975 another type of robotics was designed called Humanoid robots. This happened when Wabot-I was offered in Japan. As matter of fact, the current Wabot-III already has some minor cognitive capabilities. In 1994, another type of humanoid robot, called "Cog", was developed in MIT-AI-Lab, and in 1999 Honda's humanoid robot came around and became widely known to the public. This type is controlled remotely by humans, and it can walk autonomously. In science fiction, robots are already human's best friends, but in reality we will only see robots for specific jobs as universal programmable machine slaves in the near future (which leads to interesting questions, see [1-11]).

History of arm robots

In 1954, the first patent for robotics was received by George Devol, and in 1956, F. Engelberger had the first company, and Devol's original patents were used for this establishment. It is worthy to mention that a Unimation robot was used for transferring different objects from a point A to a point B for less than a dozen feet. Hydraulic effectors were used and they were programming in joint coordinates. Cincinnati Milacron Inc. of Ohio was a competitor for some time in Unimation robots. However, this was not continued and in late 1970's several big Japanese conglomerates began to generate the same industrial robots. While in the US Unimation robots had been patented; nevertheless, this was not the case in Japan, who rejected being obliged to international patent laws; as a result, their design was copied [12].

The Stanford arm was created by Victor Scheinman in 1969 at Stanford University. This is an all-electric, six axis articulated robot to permit an arm solution. This gave a robot an accurate feature to follow arbitrary paths in space and widened the potential of robots to more progressive uses; e.g. assembly and arc welding. Afterward, the MIT AI Lab was created by Scheinman as a second arm, called the MIT arm. After that they sold his architecture design to Unimation, who later on, with help of General Motors, developed and then sold it as the programmable Universal Machine for Assembly (PUMA) [12].

***Corresponding author:** Tarik Rashid, Software Engineering Department, College of Engineering, Salahaddin University, Hawler, Kurdistan, Iraq
E-mail: tarikrashid4@gmail.com

FAMULUS is the first industrial robot which was designed by KUKA in 1973. This type of design had six electromechanically driven axes. In late 1970s the interest in industrial robotics swelled as many different companies had entered the field, including big ones such as General Electric and General Motors, who formed a joint venture with FANUC LTD of Japan, called FANUC Robotics. US start-ups included Automatix and a dept Technology, Inc. At the height of the robot boom in 1984, Unimarion was acquired by Westinghouse Electric Corporation for 107 million US dollars. Unimation was sold to Staubli Faverges SCA by Westinghouse in 1988. Articulated robots for general and clean room application were still made by Staubli in 2004, and the robotic division of Bosch in late 2004. Finally the myopic vision of the US industry was superseded by the financial resources and strong domestic market enjoyed by the Japanese manufactures. Only a few non-Japanese companies managed to survive in this market including Adapt Technology, Staubli-Unimation, the Swedish-Swiss Company and ABB IASEA [12].

Objectives

The main objectives of this paper are as follows:

I. Design and Assemble the Arm Robot: The main objective of any project is to design the arm robot that is able to carry out certain task. The revolute robotic arm is able to move similarly to a human arm. The arm is designed so it is able to rotate clockwise and counter clockwise (180 degrees), and able to pick up and place objects. The arm needs to be as light as possible in order to maximize burden. The material for the arm structure also needs to be strong and rigid. One possible material is aluminium as used in [12].

II. Design and Construct Controller Circuit Board: The second main objective is to design and engineer the controller circuit board that will be used to control the arm robot through connections to a personal computer (PC). The most important component that is used in the controller is the microcontroller. The circuit board will be connected to a serial port of a PC through a serial connector [12].

III. Design Robot Software and Interfacing: The third main objective is to design the robot interface application using Prolog programming language to control the controller circuit board to run the robot arm.

Artificial Intelligence and Robotic Definitions and Their Types

As we have stated above, robots can fall into the categories of manipulators, mobile robots, and humanoids. Often tasks are performed using physical agents called robots. Basically, robots have two faculties: sensors through which robots can receive perception from environment; and effectors through which robots are able to perform physical tasks [13-20].

Generally speaking, artificial intelligence (AI) is one of the important disciplines in computer science. AI has been considered as a theory and agents are hot topics in artificial intelligence. The most important part in the agent is the actor. This is recognised in software. It is clear that robots are created as hardware. The construction of artificial intelligence and robotics is that a software agent controls the robot that has sensors through which data can be read, and then the robot decides what to do, then guide the actuator to perform an action in the environment [13-20].

Definitions of a robot will decide the differences between two types of intelligent and non-intelligent robots. A simple example of a robot is the thermostat on a heater or in our living room. Of course, this definition might not satisfy the definition of intelligence in human terms. Intelligence, then, may be defined as an arbitrary quality considered as such by the human operator. The term robot is used in two ways: a machine that is directed by a human operator by remote control and a machine that makes limited decisions based on a computer program. Factory welding machines are programmed robots that can sense the position of the parts. Automatic drone aircraft are given goals and can fly themselves. The Russian space craft's are autonomous cars that use cameras and sensors and on-board computers to drive themselves and stay on the road. We believe robot machines cannot think as we humans do, although we keep on hoping to arrive at this point. There are two types of robots based on intelligence, these are:

Non-intelligent robots

One important type of robot is called a non-intelligent robot. These are robots that do not have sensors to receive information. Instead, the robot is only able to follow a fixed set of instructions, no matter what is happening around it. A robot like this can never become intelligent (Figure 1).

Intelligent robots

A robot is controlled by sets of instructions called a program, which is built into the controlling device. Intelligent robots must have sensors to collect information from the environment [12-15]. This information will help the robot to become intelligent. This information will be sent to the controlling program [21-29], where it is tested to decide what the robot should do next. Generally speaking, sensors can be in the form of a camera, a pressure pad, and a microphone that will allow the robot to view, touch, and hear. These sensors are tailored or built in the robot almost exactly like a human being has eyes to see, hands and fingers to hold objects or touch, and ears to listen (Figure 2).

Methods

The methods and design of robots mainly consist of three parts: namely, the mechanical construction of an arm robot, the electronic circuit, and the programming design using Prolog programming language. In this paper, our design is mainly focused on the programming of robots.

The Mechanical construction of an arm robot

This is the main part of the arm robot which needs three different structures; these are shoulder, elbow and wrist. The challenge in this particular part is how to affix the motor to the piece or joint to achieve the preferred turning degree level. Another challenge, notably, the

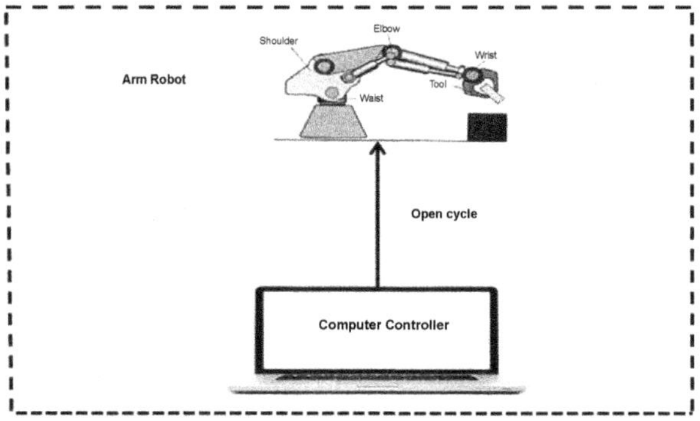

Figure 1: Shows a non-Intelligent Robot.

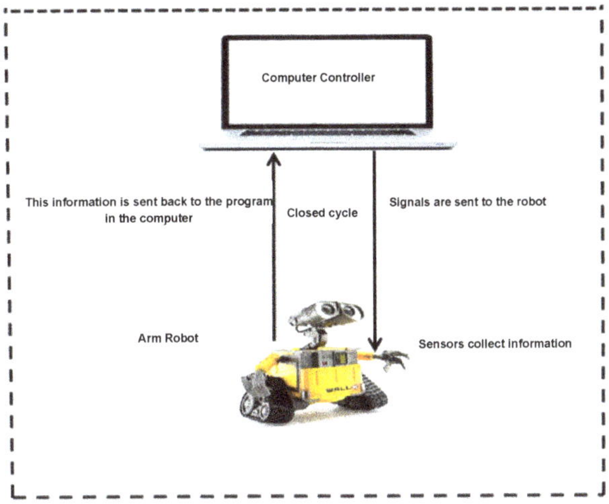

Figure 2: Shows an intelligent robot.

intended degree of freedom, can be achieved by the perfect combination of the different arm components. The mechanical design is achieved using the software Autodesk Inventor; this software is used to build and design the arm robot. Dimensions for mechanical design are a crucial issue and it is regarded as the core issue in the arm robot. They have to be very accurate, and there is no room for minor error, since a minor error in the dimensions will cause a major problem in the assembly and function of the arm structure.

Electronic circuit design

This is equally as important as the mechanical part; this part is dedicated to provide control over the movement of the arm robot. A microcontroller is used to provide this purpose.

Design the programming part using prolog language

Design a software program that enables the movement of arm robot; there must also be an interface with the software to control the arm robot.

The design of the programming part using prolog language is divided into two parts:

Plan and design: There is an arm robot with four other objects in this case: Square, Semi-Square, Circle, and Triangle. The arm is able to hold the objects and put them on each other and on the floor (Figure 3); the user of the program can control the hand robot by using some commands on a screen or menu. Additionally, the program will end whenever the user pushes on the exits command. There should be a user interface which shows the commands to the user, to wait for a user to say what to do. These important commands are:

1) Hold object
2) Put object
3) Free hand
4) Display
5) Exit

We have restricted a user with some rules that the arm robot must obey when performing tasks. These rules are clearly defined below:-

1) The arm robot can put a semi-square on a square
2) The arm robot can put a triangle on a square
3) The arm robot can put a circle on a semi-square
4) The arm robot should be free before taking hold of an object

Figure 3: Shows an arm robot with the four different objects.

5) The arm robot should be full before commanded to free arm
6) The object should be free in order to hold it
7) A blocked object cannot be held

When the user chooses the hold command, the program will ask the user about the name of the object to hold; before that, the program should check if the arm is free, if the arm is free, then the arm continues; otherwise, it writes a message to tell the user that the arm is full, and returns back to the welcome screen. Then it should check if the object is free, in other words, is not blocked or held.

When the user chooses the put command, the program should check if the arm is full or not. If the arm is full, then it continues, or else it returns back to the welcome screen; then the program should ask about the name of the target object to put it on, then the program checks if the target object is blocked or not. If the object is free then it continues, or else shows an error message to the user of "enter another target." When the user enters the free command, the program should check the arm if it is already free or not; if it's free, then the error message will be displayed to the user that says "The arm already is free!" and then return to welcome screen. If the arm is full, then free the arm by placing object that is already held, and then return back to the welcome screen.

When the user enters the Display command, the program should display the status of each object and also the arm status; it should also show which objects are blocked, and it should say which objects are already on them.

In an ideal case all the objects and arm status statements are "free".

You need a header statement here that tells your reader the following phrases are the status statements

Arm (free/full), Object (free/hold/block/another Object)

Both Arm and Objects had status, as declared above; you can see arm has two statuses, and each object has 4 types of status; the fourth one "another Object" means it is under that object, if I put triangle over square the square status will become "triangle". The flowchart is designed for the arm robot with all possibilities and functionalities indicated, (Figure 4).

Coding and implementation: After completing the design and the flowchart of the arm robot, the flowchart was implemented and coded in the prolog programming language [30-35]. Below is the full programming code for a complete arm robot system

```
Domains
x=symbol.
y=integer.
s=string.
Predicates
```

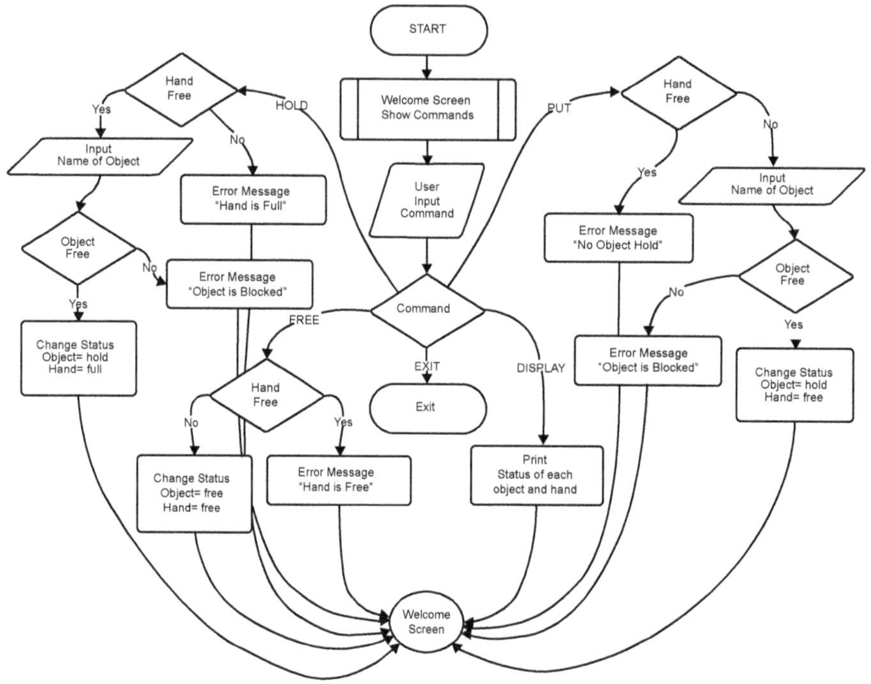

Figure 4: Shows a complete flowchart design for an arm robot system.

```
% nondetermrobothold(s,s,s,s,s,s).
% nondetermrobotput(x,x,x,x,x,s,s).
nondeterm check(s,s).
nondeterm robot(s,s,s,s,s,s,s,y).
% nondeterm call(y).
% nondeterm valid(x).
Clauses
check(“triangle”,”square”).
check(“semisquare”,”square”).
check(“circle”,”semisquare”).

robot(Square,SemiSquare,Triangle,Circle,Hand,Source,Target,0):-
nl,
write(“Press 1 to hold”),nl,
write(“Press 2 to put”),nl,
write(“Press 3 to free”),nl,
write(“Press 4 to exit”),nl,
write(“Press 7 to print”),nl,
write(“Enter Number:”),
readint(N),
robot(Square,SemiSquare,Triangle,Circle,Hand,Source,Target,N).

robot(Square,SemiSquare,Triangle,Circle,Hand,Source,Target,N):-
N<1,
write(“Enter valid input 1,2,3,4,7: “),
readint(M),
robot(Square,SemiSquare,Triangle,Circle,Hand,Source,Target,M).

robot(Square,SemiSquare,Triangle,Circle,Hand,Source,Target,N):-
N>7,
write(“Enter valid input 1,2,3,4,7: “),
readint(M),
robot(Square,SemiSquare,Triangle,Circle,Hand,Source,Target,M).

robot(_,_,_,_,_,_,_,4):-
write(“Good Bye”),nl.

robot(Square,SemiSquare,Triangle,Circle,Hand,_,Target,1):-
nl,
Hand=”free”,
write(“Enter shape to hold ‘sqaure,triangle,semisquare,circle’: “),
readln(Shape),
robot(Square,SemiSquare,Triangle,Circle,”full”,Shape,Target,6).

robot(Square,Semisquare,Triangle,Circle,”full”,Shape,Target,6):-
Square=”free”,
Shape=”square”,
robot(“hold”,Semisquare,Triangle,Circle,”full”,Shape,Target,0).

robot(Square,SemiSquare,Triangle,Circle,”full”,Shape,Target,6):-
Semisquare=”free”,
Shape=”semisquare”,
robot(Square,”hold”,Triangle,Circle,”full”,Shape,Target,0).

robot(Square,SemiSquare,Triangle,Circle,”full”,Shape,Target,6):-
Triangle=”free”,
Shape=”triangle”,
robot(Square,SemiSquare,”hold”,Circle,”full”,Shape,Target,0).

robot(Square,SemiSquare,Triangle,Circle,”full”,Shape,Target,6):-
```

```
Circle="free",
Shape="circle",
robot(Square,SemiSquare,Triangle,"hold","full",Shape,Target,0).

robot(Square,SemiSquare,Triangle,Circle,Hand,Source,Target,1):-
Hand="full",
nl,write("Error The hand is not free"),nl,
robot(Square,SemiSquare,Triangle,Circle,Hand,Source,Target,0).

robot(Square,SemiSquare,Triangle,Circle,Hand,Source,Target,7):-
nl,
write("Square: "),write(Square),nl,
write("SemiSquare: "),write(SemiSquare),nl,
write("Triangle: "),write(Triangle),nl,
write("Circle: "),write(Circle),nl,
write("Hand: "),write(Hand),nl,
write("Source: "),write(Source),nl,
write("Target: "),write(Target),nl,
robot(Square,SemiSquare,Triangle,Circle,Hand,Source,Target,0).

robot("hold",SemiSquare,Triangle,Circle,_,_,_,3):-
robot("free",SemiSquare,Triangle,Circle,"free","free","free",0).

robot(Square,"hold",Triangle,Circle,_,_,_,3):-
robot(Square,"free",Triangle,Circle,"free","free","free",0).

robot(Square,SemiSquare,"hold",Circle,_,_,_,3):-
robot(Square,SemiSquare,"free",Circle,"free","free","free",0).

robot(Square,SemiSquare,Triangle,"hold",_,_,_,3):-
robot(Square,SemiSquare,Triangle,"free","free","free","free",0).

robot(Square,SemiSquare,Triangle,Circle,"free",Source,Target,3):-
nl,write("The hand already is free "),nl,
robot(Square,SemiSquare,Triangle,Circle,"free",Source,Target,0).

robot(Square,SemiSquare,Triangle,Circle,Hand,Source,_,2):-
Hand="full",
nl,write("Enter shape to put on 'sqaure,triangle,semisquare,circle'
:"),
readln(Shape),
robot(Square,SemiSquare,Triangle,Circle,Hand,Source,Shape,5).

robot(Square,SemiSquare,Triangle,Circle,Hand,Source,Target,2):-
Hand <> "full",
nl,write("You should hold a shape before you put"),nl,
robot(Square,SemiSquare,Triangle,Circle,Hand,Source,Target,0).

robot(Square,Semisquare,Triangle,Circle,_,Source,Target,5):-
check(Source,Target),
Semisquare="free",
Circle="hold",
Source="circle",
Target="semisquare",
robot(Square,"circle",Triangle,"free","free","free","free",0).

robot(Square,SemiSquare,Triangle,Circle,_,Source,Target,5):-
check(Source,Target),
Square="free",
Triangle="hold",
Source="triangle",
Target="square",
robot("triangle",SemiSquare,"free",Circle,"free","free","free",0).

robot(Square,SemiSquare,Triangle,Circle,_,Source,Target,5):-
check(Source,Target),
Square="free",
Semisquare="hold",
Source="semisquare",
Target="square",
robot("semisquare","free",Triangle,Circle,"free","free","free",0).

robot(Square,SemiSquare,Triangle,Circle,Hand,Source,Target,5):-
nl,write("The target already blocked"),nl,
robot(Square,SemiSquare,Triangle,Circle,Hand,Source,Target,0).

robot(Square,SemiSquare,Triangle,Circle,Hand,Source,Target,5):-
Target="circle",
nl,write("Error its not possible"),
robot(Square,SemiSquare,Triangle,Circle,Hand,Source,Target,2).

robot(Square,SemiSquare,Triangle,Circle,Hand,Source,Target,5):-
Target="semisquare",
nl,write("Error its not possible"),
robot(Square,SemiSquare,Triangle,Circle,Hand,Source,Target,2).

robot(Square,SemiSquare,Triangle,Circle,Hand,Source,Target,5):-
Target="triangle",
nl,write("Error its not possible"),
robot(Square,SemiSquare,Triangle,Circle,Hand,Source,Target,2).

Goal
robot("free","free","free","free","free","free","free",0).
```

Conclusion

This paper provides a basic definition and background about robotic systems in the area of artificial intelligence. The paper refers to robotic systems with their basic facilities such as sensors and actuators. Two types of robotics are explained: these are intelligent and non-intelligent robotic systems. In this paper we have considered the non-intelligent robotic in which programming was our base to manipulate the system (program based robotics). A case study of designing an arm robot proposal is presented in which we describe possible solutions and functionalities to those problems without going deeply into theory anda firm design for solving the problem is suggested via a flowchart.

The design of flowchart is then implemented and coded. The language of the program used to design the arm robot is prolog.

References

1. Niemueller T, Widyadharma S (2003) Artificial Intelligence-An Introduction to Robotics.
2. Carlson D (1998) Synchro Drive Robot Platform.
3. Tong A (1995) Star Trek Episode: The Measure of a Man.
4. Singh S, Kumar S, Singh V, Sain C (2012) Robotic Analysis And Programming. International Journal of Advanced Scientific Research and Technology.
5. Wolf JC, Hall P, robinson P, Culverhouse P (2008) Bioloid Based Humannoid Soccer Robot Design.
6. Floreano D, Godjevac J, Martinoli A, Mondada F, Nicoud JD (2005) Design, Control And Application of Autonomous Mobile Robots. Advances in Intelligent Autonomous Systems 18: 159-186.
7. Udengaard M, Legnemma K (2007) Design of an Omnidirectional Mobile Robot for rough Terrain. IEEE International Conference on Robotics & Automation.
8. Salih JFM, Rizon M, Yaacob S, Adom AM, Mamat MR (2006) Designing Omni-Directional Mobile Robot with Mccanum Wheel. American Journal of Applied Science 3: 1831-1835.
9. Doroftei L, Grosu V, Spinu V (2007) Omnidirectional Mobile Robot Design And Implementation. Bioinspiration and Robotics: Walking and Climbing Robots.
10. Smart WD, Kaelbling LP (2008) Effective Reinforcement Learning For Mobile Robots. IEEE International Conference on Robotics & Automation 4: 3404-3410.
11. Murphy D, Challacombe B, Khan MS, Dasgupta D (2006) Robotic Technology In Urology. Postgrad Med J 82: 743-747.
12. Omar MNB (2007) Pick and Place Robotic Arm Controlled By Computer. Robotic and Automation.
13. Leang K (1999) Minibot Sonar Sensor How to.
14. Dellaert F, Fox D, Burgard W, Thrun S (1999) Monte Carlo Localization For Mobile Robots. IEEE International Conference on Robotics and Automation 2: 1322-1328.
15. De Giacomo G, Lesperance Y, Levesque H, Reiter R (2001) Indigolog Overview.
16. Grabowski B (2003) Small Robot Sensors.
17. Görz G, Rollinger CR, Schneeberger J (2000) Handbuch Der Künstlichen Intelligenz, 3. Auflage. Oldenbourg Verlag München Wien.
18. Lego (2003) Lego Mindstorms Tutorial on Correcting Course.
19. Trimble Navigation Limited (2003) Trimble-What Is Gps?
20. Helsinki University Of Technology (2001) Moving Eye-Virtual Laboratory Exercise On Telepresence, Augmented Reality, and Ball-Shaped Robotics.
21. Röbke-Doerr P (2003) Navigation Mit Satelliten. C't Magazine Fur Computer Technik 1:150-151.
22. Russell S, Norvig P (1995) Artificial Intelligence. A Modern Approach. Prentice Hall.
23. Russell S, Norvig P (2003) Artificial Intelligence. A Modern Approach. (2nd Edn), Prentice Hall.
24. Robocup Team (2003) Robocup.
25. RWTH Aachen (2003) Robocup Team. Allemaniacs.
26. Thrun S (2003) Robotic Mapping: A Survey. Robotics.
27. Weisstein EW (1999) Voronoi Diagram. Math World.
28. Welch G, Bishop G (2003) An Introduction to the Kalman Filter. University of North Carolina at Chapel Hill.
29. Nelly D, Cees VB (2007) Introduction of Robotics in Science Lessons. Comlab Conference.
30. Delson JN, West H (1994) Robot programming by human demonstration: The use of human inconsistency in improving 3D robot trajectories. Massachusetts Institute of Technology, Dept. of Mechanical Engineering, MA.
31. Edsinger A (2007) Robot Manipulation in Human Environments. Massachusetts Institute of Technology, Dept. of Electrical Engineering and Computer Science, MA.
32. Hoang H (2005) Automated construction technologies: analyses and future development strategies. Master thesis, Massachusetts Institute of Technology, Dept. of Architecture, MA.
33. Flynn AM, Brooks RA (1987) MIT Mobile Robots-What's Next? (Working Paper 302). Massachusetts Institute of Technology, Artificial Intelligence Laboratory 1: 611-617.
34. Yoon Y (2006) Modular robots for making and climbing 3-D trusses. Massachusetts Institute of Technology, Dept. of Mechanical Engineering, MA.
35. Schatz David (1983) The strategic evolution of the robotics industry. Massachusetts Institute of Technology. Sloan School of Management.

Dynamic Modelling of Differential-Drive Mobile Robots using Lagrange and Newton-Euler Methodologies: A Unified Framework

Rached Dhaouadi* and Ahmad Abu Hatab

College of Engineering, American University of Sharjah, Sharjah, UAE

Abstract

This paper presents a unified dynamic modeling framework for differential-drive mobile robots (DDMR). Two formulations for mobile robot dynamics are developed; one is based on Lagrangian mechanics, and the other on Newton-Euler mechanics. Major difficulties experienced when modeling non-holonomic systems in both methods are illustrated and design procedures are outlined. It is shown that the two formulations are mathematically equivalent providing a check on their consistency. The presented work leads to an improved understanding of differential-drive mobile robot dynamics, which will assist engineering students and researchers in the modeling and design of suitable controllers for DDMR navigation and trajectory tracking.

Keywords: Differential-drive; Mobile robot; Dynamics; Modeling; Lagrange; Newton-Euler

Introduction

In recent years, there has been a considerable interest in the area of mobile robotics and educational technologies [1-7]. For control engineers and researchers, there is a wealth of literature dealing with wheeled mobile robots (WMR) control and their applications. However, while the subject of kinematic modeling of WMR is well documented and easily understood by students, the subject of dynamic modeling of WMR has not been addressed adequately in the literature. The dynamics of WMR are highly nonlinear and involve non-holonomic constraints which makes difficult their modeling and analysis especially for new engineering students starting their research in this field. Therefore, a detailed and accurate dynamic model describing the WMR motion need to be developed to offer students a general framework for simulation analysis and model based control system design.

In the case of a differential drive mobile robot (DDMR), for example, there is no textbook available that investigates thoroughly the dynamic modeling approach taking into consideration the non-holonomic constraints in a step by step procedure. The analysis is available mainly in journals, conference papers, and technical reports [8]. Moreover, the material presented differs from one paper to another with different variables and reference frames used, and various assumptions. In addition, some papers present different results for the same DDMR used, which adds to the confusion of dynamic modeling.

For the case of DDMR, the methods used are either the Lagrangian approach [9-15] or the Newton-Euler approach [16-19]. Other formalisms such as the Kane's method have been also suggested as viable approaches of modeling DDMR [20]. Therefore, it is not clear for new engineering students and researchers which concept to use and which method offers a better physical insight on the dynamic behavior of the system and the effect of the non-holonomic constraints. Also, it is not clear if both methods will lead to the same final dynamic model.

In the Newton Euler method, one has to take into account two kinds of forces applied to a system: the given forces and the constraint forces. The given forces include the externally impressed forces by the actuators while the constraint forces are the forces of interaction between the robot platform and ground through the wheels. Moreover, in a system with interconnected elements, the components may interact with each other through gears, springs, and frictional elements. Therefore, we need to take into account all of these forces. It is clear that the Newtonian approach includes a few practical difficulties since in most cases these forces are not easily quantifiable.

The methodology developed by Lagrange overcomes these problems by expressing the forces in terms of the energies in the system, i.e., the kinetic energy and the potential energy, which are scalar quantities easily expressible in terms of the system coordinates. The derivation of the Lagrange equations requires also that the generalized coordinates be independent.

The Lagrangian approach usually provides a powerful and versatile method for the formulation of the equations of motion for holonomic systems. However, for non-holonomic systems, the usual method is to introduce the motion constraint equations into the dynamic equations using the additional Lagrange multipliers. These multipliers are not constants and are usually functions of all the generalized coordinates and often of time as well. They represent a set of unknowns whose values should be obtained as a part of the solution. To solve this computational complexity, additional methods have been suggested to remove the presence of the multipliers from the dynamic equations of the given system [21,22].

The focus of this paper is to derive simple and well-structured dynamic equations of the DDMR taking into account the non-holonomic constraints. First, the Lagrange formulation is presented. Coordinates transformation is used to cancel the Lagrange multipliers to obtain well-structured equations. Second, the Newton-Euler method is used to derive the dynamic equations of the DDMR. Major difficulties experienced in using both methods are illustrated and procedures are outlined to offer a systematic approach to the dynamic modeling of

***Corresponding author:** Rached Dhaouadi, College of Engineering, American University of Sharjah, P.O. Box 26666, Sharjah, UAE, E-mail: rdhaouadi@aus.edu

DDMR with no major mathematical complexity. It is shown that both methods reach equivalent dynamic equations for the mobile robot providing a check on their consistency.

Coordinate Systems

In order to describe the position of the WMR in his environment, two different coordinate systems (frames) need to be defined.

1. Inertial Coordinate System: This coordinate system is a global frame which is fixed in the environment or plane in which the WMR moves in. Moreover, this frame is considered as the reference frame and is denoted as $\{X_I, Y_I\}$.

2. Robot Coordinate System: This coordinate system is a local frame attached to the WMR, and thus, moving with it. This frame is denoted as $\{X_r, Y_r\}$.

The two defined frames are shown in Figure 1. The origin of the robot frame is defined to be the mid-point A on the axis between the wheels. The center of mass C of the robot is assumed to be on the axis of symmetry, at a distance d from the origin *A*.

As shown in Figure 1, the robot position and orientation in the Inertial Frame can be defined as

$$q^I = \begin{bmatrix} x_a \\ y_a \\ \theta \end{bmatrix} \tag{1}$$

The important issue that needs to be explained at this stage is the mapping between these two frames. The position of any point on the robot can be defined in the robot frame and the inertial frame as follows.

Let $X^r = \begin{bmatrix} x^r \\ y^r \\ \theta^r \end{bmatrix}$, and $X^I = \begin{bmatrix} x^I \\ y^I \\ \theta^I \end{bmatrix}$ and be the coordinates of the given point in the robot frame and inertial frame, respectively.

Then, the two coordinates are related by the following transformation:

$$X^I = R(\theta) X^r \tag{2}$$

Where $R(\theta)$ is the orthogonal rotation matrix

$$R(\theta) = \begin{bmatrix} \cos\theta & -\sin\theta & 0 \\ \sin\theta & \cos\theta & 0 \\ 0 & 0 & 1 \end{bmatrix} \tag{3}$$

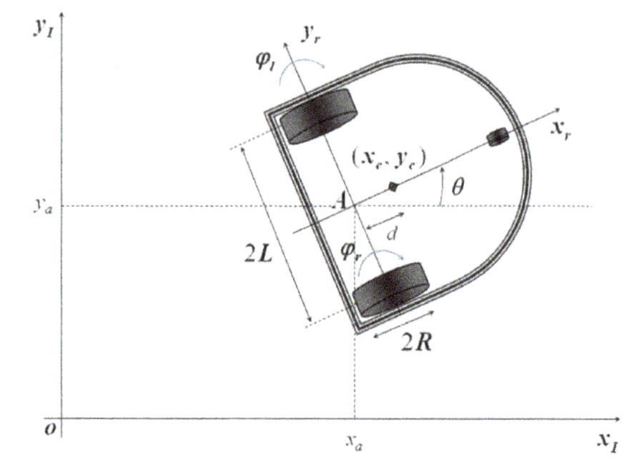

Figure 1: Differential Drive Mobile Robot (DDMR).

This transformation will enable also the handling of motion between frames.

$$\dot{X}^I = R(\theta)\dot{X}^r \tag{4}$$

It will be seen in the next section that equation (4) is very important in deriving the DDMR kinematic and dynamic models as it describes the relationship between the velocities in the Inertial Frame and the Robot Frame.

Kinematic Constraints of the Differential-Drive Robot

The motion of a differential-drive mobile robot is characterized by two non-holonomic constraint equations, which are obtained by two main assumptions:

• No lateral slip motion: This constraint simply means that the robot can move only in a curved motion (forward and backward) but not sideward. In the robot frame, this condition means that the velocity of the center-point A is zero along the lateral axis:

$$\dot{y}_a^r = 0 \tag{5}$$

Using the orthogonal rotation matrix $R(\theta)$, the velocity in the inertial frame gives

$$-\dot{x}_a \sin\theta + \dot{y}_a \cos\theta = 0 \tag{6}$$

• Pure rolling constraint:

The pure rolling constraint represents the fact that each wheel maintains a one contact point P with the ground as shown in Figure 2. There is no slipping of the wheel in its longitudinal axis (x_r) and no skidding in its orthogonal axis (y_r). The velocities of the contact points in the robot frame are related to the wheel velocities by:

$$\begin{cases} v_{pR} = R\dot{\varphi}_R \\ v_{pL} = R\dot{\varphi}_L \end{cases} \tag{7}$$

In the inertial frame, these velocities can be calculated as a function of the velocities of the robot center-point A:

$$\begin{cases} \dot{x}_{pR} = \dot{x}_a + L\theta\cos\theta \\ \dot{y}_{pR} = \dot{y}_a + L\theta\sin\theta \end{cases} \tag{8}$$

$$\begin{cases} \dot{x}_{pL} = \dot{x}_a + L\dot{\theta}\cos\theta \\ \dot{y}_{pL} = \dot{y}_a + L\dot{\theta}\sin\theta \end{cases} \tag{9}$$

Using the rotation matrix $R(\theta)$, the rolling constraint equations are formulated as follows:

$$\begin{aligned} \dot{x}_{pR}\cos\theta + \dot{y}_{pR}\sin\theta = R\dot{\varphi}_R \\ \dot{x}_{pL}\cos\theta + \dot{y}_{pL}\sin\theta = R\dot{\varphi}_L \end{aligned} \tag{10}$$

Using the contact points velocities from equation (x,y) and

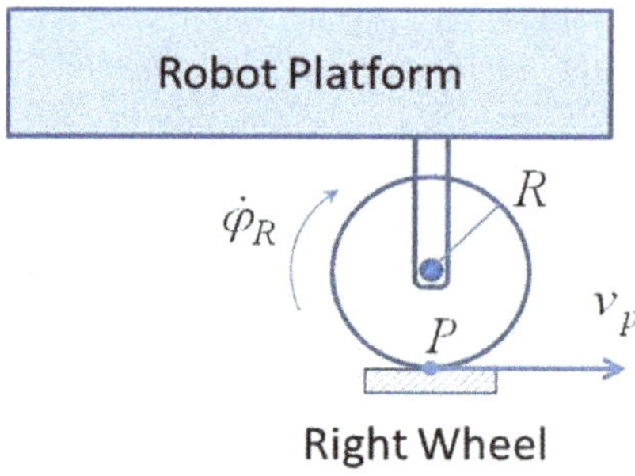

Figure 2: Pure Rolling Motion Constraint.

substituting in (x ,y), the three constraint equations can be written in the following matrix form:

$$\Lambda(q)\dot{q}=0 \tag{11}$$

Where

$$\Lambda(q)=\begin{bmatrix} -\sin\theta & \cos\theta & 0 & 0 & 0 \\ \cos\theta & \sin\theta & L & -R & 0 \\ \cos\theta & \sin\theta & -L & 0 & -R \end{bmatrix} \tag{12}$$

and

$$\dot{q}=\begin{bmatrix}\dot{x}_a & \dot{y}_a & \dot{\theta} & \dot{\varphi}_R & \dot{\varphi}_L\end{bmatrix}^T \tag{13}$$

$$\begin{cases} v_R = R\dot{\varphi}_R \\ v_L = R\dot{\varphi}_L \end{cases} \tag{14}$$

The above constraints matrix $\Lambda(q)$ will be used in the next section for the DDMR dynamic modeling.

Kinematic Model

Kinematic modeling is the study of the motion of mechanical systems without considering the forces that affect the motion. For the DDMR, the main purpose of kinematic modeling is to represent the robot velocities as a function of the driving wheels velocities along with the geometric parameters of the robot.

The linear velocity of each driving wheel in the Robot Frame is therefore, the linear velocity of the DDMR in the Robot Frame is the average of the linear velocities of the two wheels

$$v=\frac{v_R+v_L}{2}=R\frac{(\dot{\varphi}_R+\dot{\varphi}_L)}{2} \tag{15}$$

and the angular velocity of the DDMR is

$$\omega=\frac{v_R-v_L}{2L}=R\frac{(\dot{\varphi}_R-\dot{\varphi}_L)}{2} \tag{16}$$

The DDMRs velocities in the robot frame can now be represented in terms of the center-point A velocities in the robot frame as follows:

$$\begin{cases} \dot{x}_a^r = R\dfrac{(\dot{\varphi}_R+\dot{\varphi}_L)}{2} \\ \dot{y}_a^r = 0 \\ \dot{\theta}=\omega=R\dfrac{(\dot{\varphi}_R+\dot{\varphi}_L)}{2L} \end{cases} \tag{17}$$

Thus

$$\begin{bmatrix}\dot{x}_a^r \\ \dot{y}_a^r \\ \dot{\theta}\end{bmatrix}=\begin{bmatrix}\frac{R}{2} & \frac{R}{2} \\ 0 & 0 \\ \frac{R}{2L} & -\frac{R}{2L}\end{bmatrix}\begin{bmatrix}\dot{\varphi}_R \\ \dot{\varphi}_L\end{bmatrix} \tag{18}$$

The DDMR velocities can be obtained also in the inertial frame as follows:

$$\dot{q}^I=\begin{bmatrix}\dot{x}_a^r \\ \dot{y}_a^r \\ \dot{\theta}\end{bmatrix}=\begin{bmatrix}\frac{R}{2}\cos\theta & \frac{R}{2}\cos\theta \\ \frac{R}{2}\sin\theta & \frac{R}{2}\sin\theta \\ \frac{R}{2L} & -\frac{R}{2L}\end{bmatrix}\begin{bmatrix}\dot{\varphi}_R \\ \dot{\varphi}_L\end{bmatrix} \tag{19}$$

Equation (19) represents the forward kinematic model of the DDMR. Another alternative form for the kinematic model can be obtained by representing the DDMR velocities in terms of the linear and angular velocities of DDMR in the Robot frame.

$$\dot{q}^I=\begin{bmatrix}\dot{x}_a^r \\ \dot{y}_a^r \\ \dot{\theta}\end{bmatrix}=\begin{bmatrix}\cos\theta & 0 \\ \sin\theta & 0 \\ 0 & 1\end{bmatrix}\begin{bmatrix}v \\ \omega\end{bmatrix} \tag{20}$$

Dynamic Modeling of the DDMR

Dynamics is the study of the motion of a mechanical system taking into consideration the different forces that affect its motion unlike kinematics where the forces are not taken into consideration. The dynamic model of the DDMR is essential for simulation analysis of the DDMR motion and for the design of various motion control algorithms.

A non-holonomic DDMR with n generalized coordinates (q_1,q_2,… ,q_n) and subject to m constraints can be described by the following equations of motion:

$$M(q)\ddot{q}+V(q,\dot{q})\dot{q}+F(\dot{q})+G(q)+\tau_d=B(q)\tau-\Lambda^T(q)^{\lambda} \tag{21}$$

where:

$M(q)$ an nxn symmetric positive definite inertia matrix, $V(q,\dot{q})$ is the centripetal and coriolis matrix, $F(\dot{q})$ is the surface friction matrix, $G(q)$ is the gravitational vector, τ_d is the vector of bounded unknown disturbances including unstructured unmodeled dynamics, $B(q)$ is the input matrix, τ is the input vector, $\Lambda^T(q)$ is the matrix associated with the kinematic constraints, and λ is the Lagrange multipliers vector [21,22].

Lagrange dynamic approach

Lagrange dynamic approach is a very powerful method for formulating the equations of motion of mechanical systems. This method, which was introduced by Lagrange, is used to systematically derive the equations of motion by considering the kinetic and potential energies of the given system.

The Lagrange equation can be written in the following form:

$$\frac{d}{dt}\left(\frac{\partial L}{\partial \dot{q}_i}\right)+\frac{\partial L}{\partial q_i}=F-\Lambda^T(q)^{\lambda} \tag{22}$$

Where $L=T-V$ is the Lagrangian function, T, is the kinetic energy of the system, V is the potential energy of the system, q_i are the generalized coordinates, F is the generalized force vector, Λ is the constraints matrix, and λ is the vector of Lagrange multipliers associated with the constraints.

The first step in deriving the dynamic model using the Lagrange approach is to find the kinetic and potential energies that govern the motion of the DDMR. Furthermore, since the DDMR is moving in the X_I-Y_I plane, the potential energy of the DDMR is considered to be zero.

For the DDMR, the generalized coordinates are selected as

$$q=\begin{bmatrix}x_a & y_a & \theta & \varphi_R & \varphi_L\end{bmatrix}^T \tag{23}$$

The kinetic energies of the DDMR is the sum of the kinetic energy of the robot platform without wheels plus the kinetic energies of the wheels and actuators.

The kinetic energy of the robot platform is

$$T_c=\frac{1}{2}m_c v_c^2+\frac{1}{2}I_c\dot{\theta}^2 \tag{24}$$

While the kinetic energy of the right and left wheel is

$$T_{wR} = \frac{1}{2} m_w v_{wR}^2 + \frac{1}{2} I_m \dot{\theta}^2 + \frac{1}{2} I_w \dot{\varphi}_R^2 \quad (25)$$

$$T_{wL} = \frac{1}{2} m_w v_{wL}^2 + \frac{1}{2} I_m \dot{\theta}^2 + \frac{1}{2} I_w \dot{\varphi}_L^2 \quad (26)$$

where, m_c is the mass of the DDMR without the driving wheels and actuators (DC motors), m_w is the mass of each driving wheel (with actuator), I_c is the moment of inertia of the DDMR about the vertical axis through the center of mass, I_w is the moment of inertia of each driving wheel with a motor about the wheel axis, and I_m is the moment of inertia of each driving wheel with a motor about the wheel diameter.

All velocities will be first expressed as a function of the generalized coordinates using the general velocity equation in the inertial frame.

$$v_i^2 = \dot{x}_i^2 + \dot{y}_i^2 \quad (27)$$

The X_I and Y_I components of the center of mass and wheels can be obtained in terms of the generalized coordinates as follow

$$\begin{cases} x_c = x_a + d\cos\theta \\ y_c = y_a + d\sin\theta \end{cases} \quad (28)$$

$$\begin{cases} x_{wR} = x_a + L\sin\theta \\ y_{wR} = y_a + L\cos\theta \end{cases} \quad (29)$$

$$\begin{cases} x_{wL} = x_a - L\sin\theta \\ y_{wL} = y_a + L\cos\theta \end{cases} \quad (30)$$

Using equations (24)-(26) along with equations (27- (30), the total kinetic energy of the DDMR is

$$T = \frac{1}{2} m(\dot{x}_a^2 + \dot{y}_a^2) - m_c d\dot{\theta}(\dot{y}_a \cos\theta - \dot{x}_a \sin\theta) + \frac{1}{2} I_w (\dot{\varphi}_R^2 + \dot{\varphi}_L^2) + \frac{1}{2} I\dot{\theta}^2 \quad (31)$$

where the following new parameters are introduced

$m = m_c + 2m_w$ is the total mass of the robot, $I = I_c + m_c d^2 + 2m_w L^2 + 2I_m$ and is the total equivalent inertia.

Using equation (22) along with the Lagrangian function, L=T the equations of motion of the DDMR are given by

$$m\ddot{x}_a - md\ddot{\theta}\sin\theta - md\dot{\theta}^2\cos\theta = C_1 \quad (32)$$

$$m\ddot{y}_a - md\ddot{\theta}\cos\theta - md\dot{\theta}^2\sin\theta = C_2 \quad (33)$$

$$I\ddot{\theta} - md\ddot{x}_a \sin\theta + md\ddot{y}_a c\text{os}\,\theta = C_3 \quad (34)$$

$$I_w \ddot{\varphi}_R = \tau_R + C_4 \quad (35)$$

$$I_w \ddot{\varphi}_L = \tau_L + C_5 \quad (36)$$

where (C_1, C_2, C_3, C_4, C_5), are coefficients related to the kinematic constraints, which can be written in terms of the Lagrange multipliers vector λ and the kinematic constraints matrix Λ introduced in section 3.

$$\Lambda^T(q) = \begin{bmatrix} C_1 \\ C_2 \\ C_3 \\ C_4 \\ C_5 \end{bmatrix} \quad (37)$$

Now, the obtained equations of motion (32)-(36) can be represented in the general form given by equation (21) as

$$M(q)\ddot{q} + V(q,\dot{q})\dot{q} = B(q)\tau - \Lambda^T(q)^{\lambda} \quad (38)$$

Where

$$M(q) = \begin{bmatrix} m & 0 & -md\sin\theta & 0 & 0 \\ 0 & m & md\cos\theta & 0 & 0 \\ -md\sin\theta & md\cos\theta & I & 0 & 0 \\ 0 & 0 & 0 & & 0 \\ 0 & 0 & 0 & 0 & I_w \end{bmatrix},$$

$$V(q,\dot{q}) = \begin{bmatrix} 0 & -md\dot{\theta}\cos\theta & 0 & 0 & 0 \\ 0 & -md\dot{\theta}\sin\theta & 0 & 0 & 0 \\ 0 & 0 & 0 & 0 & 0 \\ 0 & 0 & 0 & 0 & 0 \\ 0 & 0 & 0 & 0 & 0 \end{bmatrix}$$

$$B(q) = \begin{bmatrix} 0 & 0 \\ 0 & 0 \\ 0 & 0 \\ 1 & 0 \\ 0 & 1 \end{bmatrix}, \text{and } \Lambda^T(q)\lambda = \begin{bmatrix} -\sin\theta & \cos\theta & \cos\theta \\ \cos\theta & \sin\theta & \sin\theta \\ 0 & L & -L \\ 0 & -R & 0 \\ 0 & 0 & -R \end{bmatrix} \times \begin{bmatrix} \lambda_1 \\ \lambda_2 \\ \lambda_3 \\ \lambda_4 \\ \lambda_5 \end{bmatrix}$$

Next, the system described by equation (38) is transformed into an alternative form which is more convenient for the purpose of control and simulation. The main aim is to eliminate the constraint term $\Lambda^T(q)\lambda$ in equation (88) since the Lagrange multipliers λ_i are unknown. This is done first by defining the reduced vector

$$\dot{\eta} = \begin{bmatrix} \dot{\varphi}_R \\ \dot{\varphi}_L \end{bmatrix} \quad (39)$$

Next, by expressing the generalized coordinates velocities using the forward kinematic model (19). Then we have

$$\begin{bmatrix} \dot{x}_a \\ \dot{y}_a \\ \dot{\theta} \\ \dot{\varphi}_R \\ \dot{\varphi}_L \end{bmatrix} = \frac{1}{2} \begin{bmatrix} R\cos\theta & R\cos\theta \\ R\sin\theta & R\sin\theta \\ \frac{R}{L} & -\frac{R}{L} \\ 2 & 0 \\ 0 & 2 \end{bmatrix} \begin{bmatrix} \dot{\varphi}_R \\ \dot{\varphi}_L \end{bmatrix} \quad (40)$$

This can be written in the form

$$\dot{q} = S(q)\eta \quad (41)$$

It can be verified that the transformation matrix $S(q)$ is in the null space of the constraint matrix $\Lambda(q)$. Therefore we have

$$S^T(q)\Lambda^T(q) = 0 \quad (42)$$

Next, taking the time derivative of equation (41) gives

$$\ddot{q} = \dot{S}(q)\eta + S(q)\dot{\eta} \quad (43)$$

Substituting equations (41) and (43) in the main equation (38) we obtain

$$M(q)\left[\dot{S}(q)\eta + S(q)\dot{\eta}\right] + V(q,\dot{q})\left[S(q)\eta\right] = B(q)\tau - \Lambda^T(q)\lambda \quad (44)$$

Next, rearranging the equation and multiplying both sides by leads to

$$S^T(q)M(q)S(q)\dot{\eta}+S^T(q)\left[M(q)\dot{S}(q)+V(q,\dot{q})S(q)\right]\eta$$
$$=S^T(q)B(q)\tau-S^T(q)\Lambda^T(q)\lambda \quad (45)$$

where the last term is identically zero. Now defining the new matrices

$$\overline{M}(q)=S^T(q)M(q)S(q)$$
$$\overline{V}=S^T(q)M(q)\dot{S}(q)+S^T(q)V(q,\dot{q})S(q),$$
$$\overline{B}=S^T(q)B(q)$$

The dynamic equations are reduced to the form

$$\overline{M}(q)\dot{\eta}+\overline{V}(q,\dot{q})\eta=\overline{B}(q)\tau \quad (46)$$

Where

$$\overline{M}(q)=\begin{bmatrix} I_w+\frac{R^2}{4L^2}(mL^2+I) & \frac{R^2}{4L^2}(mL^2-I) \\ \frac{R^2}{4L^2}(mL^2-I) & I_w+\frac{R^2}{4L^2}(mL^2+I) \end{bmatrix}$$

$$\overline{V}(q,\dot{q})=\begin{bmatrix} 0 & \frac{R^2}{2L}m_c d\dot{\theta} \\ -\frac{R^2}{2L}m_c d\dot{\theta} & 0 \end{bmatrix}, \quad \overline{B}(q)=\begin{bmatrix} 1 & 0 \\ 0 & 1 \end{bmatrix}$$

Equation (46) shows that the DDMR dynamics are expressed only as a function of the right and left wheel angular velocities $(\dot{\varphi}_R,\dot{\varphi}_L)$, the robot angular velocity $\dot{\theta}$ and the driving motor torques (τ_R,τ_L). The equations of motion (46) can be also transformed into an alternative form which is represented by the linear and angular velocities (v,w)of the DDMR. Using the kinematic model equations (15) and (16), it can be easily shown that the model equations (46) can be rearranged in the following compact form

$$\begin{cases} \left(m+\frac{2I_w}{R^2}\right)\dot{v}-m_c d\omega^2=\frac{1}{R}(\tau_R+\tau_L) \\ \left(I+\frac{2L^2}{R^2}I_w\right)\dot{\omega}+m_c d\omega v=\frac{L}{R}(\tau_R-\tau_L) \end{cases} \quad (47)$$

Newton-Euler approach

The first and most important step in Newton-Euler dynamic modeling is to draw the free body diagram of the system and to analyze the forces acting on it. The free body diagram of the differential drive mobile robot is shown in Figure 3. Using the robot local frame {xr, yr}, the following notations are introduced.

(v_u,v_w) represents the velocity of the vehicle center of mass C in the local frame; v_u is the longitudinal velocity and v_w is the lateral velocity; (a_u,a_w) represent the acceleration of the vehicle's center of mass C; (F_{u_R},F_{u_L}) are the longitudinal forces exerted on the vehicle by the left and right wheels; (F_{w_R},F_{w_L}) are the lateral forces exerted on the vehicle by the left and right wheels; θ is the orientation of the robot; ω is the angular velocity; *m* is the mass of the robot; and is the yaw moment of inertia with respect to the center of mass.

As it can be seen from the above free body diagram, the only forces acting on the robot are actuator forces acting on the robot wheels.

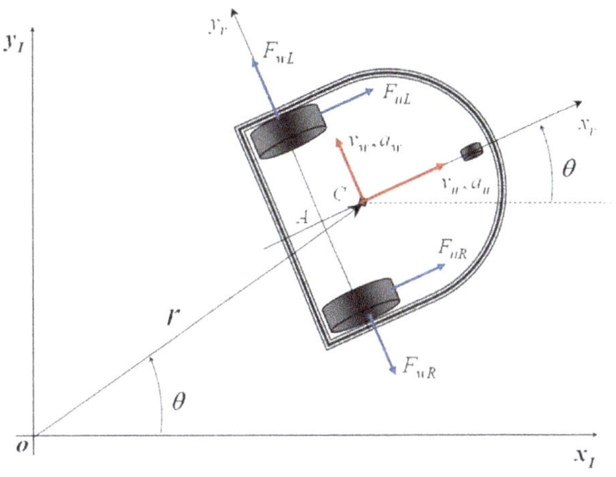

Figure 3: Robot free body diagram for Newtonian dynamic modeling.

We start the derivation by representing the robot position using polar coordinates. Assuming that the robot is a rigid body, its polar coordinates in the inertial frame can be represented using a complex vector

$$\hat{r}=re^{j\theta} \quad (48)$$

Differentiating the above position vector with respect to time will give us the velocity and acceleration of the robot in the inertial frame.

$$\dot{\hat{r}}=\dot{r}e^{j\theta}+jr\dot{\theta}e^{j\theta} \quad (49)$$
$$\ddot{\hat{r}}=\ddot{r}e^{j\theta}+2j\dot{r}\dot{\theta}e^{j\theta}+jr\ddot{\theta}e^{j\theta}-r\dot{\theta}^2e^{j\theta} \quad (50)$$

Simplifying and writing the velocity and acceleration terms in radial and tangential terms, we have

$$\dot{\hat{r}}=[\dot{r}]e^{j\theta}+[r\dot{\theta}]e^{j\left(\theta+\frac{\pi}{2}\right)} \quad (51)$$

$$\ddot{\hat{r}}=[\ddot{r}-r\dot{\theta}^2]e^{j\theta}+[2\dot{r}\dot{\theta}+r\ddot{\theta}]e^{j\left(\theta+\frac{\pi}{2}\right)} \quad (52)$$

The radial and tangential velocity and acceleration terms are defined as

$$v_u=\dot{r} \quad (53)$$
$$v_w=r\dot{\theta} \quad (54)$$
$$a_u=\ddot{r}-r\dot{\theta}^2 \quad (55)$$
$$a_w=2\dot{r}\dot{\theta}+r\ddot{\theta} \quad (56)$$

From the above four equations, we can write the following relations between the radial and tangential components of the robot velocity and acceleration

$$a_u=\dot{v}_u-v_w\dot{\theta} \quad (57)$$
$$a_w=\dot{v}_w-v_u\dot{\theta} \quad (58)$$

The above equations (57) and (58) are the fundamental acceleration equations that can be also obtained using the theorem of motion of a rigid body in a rotating reference frame [21,22].

The next step is to write the Newton's second law of motion in the robot frame and find the relationship between the forces, torques, and accelerations. The DDMR exhibits two types of motion: translations in the radial and tangential directions, and rotation around the vertical axis at the center of mass. Let M be the total mass of the robot including the wheels and actuators and J the moment of inertia with respect to the center of mass. Then the dynamic equations are

$$Ma_u = F_{uL} + F_{uR} \tag{59}$$

$$Ma_w = F_{wL} - F_{wR} \tag{60}$$

$$J\ddot{\theta} = \left(F_{uR} - F_{uL}\right)L + \left(F_{wR} - F_{wL}\right)d \tag{61}$$

Substituting the acceleration terms from (57) and (58) we get

$$\dot{v}_u = v_w\dot{\theta} + \frac{F_{uL} + F_{uR}}{M} \tag{62}$$

$$\dot{v}_w = -v_u\dot{\theta} + \frac{F_{wL} - F_{wR}}{M} \tag{63}$$

$$\ddot{\theta} = \frac{L}{J}\left(F_{uR} - F_{uL}\right) + \frac{d}{J}\left(F_{wR} - F_{wL}\right) \tag{64}$$

The absence of slipping (pure rolling) in the longitudinal direction and no sliding in the lateral direction creates independence between the longitudinal, lateral and angular velocities and simplifies the dynamic equations. These non-holonomic constraints are taken into account by defining the velocity of the center-point A in the local frame and forcing it to be zero. Using the transformation matrix $R(\theta)$, we first find the velocity of the center of mass C in the inertial frame as

$$\begin{bmatrix} \dot{x}_c \\ \dot{y}_c \end{bmatrix} = \begin{bmatrix} \cos\theta & -\sin\theta \\ \sin\theta & \cos\theta \end{bmatrix} \times \begin{bmatrix} v_u \\ v_w \end{bmatrix} \tag{65}$$

Next, using equation (28), we can find the velocity of the center-point A in the inertial frame. It can then be shown that the lateral velocity of point A in the local frame is $v_w - d\dot{\theta}$. Therefore, in the absence of lateral slippage we have $v_w = d\dot{\theta}$ (66)

Next, substituting (66) in (62), (63), and combining with (64) we obtain

$$\dot{v}_u = d\dot{\theta}^2 + \frac{1}{M}\left(F_{uL} + F_{uR}\right) \tag{67}$$

$$\ddot{\theta} = \frac{L}{Md^2 + J}\left(F_{uR} - F_{uL}\right) - \frac{Mdv_u}{Md^2 + J}\dot{\theta} \tag{68}$$

The above two equations are the dynamic equations of the robot considering the non-holonomic constraints. These equations can now easily be transformed to show the actuator torques applied to the wheels similar to the notations used in the Lagrangian approach.

$$M\dot{v}_u\dot{\theta} - Md\dot{\theta}^2 = \frac{1}{R}\left(\tau_R + \tau_L\right) \tag{69}$$

$$\left(Md^2 + J\right)\ddot{\theta} + Mdv_u\dot{\theta} = \frac{L}{R}\left(\tau_R - \tau_L\right) \tag{70}$$

Next, these two equations can be written in matrix form as follows

$$\begin{bmatrix} M & 0 \\ 0 & Md^2 + J \end{bmatrix}\begin{bmatrix} \dot{v}_u \\ \ddot{\theta} \end{bmatrix} + \begin{bmatrix} 0 & -Md\dot{\theta} \\ Md\dot{\theta} & 0 \end{bmatrix}\begin{bmatrix} \dot{v}_u \\ \dot{\theta} \end{bmatrix} = \frac{1}{R}\begin{bmatrix} 1 & 1 \\ L & -L \end{bmatrix}\begin{bmatrix} \tau_R \\ \tau_L \end{bmatrix} \tag{71}$$

As it can be observed, equation (62) is similar to equation (47), which was obtained using the Lagrangian approach. Note that in the Newton-Euler approach the mass and inertias of the wheels were not taken into consideration, and the robot is considered as one rigid body. Therefore both formulations are equivalent if the inertia and mass parameters are defined as

$$M = m_c \tag{72}$$

$$J = I_c \tag{73}$$

Next, using the forward kinematics equations (15) and (16), we can easily rewrite the general dynamic equations (71) in terms of the wheels rotational velocities and actuator torques. The leads to the following formulation

$$\left[\frac{R\left(Md^2 + J\right)}{4L^2} + \frac{MR}{4}\right]\ddot{\varphi}_R + \left[-\frac{R\left(Md^2 + J\right)}{4L^2} + \frac{MR}{4}\right]\ddot{\varphi}_L - \left(\frac{MdR^2}{4L^2}\right)\dot{\varphi}_L^2 + \left(\frac{MdR^2}{4L^2}\right)\dot{\varphi}_R\dot{\varphi}_L = \frac{1}{R}\tau_R \tag{74}$$

$$\left[\frac{R\left(Md^2 + J\right)}{4L^2} + \frac{MR}{4}\right]\ddot{\varphi}_L + \left[-\frac{R\left(Md^2 + J\right)}{4L^2} + \frac{MR}{4}\right]\ddot{\varphi}_R - \left(\frac{MdR^2}{4L^2}\right)\dot{\varphi}_R^2 + \left(\frac{MdR^2}{4L^2}\right)\dot{\varphi}_R\dot{\varphi}_L = \frac{1}{R}\tau_L \tag{75}$$

The above equations are also equivalent to those derived using the Lagrangian approach as given by equation (46).

Figure 4 shows the dynamic model of the DDMR representing the equations of motion (69) and (70). This model shows clearly the coupling between the motor torques, the linear and angular velocities of the robot, and the wheels velocities. This model can be adequately used for DDMR simulation and analysis.

Actuator Modeling

The dc motors which are generally used to drive the wheels of a differential drive mobile robot system are considered to be the servo actuators. In an armature-controlled dc motor which is the case for our DDMR system, the armature voltage v_a is used as the control input while keeping the conditions in the field circuit constant. In particular, for a permanent-magnet dc motor, we have the following equations for the armature circuit

$$\begin{cases} v_a = R_a i_a + L_a \dfrac{di_a}{dt} + e_a \\ e_a = K_b\omega_m \\ \tau_m = K_t i_a \\ \tau = N\tau_m \end{cases} \tag{76}$$

where, i_a is the armature current,(R_a,L_a) is the resistance and inductance of the armature winding respectively, e_a is the back emf, w_m is the rotor angular speed , τ_m is the motor torque, (K_t,K_b) are the torque constant and back emf constant respectively, N is the gear ratio, and is τ the output torque applied to the wheel.

Since in the DDMR the motors are mechanically coupled to the robot wheels through the gears, the mechanical equations of motion of the motors are linked directly with the mechanical dynamics of the DDMR. Therefore each dc motor will have

$$\begin{cases} \omega_{mR} = N\dot{\varphi}_R \\ \omega_{mL} = N\dot{\varphi}_L \end{cases} \tag{77}$$

The total dynamic equations of the DDMR with the actuators

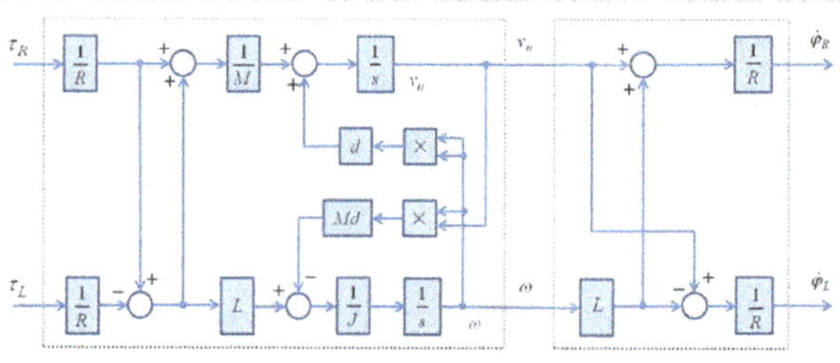

Figure 4: DDMR Dynamic Model.

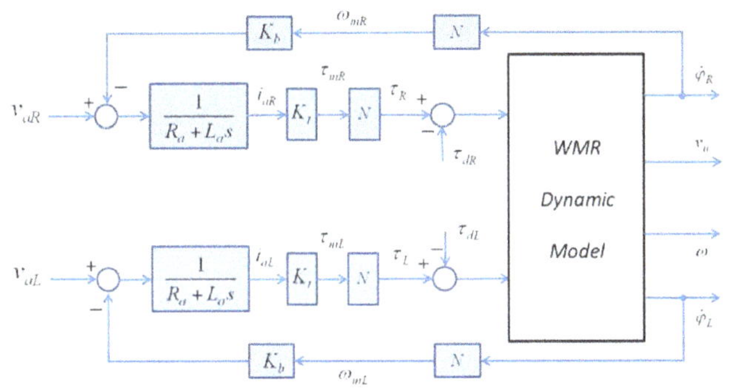

Figure 5: DDMR dynamic model with actuators.

are obtained by combining equation (76) for each motor with the mechanical dynamics of the DDMR. Additional torque disturbances acting on the wheels can be included as additive terms to the motor torques. Figure 5 shows a block diagram representation of the overall system. The forward kinematic model (19) can be added as a cascade to the dynamic model to form a complete model for simulation and analysis of the DDMR.

Conclusion

We have presented a detailed derivation of the dynamic model of a differential-drive mobile robot using the Lagrange and Newton-Euler methods. They were shown to be mathematically equivalent providing a check on their consistency. The equations of motion of DC motors actuators were also added to form the complete dynamic model of the DDMR. The insight gained in this study will assist engineering students and researchers in the modeling and design of suitable controllers for DDMR navigation and trajectory tracking.

References

1. Mitchell R, Warwick K, Browne WN, Gasson MN, Wyatt J (2010) Engaging Robots: Innovative Outreach for Attracting Cybernetics Students. IEEE Transactions on Education 53: 105-113
2. Dhaouadi R, Sleiman M (2011) Development of a modular mobile robot platform for motion-control education. IEEE Industrial Electronics Magazine 5: 35-45.
3. Shibata T, Murakami T (2012) Power-Assist Control of Pushing Task by Repulsive Compliance Control in Electric Wheelchair. IEEE Trans. Ind. Electron 59: 511-520.
4. Yang SX, Zhu A, Yuan G, Meng MQH (2012) A Bioinspired Neurodynamics-Based Approach to Tracking Control of Mobile Robots. IEEE Trans. Ind. Electron 59: 3211-3220.
5. Fang Y, Liu X, Zhang X (2012) Adaptive Active Visual Servoing of Nonholonomic Mobile Robots. IEEE Trans. Ind. Electron 59: 486-497.
6. Huang HP, Yan JY, Hu Cheng T (2010) Development and Fuzzy Control of a Pipe Inspection Robot. IEEE Trans. Ind. Electron 57: 1088-1095.
7. Li THS, Yeh YC, Da Wu J, Hsiao MY, Chen CY (2010) Multifunctional Intelligent Autonomous Parking Controllers for Carlike Mobile Robots. IEEE Trans. Ind. Electron 57: 1687-1700
8. Campion G, Bastin G, Novel BD (1996) Structural properties and classification of kinematic and dynamic models of wheeled mobile robot. IEEE Transactions on Automatic Control 12: 47-62.
9. Fukao T, Nakagawa H, Adachi N (2000) Adaptive Tracking Control of a Nonholonomic Mobile Robot .IEEE Transaction on Robotics and Automation 16: 609-615.
10. Hou ZG, Zou AM, Cheng L, Tan M (2009) Adaptive Control of an Electrically Driven Nonholonomic Mobile Robot Via Back stepping and Fuzzy Approach. IEEE Transaction on Control Systems Technology 17: 803-815.
11. Fierro R,Lewis FL (1997) Control of a nonholonomic mobile robot: back stepping kinematics into dynamics. Journal of Robotic Systems 14: 149-163.
12. Yamamoto Y, Yun X (1992) On Feedback Linearization of Mobile Robots. Technical Report No. MS- CIS-92-45, Philadelphia, PA.
13. Yamamoto Y,Yun X (1992) Coordinating Locomotion and Manipulation of a Mobile Manipulator Technical Report No. MS-CIS-92-18 Philadelphia, PA.
14. Sarkar N, Yun X, Kumar V (1994) Control of mechanical systems with rolling constraints: Application to dynamic control of mobile robots Int. J. Robot. Res 13: 55-69.
15. Yun X,Yamamoto Y (1993)Internal dynamics of a wheeled mobile robot. IEEE/RSJ International Conference on Intelligent Robots and System (IROS'93) 2: 1288-1294.
16. DeSantis RM (1995) Modeling and Path-tracking Control of a mobile Wheeled Robot with a Differential Drive. Robotica 13: 401-410.
17. de Vries TJA, van Heteren C, Huttenhuis L(1999) Modelling and control of a fast moving, highly maneuverable wheelchair. Proceedings of the International Bio mechatronics Workshop 6: 110-115.
18. Albagul A, Wahyudi A (2004) Dynamic Modelling and Adaptive Traction Control for Mobile Robots. International Journal of Advanced Robotic Systems 1: 149-154.
19. Thanjavur K, Rajagopalan R (1997) Ease of dynamic modelling of wheeled mobile robots (WMRs) using Kane's approach. IEEE International Conference on Robotics and Automation 4 : 2926-2931
20. Thanjavur K, Rajagopalan R (1997) Ease of dynamic modelling of wheeled mobile robots (WMRs) using Kane's approach. IEEE International Conference on Robotics and Automation 4: 2926-2931
21. Neĭmark JI, Fufaev NA (1972) Dynamics of Nonholonomic Systems: Translations of Mathematical Monographs. American mathematical Society
22. Bloch A, Crouch P, Baillieul J, Marsden J (2003) Nonholonomic Mechanics and Control.

A Venture to the Latest Robotic Technological Research

Prarthana BK*, Bhavana Y and Mani Shankar N

Department of Mechanical Engineering, SNIST, India

Abstract

Robots made its place in the pre 20th century era as a part of an entertaining tool who never thought it could have any sensational role in today's life. From the baby's toy to the current sensor bots and much more advanced innovations as such have made man's life much easier, luxurious and accurate outcomes. Thus this field is currently expanding and enormously growing in the way nobody can just estimate. This field has shown much of profit zone and promising outcomes which have benefited many colossal arenas. This successive generation of Robots has challenged the current scientists and is welcoming the new ideologies and technologies to improve the functioning for a better tomorrow.

This current review tries to put an effort in understanding the advancements that has taken place in the past few decades to the date. It tries to bridge up the gap and helps the current budding scientists to understand the present scenario who can challenge the enormously improving technologies and make much more advancements in this field.

Keywords: Robotic technology; Recent advancements; Latest research; End effectors; Robot programming languages

Introduction

The robotic history has its roots since ancient myths and legends. Modern concepts began when industrial revolution allowed the use of complex mechanics. After 1920s human sized robots were developed with the capacity for near human thoughts and movements. Robots were first used in industries for manufacturing tasks without the need of human assistance. Digitally controlled robots and robots making use of artificial intelligence started to develop since 1960s.

Can we ever imagine that the origin of robotics have begun from 200 BC? Archytas built a mechanical wooden bird "The Pigeon" propelled by steam around 350 BC. In 1921, Karel Capek came up with word robot for his intelligence, artificially created person for the first time in his play. Though the concept of robot was around since ancients, modern day robot was born with the arrival of computers in 1940s. The robot became a popular concept from late 1950 onwards. Industrial robots do not have human like appearance. They are computer controlled manipulators. Robots were created to help humans. Robots can do even those tasks which human cannot do. Engleberger modified the earliest robots invented by George C. Devol into industrial robots and formed a company called Unimation to produce and market the robots. For his efforts and successes, he is known as "the Father of Robotics" in the industry.

Robots can be of various types like autonomous, remote controlled or semi autonomous. Robots have replaced humans in the assistance to perform those repetitive and dangerous tasks which humans prefer not to do, or are unable to do due to size limitations, or even those such as in outer space or at the bottom of the sea where humans could not sustain the extreme environments. Robots in earlier ages were used primarily for entertainment purpose. In today's modern life, technology has contributed in many ways to comfort people's lives. Especially robotic systems with an artificial intelligence have many industrial applications and have become increasingly important for some people. From 20th century onwards the development of industrial robots changed the structure of society and allowed for safer conditions for labor. Industry has benefited drastically from the robotic work force. Automated machines have taken the dangerous jobs from humans and allowed greater productivity. Farmers have taken advantage of this new robotic technology with automated harvesters, the medical industry benefits from advancements in assisted surgical robotics. IBM runs a "lights off" factory in Texas, USA with an idea of a factory without human workers, which is staffed by fully autonomous robots making keyboards. Among the various programs launched by the currently military in the field of robotic technology, the most successful one was the predator and Reaper unmanned aerial reconnaissance vehicles which allowed only a pilot to control the robot from vast distances. The major advantage of these vehicles being high-altitude surveillance for long periods with no support to a live pilot, and during emergencies the planes can launch with small strikes on targets in zones which couldn't be operated by normal aircrafts.

As a result of artificial intelligence development as well as advances made in the robotics field the interaction between robots and humans is becoming stronger as they have become an absolutely essential tool in ensuring the quality in our lives. The advent of robots is quickly becoming an intrinsic part of our daily lives. Understanding the advancements in the current scenario of robotic technology, which is not just a subject but a vast field that is exploring within its colossal innovations. This paper tries to make an effort in the current advancements of various applications of robotic technology.

An Introduction to Advanced Sensors

The fields of robot sensing have created a boom and shocked many scientists for this innovative technology. Literature suggests that sensors are commercially available as optical, inductive, capacitive, resistive,

***Corresponding author:** Prarthana BK, Department of Mechanical Engineering, SNIST, India, E-mail: prarthanabk17@gmail.com acoustics, magnetic, piezoelectric sensing principles and experimental transducers. Recently wireless sensor networks have become forefront in scientific niche (Figure 1). The major applications of this wireless sensor networks is the effort of engineering smaller sized devices and increasing the accuracy within the small and the smallest designed equipments. The recent robot invented that behaved like human included the key features like computers, sensors, auxiliary equipments and effectors which reported the major advantages like performing tasks similar to humans in unpredictable conditions. Anyhow intelligent robots are on great demand. Intelligent robots can be manufactured by manipulating adaptive control, mobility, robot programming languages, end effectors, manufacturing process planning and mobility. The major challenge to any scientist is to manufacture robot with low cost, high efficiency with more functional applications. The scope of robotic technology is vast which includes force control, wireless communication, safe control, 3D vision, multi robot control and remote robot supervision which could be a new career for budding engineers. Model based control has now been designed to control the industrial robots which are refined accordingly to improve the functionality when required. Driving forces management could be engineered in robots to have wide applications like assembling, heavy gadgets, automobiles or other machineries or as such. Micro robots add to one more discoveries to have its application in research arena to make cost effective in small scale industries.

MARIE (mobile and autonomous robotics integration environment), has multiple applications to operate multiple machines, operating systems by mobile integrated system. Latest technology involves replacement of thousands of pressure sensors with rubber pressure sensors to reduce the cost which has application as artificial skin [1-7]. Thus decreasing labor and efficiency with much outcome in limited time is the greatest advantage in the advancements of the robotic gadgets.

Basic Improvements Concerned to Manufacturing Industry

Robots have been recently introduced into the manufacturing technology. Robots perform wide range of functions such as manipulating a tool or handling the work piece [8]. Programming and maintenance are the only human tasks associated with robot installations. An intelligent and flexible robot is considered as a general purpose machine system that may include auxiliary equipment, sensors, computers and effectors. Such a robot should be able to perform variety of tasks like a human under any unpredictable conditions [9]. The robotic system in hull grit blasting which has high quality surface preparation reduces wastage to a large extent without any emission to the environment. It is a fully automated environmental friendly system with cost effective technology [10]. The service robots such as the current floor cleaning robot were developed mainly on the human factors [11]. There are many institutions like German Aerospace Centre (DRL) Institute of Robotics and Mechatronics, KUKA Robotics GmbH and German robot manufacturer that support the researchers who are aiming to take up as a career [12]. Design approach is the fundamental key in modular robots which consists of end effectors, links and joints for performing multi-operational tasks [13]. Jean-Daniel et al. [14], in a study says that with just a language knowledge and visibility to the behavior of the robot today it has been an enormous robot community being developed.

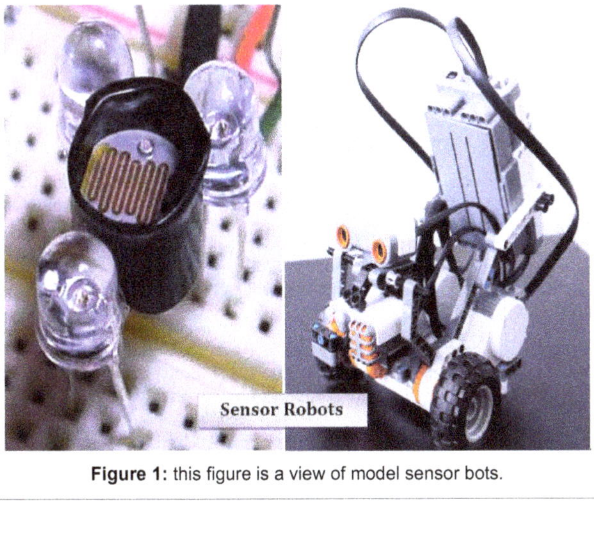

Figure 1: this figure is a view of model sensor bots.

With Reference to Computer Applications

A complex adaptive system (CAS) understands the complexity in natural systems. In this research the computer experiments led to the development of mathematical as well as computational techniques which are applicable in designing distributed control systems based complex system model which is composed of intelligent, autonomous and multiple agents [15]. Architecture and architecture styles are chosen for particular applications based on performance in previous work analysis on computer architecture design for robotics [16]. Telerobots are those service robots which keeps the human workers away from the high radiation areas. The reuse of software components has been the major advantage of teleoperation in modular architecture [17]. Experimental results conducted at Robotics and Automation Laboratory at Michigan State University and Communication and Robotics Laboratory at Oakland University reveals that internet based teleoperation systems provides various technical challenges which have profound impact on day to day human life style [18].

Current Traditions in Various Applications

The introduction of robotic technology in food industry has made the competitive and repetitive tasks simpler and also cost effective [19]. The robot takes the visual input and reasons the shape of the object and decides the best stable grasp online [20]. Robot control is the key technology that could be engineered in food industry [3]. The automatically climbing robot finds applications in industrial fields like construction and shipbuilding where maintenance work and inspection involve highly dangerous manual operations [21]. Climbing robots are developed for different applications ranging from cleaning to inspection to reach construction. Secure gripping force with light-weight mechanism is the key for proper adhesion to the surface [22]. Robots afford better human safety in workplaces where traditional vehicles are unable to reach. Gecko inspired wall climbing robots (Figure 2) which has low cost operations are used in many applications such as repair, inspection, exploration and cleaning [23]. Studies made on past and ongoing research projects in micro, macro and bio robotics reveal that these three areas of robotics have potential to provide better improvements to the present state of art of medical technology [24].

The introduction of robots with new technology has high implication such as skill requirements, investments and broader social issues. Adoption of robots in different industries varies accordingly with their applications [25]. Summary robots are used under conditions where human productivity is less or requires high quality of work. This robot has a potential application on the building site. Field robots work under any environment such as under water, on farms, in forests,

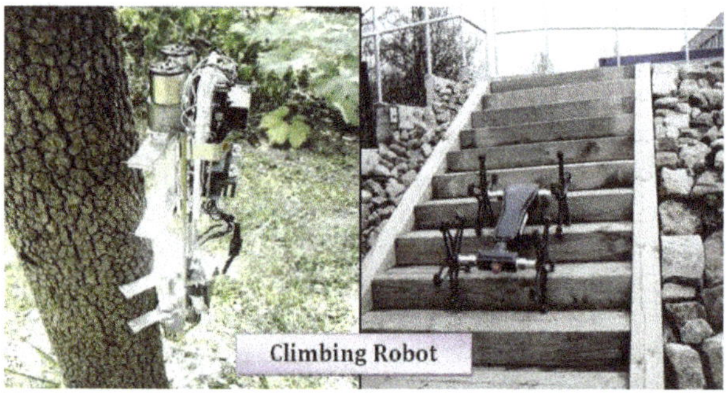

Figure 2: a glance of climbing bot.

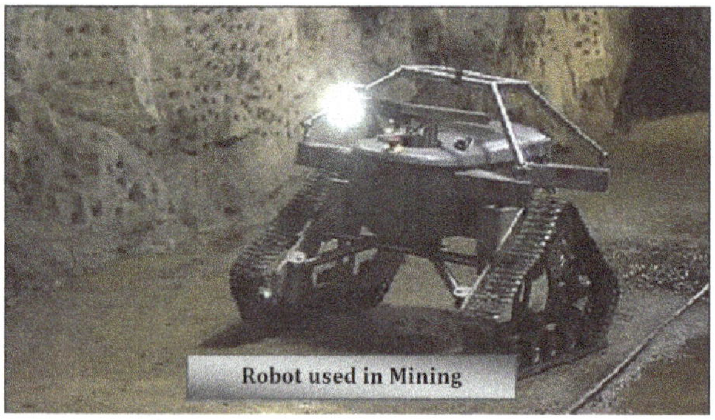

Figure 3: bot used for mining purpose.

in mines and in the air (Figure 3). A study by Chuck Thorpe Hugh Durrant-Whyte gives better ideas for design and development [26]. The space manipulation and the different class of applications like surface mobility in using polybot as it is having significant potential [27]. Miniature autonomous robots are capable of fast operations at molecular scale. Introduction of these robots in nanofactories can easily predict the exact shape and various components of nanofactories which operate at nanoscale [28].

Literature reveals that the use of robotic system in agricultural field like grass cutting on golf courses, robotic weeding in high value crops and crop scouting in cereals are more economically feasible than conventional systems [29]. Reuse of software in applications of robots is limited which is a key problem in software development of robotic system. So, software engineering techniques are implemented in industrial applications [30].

Conclusion

As the society is constantly and consistently transforming in short span is a symptom of development in the technology. Generation is growing more competitive, challenging and so the advancements are making a boom in the current scenario. Luxuries and comforts are the main cause for the new innovations with time compatibility and less labor. One such instrumental tool is robot. The current paper tries to accumulate some knowledge and information in concerned to the advancements in the robotic technology. It covers major industries and scientific areas to give a brief idea in the latest happenings of the research.

References

1. Arampatzis Th, Lygeros J, Manesis S (2005) A Survey of Applications of Wireless Sensors and Wireless Sensor Networks. Proceedings of the 2005 IEEE International Symposium on, Mediterrean Conference on Control and Automation.
2. Torgny B (2007) Present and future robot control development- An industrial perspective. Annual Reviews in Control 31: 69-79.
3. Wildberge MM (1997) Complex adaptive systems: concepts and power industry applications. Control Systems 6: 77-88.
4. Caprari G, Balmer P, Piguet R, Siegwart R (1998) The autonomous micro robot "Alice": a platform for scientific and commercial applications. Proceedings of the 1998 International Symposium on Micromechatronics and Human Science, Japan.
5. Arampatzis T, Lygeros J, Manesis S (2005) A Survey of Applications of Wireless Sensors and Wireless Sensor Networks. Proceedings of the 2005 IEEE International Symposium on Intelligent Control, Cyprus.
6. Cote C, Letourneau D, Michaud F, Valin JM, Brosseau Y, et al. (2004) Code reusability tools for programming mobile robots. (IROS 2004). Proceedings. 2004 IEEE/RSJ International Conference on Intelligent Robots and Systems, USA.
7. Someya T, Sekitani T, Iba S, Kato Y, Kawaguchi H, et al. (2004) A large-area, flexible pressure sensor matrix with organic field-effect transistors for artificial skin applications. Proceedings of the National Academy of Sciences of the United States of America, USA.
8. Edwards M (1984) Robots in industry: An overview. Applied Ergonomics 15: 45-53.
9. Torgny B (2007) Present and future robot control development-An industrial perspective. Annual Reviews in Control 31: 69-79.
10. Leung SS, Shanblatt MA (1988) Computer architecture design for robotics. Proceedings of IEEE International Conference on Robotics and Automation, USA.
11. Balaguer C, Gimenez A, Pastor JM, Pardon VM, Abderrahim M (2000) A climbing autonomous robot for inspection applications in 3D complex environments. Robotics 18: 287-297.
12. Amit G, Ning Xi, Elhajj IH (2005) Internet based robots: applications, impacts, challenges and future directions. Advanced Robotics and its Social Impacts, 2005. IEEE Workshop on 73-78.
13. Warszawski A (1986) Robots in the construction industry. Robotica 4: 181-188.
14. Jean-Daniel Dessimoz, Pierre-François Gauthey RH3-Y-Toward a Cooperating Robot for Home Applications.
15. Peter J Wallin (1997) Robotics in the food industry: an update. Trends in Food Science & Technology 8: 193-198.
16. Dario P, Guglielmelli E, Allotta B, Carrozza MC (1996) Robotics for medical applications. IEEE Robotics & Automation Magazine 3: 44-56.
17. Mark Y, Kimon R, David D, Ying Z, Craig E, et al. (2003) Modular Reconfigurable Robots in Space Applications. Autonomous Robots 14: 225-237.
18. Bernardine DM, Dani G, Anthony S. Market-Based Multirobot Coordination For Complex Space Applications
19. Iborra A, Pastor JA, Alvarez B, Fernandez C, Merono JMF (2003) Robots in radioactive environments. IEEE Robotics & Automation Magazine 10: 12-22.
20. Benhabib B, Dai MQ (1991) Mechanical design of a modular robot for industrial applications. Journal of Manufacturing Systems 10: 297-306.
21. Silva MF, Machado J, Tar JK (2008) A Survey of Technologies for Climbing Robots Adhesion to Surfaces. IEEE International Conference on Computational Cybernetics.
22. Hirzinger G, Bals J, Otter M, Stelte J (2005) The DLR-KUKA success story: robotics research improves industrial robots. IEEE Robotics & Automation Magazine 12: 16-23.
23. Fernandez-Andres C, Iborra A, Alvarez B, Pastor JA, Sanchez P, et al. (2005) Ship shape in Europe: cooperative robots in the ship repair industry. IEEE Robotics & Automation Magazine 12: 65-77.
24. Fleck J (1984) The adoption of robots in industry. Physics in Technology 15: 1.
25. Schofield M (1999) "Neither master nor slave...". A practical case study in the

development and employment of cleaning robots. Proceedings of 7th IEEE International Conference on Emerging Technologies and Factory Automation, Spain.

26. Chuck Thorpe Hugh Durrant-Whyte (2006) Field Robots 1-11.

27. Yim M, Roufas K, Duff D, Zhang Y, Eldershaw C, et al. (2003) Modular Reconfigurable Robots in Space Applications. Autonomous Robots 14: 225-237.

28. Sylvain M, Ian H (2002) Nanofactories based on a fleet of scientific instruments configured as miniature autonomous robots. Journal of Micromechatronics 2: 201-214.

29. Pedersen SM, Fountas S, Have H, Blackmore BS (2006) Agricultural robots-system analysis and economic feasibility. Precision Agriculture 7: 295-308.

30. Iborra A, Caceres D, Ortiz F, Franco J, Palma P, et al. (2009) Design of service robots. IEEE Robotics & Automation Magazine 16: 24-33.

Permissions

All chapters in this book were first published in ARA, by OMICS International; hereby published with permission under the Creative Commons Attribution License or equivalent. Every chapter published in this book has been scrutinized by our experts. Their significance has been extensively debated. The topics covered herein carry significant findings which will fuel the growth of the discipline. They may even be implemented as practical applications or may be referred to as a beginning point for another development.

The contributors of this book come from diverse backgrounds, making this book a truly international effort. This book will bring forth new frontiers with its revolutionizing research information and detailed analysis of the nascent developments around the world.

We would like to thank all the contributing authors for lending their expertise to make the book truly unique. They have played a crucial role in the development of this book. Without their invaluable contributions this book wouldn't have been possible. They have made vital efforts to compile up to date information on the varied aspects of this subject to make this book a valuable addition to the collection of many professionals and students.

This book was conceptualized with the vision of imparting up-to-date information and advanced data in this field. To ensure the same, a matchless editorial board was set up. Every individual on the board went through rigorous rounds of assessment to prove their worth. After which they invested a large part of their time researching and compiling the most relevant data for our readers.

The editorial board has been involved in producing this book since its inception. They have spent rigorous hours researching and exploring the diverse topics which have resulted in the successful publishing of this book. They have passed on their knowledge of decades through this book. To expedite this challenging task, the publisher supported the team at every step. A small team of assistant editors was also appointed to further simplify the editing procedure and attain best results for the readers.

Apart from the editorial board, the designing team has also invested a significant amount of their time in understanding the subject and creating the most relevant covers. They scrutinized every image to scout for the most suitable representation of the subject and create an appropriate cover for the book.

The publishing team has been an ardent support to the editorial, designing and production team. Their endless efforts to recruit the best for this project, has resulted in the accomplishment of this book. They are a veteran in the field of academics and their pool of knowledge is as vast as their experience in printing. Their expertise and guidance has proved useful at every step. Their uncompromising quality standards have made this book an exceptional effort. Their encouragement from time to time has been an inspiration for everyone.

The publisher and the editorial board hope that this book will prove to be a valuable piece of knowledge for researchers, students, practitioners and scholars across the globe.

List of Contributors

Yuji Ito
Graduate School of Engineering, Nagoya University, Nagoya, Japan

Youngwoo Kim
Korea Institute of Machinery & Materials (KIMM), Daegu Research Center for Medical Devices and Green Engergy, Dalseo-gu, Korea

Goro Obinata
EcoTopia Science Institute, Nagoya University, Nagoya, Japan

William R Hutchison, Betsy J Constantine and Jerry Pratt
MeMeMe Inc, 1470 Birchmount Rd, Scarborough, ON M1P 2G1, Canada

Ebrahim Mattar
College of Engineering, University of Bahrain, Kingdom of Bahrain

Ebrahim Mattar
College of Engineering, University of Bahrain, Kingdom of Bahrain

Furui Wang
Abbott Point of Care, Abbott Laboratories, Princeton, NJ, 08540, USA

Duygun Erol Barkana
Department of Electrical and Electronics Engineering, Yeditepe University, Istanbul, Turkey

Nilanjan Sarkar
Department of Mechanical Engineering, Vanderbilt University, Nashville, TN, 37212, USA

Basma El Zein
Dar Al Hekma University, Saudi Arabia

Pruthviraj RD
Department of Engineering Chemistry, Amruta institute of Engineering and Management Sciences, Bidadi, Bangalore, India

Subhash KC, Namratha K and Sushma KR
Department of Electronics and Communication Engineering, Amruta institute of Engineering and Management Sciences, Bidadi, Bangalore, India

Prasad Anipireddy and Challa Babu
Asst.Proffesor, Department of EEE, SRIST, Nellore, India

Daudi J
Department of Aerospace Engineering, School of Engineering, University of Glasgow, UK

Manolov OB
Department of Applied Informatics and Computer Technologies, European Polytechnic University, Pernik

Kamkarian P
Department of Electrical and Computer Engineering, Southern Illinois University, Carbondale, USA

Hexmoor H
Department of Computer Science, Southern Illinois University, Carbondale, USA

Yujun Wang and Can Fang
School of Computer and Information Science, Southwest University, Chongqing, China

Qimi Jiang
School of Computer and Information Science, Southwest University, Chongqing, China
Comau Inc, MI, USA

Soumya B
Department of Computer Science, University of New Mexico, USA
Ronin Institute, Montclair, USA
Complex Biological Systems Alliance, USA
Broad Institute of MIT and Harvard, USA

Zhang C and Noguchi N
Laboratory of Vehicle Robotics, Graduate School of Agriculture, Hokkaido University, Kita-9, Nishi-9, Kita-ku, Sapporo 060-8589, Japan

González-Palacios MA, Ortega-Alvarez CJ, Sandoval-Castillo JG, Cuevas-Ledesma SM and Mendoza-Patiño FJ
Division of Engineering Campus Irapuato-Salamanca, University of Guanajuato, Salamanca Carr - V of Santiago, Community Palo Blanco, Salamanca, Mexico

Jalamkar D and Selvakumar AA
School of Mechanical and Building sciences, VIT University, Chennai, India

Asif S and Webb P
School of Aerospace, Transport and Manufacturing, Cranfield University, Cranfield, MK43 0AL, UK

Abdulraheem Kinsara and Ghassan Mousa
Center of Excellence for Industrial Design and Manufacturing Research (CEIDM) Faculty of Engineering, King Abdulaziz University Jeddah, Saudi Arabia

Andrew Gunn
Industrial and Technological Development Center Mechanical and Aerospace Engineering, University of Missouri-ColumbiaColumbia, Missouri-65211, USA

Ahmed Sherif El-Gizawy
Industrial and Technological Development Center Mechanical and Aerospace Engineering, University of Missouri-ColumbiaColumbia, Missouri-65211, USA
Center of Excellence for Industrial Design and Manufacturing Research (CEIDM) Faculty of Engineering, King Abdulaziz University Jeddah, Saudi Arabia

Ahmed AIA, Cheng H and Lin X
Center for Robotics, School of Automation, University of Electronic Science and Technology of China, 611731 Chengdu, China

Omer M
School of Automation, University of Electronic Science and Technology of China

Atieno JM
School of Electronic Engineering, University of Electronic Science and Technology of China, 611731 Chengdu, China

Cindy Chan
Department of Obstetrics and Gynecology, Taipei Medical University Hospital, Taipei, Taiwan

Li-Hsuan Chiu, Ching-Hui Chen and Wei-Min Liu
Department of Obstetrics and Gynecology, Taipei Medical University Hospital, Taipei, Taiwan
Department of Obstetrics and Gynecology, School of Medicine, College of Medicine, Taipei Medical University, Taipei, Taiwan

Jebelli A, Yagoub MCE and Dhillon BS
Faculty of Engineering, University of Ottawa, Canada

Al Bandar MH, Al Sabilah J and Kim NK
Division of Colorectal Surgery, Yonsei University College of Medicine, Seoul, Republic of Korea

Lee SY and Goh BK
Department of Hepatopancreatobiliary and Transplant Surgery, Singapore General Hospital, Singapore
Duke -National University of Singapore (NUS) Medical School, Singapore

Ajiroghene O, Obiei-uyoyou O, Chuks M, Ofualagba G and Ejofodomi OA
Department of Electrical and Electronics Engineering, Federal University of Petroleum Resources (FUPRE), Nigeria

Ogaga A, Chukwumenogor O and Elvis R
Department of Mechanical Engineering, Federal University of Petroleum Resources (FUPRE), Nigeria

Udoka ED and Bright AE
Department of Marine Engineering, Federal University of Petroleum Resources (FUPRE), Nigeria

Yun Qin, Naila Rahman and Farshid Amirabdollahian
School of Computer Science, University of Hertfordshire, Hatfield, Hertfordshire, UK

Ammar H Elsheikh, Ezzat A Showaib and Abd Elwahed M Asar
Department of Production Engineering and Mechanical Design, Faculty of Engineering, Tanta University, Tanta, Egypt

Aram Azad
Computer Science and Information Technology, UTS: University of Technology, Sydney, Sydney, Australia

Tarik Rashid
Software Engineering Department, College of Engineering, Salahaddin University, Hawler, Kurdistan

Rached Dhaouadi and Ahmad Abu Hatab
College of Engineering, American University of Sharjah, Sharjah, UAE

Prarthana BK, Bhavana Y and Mani Shankar N
Department of Mechanical Engineering, SNIST, India

Index

www.ingramcontent.com/pod-product-compliance
Lightning Source LLC
Chambersburg PA
CBHW080652200326
41458CB00013B/4821
9781632407887